VOLUME FOUR HUNDRED AND THIRTY-NINE

METHODS IN ENZYMOLOGY

Small GTPases in Disease Part B

METHODS IN ENZYMOLOGY

Editors-in-Chief

JOHN N. ABELSON AND MELVIN I. SIMON

Division of Biology
California Institute of Technology
Pasadena, California

Founding Editors

SIDNEY P. COLOWICK AND NATHAN O. KAPLAN

VOLUME FOUR HUNDRED AND THIRTY-NINE

METHODS IN
ENZYMOLOGY

Small GTPases in Disease
Part B

EDITED BY

WILLIAM E. BALCH
The Scripps Research Institute
Department of Cell Biology
La Jolla, CA, USA

CHANNING J. DER
Lineberger Comprehensive Cancer Center
University of North Carolina
Chapel Hill, NC, USA

ALAN HALL
Chair, Cell Biology Program
Memorial Sloan-Kettering Cancer Center
New York, NY, USA

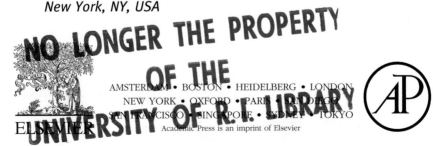

AMSTERDAM • BOSTON • HEIDELBERG • LONDON
NEW YORK • OXFORD • PARIS • SAN DIEGO
SAN FRANCISCO • SINGAPORE • SYDNEY • TOKYO
Academic Press is an imprint of Elsevier

ELSEVIER

Academic Press is an imprint of Elsevier
525 B Street, Suite 1900, San Diego, California 92101-4495, USA
84 Theobald's Road, London WC1X 8RR, UK

This book is printed on acid-free paper. ∞

For information on all Elsevier Academic Press publications
visit our Web site at www.books.elsevier.com

ISBN-13: 978-0-12-374311-4

PRINTED IN THE UNITED STATES OF AMERICA
08 09 10 11   9 8 7 6 5 4 3 2 1

Contents

Contributors

Stacey J. Adam
Department of Pharmacology and Molecular Cancer Biology, Duke University, Durham, North Carolina

Linda Van Aelst
Cold Spring Harbor Laboratory, Cold Spring Harbor, New York

Ian M. Ahearn
Departments of Medicine, Cell Biology, and Pharmacology and the Cancer Institute, New York University School of Medicine, New York

Neal M. Alto
Department of Microbiology, University of Texas Southwestern, Dallas, Texas

Richard G. W. Anderson
Department of Cell Biology, University of Texas Southwestern Medical Center, Dallas, Texas

Christelle Anguille
Universités Montpellier 2 et 1, CRBM, CNRS, UMR 5237, Montpellier, France

Francis A. Barr
University of Liverpool, Cancer Research Centre, Liverpool, United Kingdom

René Bartz
Department of Cell Biology, University of Texas Southwestern Medical Center, Dallas, Texas

Trever G. Bivona
Departments of Medicine, Cell Biology, and Pharmacology and the Cancer Institute, New York University School of Medicine, New York

Christine E. Bulawa
FoldRx Pharmaceuticals, Inc., Cambridge, Massachusetts

Paul M. Campbell
Lineberger Comprehensive Cancer Center, University of North Carolina at Chapel Hill, Chapel Hill, North Carolina

Jose A. Cancelas
Hoxworth Blood Center, University of Cincinnati College of Medicine, Cincinnati, Ohio and Division of Experimental Hematology, Cincinnati Children's Research Foundation, Cincinnati Children's Hospital Medical Center, Cincinnati, Ohio

Christopher L. Carpenter
Department of Medicine, Beth Israel Deaconess Medical Center and Harvard Medical School, Boston, Massachusetts

Linda Castaldo
OSI Pharmaceuticals, Farmingdale, New York

Ruth N. Collins
Department of Molecular Medicine, Cornell University College of Veterinary Medicine, Ithaca, New York

Natalie Cook
Department of Clinical Oncology, Addenbrookes NHS Trust, Cambridge, United Kingdom and Cancer Research UK Cambridge Research Institute, Li Ka Shing Centre, Cambridge, United Kingdom

Christopher M. Counter
Department of Pharmacology and Molecular Cancer Biology, Duke University, Durham, North Carolina

Adrienne D. Cox
Department of Radiation Oncology, Curriculum in Genetics and Molecular Biology, Department of Pharmacology, and Lineberger Comprehensive Cancer Center, University of North Carolina at Chapel Hill, Chapel Hill, North Carolina

David P. Davis
Genentech Inc., Department of Molecular Biology, South San Francisco, CA

Channing J. Der
Curriculum in Genetics and Molecular Biology, Department of Pharmacology, and Lineberger Comprehensive Cancer Center, University of North Carolina at Chapel Hill, Chapel Hill, North Carolina

Jack E. Dixon
Departments of Pharmacology, Cellular and Molecular Medicine, and Chemistry and Biochemistry, University of California, San Diego, La Jolla, California

Celine Dumont
Division of Immune Cell Biology, National Institute for Medical Research, London, United Kingdom

Marcello Ehrlich
Department of Cell Biology and Immunology, George S. Wise Faculty of Life Sciences, Tel Aviv University, Tel Aviv, Israel

James Fleming
FoldRx Pharmaceuticals, Inc., Cambridge, Massachusetts

Kristopher Frese
Cancer Research UK Cambridge Research Institute, Li Ka Shing Centre, Cambridge, United Kingdom

Gilles Gadea
Universités Montpellier 2 et 1, CRBM, CNRS, UMR 5237, Montpellier, France

Jorge E. Galán
Section of Microbial Pathogenesis, Yale University School of Medicine, Boyer Center for Molecular Medicine, New Haven, Connecticut

Andrew J. Garton
OSI Pharmaceuticals, Farmingdale, New York

Vanessa González-Pérez
Curriculum in Genetics and Molecular Biology, University of North Carolina at Chapel Hill, Chapel Hill, North Carolina

Eve-Ellen Govek
The Rockefeller University, Laboratory of Developmental Neurobiology, New York

Angela L. Groehler
Lineberger Comprehensive Cancer Center, University of North Carolina at Chapel Hill, Chapel Hill, North Carolina

Alexander K. Haas
University of Liverpool, Cancer Research Centre, Liverpool, United Kingdom

Roni Haklai
Department of Neurobiochemistry, George S. Wise Faculty of Life Sciences, Tel Aviv University, Tel Aviv, Israel

Ariella B. Hanker
Curriculum in Genetics and Molecular Biology, University of North Carolina at Chapel Hill, Chapel Hill, North Carolina

Michael A. Harding
Department of Molecular Physiology and Biological Physics, University of Virginia Health Sciences Center, Charlottesville, Virginia

Robert Henderson
Division of Immune Cell Biology, National Institute for Medical Research, London, United Kingdom

Ken-ichi Hirano
Department of Cardiovascular Medicine, Graduate School of Medicine, Osaka University, Osaka, Japan

Klaus P. Hoeflich
Genentech Inc., Department of Molecular Biology, South San Francisco, CA

Noriyuki Homma
Cardiovascular Pulmonary Research Laboratory, University of Colorado at Denver and Health Sciences Center, Denver, Colorado

Chiaki Ikegami
Department of Molecular Cardiology, Whitaker Cardiovascular Institute, Boston University School of Medicine, Boston, MA, USA

Bijay Jaiswal
Genentech Inc., Department of Molecular Biology, South San Francisco, CA

Nael Nadif Kasri
Cold Spring Harbor Laboratory, Cold Spring Harbor, New York

Chiaki Kawase
Department of Obstetrics and Gynecology, Osaka University Graduate School of Medicine, Osaka, Japan

Toshiyuki Kawashima
Division of Cellular Therapy, Institute of Medical Science, University of Tokyo, Tokyo, Japan

Vladimir Khazak
NexusPharma, Inc., Langhorne, Pennsylvania

Hong Jin Kim
Lineberger Comprehensive Cancer Center, University of North Carolina at Chapel Hill, Chapel Hill, North Carolina

Tadashi Kimura
Department of Obstetrics and Gynecology, Osaka University Graduate School of Medicine, Osaka, Japan

Toshio Kitamura
Division of Cellular Therapy, Institute of Medical Science, University of Tokyo, Tokyo, Japan

Yoel Kloog
Department of Neurobiochemistry, George S. Wise Faculty of Life Sciences, Tel Aviv University, Tel Aviv, Israel

Henry Koziel
Division of Pulmonary, Critical Care and Sleep Medicine, Department of Medicine, Beth Israel Deaconess Medical Center and Harvard Medical School, Boston, Massachusetts

Kwang M. Lee
Eppley Institute for Research in Cancer and Allied Diseases, University of Nebraska Medical Center, Omaha, Nebraska

James Leiper
BHF Laboratories, Department of Medicine, University College London, London, United Kingdom

James K. Liao
Vascular Medicine Research Unit, Brigham and Women's Hospital and Harvard, Medical School, Boston, Massachusetts

Susan L. Lindquist
Howard Hughes Medical Institute and Whitehead Institute for Biomedical Research, Cambridge, Massachusetts

Ping-Yen Liu
Vascular Medicine Research Unit, Brigham and Women's Hospital and Harvard, Medical School, Boston, Massachusetts

Pingsheng Liu
Department of Cell Biology, University of Texas Southwestern Medical Center, Dallas, Texas

Seiji Mabuchi
Department of Obstetrics and Gynecology, Osaka University Graduate School of Medicine, Osaka, Japan

Ivan F. McMurtry
Department of Pharmacology and Center for Lung Biology, University of South Alabama, Mobile, Alabama

Ken-ichirou Morishige
Department of Obstetrics and Gynecology, Osaka University Graduate School of Medicine, Osaka, Japan

Staeci Morita
Department of Pharmacology, University of North Carolina at Chapel Hill, Chapel Hill, North Carolina

Maxence V. Nachury
Department of Molecular and Cellular Physiology, Stanford University School of Medicine, Stanford, California

Matthew D. Nitz
Department of Molecular Physiology and Biological Physics, University of Virginia Health Sciences Center, Charlottesville, Virginia

Seiji Ogata
Department of Obstetrics and Gynecology, Osaka University Graduate School of Medicine, Osaka, Japan

Masahiko Oka
Cardiovascular Pulmonary Research Laboratory, University of Colorado at Denver and Health Sciences Center, Denver, Colorado

Cercina Onesto
University of North Carolina at Chapel Hill, Lineberger Comprehensive Cancer Center, Department of Pharmacology, Chapel Hill, North Carolina

Michel M. Ouellette
Eppley Institute for Research in Cancer and Allied Diseases, University of Nebraska Medical Center, Omaha, Nebraska

Tiago F. Outeiro
Institute of Molecular Medicine, Cellular and Molecular Neuroscience Unit, Lisbon, Portugal

Kenneth P. Olive
Cancer Research UK Cambridge Research Institute, Li Ka Shing Centre, Cambridge, United Kingdom

Jonathan A. Pachter
OSI Pharmaceuticals, Farmingdale, New York

Chaitali Parikh
Rosenstiel Basic Medical Sciences Research Center, Department of Biology, Brandeis University, Waltham, Massachusetts

Jayesh C. Patel
Section of Microbial Pathogenesis, Yale University School of Medicine, Boyer Center for Molecular Medicine, New Haven, Connecticut

Mark R. Philips
Departments of Medicine, Cell Biology, and Pharmacology and the Cancer Institute, New York University School of Medicine, New York

Virginie Picard
ExonHit Therapeutics, Paris, France

Joanna C. Porter
Medical Research Council Laboratory of Molecular Cell Biology, University College London, and Department of Respiratory Medicine, University College London Hospitals NHS Trust, University College London Hospital, London, United Kingdom

Jia-ying Qian
Department of Medical Biophysics, University of Toronto, Toronto, Ontario, Canada; Division of Applied Molecular Oncology, Ontario Cancer Institute, University Health Network, Ontario, Canada

Steven E. Quatela
Departments of Medicine, Cell Biology, and Pharmacology and the Cancer Institute, New York University School of Medicine, New York

Nikolina Radulovich
Division of Applied Molecular Oncology, Ontario Cancer Institute, University Health Network, Ontario, Canada

Peter B. Rahl
Department of Molecular Medicine, Cornell University College of Veterinary Medicine, Ithaca, New York

David J. Reiner
Department of Pharmacology, and Lineberger Comprehensive Cancer Center, University of North Carolina at Chapel Hill, Chapel Hill, North Carolina

Ruibao Ren
Rosenstiel Basic Medical Sciences Research Center, Department of Biology, Brandeis University, Waltham, Massachusetts

Gretchen A. Repasky
Birmingham-Southern College, Department of Biology, Birmingham, Alabama

Pierre Roux
Universités Montpellier 2 et 1, CRBM, CNRS, UMR 5237, Montpellier, France

Douglas T. Ross
Applied Genomics Inc., Burlingame, California and Huntsville, Alabama

Barak Rotblat
Department of Neurobiochemistry, George S. Wise Faculty of Life Sciences, Tel Aviv University, Tel Aviv, Israel

Masahiro Sakata
Department of Obstetrics and Gynecology, Osaka University Graduate School of Medicine, Osaka, Japan

Kenjiro Sawada
Department of Obstetrics and Gynecology, Osaka University Graduate School of Medicine, Osaka, Japan

Fabien Schweighoffer
ExonHit Therapeutics, Paris, France

Robert S. Seitz
Applied Genomics Inc., Burlingame, California and Huntsville, Alabama

Somasekar Seshagiri
Genentech Inc., Department of Molecular Biology, South San Francisco, CA

Adam Shutes
University of North Carolina at Chapel Hill, Lineberger Comprehensive Cancer Center, Department of Pharmacology, Chapel Hill, North Carolina

Mark Slack
Evotec AG, Hamburg, Germany

Vinot Stéphanie
Universités Montpellier 2 et 1, CRBM, CNRS, UMR 5237, Montpellier, France

Pamela J. Sung
Departments of Medicine, Cell Biology, and Pharmacology and the Cancer Institute, New York University School of Medicine, New York

Ayumu Tashiro
Centre for the Biology of Memory, Norwegian University of Science and Technology, Trondheim, Norway

Dan Theodorescu
Department of Molecular Physiology and Biological Physics, University of Virginia Health Sciences Center, Charlottesville, Virginia

Mrion de Toledo
Universités Montpellier 2 et 1, CRBM, CNRS, UMR 5237, Montpellier, France

Ming-Sound Tsao
Department of Laboratory Medicine and Pathobiology, and Department of Medical Biophysics, University of Toronto, Toronto, Ontario, Canada; Division of Applied Molecular Oncology, Ontario Cancer Institute, University Health Network, Ontario, Canada

David A. Tuveson
Cancer Research UK Cambridge Research Institute, Li Ka Shing Centre, Cambridge, United Kingdom

Victor L. J. Tybulewicz
Division of Immune Cell Biology, National Institute for Medical Research, London, United Kingdom

Stéphanie Vinot
Universités Montpellier 2 et 1, CRBM, CNRS, UMR 5237, Montpellier, France

Andrew Wilkins
Department of Medicine, Beth Israel Deaconess Medical Center and Harvard Medical School, Boston, Massachusetts

David A. Williams
Division of Hematology/Oncology, Childrens Hospital Harvard Medical School, Boston, MA

Beata Wojciak-Stothard
BHF Laboratories, Department of Medicine, University College London, London, United Kingdom

Yunshu Ying
Department of Cell Biology, University of Texas Southwestern Medical Center, Dallas, Texas

Shin-ichiro Yoshimura
University of Liverpool, Cancer Research Centre, Liverpool, United Kingdom

Rafael Yuste
Howard Hughes Medical Institute, Department of Biological Sciences, Columbia University, New York, New York

John K. Zehmer
Department of Cell Biology, University of Texas Southwestern Medical Center, Dallas, Texas

Zhongyan Zhang
Department of Cell Biology, Johns Hopkins University School of Medicine, Baltimore, Maryland

Yi Zheng
Division of Experimental Hematology, Cincinnati Children's Research Foundation, Cincinnati Children's Hospital Medical Center, Cincinnati, Ohio

Methods in Enzymology

VOLUME 346. Gene Therapy Methods
Edited by M. IAN PHILLIPS

VOLUME 347. Protein Sensors and Reactive Oxygen Species (Part A: Selenoproteins and Thioredoxin)
Edited by HELMUT SIES AND LESTER PACKER

VOLUME 348. Protein Sensors and Reactive Oxygen Species (Part B: Thiol Enzymes and Proteins)
Edited by HELMUT SIES AND LESTER PACKER

VOLUME 349. Superoxide Dismutase
Edited by LESTER PACKER

VOLUME 350. Guide to Yeast Genetics and Molecular and Cell Biology (Part B)
Edited by CHRISTINE GUTHRIE AND GERALD R. FINK

VOLUME 351. Guide to Yeast Genetics and Molecular and Cell Biology (Part C)
Edited by CHRISTINE GUTHRIE AND GERALD R. FINK

VOLUME 352. Redox Cell Biology and Genetics (Part A)
Edited by CHANDAN K. SEN AND LESTER PACKER

VOLUME 353. Redox Cell Biology and Genetics (Part B)
Edited by CHANDAN K. SEN AND LESTER PACKER

VOLUME 354. Enzyme Kinetics and Mechanisms (Part F: Detection and Characterization of Enzyme Reaction Intermediates)
Edited by DANIEL L. PURICH

VOLUME 355. Cumulative Subject Index Volumes 321–354

VOLUME 356. Laser Capture Microscopy and Microdissection
Edited by P. MICHAEL CONN

VOLUME 357. Cytochrome P450, Part C
Edited by ERIC F. JOHNSON AND MICHAEL R. WATERMAN

VOLUME 358. Bacterial Pathogenesis (Part C: Identification, Regulation, and Function of Virulence Factors)
Edited by VIRGINIA L. CLARK AND PATRIK M. BAVOIL

VOLUME 359. Nitric Oxide (Part D)
Edited by ENRIQUE CADENAS AND LESTER PACKER

VOLUME 360. Biophotonics (Part A)
Edited by GERARD MARRIOTT AND IAN PARKER

VOLUME 361. Biophotonics (Part B)
Edited by GERARD MARRIOTT AND IAN PARKER

VOLUME 362. Recognition of Carbohydrates in Biological Systems (Part A)
Edited by YUAN C. LEE AND REIKO T. LEE

CHAPTER ONE

HUMAN PANCREATIC DUCT EPITHELIAL CELL MODEL FOR KRAS TRANSFORMATION

Nikolina Radulovich,* Jia-ying Qian,*,† *and* Ming-Sound Tsao*,†,‡

Contents

Abstract

Mutations on the *KRAS* gene occur early during pancreatic duct cell carcinogenesis and have been identified in up to 90% of ductal adenocarcinoma. However, the functional role of *KRAS* mutations in the malignant transformation of normal pancreatic duct epithelial cells into cancer cells remains unknown. We have developed an *in vitro* model for KRAS transformation using near-normal HPV-16E6E7-immortalized human pancreatic ductal epithelial (HPDE-E6E7) cells. The expression of mutant *KRAS*^{G12V} in HPDE cells by retroviral transduction resulted in weak tumorigenic transformation, with tumors formed in 50% of immune-deficient *scid* mice implanted by these *KRAS*-transformed cells. The model provides an opportunity to dissect further the molecular and cellular mechanisms associated with human pancreatic duct cell carcinogenesis.

* Division of Applied Molecular Oncology, Ontario Cancer Institute, University Health Network, Ontario, Canada
† Department of Medical Biophysics, University of Toronto, Toronto, Ontario, Canada
‡ Department of Laboratory Medicine and Pathobiology, University of Toronto, Toronto, Ontario, Canada

Methods in Enzymology, Volume 439
ISSN 0076-6879, DOI: 10.1016/S0076-6879(07)00401-6

1. INTRODUCTION

Pancreatic cancer is the fourth leading cause of cancer death in North America. It has one of the poorest prognoses among all cancers, partly due its location that predisposes a late stage discovery and its highly metastatic and resistant phenotypes. At the time of diagnosis, only 20% of pancreatic cancer patients are eligible for surgical resection, which remains the only curative therapy available (Yeo *et al.*, 2002). More than 90% of pancreatic carcinoma is of the ductal type, which putatively arises from malignant transformation of the pancreatic duct cells.

Pancreatic duct cell carcinogenesis is a multistage event, which can be defined histopathologically in human pancreas by the putative precursor lesions termed pancreatic intraepithelial neoplasias (PanIN). PanIN evolves progressively from flat mucinous (Pan IN-1A), papillary hyperplasia without atypia (Pan IN-1B), papillary hyperplasia with atypia (Pan IN-2) to carcinoma *in situ* (Pan IN-3) (Hruban *et al.*, 2001). Molecular analyses have revealed that PanIN–carcinoma progression is associated with accumulating genetic aberrations (Hruban *et al.*, 2000). Mutations in the *KRAS* gene seem to occur early and are present in over 90% of invasive pancreatic cancers (Bardeesy and DePinho, 2002; Hilgers and Kern, 1999). The early onset and almost ubiquitous occurrence of *KRAS* oncogenic mutations in pancreatic cancers implicate a pivotal role for KRAS in the pathogenesis of this malignancy.

Recent studies have reported genes that could be promising biomarkers for early detection; however, very few have been further validated for their roles during multistage pancreatic cancer development (Grützmann *et al.*, 2004; Logsdon *et al.*, 2003; Nakamura *et al.*, 2004). To study the potential biological role of the *KRAS* oncogene in duct cell carcinogenesis, we established a dynamic model using a near diploid-immortalized epithelial cell line derived from normal human pancreatic ducts (Furukawa *et al.*, 1996, Ouyang *et al.*, 2000). In contrast to studies using pancreatic cancer cell lines that are heterogeneous genetically and biologically, a carcinogenesis model that starts from a genetically characterized clone of near normal duct cells is more likely to yield results that are common to a majority of primary ductal cancers. We showed that expression of the $KRAS^{G12V}$ oncogene in HPV16-E6E7-immortalized human pancreatic duct epithelial (HPDE) cells led to constitutive activation of the KRAS protein and increased activation of its downstream targets such as MAPK and AKT (Qian *et al.*, 2005). The tumorigenicity of $KRAS^{G12V}$-expressing HPDE cells was assessed by tumor formation in immune-deficient *scid* mice. These cells yielded tumors in nearly half of the SCID mice implanted with the cells (Qian *et al.*, 2005). This *in vitro* model may provide important mechanistic insights into functional consequences of the genetic abnormalities associated with multistage pancreatic duct cell carcinogenesis.

2. DESCRIPTION OF METHODS

2.1. Isolation of HPDE cells

2.1.1. Protocol

1. Grossly identify pancreatic ducts as tubular structures in the grossly normal-appearing part of resected pancreatectomy specimens.
2. Dissect ducts from the adjacent acinar tissue using microscissors.
3. Place dissected ducts in a 15-ml tube and wash several times with ice-cold Hanks' balanced salt solution (HBSS).
4. Transfer tissue into a 10-cm plate with 5 ml of growth medium.
5. Using a scalpel, cut the duct fragments into approximately 1-mm fragments.
6. Using pointed surgical forceps, transfer and place the dissected duct fragments onto new 6- or 10-cm-diameter tissue culture plates, with three to six fragments being distributed evenly on the plate surface.
7. Gently add drops of medium to the fragments such that the latter are just covered by the medium but remain adherent to the plastic surface. Incubate the plates at 37° in 5% CO_2 for a few hours.
8. Slowly add additional medium to allow plates to remain in the incubator undisturbed for several days without medium changes. Care should be taken not to dislodge the duct fragments from the plate surface.
9. After 3 to 6 days, a monolayer of cobblestone-appearing epithelial cells will migrate out of the duct fragments onto the plate surface. Allow further cell migration until the monolayer reaches a few millimeters in diameter.
10. The duct fragments may be removed and placed onto a new plate to repeat the explanting process, while monolayer cells can be further incubated after a fresh medium change.
11. Primary cultured cells may be subjected to the protocol for immortalization with the HPV16-E6E7 gene.

2.1.2. Complete growth media

Keratinocyte serum-free (KSF) medium (Gibco/Invitrogen, Carlsbad, CA) is supplemented with 50 mg/ml bovine pituitary extract (BPE), 5 ng/ml epidermal growth factor (EGF), and 1× antibiotic–antimycotic cocktail (Gibco/Invitrogen). Because some of the supplements are perishable, it is best to make fresh media every 2 weeks.

2.1.3. Equipment

Phase-contrast tissue culture microscope
Surgical scalpel, scissors, and forceps

2.2. Immortalizing HPDE cells by HPV16-E6E7 genes

Because normal human epithelial cells have a very limited life span in primary culture, it is advisable that the isolated HPDE cells be immortalized before the first passage. In our model, we used an amphotrophic retrovirus, LXSN16E6E7 (Halbert *et al.*, 1991), containing the E6 and E7 genes of HPV-16 to infect the duct epithelial cells in primary cultures. The *LXSN16E6E7* virus can be grown in commercially available RetroPack PT67 or AmphoPack-293 cells (Clontech, Mountain View, CA) as recommended by the manufacturer. Following retroviral infection, cells are passaged continuously (see later) until the cells appear to stop proliferating and achieve senescence. It is important that the cultures are not discarded and instead are maintained by regular medium changes. Two to 3 months later, one may notice the development of colonies of proliferating cells. These cell colonies have overcome senescence and become immortalized. The colonies can either be pooled or individual colonies can be cloned using the cloning rings.

2.2.1. Protocol

1. Prepare the retroviral cocktail (for 10-cm plate of target cells):
   - 3 ml of retroviral soup (thawed quickly at 37°)
   - 2 ml of growth medium
   - 5 μl of 1000× polybrene stock
2. Aspirate the medium and wash target cells with HBSS.
3. Add aforementioned prepared retroviral cocktail to each 10-cm plate.
4. Incubate at 37° in 5% CO_2 for 16 to 18 h.
5. Replace the virus containing medium with fresh growth medium and incubate at 37° in 5% CO_2 for an additional 48 h.
6. For neomycin resistance selection, replace the medium with fresh medium containing 400 μg/ml G-418.
7. Change the medium every 3 days until cells reach 80 to 90% confluence.
8. Proceed with the protocol for culture of HPDE cells.

   Note: A continuous selection in G-418 should be carried out for at least three subcultures.

2.2.2. Materials and reagents
Packaging cell lines RetroPack PT67 or AmphoPack-293 cells (Clontech)

2.2.2.2. Other

G418 (Clontech)
Stock polybrene solution, 4 mg/ml (Sigma, St. Louis, MO)
Cloning rings/cylinders (Fisher Scientific)

2.3. Continuous culture of immortalized HPDE cells

The following protocol is designed for a 10-cm cell culture plate format. Other plate formats will require adjustment of the volume of reagents accordingly. HPDE cells are passaged when they reach 80 to 90% confluence in a 1:3 ratio. Under normal conditions they can be passaged twice a week with one medium change between passaging.

2.3.1. Protocol

1. Aspirate the growth medium.
2. Add 10 ml of HBSS and aspirate.
3. Add 1 ml of $1 \times$ trypsin-EDTA and leave in the incubator for 5 to 10 min.

Note: For HPDE cells that have been grown on the same dish for longer periods, the cells do not detach easily. In such a case, add twice the amount of recommended trypsin and incubate for an additional 5 to 10 min.

4. Gently tap the bottom of the plate to dislodge cells from the plate surface completely.
5. Add 1 ml of 1% trypsin inhibitor and another 3 ml of growth medium to the plate.

Note: Using trypsin inhibitor is a necessary step; if this step is omitted, the cell cultures will be lost.

6. Gently pipette a few times to resuspend and transfer cells to a 15-ml centrifuge tube.
7. Spin cells at 1000 rpm for 5 min.
8. Aspirate supernatant and resuspend cells in 3 ml of growth medium.
9. Inoculate 1 ml of resuspended cells into a new 10-cm plate containing 9 ml of growth medium.
10. After overnight incubation, the culture medium should be replaced by fresh medium. This will remove any residual trypsin to which HPDE cells are sensitive.

To prepare cells for long-term storage in liquid nitrogen, follow steps 1 to 7 of the culture protocol. Then, resuspend the cell pellet in 3 ml of growth medium supplemented with 10% dimethyl sulfoxide (DMSO). Place cell suspension in cryotubes (1 ml each) and place in a Nalgene freezing container at -80°. The next day, transfer cryotubes to a liquid nitrogen tank for long-term storage.

2.3.2. Materials and reagents

Complete growth media   KSF medium (Invitrogen) supplemented with 50 mg/ml BPE, 5 ng/ml EGF, and $1 \times$ antibiotic–antimycotic (Gibco/Invitrogen). Because some of the supplements are perishable, it is best to make fresh media every 2 weeks.

Other

HBSS
1× trypsin–EDTA in HBSS (Gibco/Invitrogen)
1% soybean trypsin inhibitor in HBSS (Gibco/Invitrogen)
DMSO, molecular biology grade (Sigma; Hybri-Max)
Cryotubes (Nalgene, Rochester, NY)
Nalgene 1° freezing containers.

2.3.3. Equipment

Phase–contrast tissue culture microscope

2.4. Transformation of HPDE cells by *KRAS* oncogene (V12)

The HPDE cell line is relatively resistant to various transfection methods. On average, the transfection efficiency is lower than 10%, which is very inefficient when trying to establish stable clones. Hence, we used a retrovirus-mediated expression system to create *KRAS* oncogene-expressing HPDE cells. Retroviral vector pBabepuro-*KRAS*4B^{G12V} contains the cDNA of the human *KRAS*4B oncogene (*KRAS*$^+$) with a mutation in codon12 (GTT to GTT; Lundberg *et al.*, 2002). The puromycin resistance gene in the vector enables a selection with puromycin. This vector can be grown in the Phoenix retroviral production system, including an ecotropic packaging system (capable of delivering genes to dividing murine or rat cells) or an amphotropic packaging system (capable of delivering genes to dividing human cells). Because the *KRAS* oncogene is a biohazard requiring a biosafety level-2 plus (BL-2+) facility, we originally made the HDPE cell line expressing the murine ecotropic retroviral receptor (ecoR) and subsequently used the ecotropic packaging system to grow the pBabepuro-*KRAS*4B^{G12V} (Qian *et al.*, 2005). However, after the availability of BL-2B in our facility, we have since successfully made the pBabepuro-*KRAS*4B^{G12V} in amphotropic packaging systems. The following protocols can be adapted for use in both packaging systems. The appropriate laboratory biohazard safety procedures for the production and handling of virus capable of infecting human cells should be applied (http://bmbl.od.nih.gov/contents. htm). Construction of HPDE ecoR-expressing cell lines was published earlier (Qian *et al.*, 2005). Protocols for growing retroviruses using the Phoenix system are modified from the original protocol developed by the Nolan laboratory (http://www.stanford.edu/group/nolan/).

2.4.1. Protocol for culture of retroviral packaging cells

Phoenix ecotropic packaging cells (Phoenix-eco) and amphotropic packaging cells (Phoenix-ampho) are both from the American Type Culture Collection (Manassas, VA) and have been tested for helper virus production. Packaging cells are maintained at 37° in Dulbecco's modified Eagle's medium (DMEM) supplemented with 10% fetal bovine serum (FBS), 2 mM L-glutamine, 1× antibiotic–antimycotic (Gibco/Invitrogen). The following protocol is optimized for a 10-cm tissue culture plate.

1. Aspirate the growth medium.
2. Add 10 ml of HBSS and aspirate.

Note: Because Phoenix cells detach very easily, exercise caution when washing with HBSS not to disturb the cell monolayer.

3. Add 1 ml of 1× trypsin-EDTA and leave in the incubator for a few minutes.
4. Gently knock the plate to dislodge the cells completely.
5. Add 4 ml of growth medium to the plate.
6. Gently pipette a few times to resuspend and transfer cells to a 15-ml tube.
7. Spin cells at 1000 rpm for 5 min.
8. Aspirate supernatant and resuspend cells in 5 ml of growth medium.
9. Transfer 1 ml resuspended cells to each 10-cm plate containing 9 ml growth medium.

2.4.2. Protocol for generation of pBabe retroviruses and infection of target cells

We use the BD CalPhos mammalian transfection kit to transfect Phoenix-ampho and -ecoR cells with pBabe-puro and pBabepuro-*KRAS*4B^{G12V} constructs. The following calcium phosphate transfection protocol is optimized for a 10-cm plate format. To test for transduction efficiency, cells are transduced with the empty plasmid. A positive control includes transduction into the NIH/3T3 cells.

1. Plate two to five million Phoenix-ampho/-ecoR cells on a 10-cm plate in DMEM medium and grow overnight at 37°. The plate should be approximately 80% confluent prior to transfection.
2. Prepare solution A containing 10 μg plasmid DNA, 86.8 μl 2 M calcium solution, and dH$_2$O up to 700 μl.
3. Prepare solution B containing 700 μl 2× HBSS.
4. While gently vortexing solution B, add solution A drop wise.
5. Incubate the transfection mix at room temperature for 20 min.
6. Gently vortex the transfection solution and add solution drop wise to the plate.
7. Incubate plates at 37° for 18 h.
8. Carefully replace medium with 8 ml of fresh growth medium and incubate at 37° for another 36 h. The total transfection time is 48 h.
9. Collect supernatant and spin at 1500 rpm for 10 min to remove living cells.

10. Filter the supernatant through a 0.22-μm filter to prevent subsequent contamination.
11. To infect HPDE cells with retroviral supernatant, follow steps 1 to 5 of protocol for transduction of HPDE cells as outlined in the immortalization protocol.

Note: Retrovirus can also be transferred to freezing vials or polypropylene tubes; put on dry ice. The frozen samples are stored at $-70°$. Frozen samples are thawed quickly at $37°$ prior to use.

12. Split transduced HPDE cells in a 1:3 ratio.
13. Add growth medium supplemented with 0.5 μg/ml of puromycin.

Note: Use nontransduced HPDE cells as a control.

14. Grow cells for 1 to 2 weeks at $37°$; change media every 3 to 4 days.

Following selection in puromycin culture, pool all HPDE-KRAS+ clones and compare KRAS mRNA and KRAS protein expression and activity to that in control HPDE cells transduced with the pBABE-puro virus (HPDE-pBp).

2.4.3. Materials and reagents
Packaging cell lines

Phoenix-ampho (American Type Culture Collection)
Phoenix-eco (American Type Culture Collection)

Retroviral vectors

pBabe-puro (Addgene plasmid 1764)
pBabepuro-*KRAS*4B^{G12V} (provided by Dr. William Hahn, Dana Farber Cancer Institute, Boston, MA)

Growth medium   DMEM supplemented with 10% FBS, 2 mM L-glutamine, 1× antibiotic–antimycotic (Gibco/Invitrogen)

Other

HBSS
1× trypsin-EDTA in HBSS (Gibco/Invitrogen)
Puromycin (Sigma)
Sterile 0.22-μm filter (Pall Corporation)
Polybrene (Sigma)
BD CalPhos mammalian transfection kit

2.5. Characterization of HPDE cells expressing oncogenic KRAS^{G12V}

2.5.1. Protocol to evaluate mRNA expression in HPDE cells by reverse transcription–quantitative polymerase chain reaction (PCR)

1. Isolate RNA from HPDE-pBp and HPDE-KRAS$^+$ using the RNEasy minikit.
2. Using 4 μg of total RNA, set up the reverse transcription reaction using SuperScript II reverse transcriptase.
3. Dilute cDNA to 2 ng/μl, which is the working concentration for real-time PCR.
4. Dilute *KRAS* and RPS13 PCR primers to 20 μM.

Note: Real-time PCR primers were designed using Primer Express software (Applied Biosystems) and purchased from Invitrogen.

5. Perform real-time PCR in a total volume of 25 μl using 5 μl of the first strand cDNA synthesis mixture as template, 0.5 μl of forward primer, 0.5 μl of reverse primer, and 12.5 μl of 2XSYBR Green PCR master.

Note: Each reaction is run in duplicate and performed at least twice. A negative control reaction in the absence of template (no template control) should also be included in duplicate.

6. Set up the thermal cycling parameters as follows: 10 min activation at 95°, followed by 40 cycles of 15 s at 95° and 1 min at 60°.
7. Data analysis: The abundance of a transcript is represented by the threshold cycle of amplification (CT), which is inversely correlated to the amount of target RNA being transcribed. The relative quantification of gene expression is determined using the comparative CT method. The RPS13 mRNA is used as a reference for normalization. The CT value of RPS13 mRNA is subtracted from that of the gene of interest to obtain the ΔCT value (ΔCT = CT$_{KRAS}$ − CT$_{RPS13}$). The gene expression level in HPDE-KRAS$^+$ cells relative to HPDE-pBp cells is calculated using the following formula: $\Delta\Delta$CT = ΔCT$_{HPDE-KRAS+}$ − ΔCT$_{HPDE-pBP}$; fold change = $2^{-\Delta\Delta CT}$. Other housekeeping genes may also be used for normalization.

2.5.2. Materials and reagents
Primers *KRAS* F1 (caggctcaggacttagcaagaag), *KRAS* R1 (tgttttcgaatttctc-gaactaatgta), RPS13 F1 (gttgctgttcgaaagcatcttg), and RPS13 R1 (aatatcgagccaaacggtgaa)

Other

RNEasy minikit (Qiagen, Valencia, CA)
SuperScript II reverse transcriptase (Invitrogen)
2XSYBR Green PCR master mix (Applied Biosystems, Foster City, CA)

2.5.3. Equipment

Quantitative real-time PCR machine

2.6. KRAS protein expression, activation, and changes in RAS downstream signaling

1. Isolate protein from HPDE-pBp and HPDE-KRAS$^+$ with the protein lysis buffer.
2. Measure protein concentrations using the Bio-Rad protein assay.
3. Prepare 20 to 50 μg proteins in 6× SDS sample dye.
4. Separate proteins on a 4 to 20% gradient-ready SDS-PAGE ready gel (Bio-Rad Laboratories, Hercules, CA).
5. Transfer gel onto a polyvinylidene fluoride (PVDF) membrane at 100 V for 1 h.
6. Air dry the membrane, dehydrate in 100% methanol, and transfer to blocking buffer for 1 h at room temperature or overnight at 4°.
7. Transfer the membrane to blocking buffer containing primary antibodies and incubate overnight at 4°.
8. Wash the membrane with 1× TBST three times for 5 min each.
9. Transfer the membrane to blocking buffer containing secondary antibodies and incubate at room temperature for 1 h.
10. Wash the membrane with 1× TBST three times for 5 min each.
11. Detect membrane using the ECL Plus Western blotting detection reagent (Amersham, Piscataway, NJ).

We use the commercially available RAS activity assay kit (Upstate Biotech, Lake Placid, NY) to test KRAS oncogenic activation in HPDE-KRAS cells. The assay is performed according to the manufacturer's recommendations without any modifications.

2.6.1. Materials and reagents

Protein lysis buffer: 50 mM HEPES, pH 8.0, 10% glycerol, 1% Triton X-100, 150 mM NaCl, 1 mM EDTA, 1.5 mM MgCl$_2$, 100 mM NaF, 10 mM Na$_4$P$_2$O$_7$·H$_2$O, 5 μg/ml leupeptin, 5 μg/ml aprotinin, 100 μg/ml phenylmethylsulfonyl fluoride

Primary antibodies and dilution for western blot analyses

C-KRAS mAb (Oncogene), 1:500 dilution; total Akt (Cell Signaling, Danvers, MA), 1:500; phosphor-Akt^{Thr308} (Cell Signaling), 1:500; P44/42 MAPKinase (Cell Signaling), 1:1000; phosphor-p44/42 MAP-Kinase$^{Thr202/Tyr204}$ (Cell Signaling), 1:1000, GAPDH (AbCAM, Cambridge, MA), 1:10,000

Secondary antibodies

Horseradish peroxidase (HRP)-linked antirabbit (Cell Signaling), 1:2500
HRP-linked antimouse (Cell Signaling), 1:2500

Other

Bio-Rad Protein assay (Bio-Rad Laboratories)
6× SDS sample dye; 10 ml (3 ml of 1 M Tris-HCl, 1 g dithiothreitol, 0.06 g
  bromphenol blue,
6 ml of 100% glycerol, 1.2 g SDS, 1 ml dH_2O)
PVDF (Roche, Indianapolis, IN)
Blocking buffer (1× TBS, 0.1% Tween-20, 5% nonfat dry milk)
ECL Plus Western blotting detection reagent (Amersham)
RAS activity assay kit (Upstate Biotechnology)

2.7. Tumorigenecity in mice

The tumorigenicity of HPDE-KRAS$^+$ cells can be assayed using subcutaneous or orthotopic injection of two million cells suspended in 50 to 70 μl
of fresh medium supplemented with 10% Matrigel (BD Biosciences, San
Jose, CA). For subcutaneous implantation, cells should be injected into the
ventral abdomen and then kept for 6 to 12 months to obtain tumor
formation. Once the tumor forms, growth should be measured twice a
week. Once the tumor reaches 1 cm in diameter or becomes ulcerated, the
tumor is removed and fixed in 10% buffered formalin for histology. For
orthotopic implantation, cells are injected directly into the pancreas. The
condition of the mice is monitored twice a week over a period of 6 months,
at which time the mice are sacrificed and observed for tumor formation.
The pancreas, liver, and spleen should be removed and fixed in formalin for
routine histological processing.

3. Concluding Remarks

We have shown that the KRAS oncogene manifests weak oncogenic
activity in HPV16-E6E7-immortalized HPDE cells, with only 50% of the
animals implanted with this cell population forming tumors. Because these
cells already demonstrate aberrant p53 and Rb pathways, our model is
unsuitable for studying the transforming activity of the *KRAS* oncogene
alone in HPDE cells. Nevertheless, because p53 and p16 inactivation is
common in pancreatic ductal cancer, the model remains valuable for studying additional genes that play important roles in the full malignant transformation of HPDE cells in cooperation with *KRAS* oncogenes and

deregulated p53 and Rb pathways. This cell model may also be employed to identify relevant genes that are potentially worthy of further investigation as candidate biomarkers for early detection of pancreatic cancer.

ACKNOWLEDGMENT

This work was supported by Canadian Institutes of Health Research Grant MOP-49585.

REFERENCES

Bardeesy, N., and DePinho, R. A. (2002). Pancreatic cancer biology and genetics. *Nat. Rev. Cancer* **2**, 897–909.

Furukawa, T., Duguid, W. P., Rosenberg, L., Viallet, J., Galloway, D. A., and Tsao, M. S. (1996). Long-term culture and immortalization of epithelial cells from normal adult human pancreatic ducts transfected by the E6E7 gene of human papilloma virus 16. *Am. J. Pathol.* **148**, 1763–1770.

Grützmann, R., Pilarsky, C., Ammerpohl, O., Lüttges, J., Böhme, A., Sipos, B., Foerder, M., Alldinger, I., Jahnke, B., Schackert, H. K., Kalthoff, H., and Kremer, B. (2004)., *et al.* Gene expression profiling of microdissected pancreatic ductal carcinomas using high-density DNA microarrays*Neoplasia* **6**, 611–622.

Halbert, C. L., Demers, G. W., and Galloway, D. A. (1991). The E7 gene of human papillomavirus type 16 is sufficient for immortalization of human epithelial cells. *J. Virol.* **65**, 473–478.

Hilgers, W., and Kern, S. E. (1999). Molecular genetic basis of pancreatic adenocarcinoma. *Genes Chromosomes Cancer* **26**, 1–12.

Hruban, R. H., Adsay, N. V., Albores-Saavedra, J., Compton, C., Garrett, E. S., Goodman, S. N., Kern, S. E., Klimstra, D. S., Kloppel, G., Longnecker, D. S., Luttges, J., and Offerhaus, G. J. (2001). Pancreatic intraepithelial neoplasia: A new nomenclature and classification system for pancreatic duct lesions. *Am. J. Surg. Pathol.* **25**, 579–586.

Hruban, R. H., Wilentz, R. E., and Kern, S. E. (2000). Genetic progression in the pancreatic ducts. *Am. J. Pathol.* **156**, 1821–1825.

Logsdon, C. D., Simeone, D. M., Binkley, C., Arumugam, T., Greenson, J. K., Giordano, T. J., Misek, D. E., Kuick, R., and Hanash, S. (2003). Molecular profiling of pancreatic adenocarcinoma and chronic pancreatitis identifies multiple genes differentially regulated in pancreatic cancer. *Cancer Res.* **63**, 2649–2657.

Lundberg, A. S., Randell, S. H., Stewart, S. A., Elenbaas, B., Hartwell, K. A,, Brooks, M. W., Fleming, M. D., Olsen, J. C., Miller, S. W., Weinberg, R. A., and Hahn, W. C. (2002). Immortalization and transformation of primary human airway epithelial cells by gene transfer. *Oncogene* **21**, 4577–4586.

Nakamura, T., Furukawa, Y., Nakagawa, H., Tsunoda, T., Ohigashi, H., Murata, K., Ishikawa, O., Ohgaki, K., Kashimura, N., Miyamoto, M., Hirano, S., Kondo, S., *et al.* (2004). Genomewide cDNA microarray analysis of gene expression profiles in pancreatic cancers using populations of tumor cells and normal ductal epithelial cells selected for purity by laser microdissection. *Oncogene* **23**, 2385–400.

Ouyang, H., Mou, L., Luk, C., Liu, N., Karaskova, J., Squire, J., and Tsao, M. S. (2000). Immortal human pancreatic duct epithelial cell lines with near normal genotype and phenotype. *Am. J. Pathol.* **157**, 1623–1631.

Qian, J., Niu, J., Li, M., Chiao, P. J., and Tsao, M. S. (2005). *In vitro* modelling of human pancreatic duct epithelial cell transformation defines gene expression changes induced by *KRAS* oncogenic activation in pancreatic carcinogenesis. *Cancer Res.* **65,** 5045–5053.

Yeo, T. P., Hruban, R. H., Leach, S. D., Wilentz, R. E., Sohn, T. A., Kern, S. E., Iacobuzio-Donahue, C. A., Maitra, A., Goggins, M., Canto, M. I., Abrams, R. A., Laheru, D., *et al.* (2002). Pancreatic cancer. *Curr. Probl. Cancer* **26,** 176–275.

CHAPTER TWO

MOUSE MODEL FOR *NRAS*-INDUCED LEUKEMOGENESIS

Chaitali Parikh *and* Ruibao Ren

Contents

Abstract

Mutations that result in constitutive activation of RAS proteins are common in human hematological malignancies. In addition, functional activation of the RAS pathway can occur in leukemias, either due to mutations in genes that code for proteins upstream of RAS or due to inactivation of negative regulators of RAS. However, despite this prominent association of RAS activation with human leukemias, its precise role in leukemogenesis is not known. Previous studies have met with limited success in developing relevant animal models for leukemogenesis by oncogenic *NRAS*, the most frequently mutated *RAS* gene in human leukemias, and have suggested that oncogenic *RAS* might only act as a secondary event in leukemogenesis. This chapter describes an efficient and relevant murine model for myeloid leukemias initiated by oncogenic *NRAS* using an improved bone marrow transduction/transplantation system. This model provides a system for further studying the molecular mechanisms in the pathogenesis of myeloid malignancies and for testing targeted therapies.

Rosenstiel Basic Medical Sciences Research Center, Department of Biology, Brandeis University, Waltham, Massachusetts

Methods in Enzymology, Volume 439
ISSN 0076-6879, DOI: 10.1016/S0076-6879(07)00402-8

15

1. INTRODUCTION

RAS proteins are evolutionary conserved, small GTPases that are central regulators of cell fates (reviewed in Campbell *et al.*, 1998). They function as molecular switches by alternating between a GDP-bound inactive state and a GTP-bound active state. This process is regulated by two protein families: guanine nucleotide exchange factors (GEFs), such as SOS and RAS-GRF, which aid in the dissociation of GDP, and GTPase-activating proteins (GAPs), such as Nf1 and p120GAP, which accelerate the hydrolysis of GTP. On binding GTP, RAS undergoes a conformational change, which then allows it to bind to and activate a slew of downstream effector proteins, which are responsible for various cellular outcomes such as proliferation, differentiation, survival, or apoptosis (Shields *et al.*, 2000).

In humans, three *RAS* genes code for four highly homologous RAS proteins, namely NRAS, HRAS, and KRAS 4A and 4B. RAS proteins share 100% sequence identity in the N-terminal 80 amino acids, which includes the core effector-binding domain. However, they vary in their C-terminal, membrane-targeting domain and hence are localized to distinct microdomains of the plasma membrane and other endomembranes. It is possible that this differential subcellular localization of RAS proteins allows access to different effector proteins, which then results in distinct cellular outcomes (Hancock, 2003).

Mutations in *RAS* genes are associated with nearly 30% of human cancers, including solid tumors and hematological malignancies (Bos, 1989). Activating mutations in *RAS* genes are found in a broad range of myeloid malignancies, including acute myelogenous leukemia (AML), myelodysplastic syndromes (MDS), and myelodysplastic/myeloproliferative syndromes [MDS/MPD, including chronic myelomonocytic leukemia (CMML) and juvenile myelomonocytic leukemia (JMML)] (reviewed in Reuter *et al.*, 2000). Interestingly, the frequencies of mutations in different RAS genes vary among different types of human cancers. In myeloid malignancies, *NRAS* genes are mutated most frequently (70%), followed by *KRAS* (30%), while *HRAS* mutations are rare (Bos, 1988). Despite this prominent role of *RAS* in human leukemias, the precise molecular mechanisms and biochemical pathways that underlie the leukemogenic process are not known.

To gain a better understanding of a complex human disease such as leukemia requires the development of relevant animal models of the disease. However, previous efforts to generate mouse models for *RAS*-induced leukemias have met with limited success (summarized in Table 2.1). Expression of oncogenic *N* or *HRAS* in transgenic as well as bone marrow transduction/transplantation (BMT) models failed to recreate the myeloid leukemia phenotype associated with oncogenic RAS in human cancers. Transgenic mice expressing HRAS under the MMTV promoter developed

Table 2.1 Summary of mouse models for RAS leukemogenesis[a]

RAS isoform	Method/promoter	Disease	Reference
NRAS	Transgenic (IgH Eμ)	TLL	Haupt et al. (1992)
	Transgenic (hMRP8)	Epithelial tumors	Kogan et al. (1998)
	BMT (MoMuLV LTR)	Myeloid leukemia (long latency and incomplete penetrance)	MacKenzie et al. (1999)
	BMT (MSCV)	AML and CMML	Parikh et al. (2006)
HRAS	Transgenic (MMTV)	BLL	Sinn et al. (1987)
	BMT (MSCV LTR)	BLL and TLL	Hawley et al. (1995)
KRAS	Conditional knock in (endogenous promoter)	CMML	Braun et al. (2004); Chan et al. (2004)

[a] MoMuLV, Moloney murine leukemia virus; MMTV, mouse mammary tumor virus; TLL, T-lymphoid leukemia; BLL, B-lymphoid leukemia.

B lymphoblastic leukemia, whereas expression of *HRAS* in a bone marrow transduction/transplantation model induced B and T lymphoid leukemia/lymphoma (Hawley *et al.*, 1995; Sinn *et al.*, 1987). Transgenic mice expressing *NRAS* under the IgH Eμ enhancer or the hMRP8 promoter developed T lymphoid leukemias and epithelial tumors, respectively (Haupt *et al.*, 1992; Kogan *et al.*, 1998). In a BMT model where NRAS was expressed under the Moloney murine leukemia virus long terminal repeat (Mo-MuLV LTR), a fraction of the mice developed myeloid malignancies, but with a long disease latency and incomplete penetrance (MacKenzie *et al.*, 1999). The implication from results in transgenic as well as bone marrow transduction/transplantation studies was that oncogenic *RAS* is unable to efficiently induce myeloid malignancies in mice. The failure of previous models to induce myeloid leukemias in response to activated RAS can be attributed, at least in part, to certain drawbacks in the methods used. For instance, results from transgenic models might reflect the tissue specificities of the various promoters used rather than the precise role of oncogenic *RAS*. Another drawback of transgenic models is their inability to mimic the clonal evolution of cancers. Leukemia develops from somatic mutations in one or a few hematopoietic cells, whereas in transgenic models the onco-gene is expressed in a large population of cells (depending on the activity of the promoter) (Warner *et al.*, 2004). Additionally, the inefficient leukemo-genesis in previous models could reflect the absence of RAS expression in the relevant target cell population. For example, transcription from Mo-MuLV LTR is silenced in long-term reconstituting hematopoietic stem cells due to methylation (Challita and Kohn, 1994; Halene *et al.*, 1999). Because RAS-induced leukemia is believed to be a stem cell disease, the lack of expression of activated RAS in early progenitor cells may account for its poor leukemogenic potential in the BMT model.

These limitations can been overcome by using more efficient vectors that can target the *RAS* oncogenes to the correct target cells. Indeed, it has been shown that the expression of oncogenic *K* as well as *NRAS* in murine models can rapidly induce myeloid malignancies with complete penetrance (Braun *et al.*, 2004; Chan *et al.*, 2004; Parikh *et al.*, 2006). This chapter describes a bone marrow transduction/transplantation protocol that expresses oncogenic *NRAS* in murine bone marrow cells to develop a pathophysiolo-gically relevant mouse model for myeloid malignancies. The generation of this model provides an effective tool to further dissect the molecular pathways underlying the disease phenotype of RAS-induced myeloid malignancies. This type of research will pave the path for identifying critical drug targets, which is vital for designing rational therapies for human leukemias.

2. MODELING NRAS LEUKEMOGENESIS IN MICE USING THE BMT METHOD

2.1. Retroviral construct

The *NRAS D12* gene is amplified by polymerase chain reaction (PCR) from EST for human *NRAS* (GenBank Accession N44803). The activated mutant of *NRAS* is generated by PCR by introducing a point mutation in codon 12 (GGT to GAT), which changes glycine to aspartic acid (G12D). NRASD12 is expressed as a myc-tagged protein downstream of the IRES element, while GFP is expressed under the control of the murine stem cell virus long terminal repeat (MSCV LTR) (Fig. 2.1A).

2.2. Bone marrow transduction/transplantation protocol (Fig. 2.1)

We use 4- to 6-week-old Balb/c mice obtained from Taconic farms (Hudson, New York) as both donors and recipients of bone marrow. Mice are maintained on acid water both prior to bone marrow isolation

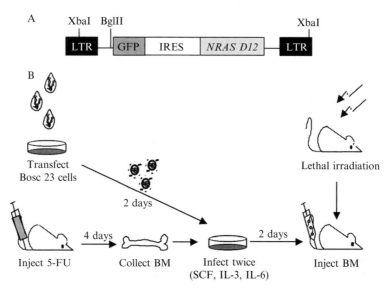

Figure 2.1 Schematic diagrams of the retroviral construct and bone marrow transduction/transplantation method. (A) A schematic diagram of the retroviral DNA construct used to transduce oncogenic NRAS. Bicistronic vectors were used that could coexpress RAS and GFP. Positions of restriction enzyme sites, as well as the probe used for Southern blot, are indicated. (B) Bosc 23 cells were transfected with the retroviral construct expressing mutant NRAS. Two days after transfection, the viral supernatant was collected and used to infect bone marrow cells isolated from 5-fluorouracil (5-FU)-treated donor mice. After culturing the cells for 2 days *in vitro*, they were injected into the tail veins of lethally irradiated recipient mice.

and after bone marrow transplantation. Acidified water helps prevent infections in these mice whose immune systems are compromised by irradiation.

Day 0

Inject 5 mg of 5-fluorouracil (5-FU) (25 mg/ml stock) into tail veins of donor mice. 5-FU is a pyrimidine analog that induces cell cycle arrest and apoptosis by inhibiting the ability of the cell to synthesize DNA (Longley *et al.*, 2003). Its role here is to enrich for the hematopoietic stem and progenitor cell population by killing the rapidly dividing bone marrow cells.

Split Bosc23 (viral packaging cell line) cells for transfections by plating 2 to 2.5 million cells per 60-mm plate (Pear *et al.*, 1993).

Day 1

Transfect Bosc23 cells by the calcium phosphate method. Add 500 μl of 2$\times$ HEPES-buffered saline (HBS) (refer to the end of this section for recipes) drop wise into a cocktail of 5 μg plasmid DNA, 62 μl 2 M CaCl$_2$, and 500 μl sterile water. After bubbling this solution for some time, add it to the plate of Bosc23 cells. Change the media in these plates 10 to 12 h after transfection.

Day 2

Split NIH/3T3 cells for infections by plating one to two million cells per 60-mm plate. These cells will be used for the determination of retroviral titers.

Day 3

Collect the viral supernatant from each plate and spin at 2500 rpm for 5 min to clear the supernatant of cell debris. The virus is relatively stable for 2 to 3 days if stored at 4°. Infect NIH/3T3 cells with serial dilutions of virus. For infections, add the appropriate volume of virus and supplement with medium to bring the volume up to 2 ml. To this add 8 μg/ml polybrene, mix, and incubate at 37°. Change the media 4 h after infection.

Day 5

a. Perform flow cytometry analysis on NIH/3T3 cells to determine percentage of infected (GFP+) cells. Determination of retroviral titers at this point is to facilitate the matching of titers for different constructs.
b. Infect NIH/3T3 cells with a normalized volume of virus, as determined for each construct from step a. This will give an approximation of the actual titers used to infect bone marrow cells.
c. Isolate bone marrow cells from 5–FU–treated mice (each step is performed in tissue culture hoods under sterile conditions). Sacrifice the mice and dissect out their long bones (femur and tibia). After stripping the muscles off the bones, extract bone marrow cells by flushing them with ice-cold media [Dulbecco's modified Eagle's medium (DMEM) + 10% fetal bovine serum (FBS) + penicillin/streptomycin]. The total bone marrow yield is determined by counting the cells in Turk's solution (the typical yield of bone marrow cells per 5–FU treated mouse is approximately 2 to 3 $\times$ 10^6). After counting, resuspend the cells at a concentration of 1 to 2 $\times$ 10^6/ml in the bone marrow cocktail medium specified later. This cocktail contains cytokines such as interleukin (IL)-3, IL-6, and stem cell factor (SCF), in addition to standard media, which are essential to maintain the stem/progenitor cells *ex vivo*. The appropriate volume of viral supernatant is added to this.

Recipe for bone marrow cocktail medium (enough for ≈10 million cells):

4.500 ml DMEM (no serum)
0.113 ml PSA mix (100$\times$ stock)(100 U/ml penicillin, 100 mg/μl streptomycin, and 0.25 mg/μl amphotericin)

0.113 ml L-glutamine $(100\times$ stock)$(2$ m$M)$
1.695 ml FBS (15% final)
0.565 ml WEHI-3B conditional medium (5% final)
4.0 μl IL-3 (7 ng/ml final, 20 μg/ml stock)
13.6 μl IL-6 (12 ng/ml final, 10 μg/ml stock)
316 μl SCF (56 ng/ml final, 2 μg/ml stock)
8.5 μl polybrene (3 μg/ml final, 4 mg/ml stock)
7.3 ml total volume

IL-3, IL-6, and SCF are available as powders and are reconstituted as follows:

IL-3: 20 μg/ml stock in 10% FBS in phosphate-buffered saline (PBS)
IL-6: 10 μg/ml stock in 10% FBS in PBS
SCF: 2 μg/ml stock in 1% bovine serum albumin in PBS

Resuspend cells in the aforementioned cocktail, plate them, and then add viral supernatant (4.0 ml, $\approx35\%$). The final volume will be approximately 11 ml. Incubate the infected bone marrow cells in a 37 °, 5% CO_2 incubator for 24 h.

Day 6

Repeat infection of bone marrow cells in the same cocktail medium specified earlier, with fresh viral supernatant, approximately 24 h after the first infection. This second round of infection serves to increase the total number of bone marrow cells transduced by the virus. Infect NIH/3T3 cells with the same volume of virus used for bone marrow infection. This will give an indication of the retroviral titers used here.

Day 7

This is the day of the actual bone marrow transplantation. Count the bone marrow cells to be transplanted by diluting in Turk's solution. Wash and resuspend the cells in $1\times$ PBS at a concentration of 2×10^6/ml. The cells should be kept on ice to prevent cell death, but while injecting, they should be at room temperature.

Lethally irradiate recipient mice (twice with a 3- to 4-h gap; 450 rads per dose). Reconstitute each recipient mouse by injecting 0.2 ml (400,000 cells) by tail vein injections. Perform flow cytometry on the first batch of infected NIH/3T3 cells to determine the titer for the first round of bone marrow infections.

Day 8

Perform flow cytometry on the second batch of infected NIH/3T3 cells to determine the titer for the second round of bone marrow infections.

2.3. Recipes for some reagents and solutions

Acid water: 20 liters of distilled water + 5 ml bleach + 3 ml of 12 N HCl

5-FU: available as a powder from Sigma–Aldrich (St. Louis, MO). Dissolve
    1 g of 5-FU in 40 ml of PBS to get a concentration of 25 mg/ml. It is
    important to maintain a pH of 9.0, as a higher pH will result in loss of
    activity.

Medium for Bosc23 cells: DMEM + 10% FBS + penicillin/streptomycin

Medium for NIH/3T3 cells: DMEM + 10% donor calf serum + penicillin/
    streptomycin

2× HEPES-buffered saline: 50 mM HEPES, pH 7.4, 10 mM KCl, 12 mM
    dextrose, 280 mM NaCl, 1.5 mM Na$_2$HPO$_4$, pH 7.05

Turk's solution: 0.1% gentian violet in 3% acetic acid

3. Characterization of Diseased *NRAS* Mice

We monitor the mice starting at 2 weeks postbone marrow transplantation by performing weekly peripheral white blood cell counts and flow
cytometry (to gauge the percentage of GFP+ cells). The mice are followed
closely and sacrificed when they are moribund (ruffled fur, cachexia, pallor).
Morphological and histopathological analyses, as well as flow cytometric
analysis, of diseased mice are performed by standard methods, following
guidelines from Kogan *et al.* (2002).

Analysis of tumor clonality is an important aspect of disease characterization, as it can shed light on the origin of the disease, specifically whether
the disease stemmed from a single or multiple transformed cells. This
provides clues to whether many oncogenic events are required for transformation or if a single event is sufficient. We determine clonality by Southern
blot analysis of tumor cells from leukemic mice. To analyze proviral
integration, we digest the genomic DNA with the enzyme *Bgl*II, which
has a unique site within the provirus, and for demonstrating the integrity of
the provirus, we digest DNA with the enzyme *Xba*I, which cuts within the
retroviral long terminal repeats (LTR) (see Fig. 2.1A). To generate a control
for single copy proviral integration, we performed single cell sorting of 32D
cells transduced with MSCV-*BCR/ABL*-IRES-*GFP*. We then tested single
cell clones by Southern blotting and isolated a cell line with a single copy
provirus. Comparison of intensity of a single copy proviral band with bands
from leukemic samples can differentiate between the occurrence of multiple
integrations within a single cell and clones of different cells with distinct
proviral integrations.

The growth and *in vitro* self-renewal capabilities of bone marrow progenitors from *NRAS D12* leukemic mice are assessed by methylcellulose

colony and serial replating assays. These assays are performed according to the manufacturer's instructions (Stem Cell Technologies, Vancouver, BC, Canada). Briefly, 1×10^4 mononuclear bone marrow cells are suspended in methylcellulose medium with and without cytokines [M3434, which contains 50 ng/ml recombinant murine (rm) SCF, 10 ng/ml rm IL-3, 10 ng/ml recombinant human (rh) IL-6, 3 units/ml rh erythropoietin, 10 μg/ml rh insulin, 200 μg/ml human transferrin, and M3231, which does not contain any cytokines, respectively; Stem Cell Technologies] and plated in 1-ml duplicate cultures. Count the individual colonies after 7 days in culture (in 5% CO_2 at 37°). For serial-replating assays, colonies are harvested every 7 days and 1×10^4 cells are replated (in duplicate) for each round.

The inheritability of leukemias can be assessed *in vivo* by performing secondary transplantation experiments. Each primary leukemic can serve as the donor for four to five secondary recipient mice. Bone marrow cells from leukemic mice are isolated as described earlier. A single cell suspension of spleen tissue from leukemic mice is prepared by crushing the spleen in PBS and repeated pipetting. Inject 1×10^6 bone marrow cells plus 1×10^6 splenic cells from primary leukemic mice (along with 1×10^5 normal bone marrow cells) into the tail veins of lethally irradiated recipient mice. Disease progression is followed as outlined previously.

ACKNOWLEDGMENT

This work was supported by grants from the National Cancer Institute (CA68008, R.R.) and the National Heart, Lung, and Blood Institute (HL083515, R.R.).

REFERENCES

Bos, J. L. (1988). Ras oncogenes in hematopoietic malignancies. *Hematol. Pathol.* **2**, 55–63.

Bos, J. L. (1989). Ras oncogenes in human cancer: A review. *Cancer Res.* **49**, 4682–4689.

Braun, B. S., Tuveson, D. A., Kong, N., Le, D. T., Kogan, S. C., Rozmus, J., Le Beau, M. M., Jacks, T. E., and Shannon, K. M. (2004). Somatic activation of oncogenic Kras in hematopoietic cells initiates a rapidly fatal myeloproliferative disorder. *Proc. Natl. Acad. Sci. USA* **101**, 597–602.

Campbell, S. L., Khosravi-Far, R., Rossman, K. L., Clark, G. J., and Der, C. J. (1998). Increasing complexity of Ras signaling. *Oncogene* **17**, 1395–1413.

Challita, P. M., and Kohn, D. B. (1994). Lack of expression from a retroviral vector after transduction of murine hematopoietic stem cells is associated with methylation *in vivo*. *Proc. Natl. Acad. Sci. USA* **91**, 2567–2571.

Chan, I. T., Kutok, J. L., Williams, I. R., Cohen, S., Kelly, L., Shigematsu, H., Johnson, L., Akashi, K., Tuveson, D. A., Jacks, T., and Gilliland, D. G. (2004). Conditional expression of oncogenic K-ras from its endogenous promoter induces a myeloproliferative disease. *J. Clin. Invest.* **113**, 528–538.

Halene, S., Wang, L., Cooper, R. M., Bockstoce, D. C., Robbins, P. B., and Kohn, D. B. (1999). Improved expression in hematopoietic and lymphoid cells in mice after transplantation of bone marrow transduced with a modified retroviral vector. *Blood* **94**, 3349–3357.

Hancock, J. F. (2003). Ras proteins: Different signals from different locations. *Nat. Rev. MolCell Biol.* **4**, 373–384.

Haupt, Y., Harris, A. W., and Adams, J. M. (1992). Retroviral infection accelerates T lymphomagenesis in E mu-N-ras transgenic mice by activating c-myc or N-myc. *Oncogene* **7**, 981–986.

Hawley, R. G., Fong, A. Z., Ngan, B. Y., and Hawley, T. S. (1995). Hematopoietic transforming potential of activated ras in chimeric mice. *Oncogene* **11**, 1113–1123.

Kogan, S. C., Lagasse, E., Atwater, S., Bae, S. C., Weissman, I., Ito, Y., and Bishop, J. M. (1998). The PEBP2betaMYH11 fusion created by Inv(16)(p13;q22) in myeloid leukemia impairs neutrophil maturation and contributes to granulocytic dysplasia. *Proc. Natl. Acad. Sci. USA* **95**, 11863–11868.

Kogan, S. C., Ward, J. M., Anver, M. R., Berman, J. J., Brayton, C., Cardiff, R. D., Carter, J. S., de Coronado, S., Downing, J. R., Fredrickson, T. N., Haines, D. C., Harris, A. W., *et al.* (2002). Bethesda proposals for classification of nonlymphoid hematopoietic neoplasms in mice. *Blood* **100**, 238–245.

Longley, D. B., Harkin, D. P., and Johnston, P. G. (2003). 5-Fluorouracil: Mechanisms of action and clinical strategies. *Nat. Rev. Cancer* **3**, 330–338.

MacKenzie, K. L., Dolnikov, A., Millington, M., Shounan, Y., and Symonds, G. (1999). Mutant N-ras induces myeloproliferative disorders and apoptosis in bone marrow repopulated mice. *Blood* **93**, 2043–2056.

Parikh, C., Subrahmanyam, R., and Ren, R. (2006). Oncogenic NRAS rapidly and efficiently induces CMML- and AML-like diseases in mice. *Blood* **108**, 2349–2357.

Pear, W. S., Nolan, G. P., Scott, M. L., and Baltimore, D. (1993). Production of high-titer helper-free retroviruses by transient transfection. *Proc. Natl. Acad. Sci. USA* **90**, 8392–8396.

Reuter, C. W., Morgan, M. A., and Bergmann, L. (2000). Targeting the Ras signaling pathway: A rational, mechanism-based treatment for hematologic malignancies? *Blood* **96**, 1655–1669.

Shields, J. M., Pruitt, K., McFall, A., Shaub, A., and Der, C. J. (2000). Understanding Ras: 'it ain't over 'til it's over.' *Trends Cell Biol.* **10**, 147–154.

Sinn, E., Muller, W., Pattengale, P., Tepler, I., Wallace, R., and Leder, P. (1987). Coexpression of MMTV/v-Ha-ras and MMTV/c-myc genes in transgenic mice: Synergistic action of oncogenes *in vivo*. *Cell* **49**, 465–475.

Warner, J. K., Wang, J. C., Hope, K. J., Jin, L., and Dick, J. E. (2004). Concepts of human leukemic development. *Oncogene* **23**, 7164–7177.

CHAPTER THREE

Inducible BRAF Suppression Models for Melanoma Tumorigenesis

Klaus P. Hoeflich,* Bijay Jaiswal,* David P. Davis,* *and* Somasekar Seshagiri*

Contents

Abstract

Somatic mutations in BRAF have been reported in 50 to 70% of melanomas. The most common mutation is a valine to glutamic acid substitution at codon 600 (V600E). ^{V600E}BRAF constitutively activates ERK signaling and promotes proliferation, survival, and tumor growth. However, although BRAF is mutated in up to 80% of benign nevi, they rarely progress into melanoma. This implicates the BRAF mutation to be an initiating event that requires additional lesions in the genome for full-blown progression to melanoma. Even though the mutations appear early during the pathogenesis of melanoma, targeted BRAF knockdown using inducible shRNA in melanoma cell lines with BRAF mutations shows that BRAF is required for growth and maintenance of tumor in xenograft models.

* Genentech Inc., Department of Molecular Biology, South San Francisco, CA

Methods in Enzymology, Volume 439
ISSN 0076-6879, DOI: 10.1016/S0076-6879(07)00403-X

1. INTRODUCTION

RAF serine/threonine kinases, ARAF, BRAF, and CRAF/Raf-1, are part of a conserved three-component kinase signaling module that regulates various cellular processes, including proliferation, apoptosis, and different-iation (Wellbrock *et al.*, 2004). RAF is activated by phosphorylation of residues in its catalytic loop upon binding the small G protein RAS through its RAS-binding domain. RAF then phosphorylates and activates the dual-specificity protein kinases MEK1 and MEK2. Active MEK1/2 in turn activates mitogen-activated protein kinases (MAPK) ERK1 and ERK2. ERK activation, depending on the cellular context, regulates a number of biological processes, including cell growth, survival, and differentiation.

The occurrence of oncogenic RAS mutations and hyperactivation of ERK in cancer is well established (Allen *et al.*, 2003; Downward, 2003). Activating oncogenic mutations in BRAF have been described in various cancers, particularly in melanoma ($\approx$70%), thyroid ($\approx$30%), colorectal ($\approx$15%), and ovarian cancers ($\approx$30%) (Davies *et al.*, 2002; Dhomen *et al.*, 2007; Wellbrock *et al.*, 2004). Among the various BRAF mutations, $\approx$95% of the missense mutations result in the substitution of valine at position 600 by a glutamic acid (^{V600E}BRAF). Further, $\approx$98% of all BRAF mutations are in the activation segment and the remaining mutations occur in the G loop.

Mutant ^{V600E}BRAF has elevated kinase activity and is a potent activator of ERK (Wan *et al.*, 2004). Furthermore, it can transform cells and promote proliferation and anchorage-independent growth (Mercer *et al.*, 2005; Wellbrock *et al.*, 2004). Studies on the occurrence of BRAF mutations during melanoma pathogenesis show that the mutations are present in over 80% of nevi (Pollock *et al.*, 2003; Yazdi *et al.*, 2003), suggesting that it is an initiating event. However, only a small proportion of these nevi progress into melanoma, suggesting the need for additional genome alterations for melanoma progression (Gray-Schopfer *et al.*, 2007).

To understand the requirement of BRAF postinitiation in tumor growth and maintenance, as well as its validity as a bona fide melanoma target, we developed a doxycycline (Dox)-inducible short hairpin RNA (shRNA) system to enable conditional BRAF silencing in ^{V600E}BRAF mutant mela-noma cell lines (Fig. 3.1). Induction of BRAF-specific shRNA results in knockdown of BRAF by 72 h postinduction, leading to reduced phos-phorylated ERK (Fig. 3.2A) and decreased proliferation (Hoeflich *et al.*, 2006). These cell lines form tumors in immunodeficient *nu/nu* mice when implanted subcutaneously. When treated with 1 to 2 mg/ml Dox plus 5% sucrose to induce the BRAF shRNA, tumor bearing mice show tumor regression in the LOX-IMVI model or stasis in the A375 model. In contrast, mice bearing tumors treated with 5% sucrose alone continued to grow

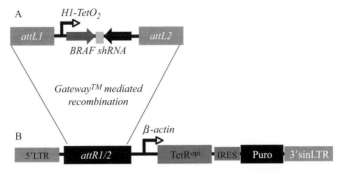

Figure 3.1   Vectors: Schematic representation of the retroviral vector system for tetra-cycline/doxycycline-inducible synthesis of shRNA. This vector system is composed of a shuttle plasmid for shRNA cloning (A) and a retroviral destination vector (B). To gener-ate BRAF targeting shRNA constructs, the sense and antisense oligonucleotides listed in Table 3.1 are annealed and cloned into the shuttle plasmid using *Bgl*II and *Hind*III. The resulting sequence-verified H1 promoter-shRNA cassette is then transferred to the viral destination vector via Gateway-mediated recombination (Invitrogen, Carlsbad CA). The Tet-repressor is constitutively expressed from the β-actin promoter. Therefore, in the absence of Dox the TetR binds to TetO$_2$ within the H1 promoter and prevents shRNA expression. Addition of Dox results in dissociation of the TetR protein from TetO$_2$ and derepression of the shRNA transcriptional unit. TetR, Tet-repressor protein; TetO$_2$, tetracycline operator 2; IRES, internal ribosomal entry site; PURO, puromycin resistance gene; LTR, long terminal repeat; attL/R- recombination sites.

(see Fig. 3.2B and C). Further, control lines expressing an inducible shRNA targeting luciferase or GFP showed no effect on tumor growth in either the presence or the absence of Dox (see Fig. 3.2B and C). These studies show the requirement for BRAF signaling for *in vivo* tumor growth in these lines and confirm BRAF to be an important therapeutic target.

Given the ability to regulate BRAF expression at will during tumor development, we sought to exploit this to understand the requirement of BRAF in tumor maintenance by cycling Dox treatment. In order to do this, we subcutaneously implanted in nude mice the engineered LOX-IVMI line described earlier. When the tumors reached ≈1500-mm³ volume, Dox was then added to the drinking water of mice bearing these large tumors to induce expression of the BRAF shRNA. These mice showed a visible decrease in tumor volume 5 days post-BRAF shRNA induction, and by 2 weeks there was a significant reduction in tumor volume (Fig. 3.3A). This suggested the requirement of BRAF for tumor maintenance, even after the tumors were effectively very large. In another experiment, mice with subcutaneous LOX-IMVI/BRAF-shRNA tumors were placed on Dox when tumors reached ≈200 mm³. The tumors regressed while under Dox treatment; however, when these mice were taken off Dox, thereby restoring BRAF expression, mice with a palpable tumor showed tumor regrowth (see Fig. 3.3B). This study indicates that restoration of BRAF signaling in these cells is important and sufficient for tumor formation.

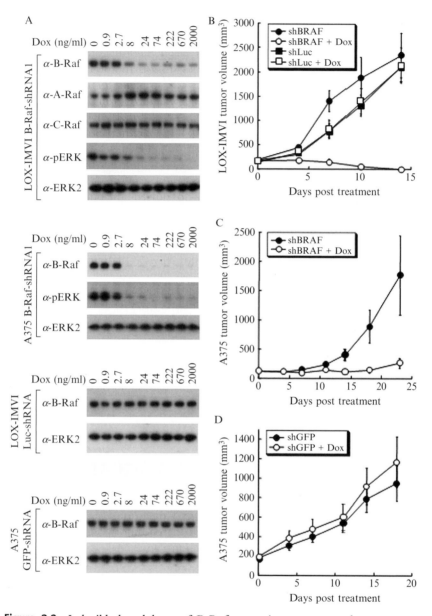

Figure 3.2 Inducible knockdown of B-Raf expression prevents melanoma tumor growth. (A) Experimental validation of B-Raf knockdown in melanoma cell lines. LOX-IMVI and A375 cell clones stably expressing B-Raf shRNA or control GFP and luciferase (Luc) shRNAs were treated with the indicated Dox concentrations for 72 h. Lysates were then analyzed by immunoblotting. (B–D) B-Raf shRNA knockdown demonstrates antitumor efficacy in xenograft models. LOX-IMVI and A375 inducible shRNA cells were implanted subcutaneously in the flank of athymic mice as described in the text. Treatment in each experiment was initiated on the day when mice had tumors

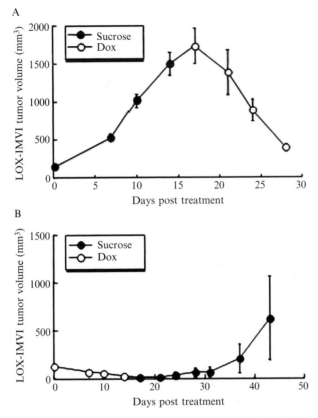

Figure 3.3 B-Raf knockdown is reversible and tightly regulated *in vivo*. (A) BRAF-dependent LOX-IMVI neoplasias were allowed to grow for 14 days before administration of Dox was initiated to knockdown BRAF-dependent signaling and tumorigenesis. (B) Dox-treated mice with regressing subcutaneous tumors that are subsequently removed from Dox at day 14 undergo tumor recurrence (adapted from Hoeflich *et al.*, 2006).

The conditional ablation of BRAF allowed us to examine the mechanism of tumor regression through the immunohistochemical analysis of tumors. We determined that cells in the LOX–IMVI model undergo apoptotic cell death following BRAF knockdown, as evidenced by increased caspase-3 levels, leading to tumor regression (Hoeflich *et al.*, 2006).

---

ranging in size from 100 to 150 mm³. Administration of 2 mg/ml Dox via drinking water produced regression in (B) LOX–IMVI or stasis in (C) A375 tumors expressing an inducible B-Raf-specific shRNA. (D) GFP or (B) Luciferase control shRNAs did not affect tumor growth kinetics. No lethality or weight loss was observed (adapted from Hoeflich *et al.*, 2006).

In an effort to understand the role of BRAF in metastatic melanoma, we engineered A375M, a metastatic melanoma line, to inducibly express BRAF using the Dox-regulated shRNA vector system (Fig. 3.4A). This line

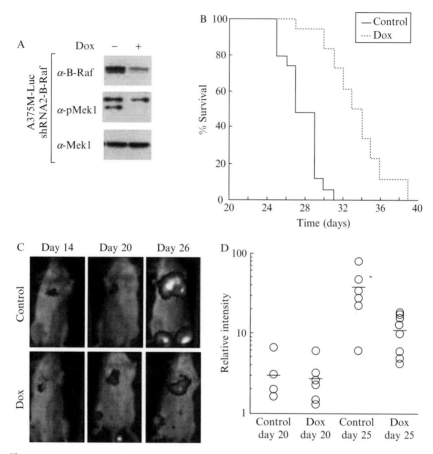

Figure 3.4 Reduction of A375M systemic tumor growth by B-Raf shRNA knockdown. (A) Western blot analysis showing expression of B-Raf and phosphorylation of Mek1 in uninduced cells (lane 1) and cells treated with 2 mg/ml Dox for 72 h (lane 2). Total Mek1 serves as an internal control to show equal loading. (B) Kaplan–Meier survival data of *scid-beige* mice injected intravenously with 4×10^5 A375M-luc/shRNA-B-Raf cells and receiving drinking water containing 5% sucrose only (control) or sucrose with 1 mg/ml Dox. Animals were monitored for tumor onset and illness until they reached a terminal stage and were euthanized. Each group consisted of at least 10 mice. The reduction in tumor growth conferred by Dox-mediated B-Raf knockdown is significant according to the log-rank test, $p < 0.0001$. (C) Representative *in vivo* bioluminescence imaging of mice and (D) quantification of tumor burden of mice receiving Dox versus sucrose-treated control mice. Homogeneous cohorts of mice with established tumor lesions were divided into treatment groups 2 weeks after injection of A375M-luc/shRNA-B-Raf cells. Bioluminescence is represented relative to intensity at day 14 for each animal (adapted from Hoeflich *et al.*, 2006). (See color insert.)

also carried a luciferase reporter gene to facilitate *in vivo* imaging. Following tail vein injection of the engineered A375M cells, mice were placed on Dox to induce BRAF shRNA and were observed for tumor formation by imaging. The overall survival of Dox-treated animals showed a statistically significant increase compared to untreated animals, indicating the requirement of BRAF in the establishment of metastastis (see Fig. 3.4B). In studies where BRAF knockdown was induced after establishment of systemic tumors, a delayed progression in metastatic tumors was evident by bioluminescent imaging and quantification (see Fig. 3.4C and D). These studies show anti-BRAF therapy as an effective strategy for treating BRAF mutant melanomas.

2. Materials and Methods

2.1. Inducible shRNA vector system

To study the effect of BRAF knockdown on the *in vivo* growth of melanoma, we created a Dox-regulated shRNA expression system (Gray *et al.*, 2007; Hoeflich *et al.*, 2006). Briefly, this vector system is composed of a shuttle plasmid for shRNA cloning (see Fig. 3.1A) and a retroviral destination vector (see Fig. 3.1B). The shuttle plasmid contains a H1 promoter (Brummelkammp *et al.*, 2002; Myslinski *et al.*, 2001) modified with a tetracycline operator 2 (TetO$_2$) (Hillen *et al.*, 1984) inserted between the TATA box and the transcriptional start site. To generate BRAF-targeting shRNA constructs, the sense and antisense oligonucleotides listed in Table 3.1 are annealed and cloned into the shuttle plasmid using *Bgl*II and *Hin*dIII. The resulting sequence-verified H1 promoter-shRNA cassette is then transferred to the appropriate viral destination vector via Gateway-mediated recombination (Invitrogen, Carlsbad, CA). As illustrated in Fig. 3.1B, the destination vector contains a second expression cassette consisting of a human β-actin promoter driving a wild-type tetracycline repressor (TetR) (Yao *et al.*, 1998) upstream of an internal ribosomal entry site (IRES) and the puromycin marker for selection. The retroviral vector backbone is derived from pQXCIP (Clontech, Palo Alto, CA).

In the "off" state, the TetR, expressed from the human β-actin promoter, binds the modified H1 promoter, thereby preventing shRNA expression. However, in the presence of a tetracycline analog (Dox), the TetR protein is released from the TetO$_2$ within the H1 promoter (Hillen *et al.*, 1984), resulting in shRNA transcription and knockdown of endogenous BRAF expression.

Table 3.1 BRAF and control RNAi oligonucleotides

Viral system	Oligonucleotide name	Oligonucleotide sequence
Retrovirus	BRAF shRNA-1 (sense)	5′-GAT CCC CAG AAT TGG ATC TGG ATC ATT TCA AGA GAA TGA TCC AGA TCC AAT TCT TTT TTT GGA AA-3′
	BRAF shRNA-1 (antisense)	5′-AGC TTT TCC AAA AAA AGA ATT GGA TCT GGA TCA TTC TCT TGA AAT GAT CCA GAT CCA ATT CTG GG-3′
	BRAF shRNA-2 (sense)	5′-GAT CCC CGC TAC AGA GAA ATC TCG ATT TCA AGA GAA TCG AGA TTT CTC TGT AGC TTT TTT GGA AA-3′
	BRAF shRNA-2 (antisense)	5′-AGC TTT TCC AAA AAA GCT ACA GAG AAA TCT CGA TTC TCT TGA AAT CGA GAT TTC TCT GTA GCG GG-3′
	Luciferase shRNA (sense)	5′-GAT CCC CCT TAC GCT GAG TAC TTC GAT TCA AGA GAT CGA AGT ACT CAG CGT AAG TTT TTT GGA AA-3′
	Luciferase shRNA (antisense)	5′-AGC TTT TCC AAA AAA CTT ACG CTG AGT ACT TCG ATC TCT TGA ATC GAA GTA CTC AGC GTA AGG GG-3′
	EGFP shRNA (sense)	5′-GAT CCC CAG ATC CGC CAC AAC ATC GAT TCA AGA GAT CGA TGT TGT GGC GGA TCT TGT TTT TTG GAA A-3′
	EGFP shRNA (antisense)	5′-AGC TTT TCC AAA AAA CAA GAT CCG CCA CAA CAT CGA TCT CTT GAA TCG ATG TTG TGG CGG ATC TGG G-3′

2.2. Generation of BRAF-inducible shRNA cell clones

Melanoma A375, A375M (American Type Culture Collection, Manassas, VA), and LOX-IMVI (NCI-60) cells are maintained at 37° and 5% CO_2 in Dulbecco's modified Eagle's medium or RPMI 1640 media, respectively, with 10% tetracycline-free fetal bovine serum, 4 mM L-glutamine, and penicillin/streptomycin. Both LOX-IMVI and A375 cell lines carry activated BRAF^{V600E} alleles and show strong BRAF/MEK/ERK signaling. Retrovirus infection is performed using Phoenix packaging cells according to the manufacturer's instructions (Orbigen, San Diego, CA). As the puromycin resistance gene encoded in the vector is under the control of a constitutive β-actin promoter, 2 to 5 μg/ml puromycin can be used to select infected cells expressing shRNA. Aliquots of the isolated stable clones should be stored in liquid nitrogen.

To ensure that any observed phenotypes result from gene silencing of the BRAF target gene and not of unintended off-target transcripts, two distinct BRAF-specific shRNAs are incorporated into these experiments to increase the confidence with which the observed changes in melanoma tumor growth can be directly linked to BRAF silencing. The first shRNA we validated for BRAF knockdown (see Fig. 3.2) corresponds to the translated sequence just following the G loop of the kinase domain (amino acids 461 to 467) in which no oncogenic mutations have been described to date. Accordingly, another hairpin specific to another region of the BRAF transcript (encoding amino acids 597 to 603 and the V600E mutation) was also selected and characterized (Hoeflich et al., 2006). Equivalent results were obtained using either shRNA construct.

2.3. In vitro characterization of BRAF-inducible shRNA cell clones

Stable clones are plated in 96-well tissue culture plates and are treated with 1 mg/ml Dox (BD Clontech) for 3 days. RNA is prepared using Turbocapture mRNA Geneplate (Qiagen, Valencia, CA), and endogenous BRAF knockdown is assessed by a quantitative reverse transcriptase polymerase chain reaction. Clones are characterized for the level of endogenous BRAF mRNA expression and extent of Dox regulation. Cell clones showing greater than 70% knockdown upon Dox treatment are further characterized by Western blot for changes in BRAF protein expression and phosphorylated ERK1/2. Cell lysates are prepared using modified RIPA buffer containing 50 mM Tris, pH 7.4, 150 mM NaCl, 1 mM EDTA, 1% Brij-35, 0.1% deoxycholate, protease inhibitors (Roche Molecular Biochemicals), and a phosphatase inhibitor cocktail (Sigma). SDS-PAGE (4 to 12% gel) is used to resolve the proteins in the lysate. After electrophoresis, the proteins are electrotransferred onto a polyvinylidene fluoride microporous membrane and immunodetected using standard procedures. Antibodies used for Western blotting are as follows: anti-ERK2, anti-p-ERK1/2 (Thr202/Tyr204), anti-MEK1, anti-p-MEK1 (Ser217/221), and anti-ARAF (Cell Signaling Technology); anti-BRAF (F-7; Santa Cruz Biotechnology); anti-RAF1 (BD Transduction Labs); anti-h-actin (Sigma Life Science); and horseradish peroxidase-conjugated secondary antibodies (Pierce Biotechnology).

For our studies, densitometry quantification of immunoblots revealed an effective BRAF protein knockdown of ≈80% and ≈98% for LOX-IMVI and A375 cell clones, respectively. The suppression of BRAF protein levels is dose dependent with a cellular IC_{50} of approximately 5 ng/ml. BRAF knockdown is reversible and time dependent, with maximal mRNA depletion detected 2 days postinduction and the corresponding protein depletion occurring at day 3. Several independent LOX-IMVI and A375 clones should be characterized to ensure against a clonal selection bias.

Following biochemical characterization of inducible shRNA clones, phenotypic cell culture assays can be performed. Upon Dox addition, LOX-IMVI and A375 cells lacking BRAF show consistent changes in two-dimensional properties as compared with control shRNA-infected cells. These phenotypes include reduced cell proliferation and a flattened epithelial-like cell morphology change. To determine the effect of BRAF ablation on cell cycle progression, A375 cells expressing either BRAF or control GFP shRNAs are cultured in 0.1% serum in the presence or absence of 1 mg/ml Dox. At 2-day intervals, viable cell counts are determined by the Trypan blue exclusion method using a Vi-Cell analyzer (Beckman Coulter) (Hoeflich *et al.*, 2006).

2.4. Role of BRAF in tumor maintenance and progression

To examine whether ablation of BRAF function in LOX-IMVI and A375 melanoma cells might affect their ability to form tumors *in vivo*, subcutaneous tumor models were established for these cell lines. In our studies, 6- to 8-week-old female *nu/nu* mice (Charles River Laboratories) are injected in the right flank with either 3×10^6 human LOX-IMVI or 1×10^7 human A375 shRNA-containing cell clones resuspended into 200 μl phosphate-buffered saline. When tumors reach a mean volume of approximately 150 mm^3, mice with similarly sized tumors are grouped into treatment cohorts. Tumor volumes are measured in two dimensions (length and width) using Ultra Cal-IV calipers (Fred V. Fowler Company) using the formula of tumor volume (mm^3) = (length × width^2) × 0.5. Between 7 and 10 mice are used for each treatment group, and results are presented as mean tumor volume ± SEM. At the end of the dosing study, or as indicated, appropriate tumor samples can be taken. In our studies, Dox-mediated knockdown of BRAF completely inhibited LOX-IMVI tumorigenesis *in vivo* and led to tumor regression (see Figs. 3.2B and 3.3), even despite the incomplete depletion of BRAF as shown *in vitro* for the selected clone. Complete responses, defined as 100% tumor regression from the initial starting tumor volume at any day during the study, were observed in 6/10 animals in the Dox treatment cohort. In A375 xenografts, BRAF shRNA induction also halted tumor progression (see Fig. 3.2C); however, the tumors did not regress as observed in studies involving the LOX-IMVI cell line. There was no discernible effect on tumor growth observed with cells expressing control shRNAs.

For oral administration of Dox, mice receive 5% sucrose only or 5% sucrose plus 1 mg/ml Dox for control and knockdown cohorts, respectively. Dox is prepared in 1-gallon carboys and stored at 4° for up to 1 month. All drinking water bottles are changed three times per week. Because Dox is light sensitive, dark-colored bottles should be used. A variety of Dox administration regimens have been tested throughout the course of these BRAF studies and we have found this to the optimal way to deliver Dox to mice for short-term xenograft studies. In-house comparison with a Dox chow diet (6 g/kg Dox in solid pellets; Bio-Serv) showed that Dox administration in drinking water results in more rapid *in vivo* knockdown and is more cost effective (data not shown).

To further explore whether inactivation of oncogenic signaling is sufficient for the elimination of well-established tumors, moribund mice with very large ($\approx$1500 mm^3) tumor volumes can be switched to Dox. An example of such a study is shown in Fig. 3.3. Within 5 days post-BRAF shRNA induction, the LOX-IMVI tumor volume had decreased visibly and after 2 weeks the tumors had grossly regressed (see Fig. 3.3). Furthermore, the effect of restoring BRAF expression in regressing tumors by discontinuing treatment in a Dox cohort can be tested. Upon Dox withdrawal, tumor recurrence was observed in mice that still had palpable subcutaneous tumors (see Fig. 3.3). This approach is useful to show that knockdown of BRAF does not lead to an irreversible cascade of molecular events in tumor cells and that prolonged BRAF suppression is necessary to eliminate tumors in preclinical models.

2.5. *In vivo* mechanism of action

Because understanding the mechanism of action of a drug target is a key step in the drug development cycle, tumor histological analysis is performed to define the spectrum of cellular responses that can be caused by targeted BRAF inhibition. LOX-IMVI tumors from mice treated with Dox for 1 to 7 days are harvested, and formalin-fixed, paraffin-embedded specimens are collected. Following routine hematoxylin and eosin evaluation of the slides, immunohistochemical staining is performed on 5-μm-thick paraffin-embedded sections using anti-Ki-67 (MIB-1, DakoCytomation), anticleaved caspase-3 (Cell Signaling Technology), and antipanendothelial cell marker (MECA-32, Pharmingen) antibodies with a standard avidin–biotin horseradish peroxidase detection system according to the manufacturer's instructions. Tissues are counterstained with hematoxylin, dehydrated, and mounted. In all cases, antigen retrieval is performed with the DAKO target retrieval kit as per the manufacturer's instructions.

Compared to xenografts from control animals, tumors from Dox-treated mice exhibited a profound decrease in Ki67-positive proliferating tumor cells and an increase in scattered apoptotic cells as determined by immuno-chemical staining for activated caspase-3 (Hoeflich *et al.*, 2006). The magnitude of these phenotypes reaches a stable maximum following 4 days for *in vivo* Dox treatment. BRAF signaling does not play a pivotal role in regulating tumor vascularization as determined by staining of endothelial cells (Hoeflich *et al.*, 2006). In summary, data obtained from this approach support the current view that BRAF signaling is important for mediating cellular proliferation and survival in tumorigenesis.

2.6. Assessment of metastatic tumor development by BRAF knockdown

Given that metastases are the predominant cause of melanoma-associated death, it is useful to provide preclinical validation for targeting BRAF in the context of an experimental metastasis model. To address this, A375M cells that have been selected for high metastatic ability can be utilized (Collisson *et al.*, 2003). Tail vein injection of 4×10^5 A375M cells (total volume of 50 μl) into female *scid-beige* mice leads to pulmonary, ovarian, and adrenal tumors after a relatively short latency. After engineering A375M cells to express BRAF-specific shRNA for Dox-regulatable knockdown of BRAF protein and signaling, mice can be monitored longitudinally for tumor onset, progression, survival, and response to BRAF knockdown. Using this approach, BRAF ablation significantly slowed tumor growth and prolonged the survival of mice ($p < 0.0001$, log-rank test), and the progression of disease was still partially inhibited when the induction of knockdown was delayed until systemic tumors were well established (see Fig. 3.4). These experiments provide evidence supporting anti-BRAF therapy as a promising strategy to inhibit certain metastatic tumors.

In our metastatic tumor model experiments, A375M cell clones were used that had been previously engineered to constitutively express a firefly luciferase protein (Ray *et al.*, 2004) and bioluminescence images could then be acquired using a cooled intensified charge-coupled device camera. Tumor progression was monitored by weekly bioluminescence imaging for luciferase and mice were monitored daily for survival. However, an alternate and more accessible method is *ex vivo* staining of tumor-bearing lungs via tracheal injection of 15% India ink. Whole lungs are soaked in water for 5 min and are fixed in Fekete's solution (70% ethanol, 10% formaldehyde, 5% glacial acetic acid) for 24 h. This method allows for bleaching of tumor colonies against the black background of the stained lungs. Superficial tumor burden can then be scored by inspection.

3. Conclusion

The inducible shRNA gene knockdown system can be used both *in vitro* and *in vivo* to understand the role of BRAF in melanoma.

ACKNOWLEDGEMENT

Parts of this work were adapted from our work published in Hoeflich *et al.* (2006).

REFERENCES

Allen, L. F., Sebolt-Leopold, J., and Meyer, M. B. (2003). CI-1040 (PD184352), a targeted signal transduction inhibitor of MEK (MAPKK). *Semin. Oncol.* **30**, 105–116.

Brummelkamp, T. R., Bernards, R., and Agami, R. (2002). A system for stable expression of short interfering RNAs in mammalian cells. *Science* **296**, 550–553.

Collisson, E. A., Kleer, C., Wu, M., De, A., Gambhir, S. S., Merajver, S. D., and Kolodney, M. S. (2003). Atorvastatin prevents RhoC isoprenylation, invasion, and metastasis in human melanoma cells. *Mol. Cancer Ther.* **2**, 941–948.

Davies, H., Bignell, G. R., Cox, C., Stephens, P., Edkins, S., Clegg, S., Teague, J., Woffendin, H., Garnett, M. J., Bottomley, W., Davis, N., Dicks, E., *et al.* (2002). Mutations of the BRAF gene in human cancer. *Nature* **417**, 949–954.

Dhomen, N., and Marais, R. (2007). New insight into BRAF mutations in cancer. *Curr. Opin. Genet. Dev.* **17**, 31–39.

Downward, J. (2003). Targeting RAS signalling pathways in cancer therapy. *Nat. Rev. Cancer* **3**, 11–22.

Gray, D. C., Hoeflich, K. P., Peng, L., Gu, Z., Gogineni, A., Murray, L. J., Eby, M., Kljavin, N., Seshagiri, S., Cole, M. J., and Davis, D. P. (2007). pHUSH: A single vector system for conditional gene expression. *BMC Biotechnol.* **7**, 61.

Gray-Schopfer, V., Wellbrock, C., and Marais, R. (2007). Melanoma biology and new targeted therapy. *Nature* **445**, 851–857.

Hillen, W., Schollmeier, K., and Gatz, C. (1984). Control of expression of the Tn10-encoded tetracycline resistance operon. II. Interaction of RNA polymerase and TET repressor with the tet operon regulatory region. *J. Mol. Biol.* **172**, 185–201.

Hoeflich, K. P., Gray, D. C., Eby, M. T., Tien, J. Y., Wong, L., Bower, J., Gogineni, A., Zha, J., Cole, M. J., Stern, H. M., Murray, L. J., Davis, D. P., *et al.* (2006). Oncogenic BRAF is required for tumor growth and maintenance in melanoma models. *Cancer Res.* **66**, 999–1006.

Mercer, K., Giblett, S., Green, S., Lloyd, D., DaRocha Dias, S., Plumb, M., Marais, R., and Pritchard, C. (2005). Expression of endogenous oncogenic V600EB-raf induces proliferation and developmental defects in mice and transformation of primary fibroblasts. *Cancer Res.* **65**, 11493–11500.

Myslinski, E., Ame, J. C., Krol, A., and Carbon, P. (2001). An unusually compact external promoter for RNA polymerase III transcription of the human H1RNA gene. *Nucleic Acids Res.* **29**, 2502–2509.

Pollock, P. M., Harper, U. L., Hansen, K. S., Yudt, L. M., Stark, M., Robbins, C. M., Moses, T. Y., Hostetter, G., Wagner, U., Kakareka, J., Salem, G., Pohida, T., *et al.* (2003). High frequency of BRAF mutations in nevi. *Nat. Genet.* **33**, 19–20.

Ray, P., De, A., Min, J. J., Tsien, R. Y., and Gambhir, S. S. (2004). Imaging tri-fusion multimodality reporter gene expression in living subjects. *Cancer Res.* **64,** 1323–1330.

Wan, P. T., Garnett, M. J., Roe, S. M., Lee, S., Niculescu-Duvaz, D., Good, V. M., Jones, C. M., Marshall, C. J., Springer, C. J., Barford, D., and Marais, R. (2004). Mechanism of activation of the RAF-ERK signaling pathway by oncogenic mutations of B-RAF. *Cell* **116,** 855–867.

Wellbrock, C., Karasarides, M., and Marais, R. (2004). The RAF proteins take centre stage. *Nat. Rev. Mol. Cell. Biol.* **5,** 875–885.

Wellbrock, C., Ogilvie, L., Hedley, D., Karasarides, M., Martin, J., Niculescu-Duvaz, D., Springer, C. J., and Marais, R. (2004). V599EB-RAF is an oncogene in melanocytes. *Cancer Res.* **64,** 2338–2342.

Yao, F., Svensjö, T., Winkler, T., Lu, M., Eriksson, C., and Eriksson, E. (1998). Tetracycline repressor, tetR, rather than the tetR-mammalian cell transcription factor fusion derivatives, regulates inducible gene expression in mammalian cells. *Hum. Gene Ther.* **9,** 1939–1950.

Yazdi, A. S., Palmedo, G., Flaig, M. J., Puchta, U., Reckwerth, A., Rutten, A., Mentzel, T., Hugel, H., Hantschke, M., Schmid-Wendtner, M. H., Kutzner, H., and Sander, C. A. (2003). Mutations of the BRAF gene in benign and malignant melanocytic lesions. *J. Invest. Dermatol.* **121,** 1160–1162.

CHAPTER FOUR

A Method to Generate Genetically Defined Tumors in Pigs

Stacey J. Adam *and* Christopher M. Counter

Contents

Abstract

As a biomedical model, pigs offer many advantages and hence have been utilized extensively for toxicology, Crohn's disease, diabetes, and organ transplantation, as well as many other research areas. However, the advantages of porcine models, particularly its large size and similarity to humans, were not exploited previously to any large degree for cancer research. One reason for this lack of porcine cancer models was the inability to induce cancer in pigs genetically. This chapter describes a rapid, reproducible, and genetically malleable method to induce large tumors in pigs.

Department of Pharmacology and Molecular Cancer Biology, Duke University, Durham, North Carolina

Methods in Enzymology, Volume 439
ISSN 0076-6879, DOI: 10.1016/S0076-6879(07)00404-1

39

1. INTRODUCTION

There is a need to develop large animal models of cancer, especially for studies that depend on large tumors that cannot be modeled in rodents (Rangarajan and Weinberg, 2003). Spontaneous tumors arising in companion animals offer on solution, although typically assembling large cohorts of animals with the appropriate cancer is difficult and can be prohibitively expensive, as well as there is no genetic control over the tumors in this approach (Dewhirst, 2000). It would therefore be of value to be able to induce tumors in a large animal genetically.

Pigs are just such an animal. Pigs are large mammals that grow in similar size and weight to humans. Their anatomy is also similar to humans; genetically, pigs bear key sequence homology to humans in xenobiotic receptors, which are divergent in mice, that are responsible for modulating the metabolism of drugs (Swanson *et al.*, 2004). Also, like humans, cancer is rare in pigs. Cancers resembling human childhood cancers occur in young pigs (Anderson and Jarrett, 1968), whereas more age-related cancers occur in adult pigs (Brown and Johnson, 1970). Moreover, because of their size, physiologically relevant sized tumors could be generated in pigs for the development of experimental therapeutics. Thus, there would be value in developing a method to gene-induced cancer in pigs.

It had been demonstrated that enforced expression of transgenes that mimic genetic changes occurring in many types of human cancers could drive normal primary human cells to a tumorigenic state. Specifically, ectopic expression of hTERT, $p53^{DD}$ (a dominant-negative truncation mutant of p53), cyclin D1, $CDK4^{R24C}$ (an activated version of cyclin-dependent kinase 4 mutant), $c\text{-}Myc^{T58A}$ (a stabilized version of the oncogene c-Myc), and $H\text{-}Ras^{G12V}$ (a constitutively active form of Ras GTPase) are together sufficient to drive a variety of human cells to form tumors when explanted into immunocompromised mice (Kendall *et al.*, 2005). These transgenes disrupt five essential cell processes involved in tumorigenesis, including telomere maintenance, cell cycle checkpoint abrogation, antigrowth signal insensitivity, mitogenic growth signal self-sufficiency, and promotion of angiogenesis (Hanahan and Weinberg, 2000). Taking advantage of this approach, we reasoned that porcine cells could similarly be converted to a tumorigenic state by expression of these transgenes and, moreover, could be engrafted into immunosuppressed pigs to form tumors, much like mice are used to grow tumors from tumor cell lines. Indeed, this has been the case for a number of porcine cell types. This chapter describes how such a genetically defined large animal model of cancer can be generated.

2. OVERVIEW

The generation of porcine cells transformed employing hTERT, p53^{DD}, cyclin D1, CDK4^{R24C}, c-Myc^{T58A}, and H-Ras^{G12V} and the use of these cells to create tumors in host pigs are described. First, establishment of primary porcine dermal fibroblasts is detailed. Second, generation of amphotrophic retroviruses containing the aforementioned genes is explained. Next, transformation of primary porcine cell strains with these viruses, detection of aberrant helper virus production, and verification of appropriate transgene expression are elucidated. Finally, injection of transformed cells into isogenic host swine and immunosuppression therapy are described.

3. DERIVATION OF CELL LINE FROM SWINE EAR NOTCHES

3.1. Materials

70% ethanol
Betadine swabs
50-ml conical tubes
Hanks' balanced salt solution (HBSS) (Sigma H-6136, 1 liter, 9.8 g/liter)
Penicillin/streptomycin 100× stock (GIBCO)
Fungizone 100× stock (GIBCO)
Sterile gauze
Sterile forceps
Sterile ear notchers
Phosphate-buffered saline (PBS), pH 7.4, 1× (GIBCO)
Cell dissociation buffer, enzyme free, PBS based (GIBCO)
Blendzyme II (Roche)
10-cm tissue culture dishes
Sterile scalpels (or razor blades)
Dulbecco's modified Eagle's medium (DMEM), 1×, liquid (GIBCO)
F10 nutrient mixture (Ham), 1×, liquid (GIBCO)
Fetal calf serum (FCS) (GIBCO)

3.2. Methods

3.2.1. Ear notch collection
Note: Solutions should be made *at least* 12 h before ear notch collection.

1. Filter deionized water through a Millipore filter, making at least 3 liters.
2. Create HBSS + 8× penicillin/streptomycin + 1× fungizone (HBSS^{++})
3. Scrub pigs ears with 70% EtOH, followed by betadine, and again with 70% EtOH.
4. Using sterile ear notchers, punch ears.
5. Using sterile forceps aseptically, drop ear notches into a sterile 50-ml conical tube containing 25 ml of HBSS^{++}.

3.2.2. Processing ear notches

1. Create stock and working solutions of Blendzyme II using a dilution buffer of cell dissociation buffer:
   Stock solution of 14 U/ml
   Working solution of 0.14 U/ml (dilution factor = 1:200, for four samples, combine 250 μl of prepared enzyme and 49.75 ml of straight PBS)

   Note: Antibiotics and fungicides inhibit enzymes.

2. Mix 50% DMEM/50% F10 nutrient mixture solution + 10% FCS in a sterile 500-ml bottle.
3. Place ear notches into a clean 50-ml conical tube. Add 5 ml of 100× penicillin/streptomycin.
4. Agitate samples and then incubate at room temperature for 10 min.
5. Pour tube contents into 10-cm tissue culture dish and dice samples using a sterile scalpel (or razor blade).
6. Collect large sample pieces and place in clean 50-ml conical tubes. Rinse 10-cm dish with 10 ml of HBSS^{++}, pouring liquid into the same 50-ml conical tube. Centrifuge samples at 1000 g for 5 min.
7. Aspirate supernatant into waste bottle containing 10% bleach. Wash sample with 50 ml of cell dissociation buffer. Centrifuge samples at 1000 g for 5 min.
8. Aspirate supernatant into waste bottle containing 10% bleach. Wash sample with 50 ml of PBS. Centrifuge samples at 1000 g for 5 min.
9. While samples spin, prepare Blendzyme II in clean 50-ml conical tube. (See step 1.)
10. Add 12 ml of Blendzyme II working solution to each sample. Incubate for 4 to 6 h at 37° (depending on tissue) to digest. (*Do not* leave samples in enzyme solution overnight.)
11. Agitate samples and add 5 ml DMEM/F10 media, QNS volume to 50 ml with HBSS^{++}. Centrifuge at 800 g for 5 min.
12. Aspirate supernatant into waste bottle containing 10% bleach. Add 50 ml of HBSS^{++} to sample. Centrifuge at 800 g for 5 min.
13. Repeat step 12 twice.

14. Aspirate supernatant into waste container. Add 10 ml of DMEM/F10 media containing 4× penicillin/streptomycin and 1× fungizone to each sample.
15. Agitate samples and let large tissue pieces settle.
16. Separate two phases into two 10-cm tissue culture dishes. One 10-cm dish should contain tissue pieces and the second should contain dispersed cells in liquid media.
17. Incubate at 37°. *Do not* disturb 10-cm dishes for at least 24 h.

Note: You may place pieces directly onto the 10-cm dish with no excess media and incubate for several hours at 37°. (This helps with attachment.) Then flood flask with ≈5 ml media, very gently, taking care not to dislodge adhered tissue pieces.

4. CREATION OF SIX GENE-TRANSFORMED CELLS

4.1. Virus creation and cell infection

Note: It is important that all virus work detailed in this manuscript be performed within a certified BSL-3 laboratory facility.

4.1.1. Materials

293T cells (ATCC)
Serum-free media (αMEM or DMEM) (GIBCO)
Porcine dermal fibroblasts established from ear notches (or other primary porcine cells)
FuGene-6 (Roche)
pCL-10A1 (viral packaging plasmid) (Imgenex)
pBabe-hTERT+p53^{DD}(Adam *et al.*, 2007)
pBabe-cyclin D1+CDK4^{R24C}(Adam *et al.*, 2007)
pBabe-c-Myc^{T58A}+H-Ras^{G12V}(Adam *et al.*, 2007)
10-cm tissue culture dishes
0.45-μm Acrodisc w/HT tufftyn membrane (VWR)
10-ml syringe
15-ml conical tubes
Sterile 1.5-ml microcentrifuge tubes
Filter-sterilized 800-μg/ml stock of polybrene (hexadimethrine bromide) (Sigma) in 1× PBS

4.1.2. Methods

Day 1 *Morning.* Seed 293T cells to 30% confluency onto a 10-cm dish in DMEM (or αMEM) supplemented with 10% FCS (no antibiotics).

Day 2 *Morning.* For the first cell transfection, combine 184 μl serum-free, antibiotic-free media, 12 μl Fugene-6, 3 μg pCL-10A1, and 3 μg pBabe plasmid (pBabe-hTERT+p53^{DD}, pBabe-cyclin D1+CDK4^{R24C}, or pBabe-c-Myc^{T58A}+H-Ras^{G12V}) in a laminar flow hood in a microcentrifuge tube, making sure to add serum-free media first and then Fugene-6. Tap the side of the tube gently to mix. Then add each plasmid DNA to the mixture and tap the side of the tube gently to mix. Incubate mixture at room temperature for 15 to 45 min. Finally, add the mixture drop wise to the aforementioned 293T cells from day 1 and incubate overnight at 37°.

Day 3 *Morning.* For the second cell transfection, repeat transfection protocol from day 2. Once the transfection mix is added to the cells, incubate for at least 8 h.

Evening. Media change cells with 6 ml media, ideally within the type of media in which primary porcine cells will be cultured.

Note: The virus collection protocol has two options. First, collect a 24-h virus and then place 6 ml back on cells for another 24 h and collect a 48-h virus, freezing the 24-h virus at −80° for later use and using the 48-h virus to infect cells. Second, the first 6 ml of media can be placed on cells and left for 48 h, collected, and used to infect cells.

Day 4 *Evening.* Collect 24-h virus if desired (see earlier discussion). Aspirate media from virus plate with a 10-ml syringe, attach a 0.45-μm filter to the syringe, and filter the virus into a 15-ml conical tube. Add polybrene to the tube and pipette to mix (800 μg/μl stock solution of polybrene, desired concentration of 4 μg/μl, 5 μl per 1 ml of media). Place the 15-ml conical tube in −80° to store virus. Split primary porcine cells to be infected to 25 to 30% confluency.

Day 5 *Evening.* Collect 48-h virus and filter virus into a 15-ml conical tube as described in the day 4 procedure. Replace media on the primary porcine cells to be infected with the pBabe-hTERT+p53^{DD} 48-h virus media. Place the 15-ml conical tubes of pBabe-cyclin D1 + CDK4^{R24C} or pBabe-c-Myc^{T58A} + H-Ras^{G12V} in −80° to store virus until required.

Note: Frozen virus appears to work as well as fresh virus.

Day 6 After 24 h of infection with the virus, media change cells with fresh, nonviral media.

Day 7 After 24 h in fresh media, allow cells to rest, infect cells with 48-h pBabe-cyclin D1 + CDK4^{R24C} virus, and let virus sit for 24 h.

Day 8 Repeat procedure from day 6.

Day 9 After 24 h in fresh media, to allow the cells to rest, infect cells with 48-h pBabe-c-Myc^{T58A} + H-Ras^{G12V} virus. Let virus sit for 24 h.

Day 10 Repeat procedure from day 6. Over the next few days, allow the plate to grow to 100% confluency.

4.2. Horizontal spread assay (for cells containing No selection markers)

4.2.1. Materials

0.45-μm Acrodisc w/HT tufftyn membrane (VWR)
10-ml syringe
15-ml conical tubes
Filter-sterilized 800-μg/ml stock of polybrene (hexadimethrine bromide) (Sigma) in 1× PBS
293T cells (ATCC)
Cell line containing puromycin resistance marker (any line verified as nonvirus producing will work; cells containing other selection markers can be used, but puromycin is usually the fastest)
10-cm tissue culture dishes

4.2.2. Methods

Day 1 Plate (or split) cells that have been infected with desired virus (es) to ≈50% confluency to be tested for helper virus.

 Note: Final viral infection of the cells must be at least 48 h prior to beginning this protocol.

Day 2 Plate (or split) cells stably expressing the desired selection marker (i.e., puromycin) to ≈50% confluency.

Day 3 Remove media from virally infected cells plated on day 1 via a syringe. Filter media from the plate through a 0.45-μm filter that attaches to the syringe into a 15-ml conical tube. Add polybrene to the tube and pipette to mix (800-μg/μl stock solution of polybrene, desired concentration of 4 μg/μl, 5 μl per 1 ml of media). Aspirate media off of cells stably expressing the desired selection marker from day 2. Use polybrene/filtered media to replace media on the desired selection marker cells from day 2.

Day 4 Plate (or split) 293T cells to ≈50% confluency.

Day 5 After 48 h, remove media that had been placed on cells stably expressing the desired selection marker during day 3 via syringe and filter through a 0.45-μm filter attached to the syringe into a 15-ml conical tube.

Add polybrene (as described on day 3). Aspirate media off 293T cells from day 4. Use polybrene/filtered media to replace media on selection marker negative cells from day 4.

Day 7   After 48 h, aspirate media off of selection marker negative cells from day 5. Replace with 10 ml of media containing appropriate growth selection agent at the appropriate concentration, i.e., DMEM + 10% FBS + puromycin 1 μg/ml for transformed cells, such as 293T cells.

Day 8   Monitor cells for death over the next few days to weeks.

Note: If no helper virus is being produced by original cells, then the entire population of selection marker negative cells should die in a relatively short amount of time.

4.3. Reverse-transcriptase polymerase chain reaction (PCR) verification of transgene expression

4.3.1. Materials

RNAzol B (Tel-Test, Inc.)
Sterile 1.5-ml microcentrifuge tubes (must be RNase free)
Sterile cell scrappers
Chloroform (stored at 4°)
75% EtOH (made with DEPC-treated H_2O)
1 mM EDTA, pH 7
Transformed cell lines for testing
Omniscript kit (Qiagen)
RNAse Out (Invitrogen)
Oligo(dT) (Invitrogen)
Red *Taq* polymerase and 5× buffer (Sigma)
2.5 mM dNTPs (diluted from Invitrogen 100 mM dNTP mix)
ddH$_2$O (PCR grade)
Gene-specific primers (see Methods for sequence)
SeaKem LE agar (Fisher Sciences)
Dimethyl sulfoxide (DMSO)

4.3.2. Methods
RNA extraction

1. Isolate RNA from one 10-cm confluent tissue culture dish, as described by the manufacturer, using RNAzol B reagent (Tel-Test, Inc.).
2. Aliquot 3 μl of RNA from step 1 into 97 μl of RNase-free deionized water to measure RNA concentration via a spectrophotometer, using standard procedures (Maniatus).

Reverse transcription reaction and PCR amplification

1. Reverse transcribe (RT) 9 μg of RNA from each sample using the Qiagen Omniscript kit, as described by the manufacturer. The RT reaction mix is as follows:
   6.0 μl 10× buffer RT
   6.0 μl 5 mM dNTPs
   0.75 μl RNase Out
   3.0 μl Omni RT
   1.5 μl oligo(dT)
   RNA (9 μg) varies according to RNA concentration
   RNase-free dH$_2$O, varies according to RNA amount
   60.0 μl final volume

   Note: A master mix can be made for everything except the RNase-free water and RNA.

2. PCR amplify cDNA products of aforementioned RT reactions for each sample to be tested, as follows:
   33.5 μl ddH$_2$O (PCR grade H$_2$O)
   5.0 μl Red *Taq* buffer
   2.0 μl 2.5 mM dNTPs
   2.0 μl 20 μM forward primer
   2.0 μl 20 μM reverse primer
   5.0 μl cDNA (from RT reaction just described)
   1.0 μl Red *Taq* polymerase
   3.5 μl DMSO

   Note: Make master mix including everything *except* primers and cDNA for number of reactions needed to be done (overestimate by approximately two reactions). For each cell line needing to be tested, seven reactions are done (β-tubulin, hTERT, p53^{DD}, CD1, CDK4^{R24C}, Myc^{T58A}, H-Ras^{G12V}).

3. Aliquot 45.0 μl of master mix into each PCR tube.
4. Add 2.0 μl of appropriate 20 μM forward primer for each reaction.

β-Tubulin F   5′-TCAGAAATACGTGCCCAGGG-3′
hTERT   5′-TGGCTGTGCCACCAAGCATT-3′
p53^{DD} 5′-GCTCACTCCAGCTACCTGAA-3′
cyclin D1   5′-AACATGGACCCCAAGGCC-3′
CDK4^{R24C} 5′-GGTGGTACCTGAGATGGA-3′
c-Myc^{T58A} 5′-ACGAGCACAAGCTCACC-3′
H-Ras^{G12V} 5′-GCACGCACTGTGGAATCT-3′

5. Add 2.0 μl of appropriate 20 μM reverse primer for each reaction.

β-Tubulin R   5′-GAAGGGGACCATGTTCACA-3′
hTERT   5′-TTTCCACACCTGGTTGC-3′

p53^{DD} 5′-ATGCCTTGCAAAATGGCG-3′
cyclin D1   5′-TTCTGCCTGCTGGGGAG-3′
CDK4^{R24C} 5′-TAGCTTGCCAAACCTACAGG-3′
c-Myc^{T58A} 5′-TTTCCACACCTGGTTGC-3′
H-Ras^{G12V} 5′-TAGCTTGCCAAACCTACAGG-3′

6. Add 5.0 μl of appropriate cDNA from each reverse transcription reaction for each cell line.
7. Cap reactions and vortex for 30 s. Centrifuge tubes 20 to 30 s to ensure that all of the reaction mix is in the bottom of each tube.
8. Using the following cycles, amplify DNA:

1 cycle: 94.0° for 5 min
25 cycles: 94.0° for 30 s
            55.0° for 1 min
            72.0° for 30 s
1 cycle: 72.0° for 10 min
Hold at 4.0°

9. Resolve 25 μl of reaction products on 2% agarose gels to confirm PCR amplification of transgenes.

5. INJECTING CELLS INTO PIGS

5.1. Tumorigenic growth assay in pigs

5.1.1. Materials

15-cm tissue culture dishes
50% DMEM/50% F10 nutrient mix media + 10% FCS, 1× (GIBCO)
0.05% trypsin (GIBCO)
PBS, pH 7.4, 1× (GIBCO)
Dulbecco's PBS with MgCl$_2$ and CaCl$_2$ (Sigma)
Cell dissociation buffer (GIBCO)
50-ml conical tubes
3-ml syringe
21-gauge needle
Transformed porcine cells to be injected
Telazol–ketamine–xylazine (TKX) anesthesia (by veterinary prescription)

5.1.2. Methods

It is important that transformed porcine cells be checked for bacterial, fungal, and mycoplasma infections before injection. Bacterial and fungal infections can be ruled out by culturing cells without penicillin/streptomycin and fungizone in culture media for 4 to 5 days. Mycoplasma should be tested for using the Gen-Probe nucleic acid hybridization procedure (Fisher Scientific) or similar assay.

1. Transfer media from 15-cm tissue culture dishes into 50-ml tubes. Centrifuge at 4° at 1000 g for 5 min.

Note: This protocol assumes that the volume in twenty 15-cm dishes will contain enough cells for injection. Cells have a tendency to no longer adhere to the tissue culture dish; however, they are still living and dividing. If needed, spin down all media.

2. Discard supernatant.
3. Add ≈8 ml of cell dissociation buffer to each 15-cm dish. Ensure full coverage of bottom of the flask.
4. Let stand for ≈5 min. Tap side of dish to dislodge cells. Transfer liquid containing cells to 50-ml conical tubes. Centrifuge at 4° at 1000 g for 5 min.

Note: If cell dissociation buffer does not cause cells to dislodge from the tissue culture dish, remove buffer and add to a 50-ml conical tube. Add 2 ml of trypsin to each dish.

5. Repeat steps 3 and 4 until all cells are removed from dishes and pelleted in tubes.
6. Resuspend each pellet in 6 ml PBS and consolidate all cell suspensions into one 50-ml conical tube per cell line. Centrifuge at 4° at 1000 g for 5 min.
7. Wash with PBS with $MgCl_2$ and $CaCl_2$. Centrifuge at 4° at 1000 g for 5 min. Discard supernatant.
8. Repeat step 7 a minimum of three times.
9. Remove 50 μl of cell suspension and place in a 1.5-ml microcentrifuge tube. Add 50 μl of Trypan blue to tube and mix.
10. Remove 10 μl of cell suspension/Trypan blue mixture and plate to a hemacytometer. Count cells.
11. Add 1×10^8 cells into each of two 50-ml conical tubes. Centrifuge at 4° at 1000 g for 5 min.
12. Aspirate PBS and resuspend each pellet in 1 ml PBS and place in a microcentrifuge tube.
13. Inject TKX anesthesia intramuscularly into the flank of each pig to be injected based on the weight of the pig (1 ml/50–75 lbs.). Wait 5 to 10 min for TKX to take effect and monitor pig's sedation according to institution IACUC approval.
14. Draw each cell suspension into a sterile 3-ml syringe. Attach a 21-gauge needle to the syringe. Inject the cell suspension subcutaneously behind the ear of the pig (one cell suspension for each ear).

Note: Cell suspensions can also be injected into the mammary fat pad of the isogenic host pig.

Over the following 15 to 30 days, track tumor progression by measuring large and small diameters and extrapolating tumor volume using the equation:

$$\text{Tumor volume} = (\text{small diameter})^2 + (\text{large diameter}/2)$$

Note: All animal work *must* be done by a trained person under an approved IACUC protocol.

5.2. Immunosuppression regime

5.2.1. Materials

Azathioprine pills (by veterinary prescription)
Prednisone pills (by veterinary prescription)
Cyclosporine A liquid (by veterinary prescription)
Standard pig chow
Chocolate syrup
Chocolate corn puff cereal (generic cocoa puffs)
Feed bowl
Rubber gloves

5.2.2. Methods

The immunosuppression regime is administered twice daily beginning 1 week prior to injection of transformed cells. Before the immunosuppression regime begins, pigs are conditioned to eat the chocolate corn puff cereal and chocolate syrup that the drugs will be administered in.

Morning dosing

1. Place azathioprine and prednisone pills (enough for 2 and 4 mg/kg, respectively) in a small dish. Crush pills into small pieces. Cover with chocolate syrup and mix with a handful of normal pig food. (Mixture should be thick enough to form into a small ball ≈3–4 in. in diameter.)
2. Add 1 ml of cyclosporine A (1 ml/50–75 lbs.) to the ball and mix in enough chocolate corn cereal to ensure full incorporation of cyclosporine A into the ball (total ≈20 ml).

Note: Ensure that food, chocolate syrup, and chocolate corn cereal are added to disguise the taste to prevent the pigs from developing an aversion to the chocolate.

3. Feed food ball mixture to pig in its designated food bowl. Make sure that the pig consumes the entire food/medicine ball.
4. Place pig back in cage and give one cup of normal feed mixed with chocolate corn cereal to reward the pig.

Evening dosing

5. Feed as described earlier with food balls made from mixing 1 ml of cyclosporine A (1 ml/50–75 lbs.), food, chocolate corn cereal, and chocolate syrup into the ball to ensure full incorporation of cyclosporine A into the ball (total ≈20 ml).

6. CONCLUDING REMARKS

The described procedure provides a rapid and reproducible method to generate genetically malleable tumors in a large mammal. Other primary porcine cell lines have already been transformed (Adam *et al.*, 2007), utilizing the six gene transformation method, and the model could be expanded still further, potentially, to any cell type that can be cultured. Similarly, the genetic makeup of these cells could also be altered to suit the specific needs of the cancer studies. In summary, the porcine tumorigenesis model opens many new avenues of research that before were limited by the lack of an affordable, genetically malleable large animal model.

REFERENCES

Adam, S. J., Rund, L. A., Kuzmuk, K. N., Zachary, J. F., Schook, L. B., and Counter, C. M. (2007). Genetic induction of tumorigenesis in swine. *Oncogene* **26,** 1038–1045.

Anderson, L. J., and Jarrett, W. F. (1968). Lymphosarcoma (leukemia) in cattle, sheep and pigs in Great Britain. *Cancer* **22,** 398–405.

Brown, D. G., and Johnson, D. F. (1970). Diseases of aged swine. *J. Am. Vet. Med. Assoc.* **157,** 1914–1918.

Dewhirst, M. W., Thrall, D., and Macewen, E. (2000). Spontaneous pet animal cancers. *In* "Tumor Models in Cancer Research" (B. A. Teicher, ed.), pp. 565–589. Humana Press, Totowa, NJ.

Hanahan, D., and Weinberg, R. A. (2000). The hallmarks of cancer. *Cell* **100,** 57–70.

Kendall, S. D., Linardic, C. M., Adam, S. J., and Counter, C. M. (2005). A network of genetic events sufficient to convert normal human cells to a tumorigenic state. *Cancer Res.* **65,** 9824–9828.

Rangarajan, A., and Weinberg, R. A. (2003). Comparative biology of mouse versus human cells: Modelling human cancer in mice. *Nat. Rev. Cancer* **3,** 952–959.

Swanson, K. S., Mazur, M. J., Vashisht, K., Rund, L. A., Beever, J. E., Counter, C. M., and Schook, L. B. (2004). Genomics and clinical medicine: Rationale for creating and effectively evaluating animal models. *Exp. Biol. Med. (Maywood)* **229,** 866–875.

CHAPTER FIVE

Tools to Study the Function of the Ras-Related, Estrogen-Regulated Growth Inhibitor in Breast Cancer

Ariella B. Hanker,* Staeci Morita,[†] Gretchen A. Repasky,[‡] Douglas T. Ross,[§] Robert S. Seitz,[§] *and* Channing J. Der*,[†],[¶]

Contents

* Curriculum in Genetics and Molecular Biology, University of North Carolina at Chapel Hill, Chapel Hill, North Carolina
[†] Department of Pharmacology, University of North Carolina at Chapel Hill, Chapel Hill, North Carolina
[‡] Birmingham-Southern College, Department of Biology, Birmingham, Alabama
[§] Applied Genomics Inc., Burlingame, California and Huntsville, Alabama
[¶] Lineberger Comprehensive Cancer Center, University of North Carolina at Chapel Hill, Chapel Hill, North Carolina

Methods in Enzymology, Volume 439
ISSN 0076-6879, DOI: 10.1016/S0076-6879(07)00405-3

Abstract

The Ras-related, estrogen-regulated growth inhibitor (Rerg) is a Ras-related small GTPase and candidate tumor suppressor. Rerg gene expression is stimulated by the estrogen receptor α (ERα), and Rerg gene expression is absent in ER-negative breast cancers. ER-negative breast cancers are highly invasive and metastastic and are typically more advanced than their ER-positive counterparts. Like Ras, Rerg binds and hydrolyzes GTP, but unlike Ras, Rerg has been shown to possess growth inhibitory activity in breast cancer cells. The precise role that Rerg loss plays in breast cancer growth and the mechanisms by which it does so are unknown. This chapter describes tools used to detect and manipulate the expression of Rerg in breast cancer cells. We validate use of an antibody to detect Rerg expression. We describe the generation of expression vectors that encode wild-type and mutants of Rerg that are altered in GDP/GTP regulation. We also describe the development of an inducible Rerg expression system and of a retrovirus-based RNA interference approach to repress Rerg expression. These tools will be invaluable in evaluating the biological function of Rerg in breast cancer.

1. INTRODUCTION

The Ras-related, estrogen-regulated growth inhibitor (Rerg) small GTPase was initially identified in a microarray screen that grouped breast tumor samples into clinically relevant subtypes (Finlin *et al.*, 2001; Sorlie *et al.*, 2001). Rerg gene expression was found to be decreased in the most aggressive, estrogen receptor (ER)-negative subtypes (Finlin *et al.*, 2001). In fact, Rerg gene expression is directly regulated by estrogen. Microarray analyses by Perou and colleagues found that Rerg expression levels are a statistically significant predictor of patient outcome, even within ER-positive breast cancers, indicating that low levels of Rerg expression correlate with breast cancer progression (personal communication, unpublished observations). Furthermore, other microarray studies have found that Rerg expression is decreased in some metastatic cancers compared to their nonmetastatic counterparts, including prostate and lung cancer (Dhanasekaran *et al.*, 2001; Lapointe *et al.*, 2004; Ramaswamy *et al.*, 2003). Rerg expression is also decreased or lost in kidney, ovary, and colon tumor tissues (Key *et al.*, 2005). These studies suggest that loss of Rerg may be an important step in tumor progression and metastasis.

Rerg belongs to the Ras branch of the Ras superfamily of small GTPases (Wennerberg *et al.*, 2005). Rerg shares ≈50% amino acid sequence homology with the Ras protooncogene proteins. Ras functions as a GTP/GDP-regulated switch that relays extracellular ligand-stimulated signals; these signals regulate cell proliferation, differentiation, and survival. Ras is mutationally activated in 30% of human cancers, but in only 5% of breast cancers. Like Ras, Rerg can bind to and hydrolyze GTP and cycles between active GTP-bound and inactive GDP-bound states (Finlin *et al.*, 2001). Whether, like Ras, the GTP-bound form of Rerg is the "active" form is not yet known. Ras binds to and activates its many downstream effectors via its core effector domain (Ras residues 32–40). Key effectors implicated in Ras-mediated oncogenesis include Raf serine/threonine kinases, phosphatidylinositol 3-kinases, and guanine nucleotide exchange factors for the Ral Ras-like small GTPases (RalGEFs) (Repasky *et al.*, 2004). Although Rerg shares significant sequence identity with the Ras effector domain, no known Ras effectors have been shown to bind to Rerg. Additionally, no Rerg-specific effectors have been identified to date. Finally, while the intrinsic biochemical properties of Rerg suggest that, like Ras, Rerg GDP/GTP cycling will be regulated by guanine nucleotide exchange factors (GEFs) and GTPase-activating proteins (GAPs), Ras GEFs and GAPs are not known to regulate Rerg and no Rerg-specific GAPs or GEFs have been identified.

A major difference between the sequences of Ras and Rerg is in their carboxyl-terminal sequences. All Ras proteins terminate in a CAAX (C = cysteine, A = aliphatic amino acid, X = terminal amino acid) tetrapeptide sequence, a substrate for farnesyl lipid modification responsible for membrane localization and critical for Ras biological activity. Rerg lacks this motif, and unlike Ras, Rerg is localized primarily to the cytosol (Finlin *et al.*, 2001).

Despite their sequence similarity, Rerg and Ras play opposing roles in human oncogenesis. While Ras functions as an oncogene and drives growth transformation and tumorigenesis, Rerg exhibits characteristics of a tumor suppressor and impedes tumor growth. Previous studies have shown that ectopic overexpression of Rerg decreases cell proliferation and tumorigenesis of Rerg-expressing, ER-positive MCF-7 breast cancer cells (Finlin *et al.*, 2001; Key *et al.*, 2005), further supporting a tumor suppressor function for Rerg in breast cancer. However, whether the absence of Rerg expression contributes to breast cancer advancement directly and whether restoring Rerg expression in ER-negative breast cancers limits their tumorigenicity and aggressiveness have not yet been determined. This chapter describes the development and validation of tools used to detect and manipulate the expression of the unique GTPase Rerg in breast cancer cells. We have developed and validated antibodies that detect ectopic and endogenous Rerg expression by Western blot and immunohistochemical staining. We have constructed mammalian expression vectors for the expression of wild-type and mutant Rerg proteins. We have generated a tamoxifen-inducible ER–Rerg fusion protein to inducibly express Rerg in ER-negative breast cancers and have used this system to study the effects of Rerg on cell growth and invasion. Finally, we have generated retrovirus vectors for interfering short hairpin RNA (shRNA) to silence endogenous Rerg expression in ER-positive breast cancers. These reagents will undoubtedly prove useful in identifying and studying the precise biological function of Rerg in normal cells, as well as in breast and other human cancers.

2. MOLECULAR CONSTRUCTS

2.1. Rerg expression constructs

Rerg wild-type, Rerg S20N, and Rerg Q64L cDNA sequences in pKH3 expression vector plasmids (generous gifts of D. Andres, University of Kentucky) are used as templates for polymerase chain reaction (PCR)-mediated DNA amplification to generate open reading frame cassettes for subcloning into the pCGN-hygro mammalian cell expression vector. Expression of the inserted cDNA is controlled from the cytomegalovirus promoter, and the vector encodes for hygromycin resistance for the selection of stably transfected cells. The cassettes contain 5' and 3' *Bam*HI sites to allow subcloning downstream of and in-frame with sequences encoding the hemaglutinin (HA)-epitope tag. Briefly, PCR-generated products and pCGN-hygro plasmid are digested with *Bam*HI to generate sticky ends. *Bam*HI-digested pCGN-hygro is then dephosphorylated with shrimp alkaline phosphatase (SAP) to prevent religation of the plasmid. The *Bam*HI-digested PCR product is then incubated with *Bam*HI-digested and SAP-treated pCGN-hygro using the rapid DNA ligation kit (Roche Applied Science, Indianapolis, IN). Competent DH5α *Escherichia coli* are then transformed with the ligation products, and positive clones are identified by restriction digest analyses. The Rerg cDNA sequences are verified for accuracy by DNA sequencing.

For the generation of retrovirus-base expression vectors, Rerg wild-type, Rerg Q64L, and Rerg S20N cDNA sequences are subcloned into the pBabe HAII retrovirus vector. This vector is a variant of the pBabe-puro retrovirus expression vector (Morgenstern and Land, 1990) that has been modified to include coding sequences for the HA epitope sequence that are added to the amino terminus of the inserted cDNA sequence (Fiordalisi *et al.*, 2001). Expression of the inserted cDNA sequence is controlled by the Moloney leukemia virus long terminal repeat promoter, and the vector encodes for puromycin resistance for the selection of stably transfected or infected cells. pCGN Rerg wild-type, Q64L, or S20N plasmid DNAs are digested with *Bam*HI, and the resulting Rerg cDNA fragments are ligated in-frame into the linearized pBabe HAII vector (as described earlier).

pBabe ER expression vectors encoding wild-type or Q64L and S20N mutants of Rerg are generated for tamoxifen-inducible expression of wild-type and mutant ER-Rerg fusion proteins. In these constructs, the cDNA sequence for the ligand-binding domain of the ERα is placed 5′ to the Rerg cDNA coding sequence. pCGN Rerg plasmid DNAs are used as templates for PCR-mediated DNA amplification to generate open reading frame cassettes containing 5′ and 3′ *Eco*RI restriction sites. The primers used are 5′-TTTTGAATTCATGGCTAAAAGTGCGG-3′ and 5′-TTTTGAATTCTTTCTAACTACTGATTTTG-3′. PCR products are digested with *Eco*RI, and Rerg fragments are ligated into the Zero Blunt Topo kit (Invitrogen, Carlsbad, CA). Rerg is then excised from the Zero Blunt Topo vector and ligated into the *Eco*RI site of the linearized pBabe ER vector (derived from the pBabe ER-H-Ras vector).

2.2. Rerg shRNA constructs

For generation of shRNA against Rerg, we selected target sequences using the Web server at http://jura.wi.mit.edu/bioc/siRNA. We chose three target sequences for the generation of shRNA Rerg expression vectors. However, subsequent analyses found that target 2 showed no knockdown activity against Rerg and is not discussed here. The target sequence for Rerg shRNA 1 corresponds to nucleotide residues 2115–2137 in the 3′-untranslated region of Rerg cDNA, whereas the Rerg shRNA 3 target sequence corresponds to nucleotide residues 427–449 in the coding sequence. The following oligonucleotides are used for insertion into the *Bgl*II and *Hin*dIII sites of the pSUPER. retro.puro vector: 5′-GATCCCCAGCGTTAGCGGCATTAATTTTCA AGAGAAATTAATGCCGCTAACGCTTTTTTGGAAA-3′ and 5′-AG CTTTTCCAAAAAAGCGTTAGCGGCATTAATTTCTCTTGAAAAT TAATGCCGCTAACGCTGGG-3′ for Rerg shRNA 1 and 5′-GATCCC CGCAACCATCGATGATGAAGTTCAAGAG A CTTCATCATCGAT GGTTGCTTTTTGGAAA-3′ and 5′-AGCTTTTCCAAAAAGCAAC CATCGATGATGAAGTCTCTTGAACTTCATCATCGATGGTTGCG GG-3′ for Rerg shRNA 3. Oligonucleotides are phosphorylated and annealed prior to ligation into the *Bgl*II and *Hind*III sites of the pSUPER.retro.puro vector (Oligoengine, Seattle, WA), as described previously (Baines *et al.*, 2005). The negative control pSUPERIOR.retro.puro GFP shRNA (Novina *et al.*, 2002) was a gift from Natalia Mitin (UNC-Chapel Hill). All inserted oligonucleotide sequences in these molecular constructs are verified by sequencing.

3. Cell Culture and Expression Vector Transfection and Infection

293T human embryonic kidney epithelial cells are maintained in Dulbecco's modified Eagle's medium (DMEM) supplemented with 10% fetal calf serum (FCS) and 100 units/ml penicillin and 100 μg/ml streptomycin in a 10% CO_2 humidified incubator at 37°. All human breast carcinoma cell lines, with the exception of SUM149 cells, are obtained from the American Type Culture Collection and cultured as described previously (Eckert *et al.*, 2004). The SUM149 basal-like breast carcinoma cell line (obtained originally from Stephen Ethier, Karmanos Cancer Institute, and provided by Carolyn Sartor, UNC-Chapel Hill) is grown in Ham's F12 medium supplemented with 5% FCS, 5 μg/ml insulin, 1 μg/ml hydrocortisone, 0.5 μg/ml fungizone, and 5 μg/ml gentamicin. All breast cancer cells are grown in a 5% CO_2 incubator at 37°.

Subconfluent cultures of each cell line are transfected with the various Rerg or shRNA expression plasmid DNAs using Lipofectamine and Plus reagents (Invitrogen) according to the manufacturer's instructions. Cells are harvested 24 h following transfection.

All retroviral infections are performed as described previously (Baines *et al.*, 2005). Briefly, 293T host cells are transfected by calcium chloride precipitation with 10 μg of the pCL10A1 virus packaging vector plasmid DNA along with 10 μg of the appropriate Rerg or shRNA expression plasmid DNA. The medium is replaced 16 h prior to infection of target cells. Two days following 293T transfection, host cells are infected with filtered virus–containing medium from 293T cells supplemented with 8 μg/ml polybrene to increase infection efficiency. Twenty-four to 48 h following infection, growth medium supplemented with puromycin is used for the selection of stably infected T-47D (2 μg/ml), MDA-MB-231 and Hs578T (1 μg/ml), and SUM149 (0.5 μg/ml) cells. After selection, multiple drug-resistant colonies (>100) are pooled together to establish mass populations of stably infected cells. All stable cell lines are maintained in medium supplemented with puromycin.

4. Cell Growth and Invasion Assays

4.1. MTT cell proliferation assay

MDA-MB-231 cells stably expressing ER–Rerg fusion proteins are seeded at 2×10^3 cells per well in growth medium. Cells are seeded in replicates of eight in 96-well plates and allowed to attach overnight. The growth medium is then removed and replaced with fresh growth medium supplemented with 1 μM 4-hydroxytamoxifen (4-OHT) to induce ER–Rerg expression. For each time point, 3-[4,5-dimethylthiazol-2-yl]-2,5-diphenyltetrazolium bromide (MTT; Sigma-Aldrich, St. Louis, MO), which stains only viable cells, is added to a final concentration of 0.5 mg/ml in media of each well, and plates are incubated for 4 h. The formazan (MTT metabolic product) is then resuspended in 100 μl dimethyl sulfoxide (DMSO), and the optical density is measured at 560 nM using an ELx100 universal microplate reader (Bio-Tek Instruments, Inc., Winooski, VT). Optical density directly correlates with cell quantity.

4.2. Matrigel invasion assay

The transwell Matrigel invasion assay using the Biocoat Matrigel invasion chamber (BD Biosciences, Bedford, MA) is performed according to the manufacturer's instructions. Briefly, MDA-MB-231 stable cell lines are seeded at 5×10^3 cells per chamber in triplicate in serum-free DMEM supplemented with 1% fatty acid–free bovine serum albumin (BSA). Medium containing 3% FCS is added to each bottom well of a 24-well plate to serve as a chemoattractant. After 2 h, growth medium supplemented with ethanol (vehicle) or 1 μM 4-OHT is added to each chamber and bottom well. After an additional 22-h incubation at 37°, noninvading cells are removed with a cotton swab and invading cells are fixed and stained using the Diff-Quick stain set (Dade Behring). Invading cells are quantified by counting five fields of view using the 20× objective lens under an inverted phase-contrast microscope.

4.3. Anchorage-independent soft agar colony formation assay

The soft agar assay used to evaluate anchorage-independent growth potential is performed as described previously (Clark et al., 1995). Briefly, 10^4 SUM149 cells stably expressing ER–Rerg or H-Ras fusion proteins are seeded in triplicate in 60-mm plates into growth medium supplemented with 0.4% bacto-agar and overlaid onto a bottom layer containing growth medium supplemented with 0.6% agar. Agar layers contain SUM149 growth medium supplemented with 10% FCS and penicillin/streptomycin. Either ethanol (vehicle) or 1 μM 4-OHT final concentration is added to both top and bottom layers. Medium containing ethanol or 1 μM 4-OHT is replaced twice weekly for 3 weeks. After 21 days, viable colonies are stained in 2 mg/ml MTT, and plates are scanned. Visible colonies are counted from scanned images on Adobe Photoshop.

5. Generation and Validation of Rerg Antibodies for Western Blotting and Immunohistochemistry

5.1. Generation of anti-Rerg antibodies

The S0222 anti-Rerg antiserum is raised in rabbits to bacterially expressed His_6-thioredoxin-Rerg fusion protein from the pET32a-Rerg *E. coli* expression vector (Finlin *et al.*, 2001). The S0068 antiserum is raised against a 21-mer peptide corresponding to the carboxyl terminus of Rerg (residues 179–199; RRSSTTHVKQAINKMLTKISS) conjugated to keyhole limpet hemocyanin (KLH) by EDC coupling. The protein or KLH conjugated peptide is mixed with Freund's adjuvant, injected, and serially boosted into two New Zealand white rabbits over a 14-week period. The boosts are made approximately 3 weeks apart. Antisera are collected and affinity purified by attaching the recombinant protein or peptide to Sepharose 4B. The antibody is eluted by a pH gradient of glycine-HCl, pH 4, decreasing to pH 2 in borate buffer.

5.2. Validation of S0222 anti-Rerg antibody for western blot analyses

In order to determine whether the S0222 anti-Rerg antibody can detect endogenous human Rerg protein expression, we first tested it against ecto-pically expressed Rerg. We determined that the optimum dilution for Western blot analyses is 1:500 (final concentration of 740 ng/ml) in 5% nonfat dry milk/TBST, incubated overnight. S0222 effectively detects expression of HA-Rerg (Figs. 5.1A and 5.2B). Furthermore, S0222 detects expression of the ER-Rerg fusion protein (Fig. 5.3). However, S0222 is not as sensitive as the HA.11 antibody (Covance, Berkeley, CA) for detection of HA-Rerg (see Fig. 5.2B) and typically requires longer exposure.

Next, we asked whether S0222 could detect endogenous Rerg expression. To test this, lysates from ER-positive and -negative breast cancer cells treated with 10 nM 17β-estradiol (Sigma-Aldrich) are generated for Western blot analysis with S0222. As shown previously (Finlin *et al.*, 2001), Rerg

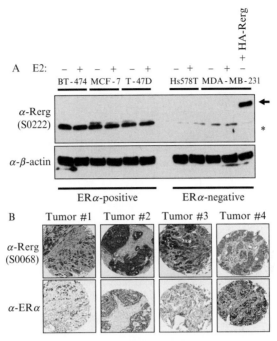

Figure 5.1 Antibody detection of endogenous Rerg expression in breast cancer cell lines and tumor tissue. (A) Detection of endogenous Rerg expression in breast carcinoma cell lines. Endogenous Rerg and ectopically expressed HA-tagged Rerg were detected in breast cancer cells by Western blot analysis using the S0222 anti-Rerg antibody. The indicated cell lines were incubated in complete growth media, either with or without 10 nM 17β-estradiol (E2, Sigma) for 24 h. A strong band at $\approx$26 kDa (asterisk) was seen in all Rerg mRNA-positive cell lines (BT-474, MCF-7, and T-47D), whereas a band at $\approx$30 kDa (arrow) was detected in MDA-MB-231 cells stably infected with the pBabe-puro HAII Rerg plasmid. Blot analysis for β-actin (Sigma) was used as a loading control. (B) Rerg expression levels correlate with ERα expression in breast cancer tissue. Rerg and ERα protein expression in primary invasive breast tissues were detected by IHC analyses. Parallel staining is shown for Rerg (S0068) and ERα from four representative samples. (See color insert.)

mRNA is detected in ER-positive, but not ER-negative, breast cancer cells. In ER-positive cell lines, S0222 detects a band at the proper molecular weight (see Fig. 5.1A). However, a weaker band at a similar weight is also seen in Hs578T and MDA-MB-231 cells (see Fig. 5.1A), two cell lines that are negative for Rerg mRNA by Northern blot analyses and reverse transcription (RT)-PCR (data not shown). It is possible that the S0222 antibody is cross-reacting with another protein of a similar size in these samples. We are fairly certain that the band seen in ER-positive samples is indeed endogenous Rerg, as knocking down Rerg gene expression by shRNA decreases the intensity of this band (Fig. 5.5C). We did not find

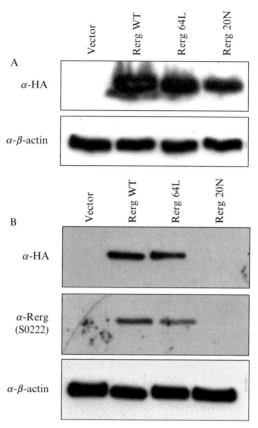

Figure 5.2 Transient and stable expression of wild-type and mutant Rerg proteins. (A) Transient expression of HA-Rerg proteins in 293T cells. 293T cells were transfected with the empty pCGN-hygro plasmid (Vector) or encoding Rerg WT, Rerg Q64L, or Rerg S20N. HA-Rerg expression was detected using the HA.11 (Covance) anti-HA epitope antibody. (B) Stable expression of HA-Rerg proteins in ER-negative Hs578T cells. Western blot analysis to verify the expression of ectopically expressed Rerg proteins was performed for Hs578T cells stably infected with the empty pBabe-puro HAII vector or encoding Rerg WT, Rerg Q64L, or Rerg S20N. Lysates were blotted with either anti-HA epitope or S0222 anti-Rerg antibody. Blot analysis with anti-β-actin was used as a loading control to verify equivalent total protein.

that treatment with estradiol stimulated an increase in Rerg expression (see Fig. 5.1A), perhaps because the cells are maintained in phenol red containing medium and are not estrogen starved. Because phenol red can stimulate the ER, Rerg may have been expressed at maximal levels even in cells not stimulated with estrogen directly. Taken together, these results suggest that the S0222 antibody can effectively detect endogenous Rerg expression by Western blot analysis.

5.3. Use of S0068 anti-Rerg antibody for immunohistochemical staining of breast tumor tissue

The tissue source and our protocols for immuohistochemical (IHC) staining of paraffin-embedded primary invasive breast tumor tissues array blocks have been described in detail elsewhere (Ring *et al.*, 2006). Briefly, prior to staining, tissue arrays are deparaffinized and dehydrated by submerging in xylene three times for 10 min each. Tissue arrays are then rinsed three times in 100% ethanol and twice in 95% ethanol and treated by microwaving and boiling for 11 min in 10 μM buffered citrate (pH 6.0). Slides are allowed to cool to room temperature and rinsed in distilled water followed by phosphate-buffered saline (PBS). Slides are dipped in 0.03% hydrogen peroxide, rinsed with PBS, and then stained using dilutions of antibodies in DAKO diluent (DAKO Cytomation Inc.) for 1 h at room temperature. The secondary antibody is applied for 1 h, and staining is visualized using the DAKO Cytomation Envision staining kit according to the manufacturer's instructions. Staining with an antibody directed against ER is performed by a commercial service (US Labs Inc.).

Anti-Rerg (S0068) and anti-ERα staining is executed in parallel using slides composed of tissue from primary invasive breast cancers. Consistent with array analyses in breast tumors and Western blot analyses of breast tumor cell lines (Finlin *et al.*, 2001), the intensity of anti-Rerg staining correlates with that of ERα. Additionally, whereas ERα staining is nuclear, Rerg staining is cytosolic, as we have seen in cell lines. These results provide validation for the use of S0068 anti-Rerg antibody for IHC analyses.

6. TRANSIENT AND STABLE EXPRESSION OF WILD-TYPE AND MUTANT RERG PROTEINS

Like Ras, Rerg cycles between GTP- and GDP-bound states (Finlin *et al.*, 2001), but whether GDP/GTP cycling is important for Rerg function is not known. In order to determine whether Rerg function depends on its GTP-bound state, mutations are generated in Rerg, based on analogous mutants of Ras proteins, that render it either constitutively GTP bound (Rerg Q64L) or preferentially bound to GDP (Rerg S20N). The Rerg Q64L mutation is analogous to the GAP-insensitive Q61L mutation in Ras and has been shown previously to possess decreased intrinsic GTPase activity *in vitro* (Finlin *et al.*, 2001). Since no GAP specific for Rerg has been identified, whether this mutant is also impaired in GAP-stimulated hydrolysis is not known. The Rerg S20N mutant has not been described previously but is based on the preferentially GDP-bound S17N mutation in Ras (Feig, 1999). Introduction of the analogous mutation has generated dominant-negative variants of a wide variety of Ras superfamily proteins. Such dominant negatives function by forming nonproductive complexes with GEFs that stimulate the activation of the wild-type GTPase. While a Rerg-specific GEF remains to be identified, an indirect evaluation of such mutants can be performed by expressing them in cells. For example, the growth suppressive activity of the Ras S17N mutant is consistent with the requirement for Ras in normal cell proliferation. Therefore, if Rerg functions as a tumor suppressor, then dominant-negative Rerg would be expected to stimulate growth in cells in which Rerg is expressed.

In order to test the expression of these mutants in the pCGN-hygro vector, 293T cells are transfected transiently with pCGN-hygro expression vectors encoding Rerg wild type, Q64L, or S20N. As shown in Fig. 5.2A, wild-type Rerg and Rerg Q64L are expressed at high levels, but the expression of Rerg S20N is reduced substantially. Similar expression patterns are observed when these constructs are stably expressed in breast cancer cells (data not shown). We also tested the expression of pBabe HA-tagged Rerg constructs stably expressed in Hs578T breast cancer cells. Wild-type Rerg and Rerg Q64L are expressed at equal levels, but Rerg S20N expression is not detected (see Fig. 5.2B). Furthermore, we consistently failed to detect transient expression of pBabe-Rerg S20N in 293T cells (data not shown), and ER-Rerg S20N expression is not induced efficiently (Fig. 5.3). Thus, we found that the Rerg S20N protein is expressed inefficiently, regardless of the expression vector or cell line used, or whether we tried transient or stable expression. We believe that the reduced expression level is most likely a consequence of decreased protein stability. The mutation in residue 20 may disrupt the Rerg protein structure. Alternatively, the mutation may decrease GDP and GTP binding, which may account for the reduced protein stability. Future identification of a Rerg GEF will be required to address these issues. Future directions may include the generation of additional Rerg mutants as candidate dominant negatives.

7. ESTABLISHMENT OF A TAMOXIFEN-INDUCIBLE RERG PROTEIN EXPRESSION SYSTEM

A major obstacle in the study of growth inhibitory proteins is that their stable expression is not well tolerated in cells. Either toxicity prevents suitable long-term expression or cells overcome growth inhibition by secondary, compensatory mechanisms. In order to evade these scenarios and to obtain a more accurate measure of Rerg growth inhibition, we generated an inducible expression system for Rerg. We constructed a chimeric gene that encodes a tamoxifen–inducible ER:Rerg fusion protein. This fusion protein is composed of a mutant form of the ER ligand-binding domain that responds only to 4-hydroxytamoxifen (4-OHT) and not to phenol red or estrogen and is fused to the N terminus of full-length Rerg. In the absence of tamoxifen, the fusion protein is inactive because it is bound by an Hsp90 complex and degraded. Upon ligand binding, the Hsp90 complex dissociates, allowing stable expression of the fusion protein (Reuter and Khavari, 2005). The use of ER fusion proteins to generate inducible protein expression has been described for Ras and many other signaling proteins (Eilers et al., 1989; Lim and Counter, 2005; Pritchard et al., 1995; Reuter and Khavari, 2005; Tarutani et al., 2003). While it is possible that the addition of ER sequences to the amino terminus of Rerg may disrupt Rerg function, previous observations with Ras and other Ras superfamily proteins indicate that the addition of sizable sequences to amino termini of small GTPases generally does not cause perturbations in function.

To determine whether ectopic expression of Rerg in ER-negative breast cancer cells decreases their tumorigenicity, two ER-negative cell lines, MDA-MB-231 and SUM149, are infected retrovirally with the empty pBabe-puro plasmid (Vector) or with plasmids encoding ER-Rerg wild type, Q64L, or S20N. To induce ER-Rerg protein expression, stably infected cells are treated with 1 μM 4-OHT for a minimum of 24 h. Figure 5.3 shows that in both cell lines, tamoxifen treatment efficiently induces the expression of wild-type Rerg and Rerg Q64L in both cell lines. In contrast, Rerg 20N is not induced efficiently. Some leakiness of expression is observed in untreated cells when the film is exposed for a longer time period (see Fig. 5.3A). Despite this leakiness, it is clear that tamoxifen treatment enhances ER-Rerg protein expression in this system greatly. Furthermore, H-Ras 12V expression is also induced efficiently by 1 μM 4-OHT in SUM149 cells (see Fig. 5.3B).

We next asked whether inducing Rerg expression decreases the growth or invasive properties of the two ER-negative breast cancer cell lines. First, an MTT assay is used to measure the anchorage-dependent cell proliferation rate of MDA-MB-231 cells stably expressing ER-Rerg wild type, Q64L, and S20N. Figure 5.4A shows that inducing wild-type or mutant Rerg with 4-OHT does not significantly alter the viability or proliferation rate of MDA-MB-231 cells. Next, the Matrigel invasion assay is used to determine whether inducing wild type or mutant Rerg decreases the invasiveness of MDA-MB-231 cells in vitro. Adding 4-OHT to cell lines stably infected with the empty pBabe-ER plasmid (Vector) or with plasmids encoding ER-Rerg wild type, Q64L, or S20N does not significantly and reproducibly affect invasion through Matrigel (see Fig. 5.4B). In addition, we did not observe significant differences in soft agar growth or

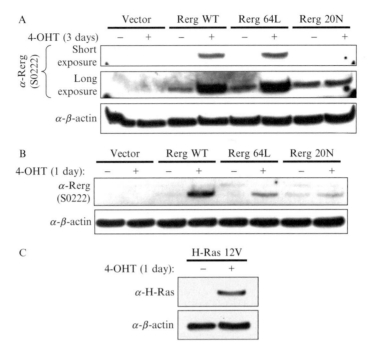

Figure 5.3 Tamoxifen-inducible expression of an ER-Rerg fusion protein. Plasmids encoding ER-Rerg fusion proteins were infected retrovirally into (A) MDA-MB-231 cells and (B) SUM149 cells. Stable cell lines were treated with either vehicle (ethanol) or 1 μM 4-hydroxytamoxifen (4-OHT) for 3 days (A) or 1 day (B). Western blot analysis using the S0222 anti-Rerg antibody detected a ≈55-kDa band corresponding to the ER: Rerg fusion protein in samples treated with 4-OHT. Similar results were obtained when cells were treated with 4-OHT for 7 days (data not shown). As a control for 4-OHT induction, SUM149 cells stably infected with an expression vector encoding the ER-H-Ras G12V protein (C) were included. Expression was induced by 1-day treatment with 4-OHT and detected by Western blot analysis with an H-Ras antibody (#146; Quality Biotech, Camden, NJ). Blot analysis for β-actin in total cell lysates was used as a loading control to verify equivalent total protein.

invasion through collagen in these cell lines (data not shown). Together, these data show that Rerg expression does not significantly affect the growth or invasive properties of ER-negative MDA-MB-231 breast cancer cells.

We also examined whether inducing Rerg expression affects the growth of SUM149 cells. A soft agar assay is used to measure the anchorage-independent growth of these cells stably expressing ER-Rerg and ER-H-Ras constructs. As expected, treating vector control cells with 4-OHT does not affect their ability to grow in soft agar. Furthermore, as expected, inducing oncogenic H-Ras G12V with 4-OHT drastically increases the ability of these cells to form colonies in soft agar (see Fig. 5.4C). However,

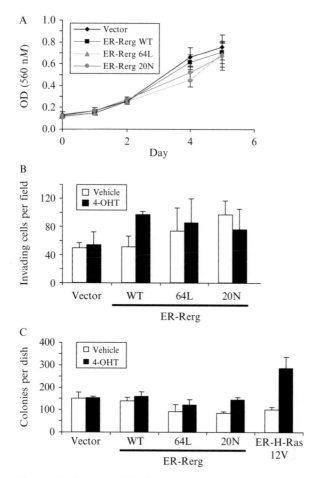

Figure 5.4 Effects of induction of ER–Rerg expression on growth and invasive properties in ER–negative breast cancer cells. (A) Evaluation of growth rate upon induction of ER–Rerg expression in MDA-MB-231 cells. Cells were incubated in growth medium supplemented with 1 μM 4-OHT to induce Rerg expression. MTT was added, and optical density was measured at the indicated time points. Results are an average of eight replicates per time point; error bars represent standard deviation. (B) Evaluation of Matrigel invasion upon induction of ER–Rerg in MDA-MB-231 cells. Stable cell lines were seeded in triplicate and allowed to invade toward growth medium supplemented with 3% FCS (chemoattractant). After 2 h, growth media supplemented with ethanol (vehicle) or 1 μM 4-OHT were added to each top and bottom well to induce Rerg expression. After incubation for 22 h, noninvading cells were removed and invading cells were fixed and stained. Invading cells were quantified by counting five fields of view using the 20× objective lens under a phase-contrast microscope. Error bars represent standard deviation of triplicate wells. (C) Evaluation of anchorage-independent growth upon induction of ER–Rerg in SUM149 cells. Cells were seeded in triplicate into 0.4% soft agar over a 0.6% bottom layer; ethanol (vehicle) or 1 μM 4-OHT was added to the top and bottom agar layers. After 21 days, viable colonies were stained in 2 mg/ml MTT, and visible colonies representing >50 cells were quantified. Results represent the average of triplicate plates; error bars represent standard deviation.

inducing wild-type Rerg or Rerg Q64L does not significantly affect their anchorage-independent growth. Inducing Rerg S20N may slightly increase the anchorage-independent growth of these cells; this may be because Rerg S20N can act as a dominant negative and may inhibit Rerg exchange factors in these cells. Thus, consistent with results obtained in the MDA-MB-231 cell line, inducing Rerg expression does not significantly alter the anchorage-independent growth of ER-negative SUM149 breast cancer cells.

We previously showed that ectopic overexpression of HA-tagged Rerg decreases cell proliferation and soft agar growth of the Rerg-expressing MCF-7 breast cancer cell line (Finlin *et al.*, 2001). However, Rerg does not show significant growth inhibition when expressed ectopically in the two ER-negative cell lines tested. This discrepancy may be because of overexpression artifacts; in previous studies, Rerg was expressed from the pCGN promoter, which drives expression much more strongly than the pBabe promoter described here. Another possibility is that the growth inhibitory function of Rerg may be cell type specific; Rerg may act as a growth inhibitor in ER-positive but not in ER-negative cells. Alternatively, it remains possible that the large ER domain fused to Rerg is disrupting its function. However, we believe this is not the case for the following reasons: (1) placing an HA tag N-terminal to Rerg does not disrupt its growth inhibitory function and (2) placing the ER ligand-binding domain N-terminal to Ras does not disrupt Ras function (Lim and Counter, 2005; Reuter and Khavari, 2005). Therefore, although we were unable to identify an assay for Rerg function, we believe that the ER-Rerg fusion proteins generated here will be very useful in the study of Rerg as other functions are identified in the future.

8. SUPPRESSION OF RERG EXPRESSION BY RNA INTERFERENCE

8.1. Evaluation of Rerg shRNA retrovirus vectors on ectopically expressed Rerg

To further define the role of Rerg in breast cancer, interfering shRNA is used to repress Rerg in Rerg-expressing cells. First, it was validated that our Rerg shRNA construct reduces the expression of ectopically expressed Rerg protein in 293T cells. Of the two constructs described here, only pSUPER. retro Rerg shRNA 3 targets a sequence in the coding region of Rerg. This construct efficiently reduces expression of pBabe HAII Rerg (Fig. 5.5A), verifying that Rerg shRNA3 does indeed target Rerg mRNA.

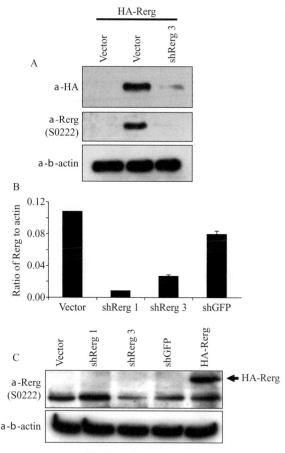

Figure 5.5   Repression of Rerg expression by shRNA. (A) Reduction of ectopically expressed Rerg by shRNA. 293T cells were transiently cotransfected with either 2.5 μg of the empty pSUPER.retro.puro vector (lanes 1 and 2) or encoding Rerg shRNA 3 (lane 3), together with 0.5 μg of the empty pBabe-puro HAII vector (lane 1) or encoding HA-Rerg (lanes 2 and 3). HA-Rerg expression was detected by immunoblotting with the anti-HA or anti-Rerg (S0222) antibodies. Blot analysis for β-actin was used to verify the equivalent total protein. (B) shRNA-induced reduction of endogenous Rerg mRNA. cDNA from T-47D cells stably infected with the empty pSUPER.retro.puro vector, Rerg shRNA 1, Rerg shRNA 3, or GFP shRNA was evaluated using quantitative real-time RT-PCR. Samples were run in duplicate. The Rerg:actin ratio was determined using the equation $2\Delta CT$, where $\Delta CT = \beta$-actin CT value − Rerg CT value. (C) shRNA-mediated reduction of endogenous Rerg protein. T-47D cells stably infected with the empty pSUPER.retro.puro vector or encoding Rerg shRNA 1, Rerg shRNA 3, or GFP shRNA, or the pBabe-puro HAII-Rerg expression vector, were lysed and analyzed by Western blot. The S0222 anti-Rerg antibody was used to detect Rerg expression. Blot analysis for β-actin was used as a loading control.

8.2. Analyzing shRNA suppression of endogenous Rerg mRNA expression by RT-PCR

Next we generated mass populations of the T-47D ER-positive cell line stably infected with either the empty pSUPER.retro.puro vector or encoding Rerg shRNA 1, Rerg shRNA 3, or GFP shRNA. After selection in growth medium supplemented with puromycin, multiple drug-resistant colonies are pooled together to establish cell lines for biological analyses. To determine whether the Rerg shRNA constructs efficiently reduce endogenous Rerg mRNA levels, quantitative real-time RT-PCR is used. RNA is isolated from each T-47D cell line using the TRIzol reagent (Invitrogen), and genomic DNA is removed by treatment with RQ1 DNase (Promega, Madison, WI). mRNA is reverse transcribed to cDNA using the SuperScript III first-strand synthesis system (Invitrogen). Rerg cDNA is amplified using Absolute SYBR Green ROX mix (ABGene House, Surrey, UK) and the following primers specific to Rerg cDNA: 5'- GGCATCTTCACCTTGCTTT-3' (sense) and 5'-GGCATGGT TAAGCTCCATT-3' (antisense). Real-time RT-PCR is performed on the ABI PRISM real-time PCR machine (Applied Biosystems, Foster City, CA) using the SYBR green detector, and data are analyzed using SDS 2.0 software. Both Rerg shRNA sequences significantly reduce the expression of Rerg mRNA (see Fig. 5.5B). Similar results are obtained with semiquantitative RT-PCR (data not shown).

8.3. Analysis of Rerg shRNA suppression of endogenous Rerg protein expression

Western blot analysis is used with the S0222 anti-Rerg antibody to determine whether the Rerg shRNA constructs could reduce the expression of endogenous Rerg protein. As shown in Fig. 5.5C, Rerg shRNA 3, but not Rerg shRNA 1, decreases the expression of endogenous Rerg protein. We cannot explain why Rerg shRNA 1 reduces the expression of Rerg mRNA but not the Rerg protein. Consistent with results from inducing Rerg expression, knocking down Rerg does not significantly affect soft agar growth or Matrigel invasion of T-47D breast cancer cells (data not shown). These studies do not support a critical role for Rerg in breast cancer growth or invasion.

9. Summary

This chapter described reagents for inducible, transient, or sustained expression of ectopic Rerg or to reduce endogenous Rerg expression in breast cancer cells. We described antibodies that detect Rerg expression by Western blot or IHC analyses. We validated expression of putative gain-of-function and loss-of-function Rerg mutants. While inducing Rerg expression or knocking down Rerg by shRNA did not significantly affect the *in vitro* growth or invasive properties of the breast cancer cell lines tested, it remains possible that Rerg expression will impair the tumorigenic, invasive, or metastatic properties of breast carcinoma cells when evaluated in mouse models. The reagents described here will prove to be useful in identifying a function for Rerg and for delineating its role in breast cancer progression.

ACKNOWLEDGMENTS

We thank Douglas Andres (University of Kentucky) for the pKH3 Rerg wild type, Rerg Q64L, and Rerg S20N plasmids; Christopher Counter (Duke University) for pBabe ER H-Ras G12V; Kevin Healy (UNC-Chapel Hill) for the pBabe ER empty vector; Natalia Mitin (UNC-Chapel Hill) for the pSUPERIOR.retro.puro GFP shRNA plasmid; Stephen Ethier (University of Michigan) and Carolyn Sartor (UNC-Chapel Hill) for the SUM149 cells; and Brian Z. Ring (Applied Genomics Inc.) for assistance in manuscript writing. This work was supported in part by a Susan G. Komen Foundation grant to C.J.D. (BCTR0402537). A.B.H. is a recipient of the UNC Cancer Cell Biology Training Grant and Department of Defense Breast Cancer Research Program Predoctoral Fellowship (BC061107).

REFERENCES

Baines, A. T., Lim, K. H., Shields, J. M., Lambert, J. M., Counter, C. M., Der, C. J., and Cox, A. D. (2005). Use of retrovirus expression of interfering RNA to determine the contribution of activated k-ras and ras effector expression to human tumor cell growth. *Methods Enzymol.* **407,** 556–574.

Clark, G. J., Cox, A. D., Graham, S. M., and Der, C. J. (1995). Biological assays for Ras transformation. *Methods Enzymol.* **255,** 395–412.

Dhanasekaran, S. M., Barrette, T. R., Ghosh, D., Shah, R., Varambally, S., Kurachi, K., Pienta, K. J., Rubin, M. A., and Chinnaiyan, A. M. (2001). Delineation of prognostic biomarkers in prostate cancer. *Nature* **412,** 822–826.

Eckert, L. B., Repasky, G. A., Ulku, A. S., McFall, A., Zhou, H., Sartor, C. I., and Der, C. J. (2004). Involvement of Ras activation in human breast cancer cell signaling, invasion, and anoikis. *Cancer Res.* **64,** 4585–4592.

Eilers, M., Picard, D., Yamamoto, K. R., and Bishop, J. M. (1989). Chimaeras of myc oncoprotein and steroid receptors cause hormone-dependent transformation of cells. *Nature* **340,** 66–68.

Feig, L. A. (1999). Tools of the trade: Use of dominant-inhibitory mutants of Ras-family GTPases. *Nat. Cell Biol.* **1,** E25–E27.

Finlin, B. S., Gau, C. L., Murphy, G. A., Shao, H., Kimel, T., Seitz, R. S., Chiu, Y. F., Botstein, D., Brown, P. O., Der, C. J., Tamanoi, F., Andres, D. A., *et al.* (2001). RERG is a novel ras-related, estrogen-regulated and growth-inhibitory gene in breast cancer. *J. Biol. Chem.* **276,** 42259–42267.

Fiordalisi, J. J., Johnson, R. L., 2nd, Ulku, A. S., Der, C. J., and Cox, A. D. (2001). Mammalian expression vectors for Ras family proteins: Generation and use of expression constructs to analyze Ras family function. *Methods Enzymol.* **332,** 3–36.

Key, M. D., Andres, D. A., Der, C. J., and Repasky, G. A. (2005). Characterization of RERG: An estrogen-regulated tumor suppressor gene. *Methods Enzymol.* **407,** 513–527.

Lapointe, J., Li, C., Higgins, J. P., van de Rijn, M., Bair, E., Montgomery, K., Ferrari, M., Egevad, L., Rayford, W., Bergerheim, U., Ekman, P., DeMarzo, A. M., *et al.* (2004). Gene expression profiling identifies clinically relevant subtypes of prostate cancer. *Proc. Natl. Acad. Sci. USA* **101,** 811–816.

Lim, K. H., and Counter, C. M. (2005). Reduction in the requirement of oncogenic Ras signaling to activation of PI3K/AKT pathway during tumor maintenance. *Cancer Cell* **8,** 381–392.

Morgenstern, J. P., and Land, H. (1990). Advanced mammalian gene transfer: High titre retroviral vectors with multiple drug selection markers and a complementary helper-free packaging cell line. *Nucleic Acids Res.* **18,** 3587–3596.

Novina, C. D., Murray, M. F., Dykxhoorn, D. M., Beresford, P. J., Riess, J., Lee, S. K., Collman, R. G., Lieberman, J., Shankar, P., and Sharp, P. A. (2002). siRNA-directed inhibition of HIV-1 infection. *Nat. Med.* **8,** 681–686.

Pritchard, C. A., Samuels, M. L., Bosch, E., and McMahon, M. (1995). Conditionally oncogenic forms of the A-Raf and B-Raf protein kinases display different biological and biochemical properties in NIH 3T3 cells. *Mol. Cell. Biol.* **15,** 6430–6442.

Ramaswamy, S., Ross, K. N., Lander, E. S., and Golub, T. R. (2003). A molecular signature of metastasis in primary solid tumors. *Nat. Genet.* **33,** 49–54.

Repasky, G. A., Chenette, E. J., and Der, C. J. (2004). Renewing the conspiracy theory debate: Does Raf function alone to mediate Ras oncogenesis? *Trends Cell Biol.* **14,** 639–647.

Reuter, J. A., and Khavari, P. A. (2005). Use of conditionally active ras fusion proteins to study epidermal growth, differentiation, and neoplasia. *Methods Enzymol.* **407,** 691–702.

Ring, B. Z., Seitz, R. S., Beck, R., Shasteen, W. J., Tarr, S. M., Cheang, M. C., Yoder, B. J., Budd, G. T., Nielsen, T. O., Hicks, D. G., Estopinal, N. C., and Ross, D. T. (2006). Novel prognostic immunohistochemical biomarker panel for estrogen receptor-positive breast cancer. *J. Clin. Oncol.* **24,** 3039–3047.

Sorlie, T., Perou, C. M., Tibshirani, R., Aas, T., Geisler, S., Johnsen, H., Hastie, T., Eisen, M. B., van de Rijn, M., Jeffrey, S. S., Thorsen, T., Quist, H., *et al.* (2001). Gene expression patterns of breast carcinomas distinguish tumor subclasses with clinical implications. *Proc. Natl. Acad. Sci. USA* **98,** 10869–10874.

Tarutani, M., Cai, T., Dajee, M., and Khavari, P. A. (2003). Inducible activation of Ras and Raf in adult epidermis. *Cancer Res.* **63,** 319–323.

Wennerberg, K., Rossman, K. L., and Der, C. J. (2005). The Ras superfamily at a glance. *J. Cell Sci.* **118,** 843–846.

CHAPTER SIX

K-Ras-Driven Pancreatic Cancer Mouse Model for Anticancer Inhibitor Analyses

Natalie Cook,*,[†] Kenneth P. Olive,[†] Kristopher Frese,[†] and David A. Tuveson[†]

Contents

Abstract

Genetically engineered mouse (GEM) models of cancer have progressively improved in technical sophistication and accurately recapitulating the cognate human condition and have had a measurable impact upon our knowledge of tumorigenesis. However, the application of such models toward the development of innovative therapeutic and diagnostic approaches has lagged behind. Our laboratory has established accurate mouse models of early and advanced ductal pancreatic cancer by conditionally expressing mutant K-ras and Trp53

* Department of Clinical Oncology, Addenbrookes NHS Trust, Cambridge, United Kingdom
† Cancer Research UK Cambridge Research Institute, Li Ka Shing Centre, Cambridge, United Kingdom

Methods in Enzymology, Volume 439
ISSN 0076-6879, DOI: 10.1016/S0076-6879(07)00406-5

alleles from their endogenous promoters in pancreatic progenitor cells. These K-Ras-dependent preclinical models provide valuable information on the cell types and pathways involved in the development of pancreatic cancer. Furthermore, they can be used to investigate the molecular, cellular, pharmacokinetic, and radiological characteristics of drug response to classical chemotherapeutics and to targeted agents. This chapter reviews the methods used to explore issues of drug delivery, imaging, and preclinical trial design in our GEM models for pancreatic cancer. We hypothesize that results of our preclinical studies will inform the design of clinical trials for pancreatic cancer patients.

1. INTRODUCTION

The primary purpose of drug efficacy testing is to predict, near the end of the preclinical development pipeline, whether a particular compound will be successful in the clinic. Two broad approaches are utilized in efficacy testing: cell-based *in vitro* systems and *in vivo* autochthonous animal models. At the interface between these two classes are tumor xenograft models in which cultured cells or tumor explants are grafted into immunodeficient mice. While such "animal culture" models are convenient to use, they generally behave differently than the corresponding human cancer. In fact, there is a poor correlation between the therapeutic activity of compounds tested in either xenografts or cell-based assays and their efficacy in humans (Johnson *et al.*, 2001). Xenograft models in particular have been used extensively in academic and industry research settings to prioritize compounds for clinical testing. Unfortunately, most drugs are found to be ineffective late in their development, with only a small percentage (<5%) of patients in phase 1 clinical trials responding to the therapies being tested (Roberts *et al.*, 2004). Such failures are costly to scientists and drug companies and are of great consequence to the patients that optimistically enroll in experimental clinical trials.

There are many reasons why preclinical studies fail to predict clinical activity, among them, differences in pharmacokinetics, pharmacodynamics, drug delivery, and metabolism. However, the basic problem is that neither cell-based studies nor xenograft models accurately reconstruct the complex interactions between tumor and host. Tumors are complicated entities, composed of mutated primary tumor cells, as well as recruited host cells that engineer a rich stromal environment. Indeed, in many tumor types, stromal cells outnumber tumor cells. This diversity is diminished and altered in xenograft systems.

In autochthonous genetically engineered mouse (GEM) models, tumor development occurs *in situ* in the appropriate tissue compartments and complex processes can be modeled. Thus it is reasonable to expect that GEM models, carrying the genetic signature of the native malignancy, could recapitulate clinical behavior, offering an alternative to traditional preclinical assays. The utility of a GEM model is dependent on a number of issues, including fidelity of the genetic lesions, tumor penetrance, kinetics of tumor initiation/progression, and ability to detect disease and perform specific interventions. To date, very few accurate GEM models have been used in well-designed preclinical drug evaluation trials.

Early efforts to model pancreatic cancer failed to reproduce the dominant histological subtype of ductal adenocarcinoma. However, substantial efforts in the molecular genetics of human pancreatic ductal adenocarcinoma (PDA) have delineated a number of common genetic alterations. Among the most common alterations in PDA are activating mutations in the K-ras protooncogene. Such mutations are found in almost 95% of human cases and have the effect of locking the K-ras protein in an active signaling conformation (Smit et al., 1998). Other common alterations include inactivation of a number of tumor suppressor genes: p16^{INK4a} (90% of cases), p53 (approximately 75% of cases), SMAD4 (55%), and BRCA2 (10%) (Bardeesy et al., 2006; Redston et al., 1994). Moreover, these alterations accumulate with increasing frequency in a series of premalignant lesions associated with pancreatic cancer, most notably pancreatic intraepithelial neoplasia (PanIN) (Hruban et al., 2000). These studies have led to the understanding that K-Ras and Ink4a alterations are early events in PDA development, whereas loss of SMAD4 and p53 tends to occur later (Maitra et al., 2003). Additionally, advances have been made in the understanding of pancreatic developmental biology and in identifying signaling pathways that are important in PDA. These include the Notch and Hedgehog signaling pathways (Miyamoto et al., 2003; Thayer et al., 2003). Further understanding of how these pathways interact with oncogenic K-ras signaling should enable us to advance therapeutic options through the development of drugs targeting key pathways and molecules.

 ## 2. Pancreatic Ductal Adenocarcinoma Models

A number of models of pancreatic disease have now been developed in genetically engineered mice. A workshop took place in 2004 to establish an internationally accepted uniform nomenclature for the pathology of genetically engineered mouse models of exocrine pancreas neoplasia (Hruban et al., 2006a,b). This resulted in the elaboration of specific proposals and goals that will allow uniform progression and hopefully success in this field.

Previous work in our laboratory has unveiled dramatic phenotypic differences between exogenous overexpression of mutant Kras and mutation of the endogenous gene at physiological levels. We therefore used homologous recombination in mouse embryonic stem cells to engineer a conditional mutant allele of Kras to enable the expression of endogenous mutant Kras exclusively in the pancreas. Mice harboring this conditional Kras mutant allele (Kras$^{LSL.G12D}$) in combination with a pancreas-specific Cre recombinase transgene (PdxCre or p48Cre) develop a full range of premalignant lesions in the pancreas, termed pancreatic intraepithelial neoplasia, before succumbing to invasive PDA and other tumors at late ages (Hingorani *et al.*, 2003). These mice are an excellent model of PanIN development and are useful for studying tumor progression. However, their use in therapeutic trials is limited because of the late onset of PDA.

For preclinical studies, a second model that also incorporates a conditional mutation in the endogenous p53 gene has proved to be more useful. These mice develop PDA with 100% penetrance and have a median survival of 4.5 months; approximately 70% of these mice present with gross metastasis (Hingorani *et al.*, 2005). Tissues not expressing Cre recombinase remain functionally hemizygous for these loci. Disease progression from preinvasive to invasive disease involved loss of the wild-type Trp53 allele, although it remains unclear at exactly what stage this occurs. The histopathology of pancreatic tumors from PDA mice is extremely similar to human PDA and also shows a high degree of chromosomal instability, another hallmark of human pancreatic cancer.

3. IMAGING

Both PanIN and PDA mice develop tumors with variable latency. Therefore, noninvasive modalities that detect and quantify tumor size are needed to properly enroll such animals in preclinical trials. Subcutaneous xenografts can be monitored easily with calipers, but most GEM models require specialized imaging modalities to monitor tumor development. Several imaging techniques are available for mice, including magnetic resonance imaging (MRI), positron emission tomography, micro–X-ray computed tomography, and high–resolution ultrasound (US). All of these methods are also used in cancer patients. Another method used exclusively in genetically engineered mice is bioluminescent imaging, relying upon transgenic reporters engineered to reflect either anatomical or functional properties *in vivo*.

The main imaging technique currently used in our PDA mice is high-resolution US. A number of factors should be considered when deciding which modality to use. These include the site of the tumor being imaged, whether the imaging modality is available within an animal barrier facility, the need for repeated imaging, session time, and invasiveness of technique. We chose to use US because it is noninvasive, readily available, requires a short session time, and is small enough to fit into our animal room. To detect tumors, mice had bimonthly US from 2 months of age and tumors were detected when they were as small as 1 mm in diameter. Animals with a defined tumor burden (3- to 5-mm-diameter tumor as measured by ultrasound) were enrolled into intervention studies along with matched littermate control mice that harbor K-ras and p53 alleles but lack Cre.

3.1. Imaging of *in situ* PDAs by high-resolution ultrasound

1. All imaging modalities require the mice to be anesthetized. Although injectable anesthetics are available, inhalation anesthetics such as isofluorane are better tolerated and therefore preferable for studies in which mice are imaged repeatedly.
2. Mice are anesthetized with a 3% isofluorane/oxygen mixture in an induction chamber and then transferred to a platform heated to 40° (VisualSonics, Inc.).
3. Core body temperature is monitored continuously using a digital rectal thermometer. Long fur is shaved from the abdomen and flanks of the mice, and a depilatory cream is applied to remove the remaining stubble. Ample washing with sterile water is used to remove the cream and prevent skin irritation.
4. We found that intraperitoneal (IP) injection of 3 to 4 ml of normal saline shortly before imaging results in a substantially improved image of the pancreas and other abdominal organs. This treatment is well tolerated, as normal saline is eliminated rapidly by diuresis.
5. Ultrasound gel, preheated to 37°, is applied to the abdomen of the mice, and scanning is performed with a 35-MHz ultrasound probe. This frequency allows for a 1-cm focal depth, which is ideal for abdominal imaging in mice. Care must be taken to minimize the introduction of bubbles into the ultrasound gel.
6. Optimal viewing of the head of pancreas is achieved through axial imaging of mice positioned supine and using the pyloris and proximal duodenum as landmarks (Figs. 6.1A and B). The tail of the pancreas is imaged axially with both supine and lateral positioning using the left kidney and the inner curvature of the spleen as landmarks (see Fig. 6.1C and D).
7. Using this protocol, mice are evaluated rapidly for the presence of pancreatic tumors with a session time of 10 to 15 min per mouse. Tumor volume is quantified by using the 3D motor (VisualSonics, Inc.)

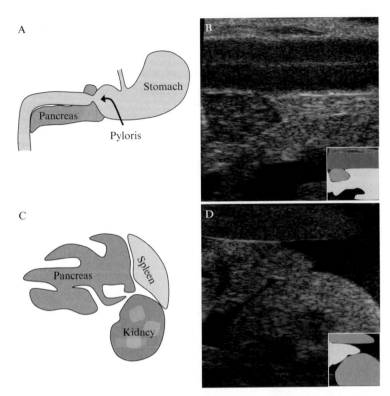

Figure 6.1 (A) Diagram showing landmarks used to locate the head of the pancreas by ultrasound. (B) Ultrasound image of the head of the pancreas showing a longitudinal section of the proximal duodenum (coded blue in the inset key), pancreas (yellow), and a axial section of a more distal segment of intestine (orange). (C) Diagram of the landmarks used to locate the tail of the pancreas by ultrasound. (D). Ultrasound image of the tail of the pancreas showing the proximity of the pancreas (coded yellow in the inset key) to the left kidney (green) and spleen (lavender). (See color insert.)

to acquire volume data and then manually outlining the tumor on frames spaced 0.25 mm apart.

8. During image acquisition, it is important to avoid obstructions in the image, such as fecal matter in the intestines and stomach.

4. DRUG DELIVERY

There are various different methods of delivering drugs in mouse models. Common routes of administration include subcutaneous (SC), IP, and oral gavage routes. Brief intravenous (IV) infusions and long-term catheterization are also possible, although the risk of infection and thrombosis can be problematic. The effect of the chosen delivery route on drug metabolism should always be taken into account.

4.1. Intraperitoneal injections

1. First sterilize the skin using an aseptic technique.
2. The needle route is usually through the abdominal wall into the peritoneal cavity. An alternative approach is to enter via the flank but this can cause retroperitoneal organ laceration.
3. The needle enters bevel up into the ventral abdominal midline with the syringe held parallel to the hind leg.
4. Initially insert the needle subcutaneously, but change the angle to almost vertical as the needle passes through the peritoneum. Take care to avoid entering the bladder and liver.
5. Repeated injections may cause adhesions between abdominal organs. The maximum amount allowed via an IP injection is usually 20 ml/kg. The frequency of injections depends on the protocol being used, for example, minimum 24 h between injections/maximum two injections within 24 h.

Another delivery method is the use of a microosmotic pump (e.g., Alzet Inc.). These have the potential to ensure long-term, continuous exposure to therapies, reduce handling and stress to laboratory animals, and can be used via IP, SC, or IV routes. This approach may have particular relevance in delivering drugs with a short half-life. Microosmotic pumps can be used with a number of different solvents, but the solubility of the drug can become an issue as the total volume of the pump is often limiting.

5. END POINT ANALYSIS

The use of GEM models for PDA to evaluate antineoplastic therapeutics requires the development of methods to measure pertinent pharmacokinetic and pharmacodynamic characteristics. This includes the availability of molecular diagnostics, including cytogenetics, genomics, and proteomics, as well as the rapid assessment of drug levels from the mouse tissues. Standard methods of tissue procurement and processing need to be established for the assessment of target inhibition in pancreatic tumors because acinar cells that occupy the majority of pancreatic tissue contain high levels of enzymes that degrade nucleic acids, proteins, carbohydrates, and fats. As high-quality human pancreatic tumor tissue is very difficult to obtain, it is likely that methods developed with GEM models may be applicable to clinical trials.

Another use of preclinical models is for the identification of biomarkers of disease progression and response to intervention. Serum proteomic screens and RNA transcriptional analyses are examples of unbiased methods used to identify correlative biomarkers for therapeutic responsiveness or resistance. Results from our PanIN mice have already shown a unique proteomic signature for mice with preinvasive pancreatic cancer (Hingorani *et al.*, 2003).

Methods for blood removal, necropsy, and RNA isolation are described next.

5.1. Removal of blood from mice (Less than 0.1 ml)

1. Asepsis should be maintained throughout so hair and superficial debris must be removed. Anesthetic is not normally needed for superficial vein bleeds.
2. The most common site for venipuncture on a mouse (and the method we use) is by puncturing the tail vein.
3. The animal should be gently restrained by an experienced animal handler and the vein raised, that is, some pressure may be needed proximal to the vein to occlude venous return and allow adequate venous sampling.
4. A sterile needle or lancet can be used to puncture the skin and underlying blood vessel.
5. Blood can be removed by a capillary tube or micropipette and plastic tip.
6. A maximum of 10% of circulating blood volume can be removed on a single occasion and a maximum of 15% of circulating blood volume in any 28-day period.
7. After blood removal, maintain firm pressure on the site for 30 s to prevent any further bleeding.
8. Possible complications associated with venipuncture include bleeding, bruising, thrombosis, and stress to the animal.

5.2. Necropsy protocol

Materials: neutral-buffered formalin, 1000 U/ml heparin solution, cryogenic freezing tubes, microcentrifuge tubes, RNALater (Qiagen), and dissection instruments

1. Euthanize the mouse using standard approaches.
2. Serum/plasma: Immediately perform cardiac puncture to collect blood.
3. Use a 21-gauge needle with a 1-ml syringe.
4. For plasma, precoat the needle and syringe with heparin by drawing a small amount into the needle and expelling, leaving some heparin in the void volume of the syringe. Collect blood into a microcentrifuge tube. For plasma, store the tube on wet ice for no more than 15 min.

Spin blood at 8000 rpm for 10 min at 4°. For serum, allow blood to clot at room temperature for 1 h. Spin at 8000 rpm for 10 min at 4°. Collect plasma or serum and usually store at −80°.

5. Following blood draw, open the abdominal cavity and immediately collect pancreatic samples, beginning with RNA samples.

6. RNA samples: Collect 50 to 250 mg of tissue into 500 μl of RNALater (Qiagen). Incubate at 4° over night. Decant RNALater and process immediately (see additional information provided later).

7. Photograph each animal with a digital camera to record anatomical position of tumors. Include the animal's identification in the photograph.

8. Snap frozen tissues: Depending on the availability of tissues, collect up to five snap frozen samples of tumor and other tissues (approximately 100 mg) into cryogenic screw-top tubes and snap freeze in liquid nitrogen.

9. Frozen sections: Collect tissue into a cryogenic mold filled with O.C.T. compound (Tissue-Tek). Place mold on dry ice and store at −80°. Do not immerse in liquid nitrogen.

10. Paraffin sections: Collect tissues into 10% formalin solution and fix for up to 24 h. An alternative is zinc-buffered formalin (Z-Fix, Anatech Ltd.). Following fixation, decant formalin and transfer tissue to 70% ethanol. Cassette and process tissue by standard protocols.

11. DNA: Collect several centimeters of tail into a microcentrifuge tube and store at −20° as an archival DNA sample from each mouse.

5.3. RNA isolation from pancreatic tissue

Note: This procedure is only used to isolate mRNA (Qiagen). Another class of RNAs that may prove diagnostically useful is micro-RNAs.

1. Always determine the correct amount of starting material to obtain optimal RNA yield and purity. In general 10 to 100 mg of fresh or frozen tissue can be processed. If there is no information about the nature of your starting material, start with 10 mg or less. For RNeasy mini-columns, use 10 mg of normal pancreatic tissue, as more than 10 mg seems to decrease the quality of the isolated RNA. Notably, Qiagen provides no information about mRNA preparation from pancreatic tissue.

2. We use the RNAeasy Protect Mini procedure (Qiagen). Always follow aseptic techniques and use sterile, RNase-free pipette tips and disposable gloves. Reagents from Qiagen are all RNase free and therefore do not require diethyl pyrocarbonate treatment.

3. Excise the tissue sample from the animal. Some method of RNA stabilization is absolutely required when isolating RNA from normal and PanIN pancreatic tissue. This is less important for tumor tissue, presumably

because there are few acinar cells present and therefore less RNase; however, if you want to compare tumor tissue to normal or PanIN tissue, they should all be treated similarly.

4. We currently use RNALater (Qiagen) for stabilization. It should be noted that RNALater-treated tissue is difficult to process for histological and immunohistochemical evaluation. We place approximately 50 to 100 mg tissue in a 1.5-m Eppendorf tube that has 10 volumes of cold RNALater and incubate overnight at 4°. The next day, RNALater can be drained and the sample is stored at −80°. For further information about RNA stabilization, please refer to the Qiagen manual

 ## 6. Model Fidelity for Preclinical Testing

The following points should be considered for pancreatic cancer GEM models used in preclinical studies.

- Mutant mice should accurately reflect the genetics and pathology of human pancreatic cancer and pathophysiological processes such as cachexia.
- The molecular alterations in murine PDA tumors should reflect those observed in human PDA, including biochemical pathways, gene expression patterns, and genomic alterations. As described previously, our PDA model accurately reflects these points, further validating our mouse model.
- Before assessing new therapies, GEM models of PDA should be treated with drugs normally used to treat patients. We refer to this as "credentialing"—the model should respond to drugs in a manner comparable to human patients (Olive and Tureson, 2006). Because the current standard therapy for pancreatic cancer, gemcitabine, has a relatively low response rate in patients with advanced disease, we would therefore predict a similar ineffectiveness in our model of PDA. Surprisingly, relatively few studies of conventional cytotoxic agents have been reported to date in GEM models.

 ## 7. Novel Therapeutics and Preclinical Trial Design

Novel therapeutics for PDA can be evaluated in the PDA GEM model once the appropriate biochemical and analytical chemical assays are established to measure pertinent pharmacokinetic and pharmacodynamic parameters. Although it is currently not possible to inhibit the K-Ras oncoprotein directly, downstream Ras effector pathways, such as RAF/MEK and PI3K/AKT, and cooperating signaling networks, including Notch, Hedgehog, and various mitogenic growth factor receptors (e.g., EGFR, Her2/neu, Met, IGFR), are potential candidates worthy of therapeutic investigation. Combinations of inhibitors blocking these pathways may be required for future effective therapies, and GEM models can be used to test synergy versus antagonism. Importantly, several candidate compounds targeting the aforementioned signaling cascades are currently undergoing early phase clinical evaluation.

Our PanIN and PDA GEM models will be used to evaluate potential therapies in a chemoprevention or intervention setting, respectively. Two study structures will be used to assess the effects of therapeutic agents on our murine PDAs (please also see Fig. 6.2).

Short-term intervention studies will be used to establish basal parameters such as serum and tumor pharmacokinetics (PK), pathway inhibition, and effects on basic tumor cell properties such as proliferation rate and apoptosis.

1. Mice are monitored by high-resolution ultrasound, MRI, or optical imaging techniques.
2. Animals with a defined tumor burden (a 3- to 5-mm-diameter tumor as measured by ultrasound) are enrolled into intervention studies.
3. Blood is collected by tail vein bleeding in order to assay serum PK levels.
4. Biweekly imaging is used to track tumor progression during the study, and daily weights are recorded.
5. Following brief pilot experiments, usually less than 2 weeks of duration, animals are euthanized and tissues harvested for tumor PK and tissue PD evaluation, including analysis of bromodeoxyuridine incorporation (injected 24 h prior to sacrifice), apoptosis, and biochemical signaling and gene expression alterations.

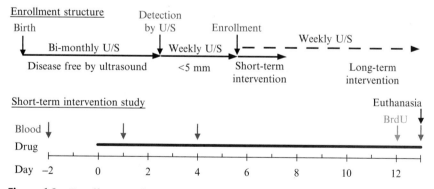

Figure 6.2 Enrollment and treatment algorithm: Mice are monitored by weekly ultrasound until the detection of measurable disease, after which they can be enrolled in a short- or long-term intervention study. Details of an example short-term intervention are provided.

Data from these studies will also advise the design of an intervention and prevention study with regards to drug dosing and schedule.

Long-term studies will directly examine the effects of drugs on survival. For these studies, drug dosages and schedules must be first evaluated for tolerability. Animals will remain on the drug treatment regimen until they require sacrifice due to demonstration of significant morbidity from pancreatic cancer. Methods of assessing the health of the mice include daily weight measurements and direct behavioral observation.

8. CONCLUSIONS

Limited progress has been made in treating pancreatic cancer over the recent years, and it is often challenging to evaluate new agents in patients due to the rapid progression of disease and the difficulty in monitoring the response to therapies. We suggest that GEM models of PDA will exceed the performance of xenograft models in predicting response to therapy due to the more accurate representation of various cell autonomous and tumor microenvironmental features in GEM models. This harbors great promise for accelerating the drug discovery process to markedly improve the success of pancreatic cancer drug development.

REFERENCES

Bardeesy, N., Aguirre, A. J., Chu, G. C., Cheng, K. H., Lopez, L. V., Hezel, A. F., Feng, B., Brennan, C., Weissleder, R., Mahmood, U., Hanahan, D., Redston, M. S., *et al.* (2006). Both p16(Ink4a) and the p19(Arf)-p53 pathway constrain progression of pancreatic adenocarcinoma in the mouse. *Proc. Natl. Acad. Sci. USA* **103**(15), 5947–5952.

Hingorani, S. R., Petricoin, E. F., Maitra, A., Rajapakse, V., King, C., Jacobetz, M. A., Ross, S., Conrads, T. P., Veenstra, T. D., Hitt, B. A., Kawaguchi, Y., Johann, D., *et al.* (2003). Preinvasive and invasive ductal pancreatic cancer and its early detection in the mouse. *Cancer Cell* **4**(6), 437–450.

Hingorani, S. R., Wang, L., Multani, A. S., Combs, C., Deramaudt, T. B., Hruban, R.H, Rustgi, A. K., Chang, S., and Tuveson, D. A. (2005). Trp53R172H and KrasG12D cooperate to promote chromosomal instability and widely metastatic pancreatic ductal adenocarcinoma in mice. *Cancer Cell* **7**, 469–483.

Johnson, L., Mercer, K., Greenbaum, D., Bronson, R. T., Crowley, D., Tuveson, D. A., and Jacks, T. (2001). Somatic activation of the K-ras oncogene causes early onset lung cancer in mice. *Nature* **410**(6832), 1111–1116.

Hruban, R. H., Adsay, N. V., Albores-Saavedra, J., Anver, M. R., Biankin, A. V., Boivin, G. P., Furth, E. E., Furukawa, T., Klein, A., Klimstra, D. S., Kloppel, G., Lauwers, G. Y., *et al.* (2006a). Pathology of genetically engineered mouse models of pancreatic exocrine cancer: consensus report and recommendations. *Cancer Res.* **66**, 95–106.

Hruban, R. H., Goggins, M., Parsons, J., and Kern, S. E. (2000). Progression model for pancreatic cancer. *Clin. Cancer Res.* **8**, 2969–2972.

Hruban, R. H., Rustgi, A. K., Brentnall, T. A., Tempero, M. A., Wright, C. V., and Tuveson, D. A. (2006b). Pancreatic cancer in mice and man: the Penn Workshop 2004. *Cancer Res.* **66,** 14–17.

Maitra, A., Adsay, N. V., Argani, P., Iacobuzio-Donahue, C., De Marzo, A., Cameron, J. L., Yeo, C. J., and Hruban, R. H. (2003). Multicomponent analysis of the pancreatic adenocarcinoma progression model using a pancreatic intraepithelial neoplasia tissue microarray. *Mod. Pathol.* **16,** 902–912.

Miyamoto, Y., Maitra, A., Ghosh, B., Zechner, U., Argani, P., Iacobuzio-Donahue, C. A., Sriuranpong, V., Iso, T., Meszoely, I. M., Wolfe, M. S., Hruban, R. H., Ball, D. W., *et al.* (2003). Notch mediates TGF alpha-induced changes in epithelial differentiation during pancreatic tumorigenesis. *Cancer Cell* **3,** 565–576.

Olive, K. P., and Tuveson, D. A. (2006). The use of targeted mouse models for preclinical testing of novel therapeutics. *Clin. Cancer Res.* **12,** 5277–5287.

Redston, M. S., Caldas, C., Seymour, A. B., Hruban, R. H., da Costa, L., Yeo, C. J., and Kern, S. E. (1994). p53 mutations in pancreatic carcinoma and evidence of common involvement of homocopolymer tracts in DNA microdeletions. *Cancer Res.* **54**(11), 3025–3033.

Roberts, T. G., Jr., Goulart, B. H., Squitieri, L., Stallings, S. C., Halpern, E. F., Chabner, B. A., Gazelle, G. S., Finkelstein, S. N., and Clark, J. W. (2004). Trends in the risks and benefits to patients with cancer participating in phase 1 clinical trials. *JAMA* **293,** 2130–2140.

Smit, V. T., Boot, A. J., Smits, A. M., Fleuren, G. J., Cornelisse, C. J., and Bos, J. L. (1998). KRAS codon 12 mutations occur very frequently in pancreatic adenocarcinomas. *Nucleic Acids Res.* **16,** 7773–7782.

Thayer, S. P., di Magliano, M. P., Heiser, P. W., Nielsen, C. M., Roberts, D. J., Lauwers, G. Y., Qi, Y. P., Gysin, S., Fernandez-del Castillo, C., Yajnik, V., Antoniu, B., McMahon, M., *et al.* (2003). Hedgehog is an early and late mediator of pancreatic cancer tumorigenesis. *Nature* **425,** 851–856.

CHAPTER SEVEN

ANALYSIS OF K-RAS PHOSPHORYLATION, TRANSLOCATION, AND INDUCTION OF APOPTOSIS

Steven E. Quatela, Pamela J. Sung, Ian M. Ahearn, Trever G. Bivona, *and* Mark R. Philips

Contents

Departments of Medicine, Cell Biology, and Pharmacology and the Cancer Institute, New York University School of Medicine, New York

Methods in Enzymology, Volume 439
ISSN 0076-6879, DOI: 10.1016/S0076-6879(07)00407-7

Abstract

K-Ras is a member of a family of proteins that associate with the plasma membrane by virtue of a lipid modification that inserts into the membrane and a polybasic region that associates with the anionic head groups of inner leaflet phospholipids. In the case of K-Ras, the lipid is a C-terminal farnesyl isoprenoid adjacent to a polylysine sequence. The affinity of K-Ras for the plasma membrane can be modulated by diminishing the net charge of the polybasic region. Among the ways this can be accomplished is phosphorylation by protein kinase C (PKC) of serine 181 within the polybasic region. Phosphorylation at this site regulates a farnesyl-electrostatic switch that controls association of K-Ras with the plasma membrane. Surprisingly, engagement of the farnesyl-electrostatic switch promotes apoptosis. This chapter describes methods for directly analyzing the phosphorylation status of K-Ras using metabolic labeling with ^{32}P, for indirectly assessing the farnesyl-electrostatic switch by following GFP-tagged K-Ras in live cells, for artificially activating the farnesyl-electrostatic switch by directing the kinase domain of a PKC to activated K-Ras using a Ras-binding domain, and for assessing apoptosis of individual cells using a YFP-tagged caspase 3 biosensor.

1. INTRODUCTION

The differential biology of Ras isoforms is generated, in large part, by distinct membrane targeting sequences. Membrane association of all Ras isoforms requires farnesylation, proteolysis, and carboxyl methylation of a C-terminal CAAX motif. Plasma membrane (PM) targeting of the principal splice variant of K-Ras also requires a unique polybasic region adjacent to the CAAX motif (Choy *et al.*, 1999; Hancock *et al.*, 1990; Jackson *et al.*, 1994). K-Ras thus falls into a broad class of proteins that are anchored to the cytoplasmic face of the PM by virtue of post-translational modification with lipids that act in conjunction with polybasic stretches of polypeptide. Whereas the lipid moieties are thought to insert into the phospholipid bilayer, the polybasic regions are believed to associate with the anionic head groups of inner leaflet phospholipids (Leventis and Silvius, 1998).

Included in this class of proteins is the myristoylated alanine–rich C kinase substrate (MARCKS) that associates with the PM via an N-terminal myristoyl modification and a polybasic region. Serine residues within the polybasic region are sites for phosphorylation, mediated by protein kinase C (PKC), that neutralize the charge and thereby cause the MARCKS protein to dissociate from the PM. The mechanism by which MARCKS is discharged from the PM through phosphorylation has been referred to as a myristoyl electrostatic switch (McLaughlin and Aderem, 1995).

Like MARCKS, the polybasic region of K-Ras harbors three potential phosphorylation sites, and this segment has been shown previously to be phosphorylated by PKC (Ballester *et al.*, 1987). We therefore tested the hypothesis that phosphorylation of the C-terminal segment of K-Ras might regulate its association with the PM and thereby constitute a farnesyl-electrostatic switch. Using green fluorescent protein (GFP)-tagged Ras proteins and live cell imaging, we have observed that PKC agonists induce a rapid translocation of K-Ras from the PM to intracellular membranes that include the endoplasmic reticulum (ER), Golgi apparatus, and, surprisingly, the outer mitochondrial membrane. Interestingly, phosphorylation and internalization of K-Ras also stimulated apoptosis. These observations show that the subcellular localization and function of K-Ras are modulated by PKC in a novel pathway resulting in cell death (Bivona *et al.*, 2006). This chapter describes assays that enabled us to determine which site(s) in the C-terminal domain of K-Ras is phosphorylated by PKC, to confirm that it is PKC phosphorylation of K-Ras that is responsible for its internalization, and to monitor the onset of apoptosis in individual cells in response to various K-Ras constructs.

2. ANALYSIS OF K-RAS PHOSPHORYLATION BY METABOLIC LABELING WITH ^{32}P

Previous studies performed by Ballester *et al.* (1987) demonstrated the presence of a PKC substrate in the C terminus of K-Ras. The C-terminal region of K-Ras harbors three potential phosphate acceptors: S171, S181, and T183. However, only serines conform to consensus phosphorylation sites, and phospho–amino acid analysis of K-Ras from cells exposed to PKC agonists revealed only phosphoserine (Ballester *et al.*, 1987). Sequence analysis can be used to determine the most likely site of phosphorylation based on the consensus PKC site, S/T-X-R/K. For K-Ras, S181 within the C-terminal membrane-anchoring region of K-Ras conforms most closely to a consensus PKC site.

Metabolic labeling of cells with ^{32}P is the standard method for monitoring the level of protein phosphorylation. Combining metabolic labeling with the expression of K-Ras constructs with mutated phosphate acceptor sites is a relatively easy way to determine the precise site of PKC phosphorylation of K-Ras or other phosphorylated proteins. One caveat is that it remains possible that mutating a phosphate acceptor site can have an indirect effect on a nearby site by altering the conformation of the region and lowering the efficiency of phosphorylation. Nevertheless, substitution of phosphate nonacceptors such as alanine for acceptors can give important information regarding protein phosphorylation. To do so, cells are transfected with expression vectors encoding the protein of interest possessing nonphosphorylatable alanine residues at potential phosphate acceptor sites. After a suitable period to allow expression, the transfected cells are labeled metabolically with ^{32}P. The cells are then stimulated with a PKC agonist, such as bryostatin-1 or phorbol myristate acetate (PMA), to induce phosphorylation and are lysed in a RIPA buffer. The lysates are immunoprecipitated for the protein of interest, which is then subjected to SDS-PAGE, and radioactive phosphate incorporation is observed by PhosphorImaging.

2.1. Transfection and metabolic labeling

Twenty-four hours before transfection, COS-1 cells are seeded into six-well plates at a density of 2.0×10^5 cells per well. The cells should be grown in 5% CO_2 at 37° in Dulbecco's modified Eagle's medium (DMEM) supplemented with 10% fetal bovine serum (FBS) and antibiotics to 70% confluence on the day of transfection. The wild-type and mutated K-Ras alleles cloned into a mammalian expression vector can be transfected into COS-1 cells with high efficiency using SuperFect (Qiagen). Lipofectamine (Invitrogen) used according to the manufacturer's instructions gives similar results. Three micrograms of each plasmid DNA is transfected with 7 μl SuperFect reagent in 100 μl unmodified DMEM. After 15 min, when transfection complexes have been allowed to form, 600 μl of DMEM supplemented with 10% FBS and antibiotics is added to bring the volume up to 700 μl. In the meantime, the medium from each well of the six-well plate is aspirated, and cells are washed with prewarmed phosphate-buffered saline (PBS). The entire transfection mixture is added to a single well, and cells are incubated for 3 h at 37° in 5% CO_2. The medium with the transfection complex is then removed, the cells are washed once with prewarmed PBS, and fresh DMEM containing serum and antibiotics is added to the cells. The cells are then grown for 24 h at 37° in 5% CO_2 to allow expression of the constructs.

Twenty-four hours after transfection, the COS-1 cells are ready for metabolic labeling. Cells should not have grown to more than 90% confluence, as phosphate uptake is maximal in rapidly growing cells. Growth medium containing serum and antibiotics is removed by aspiration. Cells are then washed three times with prewarmed phosphate-free DMEM to remove any residual phosphate-containing medium. Labeling medium is then prepared, containing 0.5 mCi/ml [^{32}P]orthophosphate (Perkin Elmer; NEM) in prewarmed phosphate-free DMEM supplemented with 10% partially dialyzed FBS. A final concentration of inorganic phosphate of 50 to 100 μM should be maintained to prevent slowing down of cellular metabolism. Serum dialyzed against phosphate-free saline can be mixed with undialyzed serum at a ratio of 1:3 to provide the required concentration of phosphate in the labeling medium. One milliliter of labeling media is added to each well of the six-well plate. The plate is then placed in a warmed Plexiglas box (to shield β radiation), and the box is put in the incubator at $37°$ in 5% CO_2 for a labeling period of 6 h. This intermediate period of labeling strikes a balance, ensuring that enough time elapses so that an equilibrium between cellular and extracellular phosphate pools is reached and ATP pools are labeled to reasonably high specific activity on the one hand, while limiting the radiation damage to cells caused by longer periods of labeling on the other.

2.2. Stimulation, immunoprecipitation, SDS-PAGE, and analysis of phosphorylation of K-Ras

At the end of the labeling period, the six-well dish is removed from the Plexiglas box and the labeling medium is removed manually using a disposable Pasteur pipette. Liquid ^{32}P waste from this and subsequent steps must be handled with caution and disposed of appropriately. The cells are then washed once with prewarmed phosphate-free DMEM. Then prewarmed phosphate-free DMEM is added with or without a PKC agonist, such as bryostatin-1 (100 nM). The six-well dish is again placed in the Plexiglas box and placed in the incubator at $37°$ with 5% CO_2 for 20 min. After the stimulation period is finished, the dish is removed from the incubator and the medium is removed from all wells with a Pasteur pipette. The cells are now washed three times with ice-cold TBS, with the wash buffer removed each time manually as before. The cells are lysed directly in the six-well plate, using 1 ml RIPA buffer [20 mM Tris-HCl, pH 7.5, 150 mM NaCl, 1% NP-40, 0.1% SDS, 0.1% Na-deoxycholate (DOC), 0.5 mM EDTA, protease inhibitor tablet (Roche), 10 nM microcystin, 50 mM NaF, and 0.2 mM Na$_2$VO$_3$] per well. The plate is left at $4°$ for 20 min to allow complete lysis, after which time the lysates are scraped from the dish and transferred to microcentrifuge tubes. The lysates are then clarified to remove detergent-insoluble material by centrifuging at 13,000 rpm for 15 min. After centrifugation, the lysate supernatant is transferred to a new microcentrifuge tube and is ready for immunoprecipitation.

To the 1 ml of clarified lysates, 30 μl agarose (1:1 slurry of beads:buffer) conjugated with Y13–259 rat monoclonal anti-Ras antibody (Santa Cruz Biotechnologies) is added and the mixture is rotated at $4°$ for 2 h. Other phosphorylated proteins can be similarly immunoprecipitated using cognate antibodies and protein A or G agarose beads. After immunoprecipitation, the agarose beads are washed three times with RIPA lysis buffer (containing protease inhibitors, microcystin, NaF, and Na$_2$VO$_3$), pelleted by centrifugation at 7000 rpm for 3 min, and eluted by boiling in 30 μl of SDS sample buffer. Lysates from parallel plates not labeled metabolically can be used to check for expression of the constructs. Eluates are then loaded onto 14% Tris-glycine gels and subjected to electrophoresis, along with the nonlabeled lysates.

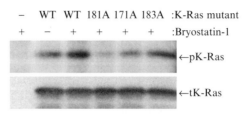

Figure 7.1 Mapping sites of K-Ras phosphorylation by metabolic labeling with ^{32}P. COS-1 cells were transfected with expression constructs for the indicated K-Ras alleles and 16 h later were labeled metabolically for 6 h with [^{32}P]orthophosphate. Ras was immunoprecipitated from cell lysates, and phosphoproteins were visualized by Phos-phorImager. Cells treated in parallel that were not exposed to [^{32}P]orthophosphate were used for Ras immunoblots (bottom) to confirm equivalent expression.

After electrophoresis has progressed to the point where the proteins of interest can be resolved, the gel is removed from the chamber and dried on Whatman 3MM filter paper for 3 h at 70°. An identical gel is run with the immunopre-cipitates of unlabeled cells, transferred to nitrocellulose, and immunoblotted with Ras10 pan-Ras monoclonal antibody (Calbiochem). The dried gel is then exposed to a PhosphorImager plate at room temperature and scanned by a laser after 10 h. Relative levels of radioactive phosphate incorporation into the various proteins are quantified by ImageQuant software.

Bryostatin-1 induced phosphorylation of K-Ras that was markedly diminished with an S181A substitution, confirming S181 as the major phosphate acceptor (Fig. 7.1) (Bivona *et al.*, 2006). However, the low level of ^{32}P incorporation into the 181A mutant suggests that, although S181 is the primary site, other minor sites of phosphorylation are likely. The diminished signal in the S171A mutant suggests that this residue might also serve as a phosphate acceptor. In contrast, a T183A substitution did not diminish ^{32}P incorporation, confirming that phosphorylation of this residue is not involved in the farnesyl–electrostatic switch of K-Ras.

3. Phoshorylation of Endogenous K-Ras in T Cells

The observation that overexpressed GFP-K-Ras in COS-1 cells responds to direct activation of PKC by diacylglycerol analogs suggested that this effect could also be observed in a more physiologic setting. To this end, we studied lymphocytes that are well known to activate both Ras (Downward *et al.*, 1990) and PKC (Valge *et al.*, 1988) following engage-ment of the T-cell receptor (TCR). Metabolic labeling of Jurkat T cells with ^{32}P followed by cross-linking of the TCR is therefore a good system with which to study phosphorylation of endogenous K-Ras following a physiologically relevant stimulus.

3.1. Labeling of Jurkat cells

Metabolic labeling of Jurkat T cells involves slight modifications from the procedure described earlier, owing to the fact that they are nonadherent cells. Moreover, because one is interested in labeling endogenous proteins in these cells, no transfection is necessary. The Jurkat T-cell culture should be maintained at a concentration between 10^5 and 10^6 cells/ml of medium, and the cells should appear healthy and double regularly. Immediately before labeling, the Jurkat cells (4×10^7) are washed three times with prewarmed phosphate-free DMEM and finally resuspended in phosphate-free DMEM supplemented with 10% FBS (undialyzed:dialyzed, 3:1) to a concentration of 10^7 cells/ml of medium. Although RPMI is the standard medium for Jurkat T cells, it contains a high concentration of phosphate and therefore phosphate-free DMEM is used instead. [^{32}P]orthophosphate is then added to the labeling medium to a final concentration of 0.5 mCi/ml. The resuspended cells in their labeling media can then be transferred to a 60-mm dish, placed in a Plexiglas box, and put in an incubator at $37°$ in 5% CO_2 for a labeling period of 6 h.

3.2. Stimulation, immunoprecipitation, SDS-PAGE, and analysis of endogenous K-Ras phosphorylation by anti-CD3

After the labeling period, Jurkat cells are washed with prewarmed DMEM as described earlier for COS-1 cells. However, because Jurkat cells grow in suspension, they must be pelleted after each wash. After the final wash, the cells are resuspended in 2 ml (2.0×10^7 cells/ml) prewarmed DMEM and are divided equally into two microcentrifuge tubes. The cells in one tube are then stimulated with 10 μg/ml anti-CD3 antibodies (Ancell, Bayport MN) and the other with buffer control. The tubes are placed in a rack in the Plexiglas box and put in the incubator for 30 min at $37°$ with 5% CO_2. After stimulation, cells are washed three times with cold TBS. The cells are finally pelleted by centrifugation at 7000 rpm for 3 min, resuspended in 500 μl TBS, and lysed with an additional 500 μl of 2$\times$ RIPA buffer. The cells are lysed for 15 min, rotating at $4°$, after which the lysates are clarified by centrifugation. Total Ras protein or other proteins of interest are immuno-precipitated as described previously. The immunoprecipitated protein is washed and eluted by boiling with SDS sample buffer, and eluates are loaded onto a 14% Tris-glycine gel and subjected to electrophoresis. The gel is then dried on Whatman 3MM filter paper and placed in a PhosphorImager plate for 3 to 6 h, and the plate is scanned in a PhosphorImager to visualize radioactive phosphate incorporation. Using this method, phosphorylation of endogenous Ras can be observed in Jurkat T cells following stimulation of the TCR, demonstrating Ras phosphorylation in response to physiological signaling.

4. ANALYSIS OF K-RAS FARNESYL-ELECTROSTATIC SWITCH BY LIVE CELL IMAGING

Having defined the phosphate acceptor sites for PKC in the C-terminal membrane targeting region of K-Ras4B by metabolically labeling as described earlier, we now determine the effects of phosphorylation at those sites on K-Ras association with the plasma membrane. This is best accomplished with live cell imaging in which GFP-tagged Ras constructs are expressed in a well-behaved cell type and then localization is scored in real time before and at various times after the addition of a PKC agonist. To identify endomembrane compartments that serve as acceptors for GFP-K-Ras or YFP-K-Ras dislodged from the plasma membrane, one can use various compartment markers.

4.1. Cloning and transfection

GFP-tagged K-Ras12V constructs are cloned in two steps. First, the K-Ras12V coding region is polymerase chain reaction (PCR) amplified with in-frame $5'$-*Hin*dIII and $3'$-*Apa*I linkers designed into the primers. The amplification product is digested with *Hin*dIII and *Apa*I and is then ligated into the pEGFP-C1 vector (Clontech), which has been linearized by double digestion with *Hin*dIII and *Apa*I and purified. It is important to use an N-terminal GFP tag to allow for proper processing of the CAAX motif of K-Ras. Second, the S171, S181, or T183 residues can be mutated to alanine or glutamic acid using QuikChange XL (Stratagene) site-directed mutagenesis according to the manufacturer's instructions. Combinations of mutations (e.g., GFP-K-Ras12V171A183A) can be created by sequential mutagenesis of each codon.

Twenty-four hours before transfection, 2.0×10^5 MDCK cells are seeded in 35-mm culture dishes that incorporate a glass coverslip in the center (MatTek) in 2 ml DMEM, supplemented with 10% FBS and antibiotics. The cells should be grown to 70% confluence on the day of transfection. Two micrograms of each GFP-tagged K-Ras12V construct is combined with 7 μl SuperFect reagent in 100 μl unmodified DMEM. After 15 min to allow transfection complexes to form, 600 μl of DMEM, supplemented with 10% FBS and antibiotics, is added to bring the volume of the transfection mixture up to 700 μl. In the meantime, the medium from each dish is aspirated, and cells are washed with prewarmed PBS. The entire transfection mixture is added to a single dish, and cells are allowed to incubate for 3 h at 37° in 5% CO_2. The medium with the transfection complex is then removed, cells are washed once with prewarmed PBS, and fresh DMEM containing serum and antibiotics is added to the cells.

4.2. Stimulation and visualization of translocation

Twenty-four hours after transfection, living cells are visualized with a Zeiss 510 inverted laser-scanning confocal microscope (LSM) using a high magnification, high NA objective (e.g., Plan-Neofluar 63x/1.25) and GFP or FITC narrow band filters. A microincubator (e.g., PDMI-2 from Harvard Apparatus) should be employed to keep the cells at 37°. A CO_2 microincubator (e.g., Incubator S-M from Zeiss) can be used, but is not necessary, as the cells will be studied for less than 30 min after removal from the incubator. Cells for imaging should be selected on the basis of morphology and intermediate fluorescent intensity. GFP-K-Ras12V in unstimulated cells should appear to localize exclusively at the PM. In a confluent monolayer of MDCK cells, PM localization should be easily discernible, as it appears as a "chicken-wire" pattern due to cell–cell contact. Whether imaging individual cells or a cell that is in a confluent patch of monolayer, one should choose a cell that appears healthy and has a structure decorated with GFP that can be unambiguously interpreted as plasma membrane. Sometimes it helps to acquire a shallow Z stack of three to five 0.45-μm slices to afford more opportunity to score plasma membrane and internal membranes. This is particularly important when the agents employed, such as bryostatin-1, can cause an acute shape change of the cell. The brightest of cells often have irregular morphology and fluorescent structures that may be artifacts and are therefore best avoided. The dimmest cells, although often of excellent morphology, may bleach below the threshold for high-quality imaging during stimulation and image acquisition such that these should also be avoided. The best cells to follow have intermediate fluorescence intensity.

Once an appropriate cell or group of cells is selected, they are stimulated while under continuous observation by adding a PKC agonist, such as 100 nM bryostatin-1 or 100 nM PMA with or without 500 ng/ml ionomycin, directly into the 35-mm MatTek plates. Thorough mixing is accomplished by pipetting 200 to 500 μl of the media two times with a 1000-μl pipette. Mixing must be done in an extremely gentle fashion so as not to lose the cell of interest from the field or focal plane. Scanned images (single or a shallow Z stack) should be acquired before and every 1 to 5 min after adding drug.

Following stimulation, GFP-K-Ras12V should translocate to internal membranes rapidly (Fig. 7.2) (<1 min onset, 3 min for maximal effect). The translocation event can also be observed with GFP-K-Ras12V171A183A, suggesting that phosphorylation of these residues is not required for the farnesyl–electrostatic switch. However, GFP-K-Ras12V181A does not translocate, and GFP-K-Ras12V with a glutamic acid at position 181 to mimic phosphorylation is internalized constitutively, confirming that serine 181 is the critical residue that controls the farnesyl–electrostatic switch. Thus, live cell imaging can confirm that serine 181 is necessary and sufficient for PKC-induced translocation (Bivona *et al.*, 2006). Because this is the same residue found to be the principal phosphate acceptor (see earlier discussion), one can deduce that phosphorylation at this residue regulates the farnesyl–electrostatic switch of K-Ras.

4.3. Colocalization of phosphorylated K-Ras with compartment markers

To determine the localization of K-Ras or other proteins of interest following translocation to intracellular compartments, one can study cells cotransfected with K-Ras tagged with a fluorescent protein (FP) and a compartment marker tagged with a different FP that can be spectrally resolved from FP-K-Ras. YFP-K-Ras is an attractive option that can be used in combination with compartment markers tagged with CFP or mCherry. To visualize organelles such as ER, Golgi, endosomes, and mitochondria, one should choose a relatively large and well-spread cell type

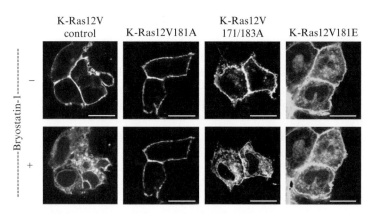

Figure 7.2 Translocation of GFP-K-Ras in response to PKC stimulation. MDCK cells were transfected with expression constructs for the indicated K-Ras alleles and 16 h later were imaged alive with a Zeiss 510 LSM before and 5 min after the addition of bryostatin-1. Constitutively active GFP-K-Ras12V translocates, as does GFP-K-Ras12V171/183A. In contrast, GFP-K-Ras12V181A remains on the plasma membrane and GFP-K-Ras12V181E is constitutively internalized. Bars indicate 10 μm.

such as COS-1. HEK293 and NIH/3T3 are not well suited for live cell imaging to determine subcellular localization, as the former are relatively small and poorly adherent and the latter are elongated. Epithelial cells such as MDCK or HeLa work well both for identifying intracellular organelles and for scoring for plasma membrane, as confluent cells grow with a semicolumnar geometry.

To mark the ER, we use the first transmembrane domain of the avian infectious bronchitis virus M1 protein (codons 1–66) with a CFP tag at the C terminus (M1-CFP). For visualization of the Golgi apparatus, we use codons 1–60 of human galactosyltransferase with a CFP tag at the C terminus (GalT-CFP). Fluorescent labeling of mitochondria can be achieved by expressing mCherry extended with the 33 C-terminal amino acids of Bcl-X$_L$ or by the addition of 25 pM MitoTracker Red CMXRos (Molecular Probes) 10 min prior to imaging. One can readily perform triple color localization using MitoTracker Red CMXRos, as it can be resolved from both the YFP-tagged Ras and the CFP-tagged ER or Golgi marker. Cotransfection is as described earlier using SuperFect using a plasmid DNA ratio of 2:1 CFP-compartment marker:YFP-Ras.

Twenty-four hours post–transfection, cells containing the YFP-tagged protein with M1-CFP, GalT-CFP, and/or MitoTracker are imaged with a Zeiss 510 LSM using a metabolic chamber (e.g., Zeiss Incubator L-M) as described previously. Images are acquired before and at various times after addition of a PKC agonist. M1-CFP is absolutely restricted to the ER (Chiu et al., 2002) and should appear as a reticular system of membranes contiguous with the nuclear envelope. In COS-1 cells, this reticular network often forms a more intense nexus in the paranuclear region also occupied by the Golgi apparatus and endosomal recycling compartment. GalT-CFP should appear as a paranuclear cluster of vesicles, tubules, and tubulovesicular structures. MitoTracker should appear as a tubulovesicular network that does not emanate from the nuclear membrane. The morphology of mitochondria often changes upon addition of a PKC agonist and these organelles can take on a more vesicular appearance as a consequence of fission. Colocalization of phosphorylated K-Ras to each compartment can be documented using confocal software to adjust the gains such that each fluorophore decorates organelles with relatively equal intensity. Under these conditions, a pseudocolor for colocalization (yellow in the case of red and green primary channels) indicates colocalization. However, one must exercise caution in not overinterpreting the yellow pseudocolor of the merge. For example, if one has a bright red mitochondrion that lies within a saturated area of green from any source, it will give a false-positive signal for colocalization on the basis of color. To avoid this, one should be careful to include morphology in the assessment of colocalization. That is, if phosphorylated YFP-Ras truly decorates mitochondria, one should see in the YFP channel a structure of identical size and shape to that made visible by MitoTracker Red CMXRos.

5. TARGETING OF PKC TO ACTIVATED K-RAS WITH THE RAS BINDING DOMAIN OF RAF-1 (RBD)

Data indicate that translocation of K-Ras is regulated by a farnesyl-electrostatic switch that is in turn regulated by phosphorylation by PKC of serine 181. PKCs are activated by recruitment via their C1 domains to membranes rich in diacylglycerol (DAG). To confirm that PKC had this ability to phosphorylate K-Ras and affect its membrane association, we devised a method to target the catalytic domain of a PKC to activated Ras in a DAG-independent fashion.

The Ras binding domain of Raf-1 binds to GTP-bound Ras with 10^4 higher affinity than it binds GDP-bound Ras and has therefore been used as a reporter for Ras activation (Chiu *et al.*, 2002). We therefore sought to determine if the catalytic domain of a PKC (PKCθ) delivered to K-Ras via a fused RBD would have the ability to phosphorylate K-Ras and alter its subcellular localization.

5.1. Construction of pCGN-HA-RBD-PKCθcat

The construction of pCGN-HA-RBD-PKCθcat can be accomplished in three steps. Vectors are available commercially with robust multiple cloning sites that add a coding sequence of interest in-frame with an HA tag. We use pCGN for this purpose, which has only a single cloning site, *Bam*HI. To facilitate cloning into this site, we first create a fusion protein in pEGFP consisting of the RBD and the catalytic domain of PKCθ (PKCθcat). First, the cDNA for the RBD is PCR amplified with 5′-*Eco*RI and 3′-*Apa*I linkers designed into the primers. The amplification product is ligated in-frame into the pEGFP-C1 vector, which has been linearized by double digestion with *Eco*RI and *Apa*I and purified. Second, the cDNA for the catalytic domain of PKCθ (PKCθcat) is PCR amplified with 5′-*Apa*I and 3′-*Bam*HI linkers designed into the primers. The amplification product is then ligated in-frame into the vector pEGFP-C1-RBD, which has been linearized by double digestion with *Apa*I and *Bam*HI and purified. Third, the coding sequence for RBD-PCKθcat is PCR amplified from the pEGFP-C1-RBD-PKCθcat vector with 5′- and 3′-*Bam*HI linkers designed into the primers. Note also that the 3′ primer has a stop codon inserted before the *Bam*HI linker. This amplification product could then be ligated into the pCGN-HA, which has been linearized by digestion with *Bam*HI and purified, to provide the desired product.

PCR primers for amplification of RBD and PKCθcat to produce pCGN-HA-RBD-PKCθcat:

Forward *Eco*RI RBD: 5′-tcgattctgccttctaagacaagcaacactatcc-3′
Reverse *Apa*I RBD: 5′-cagggccccaggaaatctacttgaagttcttctcc-3′

Forward *Apa*I PKCθcat: 5′-aacagggcccggcacaagatgttgggaaaagga-3′
Reverse *Bam*HI PKCθcat: 5′-cggtggatcccgggagcaaatgagagtctccat-3′
Forward *Bam*HI RBD-PKCθcat: 5′-tcggatccccttctaagacaagcaacactatcc-3′
Reverse *Bam*HI RBD-PKCθcat-STOP: 5′-cggtggatcccgtcaggagcaaatga-gagtctccat-3′

5.2. Transfection and analysis of localization

Twenty-four hours before transfection, 2.0×10^5 MDCK cells are seeded in 35-mm MatTek culture dishes in 2 ml DMEM, supplemented with 10% FBS and antibiotics. The cells should be grown to 70% confluence on the day of transfection. Two micrograms of pCGN-HA-RBD-PKCθcat cDNA can be cotransfected with 2 μg of either pEGFP-K-Ras12V181S or pEGFP-K-Ras12V181A plasmid DNA into MDCK cells with high efficiency using the SuperFect reagent as described earlier. After a suitable period of time has passed to ensure expression of the constructs, MDCK cells can be imaged live using a Zeiss 510 LSM. In cells expressing this construct, GFP-K-Ras12V with a wild-type C terminus is partially internalized but GFP-K-Ras12V181A is not. This result confirms that S181 is a PKC site and that its phosphorylation is necessary and sufficient for PKC-induced translocation of K-Ras to internal membranes.

6. SINGLE-CELL FLUORESCENT ANALYSIS OF APOPTOSIS: YFP-CASPASE SENSOR

We observed that activated K-Ras with a phosphomimetic substitution at residue 181, K-Ras12V181E, induced cell death that could be blocked by overexpression of Bcl-2, suggesting apoptosis. Accordingly, we sought an assay that could quantify apoptosis in cells expressing this K-Ras allele and variations thereof. Numerous assays have been developed to assess apoptosis such as annexin-V or TUNEL staining or poly (ADP-ribose) polymerase (PARP) cleavage. One disadvantage of these assays is that they measure apoptosis in a population of cells. To study apoptosis induced by expression of a transfected gene, such as K-Ras with a phosphomimetic substitution at position 181, we sought a fluorescence-based assay that would read out specifically in cells transfected with and expressing exogenous genes. A commercially available fluorescent probe for caspase-3 activation meets these criteria. Monitoring caspase-3 activation is a valuable tool for detecting apoptosis downstream of both extrinsic and intrinsic pathways of programmed cellular death.

The pCaspase3-Sensor vector (BD Clontech) encodes the enhanced yellow-green variant (EYFP) of GFP fused at the 3′ end to three copies of the nuclear localization signal (NLS) of the simian virus 40 large T-antigen. At the 5′ end, the gene contains a sequence encoding the nuclear export signal (NES) of the MAP kinase kinase (MAPKK) followed by a 36 nucleotide sequence encoding the region of PARP cleaved by caspase-3. NES is functionally dominant over the NLS such that the full-length fluorescent fusion protein distributes homogeneously to the cytosol in healthy cells. If caspase-3 is activated, it cleaves the PARP substrate, removing the NES and allowing the truncated EYFP-NLS fusion to enter the nucleus with high efficiency. Translocation of the fluorescent protein from the cytosol to the nucleus, which is detected easily by fluorescence microscopy, indicates caspase-3 activity at the single-cell level. One can cotransfect pCaspase3-Sensor along with the gene of interest in a 1:5 plasmid DNA ratio to make sure that fluorescent cells to be scored for apoptosis are cotransfected. This offers a great advantage over methods that report apoptosis over the entire population of cells, those that are transfected and those that are not.

6.1. Transfection and scoring of fluorescence

Twenty-four hours before transfection, 2.0×10^5 COS-1 cells are seeded in 35-mm glass bottom culture dishes in 2 ml DMEM, supplemented with 10% FBS and antibiotics. The cells should be grown to 70% confluence on the day of transfection. Although almost any cell type can be used in conjunction with the pCaspase3-Sensor to study the effects of a protein of interest on apoptosis, COS-1 cells were chosen for several reasons. First, COS-1 cells are transfected easily; they readily take up plasmid DNA transfected with common reagents, such as SuperFect or Lipofectamine, with typical transfection efficiencies surpassing 90%. Second, COS-1 cells generally express exogenous proteins expressed from plasmids incorporating an SV40 origin of replication to high levels, ensuring that both the experimental protein of interest and the YFP-caspase sensor are expressed to sufficient levels. Finally, and probably most critical for this assay, COS-1 cells spread well and therefore display a flat morphology that is favorable for scoring for nuclear translocation of the fluorescent probe.

The wild-type and mutated K-Ras alleles cloned into the mammalian expression vectors such as pCGN-HA could be transfected into COS-1 cells with high efficiency using the SuperFect reagent, as described earlier. Other proteins of interest in mammalian expression vectors can also be used to assess their effects on apoptosis. Two and a half micrograms of plasmid DNA for the K-Ras construct is cotransfected with 0.5 μg pCaspase3-Sensor DNA. This ratio of nonfluorescent K-Ras plasmid DNA to sensor DNA ensures that cells expressing EYFP are also highly likely to express the K-Ras construct. A 1:1 ratio of plasmid DNA to sensor DNA can be used if the protein of interest is also fluorescently tagged with a protein that can be spectrally resolved from the YFP-tagged sensor (e.g., CFP or mCherry). In this format, only cells coexpressing both the fluorescent protein of interest and the pCaspase3-Sensor are scored for caspase-3 activation. Following transfection, cells are grown for 12 to 18 h at $37°$ in 5% CO_2 to allow expression of the constructs. Optimization of the expression period is necessary to visualize cells in the early stages of apoptosis where caspase-3 is activated, but the cells have not yet undergone drastic morphological changes.

Any epifluorescence microscope equipped with a YFP or FITC filter can be used to inspect the subcellular localization of pCaspase3-Sensor EYFP fluorescence. Although confocal microscopy works well, it is not necessary to obtain a clear result. We scan various quadrants of the MatTek plate and score cells visualized as having cytosolic, nuclear, or indeterminate fluorescence. It is important to score all cells as they are encountered to avoid bias. We typically collect data until we have scored 100 cells for which the fluorescent pattern is clearly either nuclear or cytosolic. For reasons that are unknown, nucleoli are fluorescent in both healthy and apoptotic cells and can appear quite prominent so it is important to avoid scoring cells as nuclear on the basis of nucleolar fluorescence. A cell should be scored as nuclear when the nucleoplasm (excluding nucleoli) is brighter than the cytosol. For each condition, an apoptotic index can be calculated as the percentage of fluorescent cells displaying nuclear fluorescence (Fig. 7.3).

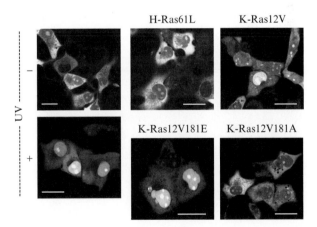

Figure 7.3   Single cell assay for apoptosis using a YFP caspase 3 sensor. COS-1 cells were transfected with pCaspase3-Sensor and imaged before and after exposure to UV radiation (left) or 12 h after cotransfection with the indicated K-Ras allele (right). Note that nucleoli are fluorescent in both healthy and apoptotic cells. Bars indicate 10 μm.

REFERENCES

Ballester, R., Furth, M. E., and Rosen, O. M. (1987). Phorbol ester- and protein kinase C-mediated phosphorylation of the cellular Kirsten ras gene product. *J. Biol. Chem.* **262,** 2688–2695.

Bivona, T. G., Quatela, S. E., Bodemann, B. O., Ahearn, I. O., Soskis, M. J., Mor, A., Miura, J., Wiener, H. H., Wright, L., Saba, S. G., Yim, D., Fein, A., *et al.* (2006). PKC regulates a farnesyl-electrostatic switch on K-Ras that promotes its association with Bcl-XL on mitochondria and induces apoptosis. *Mol. Cell.* **21,** 481–493.

Chiu, V. K., Bivona, T., Hach, A., Sajous, J. B., Silletti, J., Wiener, H., Johnson, R. L., Cox, A. D., and Philips, M. R. (2002). Ras signalling on the endoplasmic reticulum and the Golgi. *Nat. Cell Biol.* **4,** 343–350.

Choy, E., Chiu, V. K., Silletti, J., Feoktistov, M., Morimoto, T., Michaelson, D., Ivanov, I. E., and Philips, M. R. (1999). Endomembrane trafficking of ras: The CAAX motif targets proteins to the ER and Golgi. *Cell* **98,** 69–80.

Downward, J., Graves, J. D., Warne, P. H., Rayter, S., and Cantrell, D. A. (1990). Stimulation of p21ras upon T-cell activation. *Nature* **346,** 719–723.

Hancock, J. F., Paterson, H., and Marshall, C. J. (1990). A polybasic domain or palmitoylation is required in addition to the CAAX motif to loacalize p21^{ras} to the plasma membrane. *Cell* **63,** 133–139.

Jackson, J. H., Li, J. W., Buss, J. E., Der, C. J., and Cochrane, C. G. (1994). Polylysine domain of K-ras 4B protein is crucial for malignant transformation. *Proc. Natl. Acad. Sci. USA* **91,** 12730–12734.

Leventis, R., and Silvius, J. R. (1998). Lipid-binding characteristics of the polybasic carboxy-terminal sequence of K-ras4B. *Biochemistry* **37,** 7640–7648.

McLaughlin, S., and Aderem, A. (1995). The myristoyl-electrostatic switch: A modulator of reversible protein-membrane interactions. *Trends Biochem. Sci.* **20,** 272–276.

Valge, V. E., Wong, J. G., Datlof, B. M., Sinskey, A. J., and Rao, A. (1988). Protein kinase C is required for responses to T cell receptor ligands but not to interleukin-2 in T cells. *Cell* **55,** 101–112.

CHAPTER EIGHT

REGULATION OF RHoBTB2 BY THE CUL3 UBIQUITIN LIGASE COMPLEX

Andrew Wilkins *and* Christopher L. Carpenter

Contents

Abstract

The recently identified RhoBTB family is a member of the Rho GTPase family. One family member, RhoBTB2, has been implicated as a tumor suppressor in lung and breast cancer. Studies have shown that RhoBTB2 binds to the ubiquitin ligase scaffold Cul3 and that Cul3 regulates RhoBTB2 protein levels by ubiquitinating RhoBTB2 directly, leading to its degradation by the proteasome. This chapter details the cell biological and biochemical methods for analyzing the regulation of RhoBTB2 by Cul3.

Department of Medicine, Beth Israel Deaconess Medical Center and Harvard Medical School, Boston, Massachusetts

Methods in Enzymology, Volume 439
ISSN 0076-6879, DOI: 10.1016/S0076-6879(07)00408-9

1. INTRODUCTION

1.1. RhoBTB family

The RhoBTB family was first identified in 2001 (Rivero *et al.*, 2001). RhoBTB proteins are highly unusual members of the Rho GTPase family in that they are large (about 80 kDa) and are composed of an N-terminal GTPase domain followed by two BTB domains (named after the *Drosophila* proteins bric-a-brac, tramtrack and broad complex). Moreover, unlike most other Rho GTPases, RhoBTB proteins do not seem to regulate the actin cytoskeleton or bind to known Rho family effector proteins (Aspenstrom *et al.*, 2004).

The RhoBTB family is generally restricted to Metazoans, although homologues are notably absent from *Caenorhabditis elegans* (Rivero *et al.*, 2001). Mammals have three RhoBTB family members: RhoBTB1, RhoBTB2, and RhoBTB3. RhoBTB1 and RhoBTB2 exhibit a high degree of amino acid sequence identity, but RhoBTB3 is more divergent and does not seem to possess a functional GTPase domain.

1.2. RhoBTB2 is a candidate tumor suppressor

In 2002, *RHOBTB2* (it was named DBC2 for deleted in breast cancer 2) was identified as a likely tumor suppressor gene on human chromosome 8p21 (Hamaguchi *et al.*, 2002). Hamaguchi and colleagues (2002) found that *RHOBTB2* was deleted homozygously in 3.5% of primary breast cancers, gene expression was ablated in 60% of breast and 50% of lung cancer cell lines, and several somatic missense mutations in *RHOBTB2* were isolated from primary tumors and cancer cell lines. Furthermore, reintroduction of *RHOBTB2* into a breast cancer cell line (T47D) lacking endogenous *RHOBTB2* expression led to growth arrest.

1.3. RhoBTB2 is a substrate of a Cul3-based ubiquitin ligase complex

It has been shown that RhoBTB2 interacts with the ubiquitin ligase scaffold protein Cul3 (Wilkins *et al.*, 2004). This interaction requires the first BTB domain of the RhoBTB2 protein (this seems to be true for all RhoBTB family proteins tested and not a unique property of RhoBTB2) and the N-terminal region of Cul3. Cul3 is a member of the cullin family of modular ubiquitin ligase scaffolds (Petroski and Deshaies, 2005). All cullins bind the RING-domain E3, Rbx1, at their C terminus, which directly mediates the transfer of ubiquitin to the substrate. The substrate selectivity of the cullin is determined by the specific adaptor proteins, which bind to the N-terminal region of the cullin and which, in turn, recruit substrates. Each cullin has a specific set of adaptors, and several reports suggest that BTB-domain proteins function as substrate adaptors for Cul3 (Geyer *et al.*, 2003; Pintard *et al.*, 2003; Wilkins *et al.*, 2004; Xu *et al.*, 2003; Zhang *et al.*, 2006). Thus it is very likely that RhoBTB2 (and other RhoBTB family members) are substrate adaptors for a Cul3-based ubiquitin ligase complex. However, a number of lines of evidence suggest that RhoBTB2 is also a direct substrate of Cul3 (Wilkins *et al.*, 2004). First, wild-type RhoBTB2 has a short half-life, but its half-life is increased dramatically by treatment of cells with the proteasome inhibitor MG132. Second, a RhoBTB2 mutant (Y284D), identified in a lung cancer cell line that fails to bind to Cul3, has a greatly increased half-life and is not regulated by the proteasome. Third, shRNA knockdown of Cul3 protein levels leads to a significant increase in the steady-state levels of wild-type RhoBTB2 but does not affect the protein levels of Cul3-binding deficient mutants of RhoBTB2. Finally, direct ubiquination of RhoBTB2 by a Cul3-based ubiquitin ligase complex can be reconstituted *in vitro*.

This chapter details methods to investigate the regulation of RhoBTB2 protein levels by Cul3 both *in vitro* and *in vivo*.

2. METHODS

2.1. Interaction of endogenous RhoBTB2 and Cul3

2.1.1. RhoBTB2 and Cul3 antibodies

RhoBTB2 is either absent or expressed at extremely low levels in all cell types tested. Unfortunately, all the commercial anti-RhoBTB2 antibodies that are currently available (as of January 2007) are of low affinity and are unable to detect endogenous RhoBTB2. This has made investigation of the interaction of endogenous RhoBTB2 with Cul3 quite challenging. In order to visualize endogenous RhoBTB2, we have generated rabbit polyclonal antibodies against human RhoBTB2 using a bacterially expressed GST-fusion protein of amino acids 299 to 409 as the immunogen (Wilkins *et al.*, 2004). These antibodies are of high affinity and can immunoprecipitate endogenous RhoBTB2 very effectively. To detect endogenous RhoBTB2, we must immunoprecipitate RhoBTB2 prior to Western blot.

Several very good antibodies against human and mouse Cul3 are available commercially. The monoclonal antibody from BD Bioscience (611848) is by far the best antibody for Western blotting, although it cannot immuno-precipitate Cul3. A goat polyclonal antibody from Santa Cruz (C-18, sc-8556) is able to immunoprecipitate Cul3, although it is not as effective for Western blotting as the BD Bioscience monoclonal.

2.1.2. Coimmunoprecipitation of RhoBTB2 and Cul3

Because of the extremely low levels of RhoBTB2 expression, it is necessary to use a large number of cells to detect the interaction of RhoBTB2 with Cul3 (Ramos *et al.*, 2002; Wilkins *et al.*, 2004). Thus, ten 10-cm plates of HeLa cells are used for each immunoprecipitation. Also, because any RhoBTB2 that is bound to Cul3 becomes ubiquitinated and degraded by the proteasome, it is necessary to pretreat the cells with a proteasome inhibitor to maximize the amount of RhoBTB2/Cul3 complex recovered by immunoprecipitation. Therefore, 70% confluent HeLa cells are treated with a suitable proteasome inhibitor for 12 to 16 h prior to lysis. We use the proteasome inhibitor MG132. A 10 mM stock is made in dimethyl sulfoxide and applied to cells by direct addition to culture medium at a final concentration of between 10 and 50 μM. After MG132 treatment, cells are lysed in a total of 5 ml TNET [50 mM Tris, pH 7.5, 50 mM NaCl, 1 mM EDTA, 1 mM dithiothreitol (DTT), 0.1% Triton X-100, 100 μM phenylmethylsulfonyl fluoride] plus 10 μM MG132. Lysates are cleared by centrifugation at 13,000 rpm for 15 min at 4°. RhoBTB2 or Cul3 proteins are immunoprecipitated overnight at 4° with rocking using 1 μg of affinity-purified anti-RhoBTB2 antibodies or 1 μg of goat anti-Cul3 antibody (Santa Cruz C-18, sc-8556), respectively, together with 50 μl of protein G–Sepharose beads (Pharmacia). Beads and the immobilized Cul3/RhoBTB2 complexes are then washed three times with 1 ml TNET and subjected to Western blotting. Blots are probed with affinity-purified anti-RhoBTB2 antibodies or the BD-Bioscience anti-Cul3 monoclonal antibody. As a control for specificity of the antibodies, in parallel, we perform immunoprecipitations from cells that lack RhoBTB2 such as T47D or HEK293T cells.

2.2. Estimation of the half-life of RhoBTB2

The measurement of protein half-life is undoubtedly best assessed using standard pulse/chase analysis. However, the extremely low-level expression of the endogenous RhoBTB2 protein is unsuited to this assay. As an alternative, we routinely assess the half-life of transiently expressed epitope tagged-RhoBTB2. Using this method, the wild-type RhoBTB2 exhibits an apparent half-life of less than 4 h, whereas a Cul3-binding mutant of RhoBTB2 (Y284D), identified in a lung cancer cell line, has a half-life in excess of 12 h (Wilkins *et al.*, 2004). HEK293 cells in 10-cm plates are transfected with Myc-tagged RhoBTB2. Sixteen hours later the cells are harvested by trypsinization and replated into 12-well plates. Because the cells in each well come from the same initial transfected pool, there should be little well-to-well variability in the extent of transfection and protein expression. Twenty-four hours later, cells are treated with cycloheximide (100 μM final) to inhibit protein synthesis. Individual wells are harvested at various times after cycloheximide treatment and lysed in

TNET. The amount of RhoBTB2 present at each time point is assessed by Western blotting cleared lysates for Myc-RhoBTB2 using the 9E10 anti-Myc monoclonal antibody.

2.3. Effect of Cul3 RNAi knockdown on RhoBTB2 protein levels

We have demonstrated that knockdown of Cul3 protein levels by retrovirally mediated shRNA leads to a dramatic increase in the steady-state levels of endogenous RhoBTB2 protein in cells expressing wild-type RhoBTB2. There are now several published shRNA and siRNA effective at knocking down human Cul3 protein levels. Here we describe the use of retroviral-mediated shRNA to knockdown Cul3 in HeLa cells.

For each experiment, ten 10-cm plates of HeLa cells are infected twice in 24 h with shRNA-expressing retrovirus. The virus is generated by the cotransfection of HEK293T cells (any efficient transfection reagent can be used) with equal amounts of a packaging plasmid (we use the amphotrophic pCLAMPHO packaging plasmid) and the Cul3-specific shRNA plasmid. Ten 10-cm plates of HEK293 cells are required to generate sufficient virus. After 12 to 16 h, the cell culture medium is changed. Forty-eight hours after transfection, the retrovirus-containing medium is harvested and filtered through a 0.45-μm syringe filter. The viral supernatant is diluted twofold in fresh media, and polybreen (5 μg/ml final) is added to allow the virus to adhere to the surface of the target cells. We use this virus stock the same day it is prepared. The HeLa cells to be infected should be around 30% confluent on the day of infection. To infect the cells, replace the culture medium with an equal volume of the virus-containing medium (i.e., use half the virus stock). Culture cells for a further 8 h and then aspirate the medium and repeat the infection with the remaining virus stock. This double infection should result in an infection efficiency of over 90%. A GFP expressing retrovirus can be used as a control to visualize the efficiency of infection. After 48 h the cells are lysed in a total of 5 ml TNET, and lysates are cleared by centrifugation at 13,000 rpm for 15 min at 4°. The extent of Cul3 knockdown is assessed by Western blotting the cell lysate with a Cul3 monoclonal antibody (BD Bioscience). The effect of Cul3 knockdown on RhoBTB2 protein levels is assessed by immunoprecipitation and Western blotting with affinity-purified anti-RhoBTB2 antibodies as described previously.

2.4. *In Vitro* ubiquitination assay

Cul3 will directly ubiquitinate RhoBTB2 *in vitro* in a reconstitution assay using purified components. This section describes the purification of RhoBTB2, Cul3, and Rbx1 and the appropriate assay conditions.

2.4.1. Purification of Cul3/RhoBTB2/Rbx1 complexes

Because we have not been able to produce soluble, recombinant RhoBTB2 using either bacterial or baculoviral systems, we purify RhoBTB2 from mammalian cells. For ubiquitination assays, we coexpress RhoBTB2, Cul3, and Rbx1 in HEK293T cells and isolate protein complexes containing all three proteins. Approximately 2×10^7 of transfected cells (assuming around 70% transfection efficiency) are required for each individual *in vitro* assay. Cells are transfected (any efficient method of transfection is appropriate) with GST-Cul3, Myc-RhoBTB2, and HA-Rbx1 and are cultured for at least 16 h. Because RhoBTB2 is ubiquitinated by Cul3, it is necessary to inhibit the proteasome to prevent degradation of the RhoBTB2 bound to Cul3. To do this, we normally add 10 μM MG132 to the culture medium for 12 h prior to harvesting the cells. After MG132 treatment, the cells for each separate assay are lysed in a total of 1 ml of ice-cold TNET and lysates are cleared by centrifugation at 13,000 rpm for 15 min at 4°. GSH–Sepharose beads (50 μl of a 50/50 slurry in TNET) are added to the cleared supernatant and incubated for 1 h at 4° with rocking. Beads and the immobilized Cul3/Rbx1/RhoBTB2 complexes are washed three times with 1 ml TNET and then once with 1 ml of 50 mM HEPES, pH 7.6. It should be noted that even though Cul3 leads to ubiquitination of RhoBTB2 in the cell, deubiquitinating enzymes remove all of the ubiquitin chains during the immunoprecipitation such that the isolated RhoBTB2 exhibits no detectable ubiquitination.

2.4.2. *In Vitro* ubiquitination assay

In addition to purified Cul3/Rbx1/RhoBTB2, several other components are required to reconstitute ubiquitin ligase activity: a human E1 ubiquitin activating enzyme (BostonBiochem E-305), a human UbcH5a E2 ubiquitin conjugating enzyme (BostonBiochem E2–616), purified ubiquitin (Boston-Biochem U-100), ATP, and DTT. To inhibit any contaminating deubiquitinating enzymes that would interfere with the assay, ubiquitin aldehyde is added to each assay (BostonBiochem U-201).

Assays are performed as follows. An assay mixture is prepared at 4° in which each 25 μl contains 500 ng E1 (0.5 μl of a 0.5-mg/ml stock), 1 μg E2 (0.25 μl of a 4-mg/ml stock), 2 μM ubiquitin aldehyde (1 μl of a 100 μM stock), 1 mM ATP (5 μl of a 1 mM stock), 5 mM MgCl$_2$ (0.25 μl of a 1 M stock), and 11 μl of 50 mM HEPES, pH 7.6. Twenty-five microliters of the assay mixture is added to each 25 μl of packed GSH beads containing purified Cul3/Rbx1/RhoBTB2, mixed well, and incubated at 37° for 1 h. The reaction is stopped by the addition of SDS-PAGE loading buffer, and proteins are separated by SDS-PAGE and Western blotted for Myc-RhoBTB2 with anti-Myc (clone 9E10) antibody. Ubiquitination of RhoBTB2 is visible as a ladder of higher molecular weight species increasing in 8.5-kDa increments. To ensure that the assay is functioning correctly, several controls are employed. Ubiquitin is omitted from the assay to be sure that the higher molecular weight ladder is indeed due to the covalent addition of ubiquitin to RhoBTB2. As another control, RhoBTB2 is omitted from the assay to ensure that the ubiquitinated protein being visualized is RhoBTB2. To demonstrate that ubiquitination is mediated by the Cul3 complex via Rbx1 and not by other contaminating enzyme activities, the assay is performed using a truncated Cul3 mutant (residues 1–199) that can bind RhoBTB2 but cannot bind to Rbx1.

REFERENCES

Aspenstrom, P., Fransson, A., and Saras, J. (2004). Rho GTPases have diverse effects on the organization of the actin filament system. *Biochem J.* **377,** 327–337.

Geyer, R., Wee, S., Anderson, S., Yates, J., and Wolf, D. A. (2003). BTB/POZ domain proteins are putative substrate adaptors for cullin 3 ubiquitin ligases. *Mol. Cell.* **12,** 783–790.

Hamaguchi, M., Meth, J. L., von Klitzing, C., Wei, W., Esposito, D., Rodgers, L., Walsh, T., Welcsh, P., King, M. C., and Wigler, M. H. (2002). DBC2, a candidate for a tumor suppressor gene involved in breast cancer. *Proc. Natl. Acad. Sci. USA* **99,** 13647–13652.

Petroski, M. D., and Deshaies, R. J. (2005). Function and regulation of cullin-RING ubiquitin ligases. *Nat. Rev. Mol. Cell. Biol.* **6,** 9–20.

Pintard, L., Willis, J. H., Willems, A., Johnson, J. L., Srayko, M., Kurz, T., Glaser, S., Mains, P. E., Tyers, M., Bowerman, B., and Peter, M. (2003). The BTB protein MEL-26 is a substrate-specific adaptor of the CUL-3 ubiquitin-ligase. *Nature* **425,** 311–316.

Ramos, S., Khademi, F., Somesh, B. P., and Rivero, F. (2002). Genomic organization and expression profile of the small GTPases of the RhoBTB family in human and mouse. *Gene* **298,** 147–157.

Rivero, F., Dislich, H., Glöckner, G., and Noegel, A. A. (2001). The Dictyostelium discoideum family of Rho-related proteins. *Nucleic Acids Res.* **29,** 1068–1079.

Wilkins, A., Ping, Q., and Carpenter, C. L. (2004). RhoBTB2 is a substrate of the mammalian Cul3 ubiquitin ligase complex. *Genes Dev.* **18,** 856–861.

Xu, L., Wei, Y., Reboul, J., Vaglio, P., Shin, T. H., Vidal, M., Elledge, S. J., and Harper, J. W. (2003). BTB proteins are substrate-specific adaptors in an SCF-like modular ubiquitin ligase containing CUL-3. *Nature* **425,** 316–321.

Zhang, Q., Zhang, L., Wang, B., Ou, C. Y., Chien, C. T., and Jiang, J. (2006). A hedgehog-induced BTB protein modulates hedgehog signaling by degrading Ci/Gli transcription factor. *Dev. Cell* **10,** 719–729.

Characterization of EHT 1864, a Novel Small Molecule Inhibitor of Rac Family Small GTPases

Cercina Onesto,* Adam Shutes,* Virginie Picard,[†]
Fabien Schweighoffer,[†] *and* Channing J. Der*

Contents

Abstract

There is now considerable experimental evidence that aberrant activation of Rho family small GTPases promotes uncontrolled proliferation, invasion, and metastatic properties of human cancer cells. Therefore, there is considerable interest in the development of small molecule inhibitors of Rho GTPase function. However, to date, most efforts have focused on inhibitors that block Rho GTPase function indirectly, either by targeting enzymes involved in post-translational processing or downstream protein kinase effectors. We have reported the identification and characterization of the EHT 1864 small molecule as an inhibitor of Rac family small GTPases, placing Rac1 in an inert and inactive state and then impairing Rac1-mediated functions *in vivo*. Our work suggests that EHT 1864 selectively inhibits Rac1 downstream signaling and cellular transformation by a novel mechanism involving guanine nucleotide displacement. This chapter provides the details for some of the biochemical and

* University of North Carolina at Chapel Hill, Lineberger Comprehensive Cancer Center, Department of Pharmacology, Chapel Hill, North Carolina
† ExonHit Therapeutics, Paris, France

Methods in Enzymology, Volume 439
ISSN 0076-6879, DOI: 10.1016/S0076-6879(07)00409-0

biological methods used to characterize the mode of action of EHT 1864 on Rac1 and its impact on Rac1-dependent cellular functions.

 ## 1. INTRODUCTION

Rho family GTPases are molecular switches that play key roles in the modulation of a wide range of cellular processes, including cell migration, cell polarization, membrane trafficking, cytoskeleton rearrangements, proliferation, apoptosis, and transcriptional regulation (Etienne-Manneville and Hall, 2002). It is, therefore, not surprising that the aberrant functions of Rho family GTPases contribute to the generation of different human pathologies, including cancer (Boettner and Van Aelst, 2002; Sahai and Marshall, 2002). Unlike Ras proteins, activating mutations in Rho GTPases are not found in human cancers. Instead, aberrant Rho GTPase activity found in tumors is a result of alterations in Rho GTPase expression or the perturbed function of guanine nucleotide exchange factors (GEFs) or GTPase-activating proteins (GAPs) that regulate Rho GTPase function (Karnoub *et al.*, 2004).

Of the 20 members of the Rho GTPase family, proteins in particular of the Rac subfamily of small GTPases (Rac1, Rac1b, Rac2, and Rac3) have been implicated in cellular transformation and cancer progression. Rac1 is essential for transformation caused by Ras and other oncogenes, for example, promoting soft agar growth and migration of Ras-transformed cells (Ferraro *et al.*, 2006; Khosravi-Far *et al.*, 1995; Qiu *et al.*, 1995; Renshaw *et al.*, 1996; Zohn *et al.*, 1998). In addition, proteins levels of Rac1 are elevated in breast tumors (Fritz *et al.*, 1999). A splice variant of Rac1, Rac1b, is constitutively active and transforming and is found overexpressed in breast and colon cancers (Jordan *et al.*, 1999; Schnelzer *et al.*, 2000; Singh *et al.*, 2004). Mutation and overexpression of Rac3 have been seen in human brain tumors, and RNA interference demonstrated a role for Rac1 and Rac3 in human glioblastoma invasion (Chan *et al.*, 2005; Hwang *et al.*, 2005). Finally, expression of a normally hematopoietic cell-specific Rac-GEF Vav1 is upregulated in pancreatic cancers, leading to Rac1 activation, and levels of Vav1 expression correlate with patient survival rate (Denicola and Tuveson, 2005; Fernandez-Zapico *et al.*, 2005).

Because of their critical role in human oncogenesis, Rho GTPases are therefore attractive and validated targets for anticancer therapies. One approach has involved small molecule inhibitors of protein prenyltransferases (Fig. 9.1A). Those enzymes catalyze the lipid modification of Rho GTPases, which then promotes the membrane association required for Rho GTPase interaction with effectors and biological activity (Basso *et al.*, 2006; Sebti and Hamilton, 2000). However, the success of these inhibitors has been limited at best in part because of their lack of sufficient specificity to

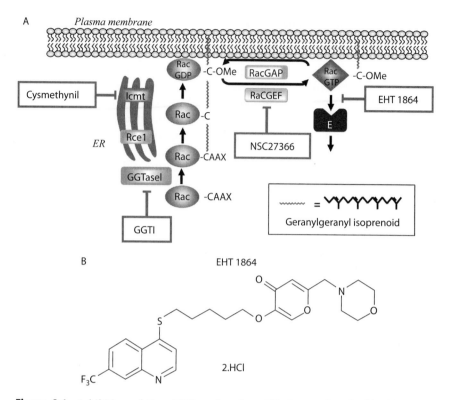

Figure 9.1 Inhibition of Rac GTPase function. (A) Approaches for blocking Rac function. Various small molecule inhibitors of Rac function have been described or considered. These include inhibitors of Rac post-translational modification. Rac terminates in a CAAX tetrapeptide sequence (C = Cys, A = aliphatic amino acid, X = Leu). This CAAX motif signals for three sequential post-translational modifications that convert the cytosolic, inactive Rac GTPase to a plasma membrane-associated protein. Geranylgeranyltransferase I (GGTaseI) catalyzes addition of the C20 geranylgeranyl isoprenoid to the Cys residue of the CAAX motif, followed by Rac converting enzyme 1 (Rce1)-catalyzed proteolytic removal of the AAX residues, and isoprenylcysteine carboxyl methyltransferase (Icmt)-catalyzed carboxyl methylation of the now terminal geranylgeranylated cysteine residue. Rac cycles between an inactive GDP-bound and an active GTP-bound state that is regulated by GTPase activating proteins (RhoGAPs) and guanine nucleotide exchange factors (RhoGEFs). Rac-GTP binds preferentially to a large spectrum of functionally diverse effectors (E) that regulate cytoplasmic signaling networks. GGTaseI inhibitors (GGTIs) block all CAAX-signaled modifications, rendering Rac cytosolic and inactive. Cysmethynil blocks the final CAAX modification step by inhibiting Icmt. NSC23766 inhibits RacGEF activation of Rac, whereas EHT 1864 impairs Rac-GTP formation and prevents Rac binding and activation of downstream effectors. (B) Structure of EHT 1864. (See color insert.)

selectively block Rho GTPase function (Reid *et al.*, 2004). Another approach has involved inhibitors of protein kinase effectors of Rho GTPases (e.g., ROCK and PAK protein kinases), but these inhibitors may not impair

Rho GTPase function effectively, as Rho GTPases utilize a multitude of downstream effectors (Bishop and Hall, 2000).

Inhibitors that antagonize Rho GTPases directly would be preferable and exhibit greater specificity and potency (see Fig. 9.1A). However, to date, there has been limited success in the identification of inhibitors that specifically interact with small GTPases. One example is the NSC23766 small molecule, which was identified as a cell-permeable compound that inhibits Rac1 binding and activation by Rac-specific RhoGEFs such as Tiam1 or Trio (Gao *et al.*, 2004). Previous studies suggested that EHT 1864 is a Rac-specific inhibitor that can inhibit association of Rac with its effector Pak, as well as a variety of downstream Rac signaling pathways (Desire *et al.*, 2005) (see Fig. 9.1B). However, the precise mechanism by which EHT 1864 can inhibit Rac signaling was unclear. We have further characterized this Rac inhibitor and showed that EHT 1864 binds to Rac1 tightly, locking the Rho GTPase in an inert and inactive state, both *in vitro* and *in vivo* (Shutes *et al.*, 2007). Moreover, EHT 1864 potently inhibited Rac1-mediated changes in cellular morphology and cellular transformation (Shutes *et al.*, 2007). We also demonstrated that in addition to Rac1, EHT 1864 binds to Rac1b and Rac2 with similar affinity, whereas binding to Rac3 is approximately 10-fold less, suggesting that EHT 1864 may form the basis for a novel class of specific Rac GTPase inhibitors. This chapter describes some of the techniques used to characterize the mechanism of action of EHT 1864 on Rac1 and its impact on Rac1-mediated cellular functions.

2. EXPERIMENTAL PROCEDURES

2.1. *In vitro* biochemical analyses of the effect of EHT 1864 on Rac GTPase nucleotide association and dissociation

A common method for monitoring the process of nucleotide exchange on small GTPases is the use of fluorescent *N*-methylanthraniloyl (mant) derivatives of guanine nucleotides. On binding to small GTPases, a change of environment around the mant group, from a solvent quenched to a hydrophobic environment, causes a significant increase in emitted fluorescence at 440 nm. The mant fluorophore is usually excited directly by 360-nm light. Rac proteins, however, contain a Trp residue buried in the nucleotide-binding pocket that can be excited at 290 nm and subsequently transfer energy directly to the mant group via fluorescence resonance energy transfer (FRET) producing emission at 440 nm. This method therefore provides a fluorescent readout of a nucleotide-bound state as an average of a population. We have observed that the compound EHT 1864 is intrinsically fluorescent (Shutes *et al.*, 2007), and this fluorescence increases on addition of Rac1 in a dose-dependent fashion, suggesting that changes in fluorescence relate to a binding event. The excitation and emission maxima of the inhibitor are extremely similar to those of mant nucleotides, and therefore direct measurement of changes in mant fluorescence is impossible. This section describes *in vitro* protocols used to analyze the interaction of EHT 1864 with Rac1.

2.1.1. Measurement of nucleotide k_{off} using Trp-mant FRET analyses

Protocols for the expression and purification of a recombinant fusion protein of glutathione *S*-transferase added to the amino terminus of human Rac1 (GST-Rac1) have been described elsewhere (Phillips *et al.*, 2003; Shutes *et al.*, 2002, 2007; Thapar *et al.*, 2002). For most *in vitro* studies, it is convenient to purify GST-Rac1 on a glutathione-Sepharose column, for example, a 5-ml GSTrap-FF column (Amersham). However, to make measurements of off rates, it is convenient to purify the protein in a batch fashion (where cell lysis supernatant and beads are incubated together), where the GST-protein remains attached to glutathione-agarose beads. An average 4-liter preparation of GST-Rac1 produces approximately 50 mg of purified recombinant protein. The bead-bound protein can then be loaded with the nucleotide by incubation in exchange buffer [20 mM Tris-HCl, 50 mM NaCl, 500 μM mant-GDP and mant-GMPPNP (Roche), 20 mM (NH$_4$)$_2$SO$_4$] for 1 min at 37°. The beads are then washed in ice-cold 20 mM Tris-HCl, 50 mM NaCl, and 1 mM MgCl$_2$ before elution in 20 mM Tris-HCl, 50 mM NaCl, 1 mM MgCl$_2$, and 0.1 mM glutathione for 10 min on ice. The beads and solution should be separated through centrifugation (13,000 rpm, 1 min) before removal of the supernatant to a fresh tube. The loaded protein is best used immediately, although it can be snap frozen in aliquots for use at a later date.

We use a SPEX Fluorolog-3 Research fluorimeter to assess off rates. Data from an example experiment are shown in Fig. 9.2A. Rac1·mant-GDP (2 μM) is incubated at 25° in 20 mM Tris-HCl (pH 7.5), 50 mM NaCl, and 1 mM MgCl$_2$ buffer (to a total volume of 300 μl) in a 1-ml quartz cuvette (with constant stirring). The addition of EHT 1864 or EDTA (to a final concentration of 50 μM or 10 mM, respectively) is done to begin the reaction. Changes in fluorescence are monitored at $\lambda_{ex} = 290$ nm and $\lambda_{em} = 440$ nm, and data are recorded by the supplied software.

2.1.2. Measurement of the effect of EHT 1864 on nucleotide association with Rac1

To examine the inhibition of nucleotide loading by EHT 1864, Rac1 is incubated with excess mant nucleotide in the presence of excess EHT 1864. Exchange is then initiated by the addition of EDTA, and the increase in fluorescence through FRET is measured on a Gemini Spectromax 96-well plate reader.

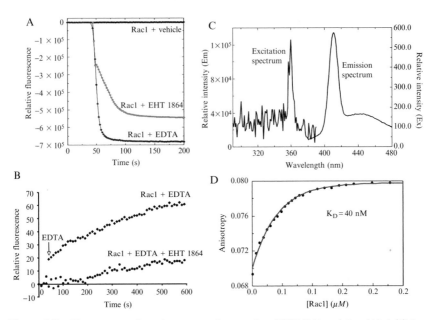

Figure 9.2 Fluorescence-based assays used to monitor EHT 1864 activity. (A) Addition of EHT 1864 to Rac1·MantGDP complexes causes loss of the bound nucleotide. Rac1 (2 μM) preloaded with mantGDP was incubated in 20 mM Tris-HCl, pH 7.5, 50 mM NaCl, and 1 mM MgCl$_2$. At the desired time, either EDTA or EHT 1864 was added to a final concentration of 10 mM and 50 μM, respectively. Changes in fluorescence were followed at $\lambda_{ex} = 290$ nm and $\lambda_{em} = 440$, where a decrease in fluorescence represents a loss in FRET between Trp 56 of Rac1 and the mant group, and therefore represents loss of mant nucleotide into solution. (B) EHT 1864 inhibits nucleotide loading at high concentrations. Incubation of 2 μM Rac1 with excess inhibitor prevents mant nucleotide loading that is stimulated by the addition of excess EDTA, as compared to in the absence of inhibitor. Exchange was followed on a SpectroMax Gemini at $\lambda_{ex} = 290$ nm, $\lambda_{em} = 440$ nm. An increase in fluorescence represents the binding of mant nucleotide to the Rho GTPase. (C) The EHT 1864 inhibitor is inherently fluorescent. Excitation and emission spectra were collected for the inhibitors, and optimal λ_{ex} and λ_{em} were found to be 360 and 440 nm, respectively. Data were collected using 10 μM inhibitor in a 20 mM Tris-HCl, pH 7.5, 1 mM MgCl$_2$, and 50 mM NaCl buffer. (D) Binding curve for the interaction of EHT 1864 and Rac1. Incremental 1-μl volumes of Rac buffer were added to a 1 μM solution of EHT 1864 (both Rac and EHT 1864 were in 20 mM Tris-HCl, pH 7.4, 50 mM NaCl, and 1 mM MgCl$_2$). The Rac solution also contained 1 μM EHT 1864. Increases in anisotropy, reflecting increases in Rac·inhibitor formation, were followed at $\lambda_{ex} = 360$ nm and $\lambda_{em} = 440$ nm. Data were fitted to a binding curve, from which a K_D can be estimated.

For these analyses, a 2× nucleotide solution containing 40 mM Tris-HCl (pH 7.5), 100 mM NaCl, 2 mM MgCl$_2$, and 4 μM mant-GDP is prepared. Fifty microliters of this solution is then placed into the desired wells of a 96-well plate (BD Falcon microtest plate). To this, Rac1·GDP is added to a final concentration of 2 μM, EHT 1864 to 50 μM, and the required amount of water to make a total volume of 100 μl within the well.

The mixture is allowed to incubate until any small disturbances in the fluorescence (λ_{ex} 290 nm, λ_{em} 440 nm) settle (usually a couple of minutes). Exchange is then initiated by the addition of 1 μl of 500 mM EDTA to provide a final concentration of 5 mM. Changes in fluorescence are then followed in a kinetic fashion so that single exponential curves can be fit to data produced (see Fig. 9.2B).

2.1.3. Measurement of EHT 1864 binding to Rac1

EHT 1864 is fluorescent in solution and this fluorescence changes on the addition of protein to this solution. This suggests that the fluorescence is affected by direct protein association. The increase in fluorescence is sufficient (30%) to provide information on inhibitor binding to various Rac isoforms, as well as Cdc42.

2.1.3.1. EHT 1864 excitation and emission spectra   A 1-ml cuvette of a SPEX Fluorolog-3 Research fluorimeter containing a 10 μM solution of EHT 1864 in 20 mM Tris-HCl (pH 7.5), 50 mM NaCl, and 1 mM MgCl$_2$ buffer is used to generate emission spectra. Emission and excitation spectra are collected by scanning through excitation and emission wavelengths to determine the values at which excitation and emission are at a maximum (see Fig. 9.2C). Optimal λ_{ex} and λ_{em} are found to be 360 and 440 nm, respectively.

2.1.3.2. Estimation of K_D of EHT/small GTPase interaction   Since we have established that binding of EHT 1864 to Rac1 causes a change in fluorescence of the inhibitor, we can monitor this change to determine a binding curve for the interaction. This can be done by titration of GTPase (at a known concentration) into a known concentration of inhibitor. The increases in fluorescence can then be followed at λ_{ex} = 360 nm and λ_{em} = 440 nm.

An example of a binding curve is shown in Fig. 9.2D. For these analyses, a 1 μM solution of EHT 1864 in a 20 mM Tris-HCl (pH 7.5), 50 mM NaCl, and 1 mM MgCl$_2$ buffer is added to a 1-ml cuvette (total volume is 300 μl) with stirring in a SPEX Fluorolog-3 Research fluorimeter in the T-format. Using a 10-μl Hamilton glass syringe, 1 μl of Rho GTPase (in GDP-bound form) is titrated into the inhibitor solution, and the increase in anisotropy (λ_{ex} = 360 nm, λ_{em} = 440 nm) is monitored. The value of increased anisotropy is read 90 s after each addition of Rho GTPase to allow for an equilibration period. Assuming a 1:1 interaction between the inhibitor and Rho GTPase, these values are plotted and fitted to a binding curve, from which a K_D is calculated. For these analyses, it is important to remember to "spike" the Rho GTPase stock solutions with inhibitor (to a final concentration of 1 μM) so that no inhibitor dilution effect occurs in the titration.

2.2. Effect of EHT 1864 on cellular Rac1 activity and cell transformation

As mentioned previously, EHT 1864 reduces the association of activated Rac1-GTP with the isolated GTPase binding domain (RBD) of its effector, Pak1 (Pak-RBD), whereas it has no effect on binding of the GTP-bound form of the related Rho GTPase, RhoA, to the isolated RBD of the RhoA effector, Rhotekin (Desire et al., 2005). We have also found that EHT 1864 treatment does not affect the interaction of Cdc42 with the Pak-RBD (Shutes et al., 2007), clearly demonstrating the specificity of interaction of EHT 1864 with Rac1 and not related Rho GTPases. By further delineating the specificity of EHT 1864 interaction with Rho family GTPases within mammalian cells, we also demonstrated that EHT 1864 is an effective inhibitor of Rac1-mediated and not RhoA- or Cdc42-mediated cellular events (Shutes et al., 2007). This section describes the techniques designed to study the impact of EHT 1864 on cellular Rac1 activity and cell transformation.

2.2.1. EHT 1864 inhibition of Rac1-mediated morphological changes

Rho GTPases are activated by extracellular stimuli, and specific Rho GTPases are regulators of distinct changes in actin cytoskeletal reorganization. Platelet-derived growth factor (PDGF) activates Rac1 and promotes Rac-mediated formation of actin-rich membrane lamellipodia (Ridley et al., 1992), whereas lysophosphatidic acid (LPA) causes activation of RhoA and RhoA-dependent formation of actin stress fibers and focal adhesions (Ridley and Hall, 1992) and bradykinin causes activation of Cdc42 and Cdc42-dependent formation of actin microspikes and finger-like membrane protrusions known as filopodia (Kozma et al., 1995). Hence, we evaluated the ability of EHT 1864 to selectively block ligand-stimulated activation of Rac1 in NIH/3T3 cells.

2.2.1.1. Cell culture NIH/3T3 mouse fibroblasts are a well-characterized cellular model used to visualize Rho GTPase-stimulated actin cytoskeleton rearrangements. NIH/3T3 cells are grown in Dulbecco's modified Eagle's medium (DMEM) supplemented with 10% calf serum (CS; Colorado Serum Company), 100 U/ml penicillin, and 10 μg/ml streptomycin and are maintained at 37° in a humidified 10% (v/v) CO_2 incubator. NIH/3T3 cells are plated onto 15-mm circular glass coverslips (Fisher) at 2×10^4 cells per well in 12-well plates. Twenty-four hours after plating, the cultures are rinsed twice with serum-free basal medium and switched to DMEM supplemented with 0.5% CS for 16 h to serum starve the cells to decrease basal Rac1 activity. For the last 4 h of this serum-starvation step, the cells are then incubated in growth medium supplemented with 5 μM of EHT 1864 or EHT 8560. EHT 8560 is a compound structurally related to EHT 1864, but is unable to inhibit Rac1 association with Pak-RBD and therefore serves as a negative control (Shutes *et al.*, 2007). At the end of the starvation period, cells are stimulated with 5 ng/ml of PDGF, 40 ng/ml of LPA, or 100 ng/ml of bradykinin (all from Sigma) for 15 min, then fixed and processed for immunofluorescence analysis as described next.

2.2.1.2. *Immunofluorescence and microscopy analysis*   All steps in the immunostaining procedure are performed at room temperature. Cells plated on coverslips are rinsed twice with room temperature phosphate-buffered saline (PBS) and are fixed for 15 min with a fresh preparation of 4% paraformaldehyde (Electron Microscopy Sciences) in PBS. After rinsing with PBS, cells are permeabilized with 0.2% Triton X-100 in PBS for 5 min, washed three times with PBS, and incubated with Alexa 568-conjugated phalloidin (1/40 dilution; Invitrogen) for 30 min. Phalloidin is a useful reagent for visualizing the distribution of F-actin in cells and hence to study actin networks at high resolution. Because phalloidin is conjugated to a light-sensitive fluorophore, coverslips should be protected from light from this step on. After two final washes in PBS and one in distilled water, coverslips are then mounted onto microscope slides with FluorSave reagent (Calbiochem) to prevent photobleaching. The slides should be maintained in the dark at room temperature for at least 1 h to allow the mounting reagent to dry and can be analyzed immediately or stored in a light-tight box at $-20°$ for a period of few months.

Although standard fluorescence microscopy with digital image acquisition is sufficient to visualize changes in the actin cytoskeleton, we use a Zeiss LSM 510 laser-scanning confocal microscope to achieve the best resolution. Images are collected using an oil immersion 63× NA 1.4 objective. Images are captured by scanning with the 543-nm spectral line of the HeNe1 laser and emission filter LP 585. As shown in Fig. 9.3A, stimulation of cells with PDGF efficiently induces membrane ruffling and lamellipodia formation, and this activity is almost completely blocked by treatment with EHT 1864 but not by the related but inactive EHT 8560 compound. EHT 1864-treated cells show an approximately 80% reduction in response to PDGF stimulation of lamellipodia formation (see Fig. 9.3B). In contrast, treatment with EHT 1864 does not affect LPA-mediated actin stress fiber formation or bradykinin-mediated filopodia formation (data not shown), clearly demonstrating that EHT 1864 is a potent and specific inhibitor of ligand-induced endogenous Rac1 activation and lamellipodia formation in NIH/3T3 cells.

2.2.2. EHT 1864 inhibition of Rac1:Pak1 complex formation

To further confirm the activity of EHT 1864 within cells, we coexpressed constitutively activated Rac1(61L) with GFP-Pak-RBD in NIH/3T3 cells. GFP-Pak-RBD is a useful tool to study the subcellular localization of

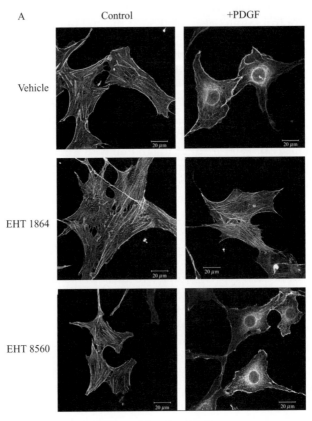

Figure 9.3 *Continued*

activated Rac small GTPases within cells because it provides a probe to detect the plasma membrane-associated, ectopically expressed, activated Rac1-GTP (Chenette *et al.*, 2006). In cells transfected with both constructs, the location of the GFP-Pak-RBD reflects its ability to associate with the activated Rac1 protein. In the absence of coexpressed activated Rac1(61L), the GFP-Pak-RBD probe is diffuse. NIH/3T3 cells plated on glass coverslips as described earlier are transfected with the relevant constructs using LipofectAmine Plus reagent (Invitrogen) and the manufacturer's protocol and are incubated in growth medium supplemented with 0.5% CS. Sixteen hours after transfection, cells are treated with 50 μM EHT 1864 for specific time periods, washed, and fixed. Cells are then examined for localization of the GFP-Pak-RBD, as well as for their overall cell morphology (see Fig. 9.3C). We use a Leica DMIRB inverted fluorescent microscope at $\lambda_{ex} = 480$ nm and $\lambda_{em} = 520$ nm, and images are captured using a black and white high-sensitivity Hamamatsu charge-coupled device camera.

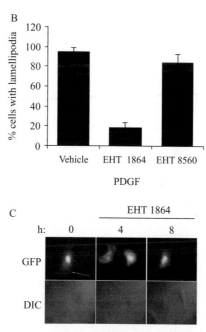

Figure 9.3   EHT 1864 is effective in specifically inhibiting PDGF-induced lamellipodia formation. (A) PDGF-induced actin reorganization is inhibited by EHT 1864. After overnight serum starvation in growth medium alone or supplemented with 5 μM EHT 1864 or EHT 8560 for the last 4 h of incubation, NIH/3T3 cultures were treated with 5 ng/ml PDGF for 15 min and then fixed, and actin filaments were visualized with Alexa-phalloidin. Scale bar represents 20 μm. Results shown are representative of three independent experiments. (B) Quantitation of data shown in A. Graphic representation of the percentage of PDGF-stimulated cells with lamellipodia in the presence or absence of EHT 1864 or 8560 and quantified on 100 cells for each condition. Results shown are the mean of three independent experiments; error bars indicate standard error of the mean. (C) EHT 1864 inhibits Rac1 interaction with Pak-RBD *in vivo*. NIH/3T3 cells were transiently cotransfected with expression constructs encoding Rac1(61L) or GFP-Pak-RBD. Cells were then maintained in growth medium supplemented with 0.5% calf serum and incubated with EHT 1864 for specific time periods. Cells were then washed and fixed for analysis on an inverted fluorescent microscope ($\lambda_{ex} = 480$ nm $\lambda_{em} = 520$ nm).

We observed that in the absence of inhibitor treatment, GFP-Pak-RBD is found enriched in cell ruffles, around the plasma membrane, as well as within a variety of internal membrane structures. This pattern corresponds to the subcellular localization of Rac1. In the presence of EHT 1864, within 4 h of treatment, membrane localization of the GFP-Pak-RBD is lost, resulting in a more cytosolic distribution (see Fig. 9.3C), demonstrating that EHT 1864 treatment disrupts Rac1 interaction with downstream effectors.

2.2.3. EHT 1864 inhibition of Rac1-mediated cellular transformation

Because previous studies showed that endogenous Rac1 activity is essential for Ras-induced growth transformation of rodent fibroblasts (Khosravi-Far *et al.*, 1995; Qiu *et al.*, 1995), we examined the ability of EHT 1864 to inhibit Ras-mediated transformation of NIH/3T3 fibroblasts.

2.2.3.1. Cell culture   NIH/3T3 mouse fibroblasts have been used extensively as a tool to study Ras-mediated morphological transformation (focus formation assays) and anchorage-independent growth (soft agar assays) (Clark *et al.*, 1995). These cells have a low level of spontaneous transformation and display density-dependent growth inhibition when maintained at confluent cell densities. However, the proper maintenance of stock cultures of NIH/3T3 cells is essential to avoid problems that can hinder reproducible, quantitative transformation assays. Because they are preneoplastic cells that are highly sensitive to the one-hit transforming actions of many different oncoproteins, they are prone to spontaneous transformation if they are allowed to persist at confluent densities. Moreover, it is important to be aware that there are many different strains of NIH/3T3 cells that may not behave similarly in transformation assays. The NIH/3T3 cells that we use routinely (sometimes referred to as the UNC strain) are propagated as described previously, following the protocols described previously (Clark *et al.*, 1995; Solski *et al.*, 2000). The source of the calf serum used for cell growth is also very critical. The frequency of spontaneous transformation, as well as oncogenic Ras focus-forming ability, can vary widely with different serum lots. Therefore, we routinely test serum lots from various vendors. We have found that calf serum from the Colorado Serum Company consistently supports excellent Ras-induced focus formation and minimum background spontaneous focus formation activity.

2.2.3.2. Establishment of oncogenic Ras-transformed NIH/3T3 fibroblasts   Several DNA transfection methods can be used for the introduction of plasmid DNA into cells. For these analyses, we chose the Lipofectamine transfection method (using Lipofectamine 2000 reagent; Invitrogen) to establish cells stably transfected with a plasmid expression vector encoding a constitutively activated, highly transforming mutant of human H-Ras [with a Gln61 to Leu mutation; designated H-Ras(61L)] into NIH/3T3 mouse fibroblasts. One day before transfection, NIH/3T3 cells are plated at a subconfluent density (4×10^5 cells per 100-mm dish) and incubated overnight to allow adhesion and cell spreading. On the next day, cultures are transfected with 4 μg of the pCGN–hygro mammalian expression vector encoding the constitutively active hemagglutinin (HA)-tagged

H-Ras(61L) protein. Parallel cultures are transfected with the empty vector to monitor the appearance of spontaneous focus-forming activity. Fresh growth medium is added 6 h after adding the DNA-Lipofectamine 2000 complexes to the cells. Two days after transfection, one-third of transfected cells are replated into a 100-mm plate containing growth medium supplemented with 200 μg/ml hygromycin B (Roche). After 14 days, several hundred drug-resistant colonies are then trypsinized and pooled together to be expanded, with one-tenth of the culture replated into another 100-mm plate containing growth medium supplemented with hygromycin B. The stable expression of HA-H-Ras(61L) within the cells is evaluated by Western blot analysis using the anti-HA antibody (clone 3F10, Roche), as shown in Fig. 9.4A. This mass population of stably transfected cells is then used in focus formation assays and soft agar assays in order to determine the impact of the compound EHT 1864 on the transforming activity of H-Ras(61L).

2.2.3.3. Focus formation assay

The NIH/3T3 focus formation assay is a widely used and well-established biological assay that measures the ability of an exogenously expressed gene to promote morphologic and growth transformation, in particular, a loss of density-dependent inhibition of growth. It is probably the most commonly used assay for examining the transforming potential of a particular oncogene in NIH/3T3 cells. There are two types of focus formation assays: primary and secondary. In the primary focus formation assay, the ability of the oncogene to transform cells is assessed by transfected cells, allowing the transfected cultures to reach confluency and to form foci of multilayered cells, which is indicative of the loss of density-dependent growth. In the secondary focus formation assay, populations of cells stably expressing a protein of interest are established first and then mixed to untransfected NIH/3T3, and the efficiency of the stably transfected cells to form foci of transformed cells is measured in confluent cultures. Here we describe results using a secondary focus formation assay and we have obtained similar results using primary focus formation assays (Shutes et al., 2007).

For this type of assay, a sufficient number of dishes are plated to perform the assay with duplicate dishes for each condition tested. A single cell suspension of 5×10^2 cells stably expressing empty pCGN-hygro (as a control) or pCGN-H-Ras(61L) is mixed with a suspension of 5×10^5 parental NIH/3T3 cells and then plated into each well of six-well plates. The ability of H-Ras(61L)-expressing cells to form multilayered colonies of cells is reflected by the appearance of foci of densely packed cells that can be visualized readily within the monolayer of untransformed, contact-inhibited NIH/3T3 cells. The cultures are fed every other day with fresh growth medium supplemented or not with 5 μM of the compound EHT 1864. Ras-induced transformed foci are visible after 10 to 14 days and are subsequently quantified. The

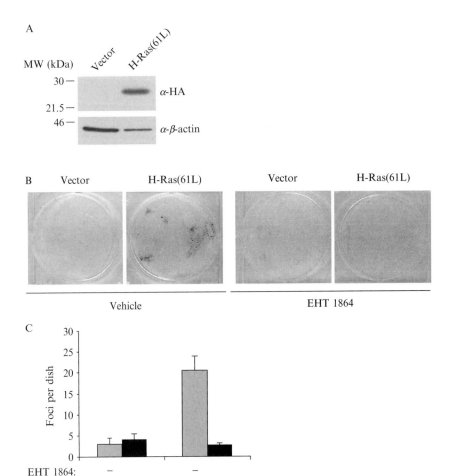

Figure 9.4 EHT 1864 blocks oncogenic Ras-stimulated cell transformation. (A) Expression level of H-Ras(61L) in stably transfected NIH/3T3 cells. The expression of HA epitope-tagged H-Ras(61L) protein was detected by blot analysis using an anti-HA antibody (clone 3F10; Roche Diagnostics). Blot analysis for β-actin (clone AC15; Sigma-Aldrich) was also done to verify equivalent total protein loading. (B) EHT 1864 inhibition of Ras-induced formation of foci of transformed cells. NIH/3T3 cells stably transfected with the empty pCGN-hygro (vector) or encoding H-Ras(61L) were plated and allowed to reach confluency. Cells were cultured in 10% serum growth medium, either alone or supplemented with 5 μM EHT 1864. The appearance of foci of transformed cells was quantitated 14 days after plating. Cells were then fixed and stained with crystal violet. (C) Quantitation of data shown in B. Data shown are representative of two independent experiments, each performed in duplicate.

most accurate quantification of foci is performed by visual inspection of live cultures under an inverted phase-contrast microscope at 4× magnification. The cultures can then be quantitated for the appearance of foci of transformed cells. The cultures can also then be subjected to fixation and staining with crystal violet for permanent storage of the cultures. For cell fixation, cells are rinsed once with room temperature PBS, and then 2 ml of a fixing solution containing 10% acetic acid and 10% methanol in water is added to each well for 10 min. The fixative is then removed, and cells are stained by adding 2 ml of crystal violet 0.4% (diluted in ethanol) per well for 5 min. Stained cells are rinsed with distilled water thoroughly but carefully, taking care to preserve the thin monolayer of cells. An example of Ras-induced foci is illustrated in Fig. 9.4B, which shows that EHT 1864 treatment causes an essentially complete inhibition of Ras-induced focus formation activity.

2.2.3.4. Soft agar assay

Colony formation in soft agar is the most widely used assay to evaluate anchorage-independent growth potential and represents one of the best *in vitro* assays that correlates strongly with *in vivo* tumorigenic cell growth potential. Like normal cells, untransformed NIH/3T3 cells require adherence and spreading onto a solid substratum in order to remain viable and proliferate. In contrast, transformed NIH/3T3 cells lose this requirement and therefore can form proliferating colonies of cells when suspended in a semisolid agar medium. To evaluate the impact of EHT 1864 treatment on Ras-mediated colony formation in soft agar, we used a protocol that has been described previously (Solski *et al.*, 2000). We performed this assay in six-well plates using triplicate wells for each condition evaluated. Briefly, a 0.6% Bacto agar (BD Biosciences) bottom layer (prepared in NIH/3T3 growth medium supplemented or not with 5 μM EHT 1864) is poured first and allowed to harden (2.5 ml per well). NIH/3T3 cells stably transfected with the empty pCGN-hygro plasmid (Vector) or encoding H-Ras(61L) are trypsinized to generate a single cell suspension. It is important at this step to confirm that cells are a uniform single-cell suspension. Cells (5×10^3 per well) are resuspended in 0.4% agar supplemented with complete growth medium or are supplemented additionally with 5 μM EHT 1864 to form the top layer (0.25 ml per well). After overnight solidification of the agar top layer, 0.25 ml of growth medium supplemented or not with 5 μM EHT 1864 is added to the top of each well. The medium is then changed every other day to continuously maintain the cells in the presence of fresh EHT 1864. Whereas untransformed NIH/3T3 cells do not form colonies in soft agar, Ras-transformed NIH/3T3 cells form colonies of proliferating cells that can be detected readily within a week. When the colonies are large enough to be stained (typically after 16 days), the medium is removed, and 0.25 ml per well of a 2-mg/ml solution of thiazolyl blue tetrazolium bromide (MTT) (Sigma) is added to the cells and incubated at least 1 h at 37° before the plates are scanned. Colonies are scored by counting the whole well for each condition under a microscope. As shown in Fig. 9.5. the EHT 1864 compound partially inhibits oncogenic Ras-induced anchorage-independent growth

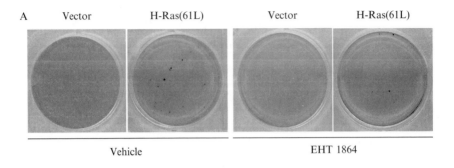

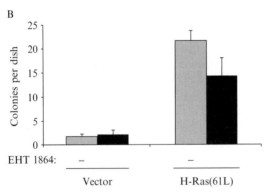

Figure 9.5 EHT 1864 partially inhibits oncogenic Ras-induced anchorage-independent growth of NIH/3T3 fibroblasts. (A) Single-cell suspensions of NIH/3T3 cells stably transfected with the empty pCGN-hygro (vector) plasmid or encoding H-Ras(61L) were cultured in a soft agar medium in the presence or absence of 5 μM EHT 1864, and the appearance of colonies of proliferating cells was monitored 16 days later. (B) Quantitation of data shown in A. Data shown are representative of two independent experiments, each performed in triplicate.

of NIH/3T3 cells. Therefore, results obtained with both focus formation assays and soft agar assays suggest that EHT 1864 impairs Ras-mediated cellular transformation.

3. CONCLUDING REMARKS

This chapter presented methods to characterize Rho GTPase specificity and biochemical and cellular mechanisms of action of EHT 1864, a lead compound that specifically inhibits Rac GTPases by binding directly to Rac1 (and potentially to other members of the Rac family) and placing it in an inert and inactive state. We also found that EHT 1864 has also been effective in modulating Rac1-mediated cellular functions, such as morphological changes and malignant transformation. Future analyses of EHT 1864 activity will include studies of Rac function inhibition in human tumor cell lines with validated activated Rac, using both cell culture and mouse models. In summary, the current lack of identification of successful inhibitors that selectively target small GTPases is surprising given the amount of current knowledge on their structure and functions. Because of its specificity of action toward Rac GTPases, EHT 1864 may represent in the future a useful tool for modulating aberrant Rac activities in therapeutics.

ACKNOWLEDGMENTS

We thank Wendy Salmon and Michael Chua at the UNC Michael Hooker Microscopy Facility for their assistance with image acquisition and Misha Rand for assistance with the manuscript and figure preparation. This work was supported by grants from the National Institutes of Health to C.J.D (CA67771, CA92240, and CA063071). A.S. was supported by a fellowship from the Susan G. Komen Breast Cancer Foundation.

REFERENCES

Basso, A. D., Kirschmeier, P., and Bishop, W. R. (2006). Lipid posttranslational modifications. Farnesyl transferase inhibitors. *J. Lipid Res.* **47,** 15–31.

Bishop, A. L., and Hall, A. (2000). Rho GTPases and their effector proteins. *Biochem. J.* **348** (Pt 2), 241–255.

Boettner, B., and Van Aelst, L. (2002). The role of Rho GTPases in disease development. *Gene* **286,** 155–174.

Chan, A. Y., Coniglio, S. J., Chuang, Y. Y., Michaelson, D., Knaus, U. G., Philips, M. R., and Symons, M. (2005). Roles of the Rac1 and Rac3 GTPases in human tumor cell invasion. *Oncogene* **24,** 7821–7829.

Chenette, E. J., Mitin, N. Y., and Der, C. J. (2006). Multiple sequence elements facilitate Chp Rho GTPase subcellular location, membrane association, and transforming activity. *Mol. Biol. Cell* **17,** 3108–3121.

Clark, G. J., Cox, A. D., Graham, S. M., and Der, C. J. (1995). Biological assays for Ras transformation. *Methods Enzymol.* **255,** 395–412.

Denicola, G., and Tuveson, D. A. (2005). VAV1: A new target in pancreatic cancer?. *Cancer Biol. Ther.* **4,** 509–511.

Desire, L., Bourdin, J., Loiseau, N., Peillon, H., Picard, V., De Oliveira, C., Bachelot, F., Leblond, B., Taverne, T., Beausoleil, E., Lacombe, S., Drouin, D., *et al.* (2005). RAC1 inhibition targets amyloid precursor protein processing by gamma-secretase and decreases Abeta production in vitro and in vivo. *J. Biol. Chem.* **280,** 37516–37525.

Etienne-Manneville, S., and Hall, A. (2002). Rho GTPases in cell biology. *Nature* **420,** 629–635.

Fernandez-Zapico, M. E., Gonzalez-Paz, N. C., Weiss, E., Savoy, D. N., Molina, J. R., Fonseca, R., Smyrk, T. C., Chari, S. T., Urrutia, R., and Billadeau, D. D. (2005). Ectopic expression of VAV1 reveals an unexpected role in pancreatic cancer tumorigenesis. *Cancer Cell* **7,** 39–49.

Ferraro, D., Corso, S., Fasano, E., Panieri, E., Santangelo, R., Borrello, S., Giordano, S., Pani, G., and Galeotti, T. (2006). Pro-metastatic signaling by c-Met through RAC-1 reactive oxygen species (ROS). *Oncogene* **25**, 3689–3698.

Fritz, G., Just, I., and Kaina, B. (1999). Rho GTPases are over-expressed in human tumors. *Int. J Cancer* **81**, 682–687.

Gao, Y., Dickerson, J. B., Guo, F., Zheng, J., and Zheng, Y. (2004). Rational design and characterization of a Rac GTPase-specific small molecule inhibitor. *Proc. Natl. Acad. Sci. USA* **101**, 7618–7623.

Hwang, S. L., Chang, J. H., Cheng, T. S., Sy, W. D., Lieu, A. S., Lin, C. L., Lee, K. S., Howng, S. L., and Hong, Y. R. (2005). Expression of Rac3 in human brain tumors. *J. Clin. Neurosci.* **12**, 571–574.

Jordan, P., Brazao, R., Boavida, M. G., Gespach, C., and Chastre, E. (1999). Cloning of a novel human Rac1b splice variant with increased expression in colorectal tumors. *Oncogene* **18**, 6835–6839.

Karnoub, A. E., Symons, M., Campbell, S. L., and Der, C. J. (2004). Molecular basis for Rho GTPase signaling specificity. *Breast Cancer Res. Treat.* **84**, 61–71.

Khosravi-Far, R., Solski, P. A., Clark, G. J., Kinch, M. S., and Der, C. J. (1995). Activation of Rac1, RhoA, and mitogen-activated protein kinases is required for Ras transformation. *Mol. Cell. Biol.* **15**, 6443–6453.

Kozma, R., Ahmed, S., Best, A., and Lim, L. (1995). The Ras-related protein Cdc42Hs and bradykinin promote formation of peripheral actin microspikes and filopodia in Swiss 3T3 fibroblasts. *Mol. Cell. Biol.* **15**, 1942–1952.

Phillips, R. A., Hunter, J. L., Eccleston, J. F., and Webb, M. R. (2003). The mechanism of Ras GTPase activation by neurofibromin. *Biochemistry* **42**, 3956–3965.

Qiu, R. G., Chen, J., Kirn, D., McCormick, F., and Symons, M. (1995). An essential role for Rac in Ras transformation. *Nature* **374**, 457–459.

Reid, T. S., Terry, K. L., Casey, P. J., and Beese, L. S. (2004). Crystallographic analysis of CaaX prenyltransferases complexed with substrates defines rules of protein substrate selectivity. *J. Mol. Biol.* **343**, 417–433.

Renshaw, M. W., Lea-Chou, E., and Wang, J. Y. (1996). Rac is required for v-Abl tyrosine kinase to activate mitogenesis. *Curr. Biol.* **6**, 76–83.

Ridley, A. J., and Hall, A. (1992). The small GTP-binding protein rho regulates the assembly of focal adhesions and actin stress fibers in response to growth factors. *Cell* **70**, 389–399.

Ridley, A. J., Paterson, H. F., Johnston, C. L., Diekmann, D., and Hall, A. (1992). The small GTP-binding protein rac regulates growth factor-induced membrane ruffling. *Cell* **70**, 401–410.

Sahai, E., and Marshall, C. J. (2002). RHO-GTPases and cancer. *Nat. Rev. Cancer* **2**, 133–142.

Schnelzer, A., Prechtel, D., Knaus, U., Dehne, K., Gerhard, M., Graeff, H., Harbeck, N., Schmitt, M., and Lengyel, E. (2000). Rac1 in human breast cancer: Overexpression, mutation analysis, and characterization of a new isoform, Rac1b. *Oncogene* **19**, 3013–3020.

Sebti, S. M., and Hamilton, A. D. (2000). Farnesyltransferase and geranylgeranyltransferase I inhibitors and cancer therapy: Lessons from mechanism and bench-to-bedside translational studies. *Oncogene* **19**, 6584–6593.

Shutes, A., Onesto, C., Picard, V., Leblond, B., Schweighoffer, F., and Der, C. J. (2007). Specificity and mechanism of action of EHT 1864, a novel small molecule inhibitor of Rac family small GTPases. *J. Biol. Chem.* **282**, 35666–35678.

Shutes, A., Phillips, R. A., Corrie, J. E., and Webb, M. R. (2002). Role of magnesium in nucleotide exchange on the small G protein rac investigated using novel fluorescent guanine nucleotide analogues. *Biochemistry* **41**, 3828–3835.

Singh, A., Karnoub, A. E., Palmby, T. R., Lengyel, E., Sondek, J., and Der, C. J. (2004). Rac1b, a tumor associated, constitutively active Rac1 splice variant, promotes cellular transformation. *Oncogene* **23,** 9369–9380.

Solski, P. A., Abe, K., and Der, C. J. (2000). Analyses of transforming activity of Rho family activators. *Methods Enzymol.* **325,** 425–441.

Thapar, R., Karnoub, A. E., and Campbell, S. L. (2002). Structural and biophysical insights into the role of the insert region in Rac1 function. *Biochemistry* **41,** 3875–3883.

Zohn, I. M., Campbell, S. L., Khosravi-Far, R., Rossman, K. L., and Der, C. J. (1998). Rho family proteins and Ras transformation: The RHOad less traveled gets congested. *Oncogene* **17,** 1415–1438.

CHAPTER TEN

ANALYSIS OF RHO-GTPASE MIMICRY BY A FAMILY OF BACTERIAL TYPE III EFFECTOR PROTEINS

Neal M. Alto* *and* Jack E. Dixon

Contents

Abstract

Microbial pathogens may hijack the actin cytoskeleton machinery to facilitate actin-based motility, cellular invasion, and intracellular trafficking through the endocytic pathway. Of particular interest has been a large class of virulence

Departments of Pharmacology, Cellular and Molecular Medicine, and Chemistry and Biochemistry, University of California, San Diego, La Jolla, California
* Department of Microbiology, University of Texas Southwestern, Dallas, Texas

Methods in Enzymology, Volume 439
ISSN 0076-6879, DOI: 10.1016/S0076-6879(07)00410-7

factors collectively termed the type III "effector" proteins. Following contact with a eukaryotic cell, Gram-negative bacterial pathogens employ a dedicated protein secretion apparatus termed type III to translocate effector proteins into the cellular cytoplasm where they initiate a pathogenic response. Interestingly, these effectors can mimic the structure or function of eukaryotic signaling proteins and often target components of the actin cytoskeleton. This chapter describes methodologies to examine the cellular role of the recently defined WxxxE family of effector proteins that functionally mimic Rho family small G proteins.

 ## 1. INTRODUCTION

Because of their ubiquitous role in cell biology, Ras-like small G proteins are a common target for virulence factors secreted by bacterial pathogens (Boquet, 2000). At least three key features of small G proteins make them good targets for bacterial toxins and effector proteins. First, they act as molecular switches cycling between GTP-bound (active) and GDP-bound (inactive) conformations. Second, they transmit signaling events in a nucleotide-dependent manner by activating and/or recruiting downstream target proteins to their sites of action. Finally, they are targeted to specific subcellular compartments through the addition of isoprenoid lipid moieties at their carboxyl terminus. Because each feature of the small G protein requires extensive regulation in the host cell, pathogens have designed sophisticated mechanisms to subvert these signaling pathways.

Virulence associated with several Gram-negative bacterial pathogens requires the translocation of "effector" proteins from bacteria into host cells through a dedicated protein translocation apparatus termed the type III secretions system (TTSS). For example, *Salmonella typhi*, *Shigella flexneri*, and enteropathogenic *Escherichia coli* (EPEC) utilize type III effector proteins to facilitate human disease progression by allowing each pathogen to limit antibacterial host immune responses or generate a cellular architecture that is advantageous to bacterial replication. Importantly, each pathogen secretes a distinct set of effector proteins that propagate their complex life cycles within the host organism (Cossart and Sansonetti, 2004).

We have described the biological function of three members of a 24-member type III effector protein family (Alto *et al.*, 2006). IpgB1 and IpgB2 from *Shigella*, as well as Map from EPEC, function in a similar manner by mimicking the GTP-active form of Rho family GTPases. Subsequent studies have further elucidated the function of IpgB1 during *Shigella* invasion of host cells (Handa *et al.*, 2007). Although these important bacterial proteins seem to lack the structural and biochemical attributes of GTPases, they induce actin cytoskeletal dynamics in the host cell by stimulating Rho family signaling pathways. Moreover, the signaling specificities of these type III effector proteins are governed in part by eukaryotic-like targeting motifs (Alto *et al.*, 2006; Boucrot *et al.*, 2003). This chapter describes several cell biological and yeast genetic assays used to distinguish the function of this family of bacterial small G-protein mimics.

2. Analysis of WxxxE Effectors as Rho-GTPase Mimics in Mammalian Cells

2.1. Identification of bacterial pathogen families with low sequence homology

The high rate of bacterial replication and the possibility of horizontal gene transfer between organisms have led to remarkable diversity among type III effector proteins. In fact, it is now apparent that pathogens can share common virulence factors that have evolved to exploit unique host cell niches. We and others have used bioinformatics database search tools to identify new effector protein families and to discover novel biochemical activities. While standard BLAST searches are useful to identify members of an effector protein family that share a high degree of sequence identity, a more powerful approach is to use multiple rounds of the PSI-BLAST algorithm to reveal new family members with much lower sequence identity. For example, using the type III bacterial effector protein Map from enterohaemorrhagic *E. coli* (EHEC 0157:H7) as an index protein, we performed multiple PSI-BLAST iterations to identify a 24-member protein family that shared significant sequence identity over ≈150 residue domain (Fig. 10.1A). Clustal alignment revealed several new aspects of these proteins. First, Map shared sequence homology to type III effector proteins from unique organisms, including *Shigella* (IpgB1 and IpgB2) and *Salmonella* (SifA and SifB). Second, there were only two invariant amino acids, a Trp and Glu residues found in a WxxxE motif common between all 24 proteins. Finally, the C terminus of several of these bacterial effectors harbored putative eukaryotic targeting motifs, including PDZ ligands and CaaX box motifs (data not shown). Thus, we propose that each of these type III effector proteins shares a common and unidentified biochemical activity to promote bacterial pathogenesis.

2.2. Expression constructs and mutagenesis

Because the biological function of WxxxE effectors is unknown, we first explored a strategy of overexpressing Map, IpgB1, and IpgB2 in eukaryotic cells. The coding region of EHEC Map (accession #AP002566) and *S. flexneri* IpgB2 (accession #AF348706) was polymerase chain reaction (PCR)

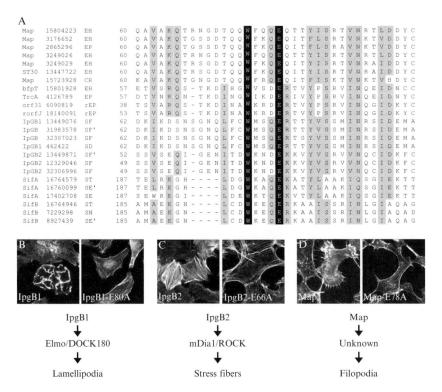

Figure 10.1 (A) Clustal alignment comparing the WxxxE region (black) of EHEC Map and 23 other family members. GenBank ID # is listed. Residues with similar chemical properties are in gray. Immunofluorescence microscopy of HEK293A cells transfected with Tap-IpgB1 or Tap-IpgB1-E80A (B), EGFP-IpgB2 or EGFP-IpgB2-E66A (C), and EGFP-Map or EGFP-Map-E78A (D). Cells were stained with rhodamine-phalloidin to detect actin. Arrows indicate transfected cells. Scale bar: 15 mm. Below each image is the cellular pathway regulated by IpgB1, IpgB2, and Map. Modifications of these figures were reprinted from Alto *et al.* (2006) with permission.

amplified and ligated in-frame into pEGFP-C2 (BD Clontech) with restriction sites *Eco*RI and *Xho*I. Similarly, IpgB1 (accession #CAA33379) was cloned into a modified Tandem affinity tag (Tap) vector termed pSurf-flag with restriction sites *Eco*RV and *Xho*I. These cloning efforts led to EGFP-Map and EGFP-IpgB2, and Tap-IpgB1 chimeric proteins with amino-terminal enhanced green fluorescent protein (EGFP) and Tap tags (protein A-flag tag).

Generation of nonfunctional mutants is a convenient strategy for the analysis of Map, IpgB1, and IpgB2 in cells. We substituted the conserved glutamic acid (E) in the WxxxE motif to alanine (A) using the Quickchange site-directed mutagenesis kit (Stratagene). Primers used for PCR-based point mutagenesis include Map-E78A (5′- CAA TGG TTC CAG CAG <u>GCG</u> CAG ACG ACT TAT ATA TCC-3′), IpgB1 E80A (5′-TGT TGG ATG AGC CAA <u>GCA</u> CGA ACC ACT TAT GTC-3′) and IpgB2-E66A (5′-ACA GAT TGG AAA AAT GAT <u>GCA</u> AAA AAA GTC TAC GTA TCC-3′) with the underlined codon switch to alanine. All constructs were verified by DNA sequencing.

2.3. Effects of IpgB1, IpgB2, and Map expression in eukaryotic cells

To characterize the properties of these bacterial effectors in eukaryotic cells, we used a transient transfection model system using either Hek293A or HeLa cells. The extent of the observed phenotypes described here can be cell dependent, most likely because of different eukaryotic ligand concentrations or the subcellular localization patterns of WxxxE effectors in distinct cell types. However, robust effects were observed for all WxxxE effectors analyzed in Hek293A cells, a clonal cell line that was specifically selected for strong adherent properties to polystyrene dishes (see Figs. 10.1B–10.1D).

Immunocytochemistry experiments are used to determine the effects of WxxxE type III effectors on the actin cytoskeleton. We routinely culture Hek293A cells at 37° in 5% CO_2 in Dulbecco's modified Eagle's medium (DMEM) containing penicillin streptomycin and 5% fetal bovine serum (GIBCO). For microscopy studies, cells are seeded onto fibronectin-coated coverslips (Becton Dickinson) at ≈50% confluence. The next day, cells are transiently transfected with 1 μg of plasmid DNA using FuGene 6 (Roche Applied Science). Cells are grown for 16 h and fixed the next day with 3.7% paraformaldehyde in phosphate-buffered saline (PBS; 138 mM NaCl, 2.7 mM KCl, 8.2 mM Na_2HPO_4, 1.5 mM KH_2PO_4, 25 mM glacial acetic acid, pH 7.2). Transfected coverslips are gently washed two times in PBS and fixed for 10 min with 3.7% paraformaldehyde in PBS. Fixed cells are permeabilized for 5 min with 0.1% Triton-X 100 (Sigma) in PBS and then blocked with buffer-supplemented 1% bovine serum albumin for 30 min. A 90-min incubation with the primary antibody to flag (M2 Sigma) can be used to detect TAP-IpgB1 detection (1:200). Secondary antibody incubation for 60 min (1:500, Alexa 488) is used to detect TAP-IpgB1. EGFP-IpgB2 or EGFP-Map is detected by indirect fluorescence. Filamentous actin is stained with rhodamine-phalloidin (Molecular Probe) for at least 30 min during the secondary antibody incubation. Cellular phenotypes are examined with a Zeiss Axiovert 200M microscope using the appropriate filter sets and Metafluor software. Examples of cellular effects of Map, IpgB1, and IpgB2 on the cytoskeleton of Hek293A cells are shown in Figs. 10.1B and 10.1C Map induces cell surface filopodia, and IpgB1 induces dorsal membrane ruffles and lamellipodia, whereas IpgB2 induces actin stress fibers.

In addition, mutations in the conserved glutamic acid abolished the cellular phenotypes in each WxxxE effector protein, suggesting that each bacterial protein utilizes a conserved molecular mechanism. The actin phenotypes observed with IpgB1, IpgB2, and Map are similar to those produced by Rac1, RhoA, and Cdc42, respectively (Hall, 1998).

2.4. Epistatic analysis of Rho mimicry by IpgB2 using the C3-exoenzyme and the Y27632 ROCK inhibitor

Botulinum C3 exoenzyme is the prototype member of Rho-ADP-ribosy-lating toxins that irreversibly inactivates the Rho subfamily of GTPases such as RhoA, RhoB, and RhoC (Aktories et al., 2004). Therefore, this toxin is an appropriate tool test if IpgB2 is epistatic to RhoA. We cotransfected EGFP-IpgB2 or HA-RhoA with pEF-C3-exoenzyme at threefold excess concentration of C3-exoenzyme DNA (Figs. 10.2A–10.2D). This ensures that the majority of IpgB2 or RhoA expressing cells also express C3 or the pEF vector as an appropriate control. Transfected cells are incubated 16 h and processed for immunofluorescence microscopy. The functional expression of the C3-exonenzyme can be determined by morphological changes in cells expressing RhoA, which include the loss of actin-stress fibers and severe retraction fibers (see Figs. 10.2A–10.2D). Unlike expression of RhoA, the C3-exoenzyme has no effect on cell expression EGFP-IpgB2 (see Figs. 10.2A–10.2D), indicating that IpgB2 can induce stress fiber formation independent of any Rho family member activity.

Y-27632 (Ishizaki et al., 2000), a pharmacological inhibitor of Rho kinases, is also used to inhibit the RhoA target proteins ROCKI and ROCKII required for stress fiber formation (Amano et al., 1997). Sixteen hours after EGFP-IpgB2 or HA-RhoA transfection, 10 μM Y27632 (Calbiochem) is added to cells for 30 min and fixed and processed for immunofluourescence microscopy. Y27632 blocks stress fiber formation induced by EGFP-IpgB2, indicating that IpgB2 is epistatic to RhoA and behaves as a functional Rho mimic in tissue culture cells. Together, the use of well-characterized bacterial toxins and specific pharmacological inhibitors provides a useful strategy to dissect the signaling pathways of type III effector protein that function as small G-protein mimics.

2.5. Cellular interaction between IpgB2 and mDiaphanous

It is well established that RhoA induces stress fiber assembly by stimulating the mammalian homologue of Diaphanous (mDia1) (Watanabe et al., 1999). mDia1 is composed of an N-terminal Rho binding sequence (GBD) followed by actin nucleating Formin homology domains 1 and 2 (see Fig. 10.2E). We have performed coimmunoprecipitation studies to demonstrate that IpgB2 interacts with the GBD region of mDia1

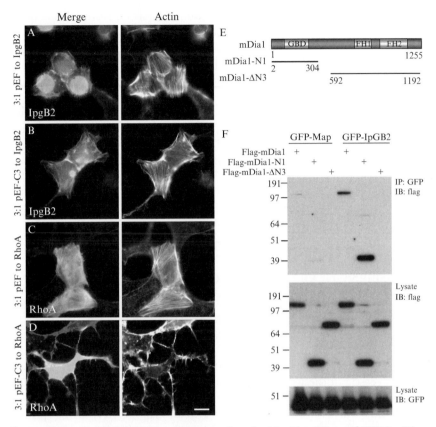

Figure 10.2 (A–D) HEK293A cells cotransfected with either 0.5 μg of GFP-IpgB2 or HA-RhoA and threefold excess (1.5 μg) pEF control vector or pEF-C3-exoenzyme as indicated to the left of each image. Merged image of IpgB2 or RhoA and actin (left) and actin morphology (right). Scale bar: 15 μm. (E) Diagram of mDia1 and truncation mutants used in this study. (F) HEK293A cells were cotransfected with GFP-Map or GFP-IpgB2 and the indicated flag-mDia1 plasmids. Anti-GFP immunoprecipitations (IP) were probed by flag immunoblot (IB) (top). Modifications of these figures were reprinted from Alto *et al.* (2006) with permission.

(see Fig. 10.2F). A 100-mm dish of Hek293A cells at 70 to 80% confluency is transfected with 2.5 μg of each plasmid DNA using FuGENE 6 (Roche Applied Science). Twenty-four hours after transfection, cells are washed with PBS and lysed in 1 ml of ice-cold lysis buffer (50 mM Tris, pH 7.5, 150 mM NaCl, 0.5% Triton X-100, 100 mM NaF, 1 mM phenylmethyl-sulfonyl fluoride, 1 μg/ml leupeptin, 10.5 μg/ml aprotinin, and 1 μg/ml pepstatin). Lysates are vortexed and cleared by centrifugation (17,000 g for 15 min at 4°). EGFP-tagged proteins are immunoprecipitated using the anti–EGFP polyclonal antibody and protein A-agarose. Immunoprecipitations are incubated for 2 h at 4° and subsequently washed twice with 1 ml of lysis buffer, three times with 1 ml of lysis buffer containing 0.5 M NaCl, and once more with 1 ml of lysis buffer. Immunoprecipitates are suspended in NuPAGE LDS sample buffer (Invitrogen). All immunoprecipitates and lysates are resolved in 4 to 12% NuPAGE Bis–Tris gels (Invitrogen), transferred to polyvinylidene difluoride membranes. Flag-tagged mDia1 onstructs are detected by anti–FLAG-M2 and analyzed by immunoblotting. These studies demonstrate that IpgB2 interacts with the GBD region of mDia1, a protein interaction that is similar to that of RhoA.

3. Analysis of Small G-Protein Mimicry Using Yeast as a Model System

3.1. Using yeast genetics to study small G-protein mimicry by WxxxE effector proteins

Yeast can be used as a powerful genetic tool to study signaling pathways in eukaryotic cells. Yeast cells possess Rho family small G proteins involved in gene transcription responses, cytoskeletal changes, and cell polarity. We found that the expression of certain WxxxE family members can be toxic to yeast. For these studies, the coding region of IpgB2 and single point mutations are first PCR cloned into the Gateway recombination vector pENTR/D (Invitrogen, pENTR/D-Directional TOPO) using primer pairs (forward 5′-CAC CAT GCT TGG AAC ATC TTT TAA TAA TTT TGG AAT C and reverse 5′-GAA AGG CGA TTC TAA ATT TGT AAT ATA GTC ACT CAT ATT CCC). Next, IpgB2 is Gateway recombined (Invitrogen) into the yeast expression vector pYes-DEST52, which possesses a galactose inducible promoter and *URA3* selection marker. Constructs are transformed into the yeast strain Y7092 (*MATα can1Δ::STE2pr-Sp Δhis5 his3Δ1 leu2Δ0 ura3Δ0 met15Δ0 lyp1Δ*) using standard lithium acetate methods, and cultures are plated on SD-Ura (synthetic dextrose medium minus Uracil) containing 20% glucose. The lack of uracil allows the selection of yeast transformed with the pYes-Dest52 plasmid. Glucose (20%) is sufficient to repress IpgB2 expression from the *GAL1* promoter (Fig. 10.3A). Yeast plates are grown for 3 days at 30°. Single colonies are streaked to SGR/-Ura with 10% galactose and 10% rafinose as carbon sources to express IpgB2 (see Fig. 10.3A). Control experiment should also be conducted with SD/-Ura minimal medium with 20% galactose. Yeast plates are incubated for 48 h at 30° and subsequently scored for growth phenotypes as shown in Fig. 10.3B.

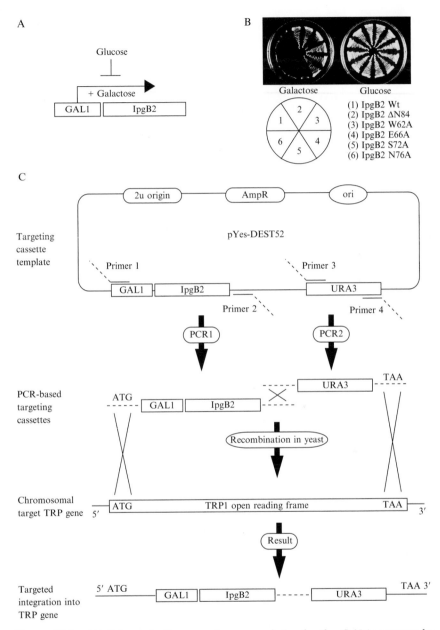

Figure 10.3 (A) Schematic diagram of gene regulation by the *GAL1* promoter by galactose or glucose carbon source. (B) Yeast transformed with *GAL1-IpgB2* or the indicated point mutations were streaked on galactose or glucose containing minimal medium as indicated. (C) Schematic diagram of the PCR-based yeast integration procedure of IpgB2.

3.2. Integrating IpgB2 into yeast for PGA screening procedure

To gain additional insights into the mechanism of IpgB2 yeast toxicity, we designed a whole genome screening strategy modeled after the synthetic genetic array (SGA) (Tong *et al.*, 2001). The detailed methodology of the SGA screening protocol has been reported previously and is not expounded on here (Tong and Boone, 2006). However, our modified type III effector screen is referred to as a pathogenic genetic array (PGA) because it allows an ordered array of ≈5000 viable yeast deletion mutants to be assayed for growth defects associated with conditional expression of any pathogen protein. For the PGA, it is critical that IpgB2 is integrated into the genome of Y7092 to aid in stability of the toxic gene, as well as to ensure that a single copy of IpgB2 is expressed in all ≈5000 yeast clones. We have designed a universal integration/selection/screening strategy for the targeted insertion of IpgB2 into the Trp1 locus of Y7092 and this method can be easily adapted to any pathogen gene (see Fig. 10.3B).

3.3. General approach

Two PCR products, one coding for *GAL1*-IpgB2 and the second coding for the *URA3* selectable marker, are transformed into yeast and screened for homologous recombination into the *TRP1* genetic locus. We use the pYes-Dest52 plasmid as a template for both the *GAL1*-IpgB2 and the *URA3* gene PCR with the specific primer sets shown (see Fig. 10.3C).

PCR primer 1: 5′-<u>ATG TCT GTT ATT AAT TTC ACA GGT AGT TCT GGT CCA TTG GTG AAA GTT TGC GGC TAG</u> **ACG GAT TAG AAG CCG CCG AGC GGG TGA CAG CCC TCC**-3′. Underlined is the 54-bp 5' sequence from the *TRP1* open reading frame (ATG is the start codon) and bold is the 5′ sequence of the *GAL1* promoter (see Fig. 10.3C).

PCR primer 2: 5′-<u>TGA TGC CTC CAC TCC CAT CGG AGT C</u>**CT AGG CTT ACC TTC GAA GGG CCC**-3′. Underlined is the 25 nucleotide unique overlap sequence and bold sits down on the pYes-Dest52 sequence 3′ to the IpgB2 gene (see Fig. 10.3C).

PCR primer 3: 5′-<u>GAC TCC GAT GGG AGT GGA GGC ATC A</u>**GT GGC TGT GGT TTC AGG GTC CAT AAA GCT TTT CAA TTC ATC**-3′. Underlined is the 25 nucleotide unique overlap sequence and bold sits down at the 5′ end of the *URA3* gene promoter (see Fig. 10.3C).

PCR primer 4: 5′-<u>CTA TTT CTT AGC ATT TTT GAC GAA ATT TGC TAT TTT GTT AGA GTC TTT TAC ACC</u> **TTA GTT TTG CTG GCC GCA TCT TCT CAA ATA TGC TTC CCA GCC**

TGC-3′. Underlined is the 54-bp 3′ sequence from the TRP1 open reading frame (the 5′ CTA is the reverse and complemented TAG stop codon in Trp1) and bold sits down at the 3′ end of the *URA3* gene (see Fig. 10.3C).

3.4. PCR-based yeast integration procedure

1. Two separate PCR reactions, one with primer sets 1 and 2 and the other with primer sets 3 and 4, are performed on the pYes–Dest-52 template coding for IpgB2. We have successfully PCR amplified templates using long primer pairs with hot start *Taq* polymerase Platinum PCR Super-Mix high fidelity (Invitrogen). In addition, high-fidelity polymerases are essential to guard against the introduction of unwanted mutations.
2. The entire 50-μl PCR products are gel purified with the QIAquick gel extraction kit (Qiagen).
3. The yeast strain Y7092 is transformed with both PCR products using the standard lithium acetate method. We routinely heat shock the Y7092 strain for 16 min at 42° in dimethyl sulfoxide.
4. Transformed Y7092 are plated onto SD/-Ura agar plates with 20% glucose to select for recombination of the *URA3* gene.
5. Large colonies that appear after 3 days are further screened for insertion and expression of the toxic IpgB2 gene by picking colonies onto SGR–Ura supplemented with 10% galactose and 10% raffinose carbon source. Yeast colonies that grow well on SD-Ura glucose but not on SD-Ura galactose selectively express the toxic IpgB2 gene that is incorporated into the yeast genome.
6. Finally, to confirm that the yeast integration cassette is targeted to the *TRP1* locus, strains are plated on SD-Ura-Trp to screen for the deletion of *TRP1*. *TRP1*-targeted gene deletions will not grow on SD-Ura-Trp agar plates. A PCR-based strategy can also be used to further confirm the site of IpgB2 integration.
7. The final yeast strain is Y7092 (*MATα can1Δ::STE2pr-Sp his5 his3Δ1 leu2Δ0 ura3Δ0 met15Δ0 lyp1Δ trp1Δ::GAL1-IpgB2-URA3*).

3.5. Pathogenic genetic array

The PGA analysis is conducted using the SGA methodology as described in Tong *et al.* (2001) with the following modifications: double mutant meiotic progeny are selected on medium containing glucose, moved to medium containing raffinose, and then transferred to and scored on galactose medium. We introduced Y7092 carrying the integrated *GAL1-IpgB2* into ≈5000 viable gene deletion mutants through a mating procedure (Tong and Boone, 2006) and examined their growth rates on medium containing galactose (Fig. 10.4A). Because expression of IpgB2 is toxic to wild-type yeast, any gene deletion mutant that grows should identify the corresponding

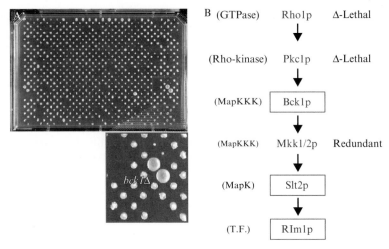

Figure 10.4 (A) Pathogenic genetic array showing an example of a single 24 × 16 array plate carrying ≈348 individual gene deletion yeast strains that express *GAL1-IpgB2* (out of 16 plates and ≈5000 strains). The duplicate *bck1Δ* yeast strain expressing *GAL1-IpgB2* is shown (enlarged image). (B) Diagram of the Rho1p-activated Pkc1p/MAPK module. The function of each protein is in parentheses (T.F. denotes transcription factor). Gene deletions that allow IpgB2 expressing yeast to survive are boxed. Modifications of these figures were reprinted from Alto *et al.* (2006) with permission.

gene as a potential target for IpgB2-induced toxicity. Out of the ~5000 strains tested, only gene deletions in *BCK1*, *SLT2*, and *RLM1* resulted in a robust growth phenotype. Remarkably, each of these genes encodes components of the cell wall integrity Rho1p/MAP kinase (MAPK) signaling module (Heinisch *et al.*, 1999) illustrated in Fig. 10.4B. Although we used these genome-screening procedures as an unbiased conformation of the Rho mimicry by IpgB2, the PGA screen may serve as a useful tool to help identify the functions of other pathogen effector proteins or toxins that target specific substrates in yeast.

 ## 4. Concluding Remarks

We have used the *Shigella* effectors IpgB1 and IpgB2, as well as the *E. coli* effector Map, as experimental prototypes to demonstrate that several WxxxE family members functionally mimic the activated forms of distinct Rho family small G proteins. These effector proteins appear to be a bacterial "invention," as there are no WxxxE motifs in eukaryotic small G proteins. In addition, IpgB1, IpgB2, and Map do not bind guanine nucleotides or display sequences, suggesting that they have GTPase activity. Thus results further highlight the remarkable ability of microbial pathogens to subvert small G-protein signaling pathways through molecular mimicry of eukaryotic proteins. In addition, we highlighted methods demonstrating how yeast can serve as an excellent model organism to study bacterial effector proteins and toxins *in vivo*.

ACKNOWLEDGMENTS

We thank Michael Costanzo, Cheri Lazar, and members of the Dixon laboratory for help in preparation of this manuscript. Grants from the NIH, the Walther Cancer Institute, and the Ellison Foundation supported this research (J.E.D.). N.M.A. was supported by Hematology and Diabetes NIH training grants.

REFERENCES

Aktories, K., Wilde, C., and Vogelsgesang, M. (2004). Rho-modifying C3-like ADP-ribosyltransferases. *Rev. Physiol. Biochem Pharmacol.* **152,** 1–22.

Alto, N. M., Shao, F., Lazar, C. S., Brost, R. L., Chua, G., Mattoo, S., McMahon, S. A., Ghosh, P., Hughes, T. R., Boone, C., and Dixon, J. E. (2006). Identification of a bacterial type III effector family with G protein mimicry functions. *Cell* **124,** 133–145.

Amano, M., Chihara, K., Kimura, K., Fukata, Y., Nakamura, N., Matsuura, Y., and Kaibuchi, K. (1997). Formation of actin stress fibers and focal adhesions enhanced by Rho-kinase. *Science* **275,** 1308–1311.

Boquet, P. (2000). Small GTP binding proteins and bacterial virulence. *Microbes Infect.* **2,** 837–843.

Boucrot, E., Beuzon, C. R., Holden, D. W., Gorvel, J. P., and Meresse, S. (2003). *Salmonella typhimurium* SifA effector protein requires its membrane-anchoring C-terminal hexapeptide for its biological function. *J. Biol. Chem.* **278,** 14196–14202.

Cossart, P., and Sansonetti, P. J. (2004). Bacterial invasion: The paradigms of enteroinvasive pathogens. *Science* **304,** 242–248.

Hall, A. (1998). Rho GTPases and the actin cytoskeleton. *Science* **279,** 509–514.

Handa, Y., Suzuki, M., Ohya, K., Iwai, H., Ishijima, N., Koleske, A. J., Fukui, Y., and Sasakawa, C. (2007). Shigella IpgB1 promotes bacterial entry through the ELMO-Dock180 machinery. *Nat. Cell Biol.* **9,** 121–128.

Heinisch, J. J., Lorberg, A., Schmitz, H. P., and Jacoby, J. J. (1999). The protein kinase C-mediated MAP kinase pathway involved in the maintenance of cellular integrity in *Saccharomyces cerevisiae*. *Mol. Microbiol.* **32,** 671–680.

Ishizaki, T., Uehata, M., Tamechika, I., Keel, J., Nonomura, K., Maekawa, M., and Narumiya, S. (2000). Pharmacological properties of Y-27632, a specific inhibitor of rho-associated kinases. *Mol. Pharmacol.* **57,** 976–983.

Tong, A. H., and Boone, C. (2006). Synthetic genetic array analysis in *Saccharomyces cerevisiae*. *Methods Mol. Biol.* **313,** 171–192.

Tong, A. H., Evangelista, M., Parsons, A. B., Xu, H., Bader, G. D., Page, N., Robinson, M., Raghibizadeh, S., Hogue, C. W., Bussey, H., Andrews, B., Tyers, M., and Boone, C. (2001). Systematic genetic analysis with ordered arrays of yeast deletion mutants. *Science* **294,** 2364–2368.

Watanabe, N., Kato, T., Fujita, A., Ishizaki, T., and Narumiya, S. (1999). Cooperation between mDia1 and ROCK in Rho-induced actin reorganization. *Nat. Cell Biol.* **1,** 136–143.

CHAPTER ELEVEN

INVESTIGATING THE FUNCTION OF RHO FAMILY GTPASES DURING *SALMONELLA*/HOST CELL INTERACTIONS

Jayesh C. Patel *and* Jorge E. Galán

Contents

Abstract

Salmonella enterica comprise a family of pathogenic gram-negative bacteria that have evolved sophisticated virulence mechanisms to enter non-phagocytic cells. The entry event is the result of a carefully orchestrated modulation of Rho family GTPase activity within the host cell, which in turn triggers localized remodeling of the actin cytoskeleton. These cytoskeletal rearrangements drive profuse membrane ruffling and lamellipodial extensions that envelop bacteria and trigger their internalization. This chapter describes a number of

Section of Microbial Pathogenesis, Yale University School of Medicine, Boyer Center for Molecular Medicine, New Haven, Connecticut

Methods in Enzymology, Volume 439
ISSN 0076-6879, DOI: 10.1016/S0076-6879(07)00411-9

methods used to investigate the role of Rho family GTPases during *Salmonella/* host cell interactions. In particular, we detail a variety of complementary techniques, including affinity pull-down assays and bacterial-induced membrane ruffling and internalization assays to show that *Salmonella*-induced actin remodeling and entry require the Rho family members Rac and RhoG.

1. INTRODUCTION

Manipulation of the actin cytoskeleton is a prevailing theme in host/ pathogen interactions (Cossart and Sansonetti, 2004; Munter *et al.*, 2006; Rottner *et al.*, 2005). A prime example is *Salmonella enterica*, gram-negative bacteria commonly associated with food-borne illness in humans. A hallmark of its close association with vertebrate hosts, *Salmonella* has evolved very sophisticated strategies to enter cells that are not normally phagocytic (Patel and Galan, 2005). Following host cell contact, *Salmonella* employs a specialized organelle termed the type III secretion system to inject multiple bacterial effector proteins directly into the host cell cytosol (Galan, 2001). A specific subset of these effector proteins taps into the eukaryotic signaling networks responsible for regulating actin dynamics and together they orchestrate profuse remodeling of the cytoskeleton at the site of entry (Patel and Galan, 2005). This reprogramming of the cytoskeletal machinery drives localized membrane ruffling and lamellapodial extensions that envelop bacteria and promote their internalization into membrane-bound vacuoles.

As key modulators of cytoskeletal remodeling in eukaryotes, the Rho family of small GTPases (Bustelo *et al.*, 2007; Etienne-Manneville and Hall, 2002) presents ideal targets for bacterial effector proteins to promote internalization into non–phagocytic cells. Like all small GTPases, the Rho family cycles between an active (GTP bound) state, competent for downstream signaling, and an inactive (GDP bound) state. This binary exchange of guanine nucleotides is largely controlled by two classes of regulatory proteins: guanine nucleotide exchange factors (GEFs), which activate Rho proteins by catalyzing the exchange of GDP for GTP, and GTPase-activating proteins (GAPs), which accelerate the intrinsic GTPase activity inherent to Rho family members, leading to their inactivation (Moon and Zheng, 2003; Rossman *et al.*, 2005). *Salmonella* delivers three distinct effector proteins that functionally converge at the level of Rho family GTPases to elicit host cell entry. SopE and its homolog SopE2 act as bona fide GEFs for the Rho family members Cdc42, Rac and RhoG (Bakshi *et al.*, 2000; Hardt *et al.*, 1998; Stender *et al.*, 2000). A second effector protein, SopB, encodes a phosphoinositide phosphatase that feeds into the GTPase cycle by indirectly activating a host-encoded exchange factor, SGEF, specific for

RhoG (Patel and Galan, 2006). A third effector protein, SptP, antagonizes the actions of SopE/E2 and SopB by functioning as a GAP for several Rho family members (Fu and Galan, 1999). This limited repertoire of effectors proteins provides *Salmonella* the capacity to reversibly modulate the Rho GTPase switch and thereby initiate the transient cytoskeletal remodeling events necessary for internalization. Moreover, Rho family GTPases play an essential role beyond bacterial entry, driving the accompanying *Salmonella*-induced macropinocytosis events and reprogramming of host gene expression (Chen *et al.*, 1996; Patel and Galan, 2006). The manipulation of GTPase function is thus essential to *Salmonella* pathogenesis, and the ability to fine-tune their activity ensures against overt cellular damage to the host.

This chapter describes a number of protocols used to investigate the differential activation and function of Rho family GTPases during *Salmonella*/host cell interactions. The different methods include affinity pull-down assays and *Salmonella*-induced membrane ruffling and internalization assays. Collectively, they demonstrate that *Salmonella* infection triggers activation of three distinct Rho family GTPases, Cdc42, Rac, and RhoG, and that *Salmonella*-induced actin remodeling and entry requires Rac and RhoG.

2. Determining *Salmonella*-Induced Rho GTPase Activation

The use of standard affinity pull-down assays to measure Rho GTPase activation in response to different cellular stimuli has been described extensively (Benard and Bokoch, 2002; Prieto-Sanchez *et al.*, 2006; Ren and Schwartz, 2000). These assays exploit downstream binding partners that specifically bind to the activated, GTP-bound form of their cognate GTPase. The binary complex can be isolated by coupling the GTPase-binding protein to glutathione *S*-transferase (GST) and using affinity precipitation. We utilize pull-down assays based on the Rac and Cdc42-binding partner PAK CRIB and the RhoG-binding target ELMO to evaluate GTPase activation in response to *Salmonella* infection.

2.1. Reagents

COS 1 cells: a monkey kidney fibroblast cell line (available at ATCC, Rockville, MD) used as host cells for *Salmonella* infection. Cells are cultured in Dulbecco's modified Eagle's medium (DMEM; Invitrogen) supplemented with 10% heat-inactivated fetal calf serum (Gemini) and 100 μg/ml penicillin/streptomycin (Invitrogen) at 37°, 5% CO_2.

Bacterial strains: the wild-type *S. enterica* serovar *typhimurium* (*S. typhimurium*) strain SL1344 is cultured to an invasion-competent state by inoculating 3 ml Luria broth (LB) with a freshly streaked colony and growing overnight ($\approx$16 h) at 37° on a rotating wheel ($\approx$20–30 rpm). Cultures are diluted in 2.5 ml LB containing 0.3 M NaCl, to an $OD_{600 \text{ nm}}$ of $\approx$0.125 and are grown on a rotating wheel at 37° until an $OD_{600 \text{ nm}}$ of 0.9, prior to immediate use for infection.

GST-PAK CRIB: the Cdc42/Rac-binding domain of PAK (amino acids 67–150) fused to GST is expressed and purified by binding to glutathione agarose beads as described previously (Benard and Bokoch, 2002).

GST-ELMO: a RhoG-binding fragment of ELMO-2 (amino acids 1–362) fused to GST is expressed and purified by binding to glutathione agarose beads as described previously (Prieto-Sanchez *et al.*, 2006).

PAK CRIB pull-down lysis buffer (PAK LB): 50 mM Tris-HCl, pH 7.6, 200 mM NaCl, 1% (v/v) Triton X-100, 5% (v/v) glycerol, 10 mM MgCl$_2$, 1 mM phenylmethylsulfonyl fluoride (PMSF), 1× complete protease inhibitor cocktail (Roche).

ELMO pull-down lysis buffer (ELMO LB): 20 mM Tris-HCl, pH 7.6, 150 mM NaCl, 0.5% (v/v) Triton X-100, 5 mM MgCl$_2$, 1 mM dithiothreitol (DTT), 5 mM β-glycerophosphate, 1× complete protease inhibitor cocktail.

Wash buffer: 50 mM Tris-HCl, pH 7.6, 150 mM NaCl, 1% (v/v) Triton X-100, 1 mM PMSF, 1× complete protease inhibitor cocktail.

Antibodies: endogenous levels of Cdc42 and Rac1 protein are probed using rabbit anti-Cdc42 (Santa Cruz Biotechnology) and mouse anti-Rac1 (clone 23A8, Upstate Biotechnology) antibodies. Because of a lack of suitable antibodies for endogenous RhoG, activation of this GTPase is determined by assaying levels of ectopically expressed flag-tagged RhoG using mouse anti-Flag M2 (Sigma).

2.2. Procedure

To assay activation of Cdc42 and Rac upon *Salmonella* infection, COS cells are seeded in 10-cm dishes overnight to $\approx$80% confluency. In general, one 10-cm dish is sufficient to analyze activation of both GTPases at a selected time point postinfection. Cells are washed twice with 5 ml prewarmed Hanks buffered saline solution (HBSS; Invitrogen) and allowed to equilibrate for 10 min in 5 ml warm HBSS at 37°, 5% CO_2 prior to infection with invasion-competent *S. typhimurium*. Bacteria, resuspended in warm HBSS, are added to cells at a multiplicity of infection (MOI) of $\approx$50 bacteria/cell, and infection is allowed to proceed by incubation at 37°, 5% CO_2. At desired time points, cell extracts are prepared by washing cells twice with ice-cold HBSS and lysing cells on ice by scraping into 1 ml PAK LB. Crude lysates are clarified by centrifugation at 16,000 g for 10 min at 4°.

An aliquot of the cleared lysate ($\approx$50 μl) is saved to assess total GTPase levels while the remainder is divided equally into two aliquots to measure levels of active Cdc42 and Rac. Cleared cell lysates should be used immediately for affinity precipitation by adding $\approx$20 μg GST-PAK CRIB glutathione agarose beads and rotating for 1 h at 4°. If multiple dishes are assayed for different time points, then it is important to maintain an equal volume for each pull down, diluting samples with PAK LB if necessary. Beads are collected by centrifugation at 1500 g, 4° for 2 min and washed gently four times with 1 ml cold wash buffer. Following the final wash, the wash buffer is removed carefully and the agarose pellet is resuspended in $\approx$20 μl Laemmli sample buffer containing 100 mM DTT. Samples are heated at 95° for 10 min and resolved by SDS-PAGE on a 13% polyacrylamide gel, together with aliquots of total lysate to determine both active and total GTPase levels, respectively. The separated proteins are transferred to polyvinylidene difluoride (PVDF) membranes for immunodetection with anti-Cdc42 or anti-Rac1 antibodies. Immunoblots are visualized using chemiluminescence reagents (Pierce) and can be quantified by scanning X-ray blots, provided that the detection system is in the linear range (this can be verified by ensuring linearity of twofold sample dilutions).

Salmonella-induced RhoG activation is determined using a similar affinity pull-down approach, with minor modifications. First, given the lack of a suitable antibody to detect endogenous RhoG, COS cells are transfected using Lipofectamine 2000 (Invitrogen) 1 day prior to infection with an expression plasmid encoding flag-tagged RhoG. Furthermore, activated RhoG is isolated by lysing infected cells in ELMO LB and affinity precipitation using GST-ELMO glutathione agarose as the affinity probe.

Salmonella-induced activation of Cdc42, Rac, and RhoG is evident as early as 10 min, reaching its peak 20–30 min postinfection (Fig. 11.1). In the case of RhoG, activation persists for up to 45 min after infection. The precise kinetics of activation depends greatly on the "invasion competency" of the bacteria, which is a function of the expression of the invasion-associated type III secretion system. To monitor infection and *S. typhimurium* "invasion competency," cells can be observed *briefly* during the assay by phase-contrast microscopy to ensure that the bacteria induce membrane ruffling.

2.3. Troubleshooting

A number of variables may affect the magnitude of GTPase activation induced by *S. typhimurium*. In addition to the invasion competence of bacteria, a major obstacle to this assay is excessive GAP activity in cell extracts. It is recommended that samples are processed rapidly and cell lysates are maintained at 4° throughout the pull down to minimize GTP hydrolysis of the target GTPase. If necessary, for example, during a time course experiment, a shorter centrifugation time (2 min) may be used to

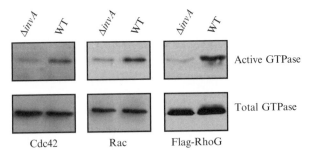

Figure 11.1 Activation of Cdc42, Rac, and RhoG upon infection with *S. typhimurium*. COS-1 cells were infected with wild-type (WT) *S. typhimurium* or its isogenic *invA* mutant (defective in type III secretion and thus unable to induce actin cytoskeletal rearrangements or bacterial entry) for 20 min (Cdc42 and Rac) or 30 min (RhoG). (Top) Relative levels of activated endogenous Cdc42 and Rac and ectopically expressed flag tagged RhoG as determined by affinity pull-down assays using GST-PAK CRIB (Cdc42 and Rac) or GST-ELMO (RhoG), respectively. (Bottom) Total GTPase levels in cell lysates.

clarify crude lysates and make processing of multiple samples easier. Another important consideration is the detergent composition of the lysis buffer. Different cell types may require different detergents to achieve sufficient protein extraction and solubilization. Moreover, caution should be observed to ensure that detergents do not disrupt specific binding of GTPases during affinity precipitation (Benard and Bokoch, 2002).

3. DETERMINING THE ROLE OF RHO GTPASES IN *S. TYPHIMURIUM*-INDUCED ACTIN REMODELING

The role of Rho family GTPases in diverse cellular processes traditionally entails blocking their function by overexpression of dominant interfering mutants within the cell (Feig, 1999). Although this approach is very powerful and yields significant functional data regarding GTPase activity *in vivo*, concerns regarding its specificity have arisen (Czuchra *et al.*, 2005; Prieto-Sanchez and Bustelo, 2003; Wennerberg *et al.*, 2002). Indeed, as shown previously (Patel and Galan, 2006), dominant interfering mutants of Cdc42 (^{N17}Cdc42) drastically reduce *Salmonella*-induced Rac activation by affinity pull-down assays. The advent of RNA interference (RNAi) technology has offered a viable alternative to selectively deplete and hence block endogenous GTPase function, given that appropriate controls to assess potential "off-target" effects are observed. The principles of RNAi, together with rationale for experimental design and controls, have been reviewed extensively elsewhere (Endo *et al.*, 2006; Pei and Tuschl, 2006). The following protocol describes a microscopy-based assay using RNAi to examine the roles of Cdc42, Rac1, and RhoG in *Salmonella*-induced actin remodeling.

3.1. Reagents

Henle-407 cells: a human intestinal epithelial cell line (available at ATCC) used as host cells for *S. typhimurium* infection. Cells are cultured in DMEM supplemented with 10% heat-inactivated bovine calf serum (BSA; Hyclone) and 100 μg/ml penicillin/streptomycin at 37°, 5% CO_2.

Bacterial strains: *S. typhimurium* cultured as described earlier.

RNAi constructs: small interfering RNA (siRNA) or short hairpin RNA (shRNA) constructs engineered to silence the expression of endogenous Rho family GTPases are readily available commercially or have been described extensively. We employ the shRNA approach to deplete endogenous Cdc42 using a hairpin sequence targeting nucleotides 296–318 of human Cdc42. Oligonucleotide pairs A (ACAGTGGTGAGTTATCT-CAGGAAGCTTGCTGAGATAACTCACCACTGTCCAT TTTTT) and B (GATCAAAAAATGGACAGTGGTGAGTTATCTCAGCA AGCTTCCTGAGATAACTCACCACTGTCG) are annealed and ligated into the vector pSHAG (a gift from G. Hannon, Cold Spring Harbor Laboratory, Cold Spring Harbor, NY; Hannon and Conklin, 2004). Attempts to knockdown endogenous Rac1 and RhoG using a similar vector-based approach proved unsuccessful. Therefore, to silence Rac1 and RhoG gene expression, we use synthetic SMARTpools, each comprising four proprietary siRNA sequences (Dharmacon). A control nontargeting siRNA pool (siCONTROL; Dharmacon) is used as a negative control.

Lipofectamine 2000: transfection reagent (Invitrogen).

Microscope: we utilize an inverted Nikon Eclipse TE2000-U microscope equipped with oil immersion plan Apo 60× NA 1.4, 100× NA 1.4 or dry plan Fluor 40× NA 0.6 objectives and a charge-coupled device (CCD) camera (MicroMAX RTE/CCD-1300Y, Princeton Instruments). Acquisition and analysis are performed using MetaMorph imaging software (version 6.1, Universal Imaging Corp.).

3.2. Procedure

One day prior to transfection, Henle cells are seeded onto glass coverslips (13 mm) in a 24-well plate (Falcon) at a density of 5×10^4/well in culture media without antibiotics. The absence of antibiotics is critical to increase transfection efficiency. Cells at ≈60 to 70% confluency are cotransfected with a GTPase-specific RNAi construct and an expression plasmid encoding GFP to identify transfected cells by microscopy. When working on a 24-well format, we routinely use 200 ng shRNA plasmid DNA or 1 pmol siRNA + 40 ng pEGFP diluted in 50 μl serum-free media (SFM) per single well. Lipofectamine 2000 is diluted (1 μl in 50 μl SFM) prior to mixing gently with the RNAi/GFP solution. Once mixed, samples are incubated at room temperature for 30 min to generate transfection complexes, before adding drop wise to cells.

Cells are maintained at 37°, 5% CO_2 for a further 36 to 48 h to promote silencing and turnover of endogenous target GTPases. The RNAi efficacy for knockdown is verified at gene and protein levels by quantitative real-time polymerase chain reaction (qRT-PCR) and Western blotting (Fig. 11.2C). To perform a *S. typhimurium*-induced cytoskeletal remodeling assay, cells are equilibrated into warm HBSS for 10 min at 37°, 5% CO_2 prior to infection with invasion-competent bacteria. *S. typhimurium* is added in warm HBSS at a MOI of 30, and infection is allowed to proceed at 37°

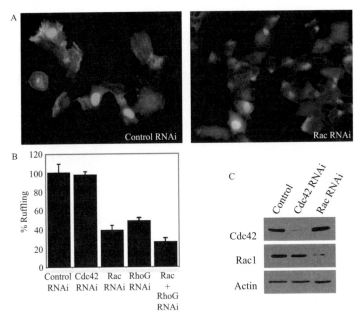

Figure 11.2 *Salmonella*-induced actin remodeling requires Rac and RhoG. (A) Henle-407 cells were cotransfected with GFP (to detect transfected cells) and RNAi constructs targeting Cdc42, Rac, or RhoG as indicated. Two days post-transfection, cells were infected with wild-type *S. typhimurium* for 30 min and stained with TRITC-conjugated phalloidin to visualize remodeling of the actin cytoskeleton. (B) The percentage of transfected (GFP positive) cells displaying actin rearrangements as a consequence of bacterial infection was enumerated and standardized considering the number of transfected cells exhibiting actin-rich ruffles in RNAi control transfections to be 100%. Results represent the mean ± SD of three independent experiments. (C) Western blot showing RNAi efficacy. Cell lysates were prepared 2 days post-transfection with RNAi constructs or a mock control, and protein levels were analyzed using antibodies directed against endogenous Cdc42, Rac, and actin. (See color insert.)

for 30 min. This is sufficient time for bacteria to initiate cytoskeletal remodeling, clearly evident by phase-contrast microscopy as membrane ruffling. The assay is terminated by washing cells twice with 1 ml cold phosphate-buffered saline (PBS) and fixing with freshly prepared 4% paraformaldehyde (PFA)/PBS for 20 min. To visualize actin remodeling by fluorescence microscopy, cells are permeabilized with 0.1% Triton X-100 for 5 min at room temperature followed by incubation with ammonium chloride (2 mg/ml PBS) for 10 min to minimize autofluorescence. Cells are washed three times with 1 ml PBS and processed for immunofluorescence staining.

We routinely employ tetramethylrhodamine isothiocyanate (TRITC)-conjugated phallodin (Sigma) to visualize polymerized F-actin and 4′,6-diamidino-2-phenylindole (DAPI; Sigma) to detect bacteria. The latter also stains cellular nuclei, which, if necessary, may be avoided by specifically labeling bacteria with rabbit anti-*S. typhimurium* lipopolysaccharide (Difco Laboratories) followed by Alexa350-conjugated anti-rabbit IgG. Indeed, the only restriction when staining is to avoid the green ($\approx$488-nm emission) channel, which is used to detect the GFP expressing RNAi-transfected cells. All incubations are carried out at room temperature, in the dark, for 45 min, by inverting coverslips onto 75-μl antibody solution (diluted in 1% BSA/PBS) drops. Coverslips are washed between each staining step by sequentially dipping three times into PBS, each time draining excess PBS. Following the final PBS wash, coverslips are further washed by dipping twice into ddH$_2$O and are mounted by inverting onto ProLong Gold antifade mounting media (Molecular Probes) on microscopy slides.

The actin remodeling elicited by *S. typhimurium* is manifested as a dense accumulation of phalloidin staining (F-actin), specifically at the sites of bacteria and host cell contact (see Fig. 11.2A). Under closer inspection, for instance by confocal microscopy, bacteria are evident at the base of an actin rich "ruffle," which protrudes from the apical cell surface. In order to quantify the cytoskeletal remodeling phenotype, the percentage of transfected cells, identified by GFP expression, displaying actin rearrangements as a consequence of bacterial infection is determined and standardized using the values obtained in control cells transfected with nontargeting siRNA (see Fig. 11.2B). By this measure, cells depleted of Cdc42 following RNAi are indistinguishable from control cells. In contrast, depletion of Rac1 or RhoG significantly reduces bacterial-induced actin remodeling, with a more pronounced effect seen upon simultaneous depletion of both these GTPases.

3.3. Troubleshooting

As in other infection experiments, it is essential to ensure the invasion competence of the bacteria. Moreover, the confluency of host cells at the time of infection should ideally be $\approx$80 to 90%. This may require some trial and error, as transfection and RNAi requirements dictate that cells are seeded ≈3 days prior to infection. It is important to proceed rapidly once *Salmonella* are ready for infection and maintain cells at 37° throughout the infection period. In general, *Salmonella*-induced actin ruffling should be clearly seen on ≈70 to 80% of control transfected cells. If necessary, an increased MOI can be used to promote increased actin remodeling.

When discerning Rho GTPase function, it is critical to verify the efficacy of their RNAi-mediated depletion (see Fig. 11.2C). This is a function of both the transfection efficiency, which may vary greatly between cell types, confluency and passage number of cells, and the efficacy of individual RNAi constructs. Whenever possible, cells should be seeded in duplicate to directly correlate RNAi knockdown of target GTPases by qRT-PCR and Western blotting with *S. typhimurium*-induced ruffling.

4. Determining the Role of Rho GTPases in *S. typhimurium* Internalization

The actin cytoskeletal rearrangements triggered by *Salmonella* are an essential prerequisite to bacterial internalization. Because the two events are tightly coupled, bacterial entry can serve as an alternative assay to complement and validate results gained from a ruffling assay. A variety of techniques have been developed to determine bacterial entry. The majority of these rely on differential staining protocols to directly discriminate internalized bacteria from those that are cell associated. The following protocol describes a variation on this theme, in which *Salmonella* carrying an inducible fluorescent marker are used to infect host cells. The assay involves stimulating fluorescence selectively within internalized bacteria following treatment of cells with gentamicin, a membrane-impermeable antibiotic that kills external bacteria by inhibiting protein synthesis. This strategy strongly correlates with the traditional approach of inside/outside staining of bacteria pre- and postcell permeabilization with different fluorophores (Van Putten *et al.*, 1994) and benefits from requiring only a single fluorescent channel. Together with RNAi, it can be used as a microscopy-based assay to determine the requirements for Rho family GTPases in *Salmonella* entry.

4.1. Reagents

Cell lines, *S. typhimurium*, and RNAi constructs are used as described earlier. To label *S. typhimurium* fluorescently, a bacterial strain is used that carries the plasmid pBAD DsRED T3.S4T (a generous gift from D. Bumann, Max-Planck Institute, Berlin, Germany) (Sorensen *et al.*, 2003), an expression vector that encodes the bacterially optimized fluorescent protein DsRED under the control of an arabinose-inducible promoter. Overnight but *not* subcultures of *Salmonella* are grown in the presence of ampicillin (100 μg/ml) to select for the plasmid-encoded marker.

Gentamicin: purchased from Sigma. Prepare fresh as a 100-mg/ml stock solution in ddH$_2$O.

L-(+)-Arabinose: purchased from Sigma. Prepare fresh as a 10% stock solution in ddH$_2$O.

4.2. Procedure

Henle cells are seeded onto coverslips and transfected with RNAi constructs as described previously. Following sufficient incubation (36–48 h) to deplete target GTPases, cells are equilibrated into warm HBSS and infected at an MOI of 20 with invasion-competent *S. typhimurium*/pBAD DsRED for 45 min at 37°, 5% CO$_2$. A longer incubation time than that used for cytoskeletal remodeling is used to ensure efficient bacterial entry. Cells are washed twice with 1 ml warm HBSS, and noninternalized bacteria are killed by incubating cells for 45 min in culture media containing 100 μg/ml gentamicin. To selectively visualize internalized bacteria, the culture medium is replaced with media containing 100 μg/ml gentamicin + 0.1% L-arabinose, and infection is allowed to proceed for a further 2 h. The arabinose chase may be shortened if necessary, as DsRED expression and detection of internalized bacteria are evident >1 h postinduction. However, extended chase times (>3–4 h) should be avoided to ensure against excessive bacterial replication. Finally, cells are washed twice with 1 ml cold PBS, fixed with freshly prepared 4% PFA/PBS for 20 min, and mounted on slides for analysis by immunofluorescence microscopy.

As with the actin-remodeling assay, RNAi-transfected cells are identified by coexpression of GFP. Therefore, the entry phenotype is scored based on the proportion of GFP expressing cells displaying internalized, that is, DsRED positive, *Salmonella* (Fig. 11.3A). As before, values are standardized considering the percentage of internalized bacteria in control siRNA-transfected cells to be 100% (see Fig. 11.3B). Depletion of Cdc42 has little effect on bacterial internalization, whereas depletion of Rac1 and RhoG by RNAi impairs *Salmonella* entry significantly.

4.3. Troubleshooting

This assay is based on the ability to score internalized bacteria unambiguously. A failure to kill external bacteria completely prior to inducing DsRED expression with arabinose can compromise its validity severely. Hence, it is important to ensure that gentamicin and arabinose solutions are freshly prepared. Another variable that may affect DsRED induction within internalized bacteria is excessive cell confluency. Thus, it is good

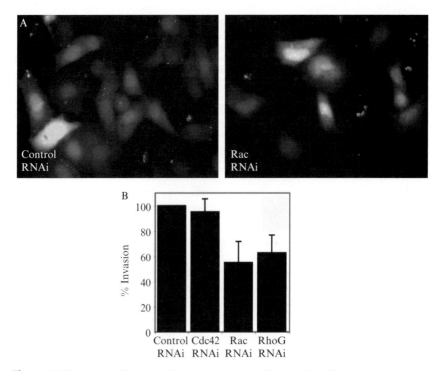

Figure 11.3 *Salmonella* internalization into non-phagocytic cells requires Rac and RhoG. (A) Intestinal Henle-407 cells were cotransfected with GFP (to detect transfected cells) and RNAi constructs targeting Cdc42, Rac, or RhoG as indicated. Two days post-transfection, cells were infected with wild-type DsRED *S. typhimurium* as described in the text. DsRED-positive *Salmonella* represent internalized bacteria only, as the expression of DsRED is induced once external bacteria are killed by gentamicin treatment. (B) The percentage of transfected (GFP positive) cells displaying internalized DsRED bacteria was enumerated and standardized considering the number of transfected cells with internalized bacteria in RNAi control transfections to be 100%. Results represent the mean ± SD of three independent experiments. (See color insert.)

practice to initially ensure that results parallel those seen by an alternative labeling method, such as inside/out staining. If necessary, variables, such as the duration of the gentamicin and arabinose chase, may need to be optimized for different conditions. This assay may also be coupled to fluorescence-activated cell sorting to simplify quantitation.

 ## 5. SUMMARY

Rho family GTPases play a critical role during *Salmonella* infection. This chapter focused on the Rho family GTPase-dependent events that lead to bacterial internalization and described several complementary methods to showcase the remarkable adaptations that *Salmonella* has evolved to carefully modulate the Rho GTPase switch.

ACKNOWLEDGMENT

This work was supported by Public Health Service Grant AI 055472 from the National Institutes of Health to J.E.G.

REFERENCES

Bakshi, C. S., Singh, V. P., Wood, M. W., Jones, P. W., Wallis, T. S., and Galyov, E. E. (2000). Identification of SopE2, a *Salmonella* secreted protein which is highly homologous to SopE and involved in bacterial invasion of epithelial cells. *J. Bacteriol.* **182,** 2341–2344.

Benard, V., and Bokoch, G. M. (2002). Assay of Cdc42, Rac, and Rho GTPase activation by affinity methods. *Methods Enzymol.* **345,** 349–359.

Bustelo, X. R., Sauzeau, V., and Berenjeno, I. M. (2007). GTP-binding proteins of the Rho/Rac family: Regulation, effectors and functions *in vivo*. *Bioessays* **29,** 356–370.

Chen, L. M., Hobbie, S., and Galan, J. E. (1996). Requirement of CDC42 for *Salmonella*-induced cytoskeletal and nuclear responses. *Science* **274,** 2115–2118.

Cossart, P., and Sansonetti, P. J. (2004). Bacterial invasion: The paradigms of enteroinvasive pathogens. *Science* **304,** 242–248.

Czuchra, A., Wu, X., Meyer, H., van Hengel, J., Schroeder, T., Geffers, R., Rottner, K., and Brakebusch, C. (2005). Cdc42 is not essential for filopodium formation, directed migration, cell polarization, and mitosis in fibroblastoid cells. *Mol. Biol. Cell* **16,** 4473–4484.

Endo, Y., Even-Ram, S., Pankov, R., Matsumoto, K., and Yamada, K. M. (2006). Inhibition of rho GTPases by RNA interference. *Methods Enzymol.* **406,** 345–361.

Etienne-Manneville, S., and Hall, A. (2002). Rho GTPases in cell biology. *Nature* **420,** 629–635.

Feig, L. A. (1999). Tools of the trade: Use of dominant-inhibitory mutants of Ras-family GTPases. *Nat. Cell Biol.* **1,** E25–E27.

Fu, Y., and Galan, J. E. (1999). A *Salmonella* protein antagonizes Rac-1 and Cdc42 to mediate host-cell recovery after bacterial invasion. *Nature* **401,** 293–297.

Galan, J. E. (2001). *Salmonella* interactions with host cells: Type III secretion at work. *Annu. Rev. Cell Dev. Biol.* **17,** 53–86.

Hannon, G. J., and Conklin, D. S. (2004). RNA interference by short hairpin RNAs in vertibrate cells, *Methods Mol. Biol.* **257,** 255–256.

Hardt, W. D., Chen, L. M., Schuebel, K. E., Bustelo, X. R., and Galan, J. E. (1998). *S. typhimurium* encodes an activator of Rho GTPases that induces membrane ruffling and nuclear responses in host cells. *Cell* **93,** 815–826.

Moon, S. Y., and Zheng, Y. (2003). Rho GTPase-activating proteins in cell regulation. *Trends Cell Biol.* **13,** 13–22.

Munter, S., Way, M., and Frischknecht, F. (2006). Signaling during pathogen infection. *Sci. STKE* **2006,** re5.

Patel, J. C., and Galan, J. E. (2005). Manipulation of the host actin cytoskeleton by *Salmonella*: All in the name of entry. *Curr. Opin. Microbiol.* **8,** 10–15.

Patel, J. C., and Galan, J. E. (2006). Differential activation and function of Rho GTPases during *Salmonella*-host cell interactions. *J. Cell Biol.* **175,** 453–463.

Pei, Y., and Tuschl, T. (2006). On the art of identifying effective and specific siRNAs. *Nat. Methods* **3,** 670–676.

Prieto-Sanchez, R. M., Berenjeno, I. M., and Bustelo, X. R. (2006). Involvement of the Rho/Rac family member RhoG in caveolar endocytosis. *Oncogene* **25,** 2961–2973.

Prieto-Sanchez, R. M., and Bustelo, X. R. (2003). Structural basis for the signaling specificity of RhoG and Rac1 GTPases. *J. Biol. Chem.* **278,** 37916–37925.

Ren, X. D., and Schwartz, M. A. (2000). Determination of GTP loading on Rho. *Methods Enzymol.* **325,** 264–272.

Rossman, K. L., Der, C. J., and Sondek, J. (2005). GEF means go: Turning on RHO GTPases with guanine nucleotide-exchange factors. *Nat. Rev. Mol. Cell Biol.* **6,** 167–180.

Rottner, K., Stradal, T. E., and Wehland, J. (2005). Bacteria-host-cell interactions at the plasma membrane: Stories on actin cytoskeleton subversion. *Dev. Cell* **9,** 3–17.

Sorensen, M., Lippuner, C., Kaiser, T., Misslitz, A., Aebischer, T., and Bumann, D. (2003). Rapidly maturing red fluorescent protein variants with strongly enhanced brightness in bacteria. *FEBS Lett.* **552,** 110–114.

Stender, S., Friebel, A., Linder, S., Rohde, M., Mirold, S., and Hardt, W. D. (2000). Identification of SopE2 from *Salmonella typhimurium*, a conserved guanine nucleotide exchange factor for Cdc42 of the host cell. *Mol. Microbiol.* **36,** 1206–1221.

Van Putten, J. P., Weel, J. F., and Grassme, H. U. (1994). Measurements of invasion by antibody labeling and electron microscopy. *Methods Enzymol.* **236,** 420–437.

Wennerberg, K., Ellerbroek, S. M., Liu, R. Y., Karnoub, A. E., Burridge, K., and Der, C. J. (2002). RhoG signals in parallel with Rac1 and Cdc42. *J. Biol. Chem.* **277,** 47810–47817.

CHAPTER TWELVE

CONTRIBUTION OF CDC42 TO CHOLESTEROL EFFLUX IN FIBROBLASTS FROM TANGIER DISEASE AND WERNER SYNDROME

Ken-ichi Hirano,[*] Chiaki Ikegami,[†] and Zhongyan Zhang[‡]

Contents

Abstract

Atherosclerotic cardiovascular disease is a life-threatening disorder. Cholesterol efflux from the cells is the rate-limiting step in regulating the intracellular cholesterol content as well as raft structure in the plasma membrane. The defect of cholesterol efflux leads to the development of atherosclerosis. Tangier disease (TD), a hereditary high-density lipoprotein deficiency, is characterized by the presence of defective cellular cholesterol efflux. Using the cDNA subtraction technique, we found that expression of Cdc42 was decreased markedly in fibroblasts and macrophages from patients with TD. Madin–Darby canine

* Department of Cardiovascular Medicine, Graduate School of Medicine, Osaka University, Osaka, Japan
† Department of Molecular Cardiology, Whitaker Cardiovascular Institute, Boston University School of Medicine, Boston, MA, USA
‡ Department of Cell Biology, Johns Hopkins University School of Medicine, Baltimore, Maryland

Methods in Enzymology, Volume 439
ISSN 0076-6879, DOI: 10.1016/S0076-6879(07)00412-0

kidney cells expressing the dominant-negative form of Cdc42 had a reduced lipid efflux; inversely, cells expressing the active form had increased efflux. Furthermore, we found that cellular lipid efflux was defective and Cdc42 was reduced in fibroblasts from a premature aging disorder, Werner syndrome. Complementation experiments using an adenovirus carrying wild-type Cdc42 successfully corrected impaired lipid efflux in Werner syndrome cells. We concluded that Cdc42 may play important roles in cellular cholesterol efflux and that dysregulation of this type of RhoGTPase might lead to the development of atherosclerotic cardiovascular disease.

1. INTRODUCTION

Atherosclerotic cardiovascular disease, including myocardial infarction and stroke, is one of the major causes of death in well-developed countries. Reverse cholesterol transport is the formulated concept for the protective system against atherosclerosis. In reverse cholesterol transport, high–density lipoproteins (HDL) are thought to play an important role as shuttles carrying cholesterol from lipid-laden cells in the arterial walls to the liver, a terminal of reverse cholesterol transport, where cholesterol is catabolized (Hirano *et al.*, 2004). The initial step of reverse cholesterol transport is called "cholesterol efflux," where small and lipid-poor HDL or free apolipoprotein (apo) AI removes cholesterol from the cells.

Cholesterol efflux is a rate-limiting step of intracellular cholesterol as well as cholesterol on the cell surface. Many studies have indicated that the loss of cholesterol efflux causes the accumulation of cholesterol esters and foam cell formation. Tangier disease (TD), of which cause is mutations in the ATP-binding cassette transporter-A1 (ABCA1) gene, is a kind of model for impaired cholesterol efflux. Patients with TD suffer from atherosclerotic cardiovascular diseases. We have reported that cells from TD patients and ABCA1 knockout mice show an accumulation of lipid rafts on the cell surface, which may lead to the increased secretion of tumor necrosis factor-α (Koseki *et al.*, 2007).

Enhancement of cholesterol efflux from cells is one of the novel strategies for antiatherosclerotic treatment, which would make it possible to regress and stabilize cholesterol-rich plaque in atherosclerotic lesions. However, little is known about the molecular mechanism for cholesterol efflux. Therefore, we have been trying to elucidate the detailed mechanism for cholesterol efflux by patient- and disease-oriented approaches. During these studies, we have found a contribution of Cdc42, a RhoGTPase, to cellular cholesterol efflux in some disease and pathological conditions, including Tangier disease, a hereditary HDL deficiency, and Werner syndrome (WS), a premature aging disorder.

2. cDNA Subtraction Technique Reveals Decreased Expression of Cdc42 in Macrophages from a Patient with TD

As mentioned earlier, TD, a rare lipoprotein disorder, may be a model for the impairment of cholesterol efflux. In order to elucidate the pathophysiology of TD, we performed cDNA subtraction using a polymerase chain reaction-select subtraction kit (Clontech, Palo Alto, CA). We used monocyte-derived macrophages as a source for RNA.

2.1. Isolation of human monocyte-derived macrophages and cell culture

1. Obtain the approval of the ethical committee at the university.
2. Obtain informed consent from the patient and control subjects.
3. Prewarm lymphocyte separation solution (Nacalai Tesque, Kyoto, Japan) and RPMI (Nacalai Tesque).
4. Draw 60 ml of whole blood and isolate mononuclear cells as follows.
5. Add 15 ml of lymphocyte separation solution in two 50-ml tubes.
6. Overlay 30 ml of whole blood very gently by pipetting into each tube.
7. Centrifuge the tube (400 g, 20 to 30 min).
8. Obtain the layer containing lymphocytes, monocytes, and platelets by gentle pipetting and put them into a 50-ml tube containing 40 ml of RPMI only.
9. Mix gently by pipetting (do not vortex).
10. Centrifuge the tube (400 g, 5 min).
11. Discard the supernatant, add another 50 ml of RPMI only, and mix by gentle pipetting.
12. Repeat steps 10 and 11 again.
13. Suspend the cells with RPMI/10% human AB serum.
14. Plate cells onto the appropriate culture dishes (multiwell Primaria 24 well or 6 well, Becton Dickinson) and incubate cells at 37 ° for 1 h.
15. After incubation, discard the medium by pipetting (do not aspirate, as the attachment of cells is still weak).
16. Add the appropriate amount of RPMI only and discard in order to avoid cells that do not attach to the bottom.
17. Repeat step 16 again.
18. Add RPMI/10% AB serum and incubate cells at 37° in a CO_2 incubator for 5 to 10 days.

After a week of culture, cells are harvested and total RNA is extracted. After evaluating the quality of total RNA obtained, cDNA subtraction is performed according to the manufacturer's protocol. After subtraction, sequencing analyses revealed that Cdc42 may be decreased in cells from the TD patient. We confirmed the mRNA and protein expression levels of Cdc42 in macrophages by Northern and Western blot analyses, respectively. Furthermore, we analyzed the reduced expression in fibroblasts from TD (Hirano *et al.*, 2000). Diederich *et al.* (2001) also reported that Cdc42 was decreased in macrophages obtained from their TD patients.

In addition, levels of RhoA, RhoB, RhoG, and Rac1, which are other RhoGTPases, appeared to be elevated in fibroblasts from TD patients (Utech *et al.*, 2001). The underlying mechanism and significance still remain to be investigated.

3. Cdc42 Plays Some Role in Cholesterol Efflux from Cells

In order to know whether Cdc42 plays some role in cholesterol efflux, we tested cellular cholesterol efflux from Madin–Darby canine kidney (MDCK) cells expressing dominant-active (V12Cdc42) or -negative (N17Cdc42) forms of Cdc42, developed by Kodama *et al.* (1999).

3.1. Cholesterol and phospholipid efflux assay

1. Culture Cos7, MDCK, or HEK293 cells in Dulbecco's modified Eagle's medium supplemented with 10% fetal calf serum (FCS), 100 units/ml of penicillin, and 100 μg/ml of streptomycin. Culture fibroblasts in modified Eagle's medium (MEM). Incubate human monocyte-derived macrophages in RPMI 1640 with 10% human type AB serum.
2. Label subconfluent passaged cells with 1 μCi/ml [^{3}H]cholesterol (NEN Life Science Products, Inc., Boston, MA) or 4 μCi/ml [*methyl*-^{3}H] choline-chloride (Amersham Biosciences, Uppsala, Sweden) (Zanotti *et al.*, 2006) for 18 to 24 h with 2 mg/ml of an acyl CoA:cholesterol acyltransferase (ACAT) inhibitor, F-1349 (Fujirebio, Tokyo, Japan), in corresponding medium supplemented with 5% FCS, at 37° in a CO_2 incubator. Incubate human monocyte-derived macrophages for 48 h in RPMI 1640 containing 10% human type AB serum with or without 50 μg/ml AcLDL (Matsuura *et al.*, 2005).
3. After labeling, wash cells three to five times with phosphate-buffered saline (PBS) and equilibrate at 37° for 2 or 18 h in phospholipid efflux (Zanotti *et al.*, 2006), with medium plus 0.1% essential fatty acid-free bovine serum albumin (BSA) (Hou *et al.*, 2007; Nishida *et al.*, 2002; Zanotti *et al.*, 2006).

4. In the case of Cos7 and MDCK cells, replace the medium with fresh medium/0.1% BSA with or without the presence of 20 μg/ml of apoA1 or 100 μg/ml of HDL3. Incubate cells for 4 to 6 h at 37°. For fibroblasts and human macrophages, add 10 μg/ml of apoA1 or 50 μg/ml of HDL3 to the medium and incubate cells overnight (Zhang *et al.*, 2005).

5. After incubation, filter the acceptor-containing medium through a 0.45-μm filter (Millex-HV) to remove floating cells and debris and then measure for radioactivity by liquid scintillation counting. Dissolve the remaining cells in 0.1 *N* NaOH, and determine the residual radioactivity in cells (Hirano *et al.*, 2000; Ohama *et al.*, 2002).

6. Percentage efflux is the ratio of counts in the medium to the sum of radioactivities in the cells plus medium.

7. In all experiments, fractional efflux is corrected for the small amount of radioactivity released to medium/0.1% BSA without an acceptor (Zhang *et al.*, 2001). Carry out each experiment in quadruplicate.

As a result, cholesterol efflux is increased in MDCK cells expressing the dominant-active form and decreased in cells expressing the negative form (Hirano *et al.*, 2000).

4. Cdc42 Binds with a Gate Player for Cholesterol Efflux, ABCA1

In order to know the mechanism of how Cdc42 plays a role in cholesterol efflux, we examined the intracellular events though Cdc42 signaling. First we assessed the hypothesis that Cdc42 binds with ABCA1, an essential gate player for cholesterol efflux.

1. Transfect plasmids containing myc-Cdc42 and/or FLAG-tagged-ABCA1 into Cos-7 cells using the Lipofectamine 2000 reagent according to the manufacturer's protocol.

2. Two days after transfection, lyse cells with lysis buffer containing 150 m*M* NaCl, 40 m*M* Tris-HCl (pH 7.4), 5 m*M* $MgCl_2$, protease inhibitors, and detergent.

3. Rotate cell lysates overnight at 4° to solubilize the protein.

4. Perform immunoprecipitation using the anti-FLAG immunoprecipitation kit (Sigma).

5. Briefly, centrifuge cell lysates at 800 *g* for 20 min to remove detergent-insoluble fractions.

6. Mix agarose beads covalently conjugated with the anti-FLAG M2 antibody.

7. Rotate the mixture for 2 h at 4°.

8. Wash beads four times with Tris-buffered saline supplemented with 5 mM MgCl$_2$.
9. Eluate proteins with elution buffer containing 3xFLAG peptides.
10. Analyze samples before and after immunoprecipitation by Western blot.

As a result, myc-Cdc42 is coimmunoprecipitated with the anti-FLAG M2 antibody only when it is cotransfected with ABCA1-FLAG. Furthermore, we demonstrated the colocalization of Cdc42 and ABCA1 in overexpressing Cos-7 cells (Tsukamoto *et al.*, 2001). These results suggest that human ABCA1 can possibly interact with Cdc42.

Nofer *et al.* (2003, 2006) reported that Cdc42 is involved in apo AI-induced cholesterol efflux by the following mechanism. Apo AI activates Cdc42 and a stress kinase, JNK. The binding of apo AI with ABCA1 is required for this cellular event.

4.1. ABCA1 induces rearrangement of actin cytoskeleton possibly through Cdc42/neural Wiskott–Aldrich syndrome protein (N-WASP) pathway

We and others have reported that cells from patients with Tangier disease have abnormal cell shape as follows: larger in size, coarse actin fibers, and less filopodia formation (Drobnik *et al.*, 1998; Hirano *et al.*, 2000). Therefore, we assessed the hypothesis that ABCA1 may regulate the arrangement of actin cytoskeletons. Cells overexpressing ABCA1 were divided into the following two groups by distinct morphology with altered actin cytoskeletons: one had increased formation of filopodia and the other had long protrusions. When dominant-negative forms of Cdc42 and N-WASP lacking the verprolin homology domain were transfected, ABCA1-induced morphological changes were inhibited significantly. These data suggest that ABCA1 regulates actin organization through a possible interaction with Cdc42 (Tsukamoto *et al.*, 2001).

5. Role of Cdc42 in Intracellular Lipid Transport

From data given earlier, it was demonstrated that Cdc42 mediates cholesterol efflux from cells, possibly though apo AI-induced activation of Cdc42 and its related downstream effector molecules such as N-WASP and JNK. Next, we examined whether Cdc42 mediates intracellular lipid transport using fluorescent recovery after the photobleaching (FRAP) technique. FRAP is a powerful technique used to investigate the intracellular transport of lipids and proteins in living cells.

1. Plate cells on glass bottom culture dishes (35-mm dish, poly-D-lysine coated, MatTek Corporation).
2. After growing for 2 days, perform FRAP using laser-scanning microscopy (LSM510, Carl Zeiss).
3. Dissolve 5 mg of C6-NBD-ceramide (Molecular Probes) with 150 μl of distilled water (protected from light). Make one to hundred dilution of the ceremide by MEM only.
4. Wash cells with ice-cold PBS twice.
5. Add the diluted ceramide and incubate for 30 min on ice.
6. Wash cells with ice-cold PBS twice.
7. Add the complete growth medium and incubate at 37° for 30 min.
8. Subject cells to FRAP.
9. Focus a beam of light using 488-nm laser lines on the indicated part of C6-ceramide-positive regions in the living cells.
10. Typically, 20 to 25 iterations are required for almost complete photobleaching.
11. After the appropriate bleach pulse, monitor FRAP at the bleach area until 30 s after bleaching.
12. Measure relative fluorescence by dividing the fluorescence in the FRAP area by that in the reference spot.
13. The recovery is reasonably fit by a single exponential function.
14. After the values of relative fluorescence are plotted, calculate the time constant.

Orso et al. (2000) reported that the intracellular kinetics of Golgi-associated lipids is retarded in Tangier fibroblasts with reduced expression of Cdc42. We reported that fibroblasts from aged human subjects had a lower expression of Cdc42 and retarded intracellular kinetics of C6-ceramide. We also examined the effect of transfection of dominant active, negative, and wild-type Cdc42 on intracellular lipid transport detected by FRAP in fibroblasts. Results showed that the introduction of wild type and dominant active forms of Cdc42 recovered retarded FRAP in aged fibroblasts. Conversely, control fibroblasts infected with dominant negative Cdc42 exhibited significantly retarded FRAP (Tsukamoto et al., 2002).

The ceramide used in our study, C6-NBD-ceramide, is believed to be metabolized and accumulated in the Golgi apparatus and to be sorted and transported to the plasma membrane via vesicular transport. It is also thought that the kinetics of C6-NBD-ceramide is closely correlated with that of cholesterol for the following reasons: this kind of ceramide is an efficient substrate for glucosylsphingolipid and sphingolipids, both of which are known to be assembled together with cholesterol to form a cholesterol-sphingolipid microdomain or raft. The FRAP of ceramide can be affected by cholesterol depletion and repletion.

6. THE SIGNIFICANCE OF CDC42 IN WERNER SYNDROME

Werner syndrome is an autosomal recessive disorder that belongs to a category of diseases called human premature aging disorders (Brosh and Bohr, 2002; Chen and Oshima, 2002). Patients with WS suffer from malignant neoplasmas and premature atherosclerosis in their 40s. Cdc42, a member of the RhoGTPase family, has been shown to play an important role in intracellular cholesterol trafficking and its export from the cells, a terminal of reverse cholesterol transport, which is a major protective system against atherosclerosis (Zhang *et al.*, 2005). Cdc42 was originally identified as a molecule responsible for the budding of yeast, as well as regulating actin dynamics, cell cycle, transformation, and vesicular transportation (Hall, 1998; Takai *et al.*, 2001).

It has been demonstrated that abnormal intracellular lipid transport, marked reduction of apoA1- and HDL3-mediated cholesterol efflux, and increased cholesterol accumulation in WS fibroblasts go along with the decreased expression of Cdc42, which may be one of the possible crucial phenotypes for atherosclerosis. This phenotype was completely corrected by the introduction of wild-type Cdc42 (Zhang *et al.*, 2005).

The reduction of Cdc42 may cause abnormal intracellular lipid transport and lipid efflux, subsequently leading to the accumulation of cellular cholesterol. High levels of cholesterol could activate cholesterol esterification (Soccio and Breslow, 2004) and process the formation of foam cells, which form the initial morphological lesion, the fatty streak, of atherosclerosis. Cholesterol accumulation could increase the susceptibility of lipid peroxidation (Joseph *et al.*, 1996), the products of which might be able to damage cells long before the typical signs of atherosclerosis and aging (Cutler *et al.*, 2004; Johnson *et al.*, 1999; Tahara *et al.*, 2001). Therefore, it is possible that the decreased expression of Cdc42 could be involved in the expression of premature aging phenotypes in WS. As a result, the reduction of Cdc42 protein may be a general problem in cellular senescence.

7. SUMMARY

This chapter demonstrated that the expression of Cdc42 was reduced in both TD and WS cells and that Cdc42 played some role in intracellular lipid transport and cholesterol efflux. As shown in Fig. 12.1, the binding of apo AI with ABCA1 on the cell surface activates Cdc42. The GTP form of Cdc42 triggers intracellular events, such as the rearrangement of actin cytoskeleton and activations of some kinases. Although we still do not

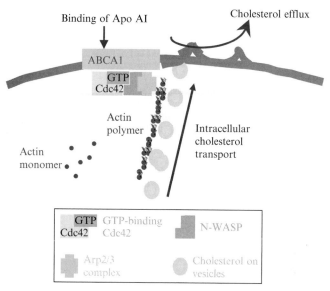

Figure 12.1 Hypothetical scheme for Cdc42-mediated cholesterol efflux from cells. The binding of apo AI with ABCA1 on the cell surface activates Cdc42. The GTP form of Cdc42 triggers intracellular events, such as assembly and rearrangement of the actin cytoskeleton and activations of some kinases. Although we still do not know the detailed mechanism, we believe that cholesterol moves as vesicles from the intracellular site to the plasma membrane along with arrangement of the actin cytoskeleton. On the cell surface, ABCA1 assembles apo AI and cholesterol for its efflux from cells. (See color insert.)

know the detailed mechanism, we think that cholesterol moves as vesicles from the intracellular site to the plasma membrane along with the arrangement of the actin cytoskeleton. On the cell surface, ABCA1 assembles apo AI and cholesterol for its efflux from cells.

ACKNOWLEDGMENTS

We thank Drs. Yoshimi Takai (Department of Molecular Biology and Biochemistry, Osaka University, Japan) and Kazumitsu Ueda (Kyoto University, Kyoto, Japan) for providing plasmids for Cdc42 and anti-ABCA1 antibody, respectively. We also thank Dr. Kosuke Tsukamoto (Sakai Municipal Hospital, Sakai, Osaka, Japan) for helpful discussions and comments. We thank Mr. Akira Yaguchi for technical advice on FRAP.

REFERENCES

Brosh, R. M., Jr., and Bohr, V. A. (2002). Roles of the Werner syndrome protein in pathways required for maintenance of genome stability. *Exp. Gerontol.* **37,** 491–506.
Chen, L., and Oshima, J. (2002). Werner syndrome. *J. Biomed. Biotechnol* **2,** 46–54.

Cutler, R. G., Kelly, J., Storie, K., Pedersen, W. A., Tammara, A., Hatanpaa, K., Troncoso, J. C., and Mattson, M. P. (2004). Involvement of oxidative stress-induced abnormalities in ceramide and cholesterol metabolism in brain aging and Alzheimer's disease. *Proc. Natl. Acad. Sci. USA* **101,** 2070–2075.

Diederich, W., Orso, E., Drobnik, W., and Schimitz, G. (2001). Apolipoprotein AI and HDL3 inhibit spreading of primary human monocytes through a mechanism that involves cholesterol depletion and regulation of Cdc42. *Atherosclerosis* **159,** 313–324.

Drobnik, W., Liebisch, G., Biederer, C., Trumbach, B., Rogler, G., Muller, P., and Schmitz, G. (1998). Growth and cell cycle abnormalities of fibroblasts from Tangier disease patients. *Arterioscler. Thromb. Vasc. Biol.* **19,** 28–38.

Hall, A. (1998). Rho GTPases and the actin cytoskeleton. *Science* **279,** 509–514.

Hirano, K., Matsuura, F., Tsukamoto, K., Zhang, Z., Matsuyama, A., Takaishi, K., Komuro, R., Suehiro, T., Yamashita, S., Takai, Y., and Matsuzawa, Y. (2000). Decreased expression of a member of the Rho GTPase family, Cdc42Hs, in cells from Tangier disease: The small G protein may play a role in cholesterol efflux. *FEBS Lett.* **484,** 275–279.

Hirano, K., Yamashita, S., Sakai, N., and Matsuzawa, Y. (2004). Low and high HDL syndrome. *In* "Encylopedia of Endocrine Disease," Vol. 3, pp. 199–205. Elsevier, San Diego.

Hou, M., Xia, M., Zhu, H., Wang, Q., Li, Y., Xiao, Y., Zhao, T., Tang, Z., Ma, J., and Ling, W. (2007). Lysophosphatidylcholine promotes cholesterol efflux from mouse macrophage foam cells via PPARgamma-LXRalpha-ABCA1-dependent pathway associated with apoE. *Cell Biochem. Funct.* **25,** 33–44.

Johnson, F. B., Sinclair, D. A., and Guarente, L. (1999). Molecular biology of aging. *Cell* **96,** 291–302.

Joseph, J. A., Villalobos-Molina, R., Denisova, N., Erat, S., Jimenez, N., and Strain, J. (1996). Increased sensitivity to oxidative stress and the loss of muscarinic receptor responsiveness in senescence. *Ann. N.Y. Acad. Sci.* **786,** 112–119.

Kodama, A., Takaishi, K., Nakano, K., Nishioka, H., and Takai, Y (1999). Involvement of Cdc42 small G protein in cell-cell adhesion, migration, and morphology of MDCK cells. *Oncogene* **18,** 3996–4006.

Koseki, M., Hirano, K., Masuda, D., Ikegami, C., Tanaka, M., Ota, A., Sandoval, J. C., Nakagawa-Toyama, Y., Sato, S. B., Kobayashi, T., Shimada, Y., Ohno-Iwashita, Y., *et al.* (2007). Increased lipid rafts and accelerated lipopolysaccharide-induced tumor necrosis factor-alpha secretion in Abca1-deficient macrophages. *J. Lipid Res.* **48,** 299–306.

Matsuura, F., Hirano, K., Koseki, M., Ohama, T., Matsuyama, A., Tsujii, K., Komuro, R., Nishida, M., Sakai, N., Hiraoka, H., Nakamura, T., and Yamashita, S. (2005). Familial massive tendon xanthomatosis with decreased high-density lipoprotein-mediated cholesterol efflux. *Metabolism* **54,** 1095–1101.

Nishida, Y., Hirano, K., Tsukamoto, K., Nagano, M., Ikegami, C., Roomp, K., Ishihara, M., Sakane, N., Zhang, Z., Tsujii Ki, K., Matsuyama, A., Ohama, T., *et al.* (2002). Expression and functional analyses of novel mutations of ATP-binding cassette transporter-1 in Japanese patients with high-density lipoprotein deficiency. *Biochem. Biophys. Res. Commun.* **290,** 713–721.

Nofer, J. R., Feuerborn, R., Levkau, B., Sokoll, A., Seedorf, U., Assmann, G. (2003). Involvement of Cdc42 signaling in apo A-I-inducedcholesterol efflux. *J. Biol. Chem.* **278,** 53055–53062.

Nofer, J. R., Remaley, A. T., Feuerborn, R., Wolinnska, I., Engel, T., von Eckardsterin, A., Assmann, G. (2006). Apolipoprotein A-I activates Cdc42 Signalling through the ABCA1 transporter. *J. Lipid Res.* **47,** 794–803.

Ohama, T., Hirano, K., Zhang, Z., Aoki, R., Tsujii, K., Nakagawa-Toyama, Y., Tsukamoto, K., Ikegami, C., Matsuyama, A., Ishigami, M., Sakai, N., Hiraoka, H., *et al.* (2002). Dominant expression of ATP-binding cassette transporter-1 on basolateral surface of Caco-2 cells stimulated by LXR/RXR ligands. *Biochem. Biophys. Res. Commun.* **296,** 625–630.

Orso, E., Broccardo, C., Kaminski, W. E., Böttcher, A., Liebisch, G., Drobnik, W., Götz, A., Chambenoit, O., Diederich, W., Langmann, T., Spruss, T., Luciani, M. F., *et al.* (2000). Transport of lipids from Golgi to plasma membrane is defective in Tangier disease patients and Abc1-deficient mice. *Nat. Genet.* **24,** 192–196.

Soccio, R. E., and Breslow, J. L. (2004). Intracellular cholesterol transport. *Arterioscler. Thromb. Vasc. Biol.* **24,** 1150–1160.

Tahara, S., Matsuo, M., and Kaneko, T. (2001). Age-related changes in oxidative damage to lipids and DNA in rat skin. *Mech. Ageing Dev.* **122,** 415–426.

Takai, Y., Sasaki, T., and Matozaki, T. (2001). Small GTP-binding proteins. *Physiol. Rev.* **81,** 153–208.

Tsukamoto, K., Hirano, K., Tsujii, K., Ikegami, C., Zhongyan, Z., Nishida, Y., Ohama, T., Matsuura, F., Yamashita, S., and Matsuzawa, Y. (2001). ATP-binding cassette transporter-1 (ABCA1) induces rearrangement of actin cytoskeleton through possible interaction between ABCA1 and Cdc42. *Biochem. Biophys. Res. Commun.* **287,** 757–765.

Tsukamoto, K., Hirano, K., Yamashita, S., Sakai, N., Ikegami, C., Zhang, Z., Matsuura, F., Hiraoka, H., Matsuyama, A., Ishigami, M., and Matsuzawa, Y. (2002). Retarded intra-cellular lipid transport associated with reduced expression of a member of Rho-GTPases, Cdc42, in human aged skin fibroblasts: a possible function of Cdc42 to mediated intracellular lipid transport. *Arterioscler. Thromb. Vasc. Biol.* **22,** 1899–1904.

Utech, M., Höbbel, G., Rust, S., Reinecke, H., Assmann, G., and Walter, M. (2001). Accumulation of RhoA, RhoB, RhoG, and Rac1 in fibroblasts from Tangier disease subjects suggests a regulatory role of Rho family proteins in cholesterol efflux. *Biochem. Biophys. Res. Commun.* **280,** 229–236.

Zanotti, I., Potì, F., Favari, E., Steffensen, K. R., Gustafsson, J. A., and Bernini, F. (2006). Pitavastatin effect on ATP binding cassette A1-mediated lipid efflux from macrophages: evidence for liver X receptor (LXR)-dependent and LXR-independent mechanisms of activation by cAMP. *J. Pharmacol. Exp. Ther.* **317,** 395–401.

Zhang, Z., Hirano, K., Tsukamoto, K., Ikegami, C., Koseki, M., Saijo, K., Ohno, T., Sakai, N., Hiraoka, H., Shimomura, I., and Yamashita, S. (2005). Defective cholesterol efflux in Werner syndrome fibroblasts and its phenotypic correction by Cdc42, a RhoGTPase. *Exp. Gerontol.* **40,** 286–294.

Zhang, Z., Yamashita, S., Hirano, K., Nakagawa-Toyama, Y., Matsuyama, A., Nishida, M., Sakai, N., Fukasawa, M., Arai, H., Miyagawa, J., and Matsuzawa, Y. (2001). Expression of cholesteryl ester transfer protein in human atherosclerotic lesions and its implication in reverse cholesterol transport. *Atherosclerosis* **159,** 67–75.

CHAPTER THIRTEEN

RAC AND NUCLEAR TRANSLOCATION OF SIGNAL TRANSDUCERS AND ACTIVATORS OF TRANSCRIPTION FACTORS

Toshiyuki Kawashima *and* Toshio Kitamura

Contents

Abstract

Signal transducers and activators of transcription (STATs) are tyrosine phosphorylated upon cytokine stimulations, and tyrosine-phosphorylated STATs (p-STATs) enter the nucleus to activate a variety of target genes. However, how activated STATs are transported to the nucleus has remained unclear. We have demonstrated that the small Rho GTPase Rac1 and the GTPase activating protein MgcRacGAP promote nuclear translocation of p-STATs via the importin α/β pathway through mediating a complex formation of p-STATs with importin α. This chapter discusses the use of an *in vitro* nuclear transport assay with Flag-tagged purified proteins, providing detailed protocols regarding such experiments. Although this chapter focuses on STATs, Rac, and MgcRacGAP the methods described here can be adapted easily to other nuclear factors and their regulators.

Division of Cellular Therapy, Institute of Medical Science, University of Tokyo, Tokyo, Japan

Methods in Enzymology, Volume 439
ISSN 0076-6879, DOI: 10.1016/S0076-6879(07)00413-2

1. INTRODUCTION

Nuclear transport is a fundamental physiological process occurring in eukaryotic cells; it is highly controlled, with regulatory mechanisms at each stage of nuclear transport. Molecular trafficking between the nucleus and the cytoplasm occurs through nuclear pore complexes (NPC). To enter the nucleus, proteins larger than $\approx$40 kDa usually require mono- or bipartite polybasic nuclear localization signals (NLS). Proteins carrying NLS are usually recognized by importin α-β heterodimers, and importin β docks the ternary complex to the nuclear pore, thus inducing migration of the complex into the nucleus. Then the GTP-bound form of the small GTPase Ran binds directly to importin β in the complex, causing disassembly of the complex inside the nucleus (Gorlich and Kutay, 1999; Mattaj and Englmeier, 1998). Some nuclear imports can be regulated by modifying the NLS. For example, the NLS can be masked or unmasked by the phosphorylation or dephosphorylation of adjacent amino acids (Heist *et al.*, 1998; Mattison and Ota, 2000; Tagawa *et al.*, 1995). Exposure of the NLS may also occur after proteolytic cleavage of the protein (Ghoda *et al.*, 1997). However, some proteins lacking functional NLS can associate NLS-containing molecules to enter the nucleus by piggyback import (Leslie *et al.*, 2004; Weil *et al.*, 1999). Therefore, a part of importin-mediated nuclear translocations is tightly regulated by kinases, phosphatases, and NLS-containing chaperones. Other regulations or transport systems are likely to exist and would be of great interest.

The STATs (STAT1-4, 5A, 5B, and 6) are activated by cytokine stimulations through tyrosine phosphorylation, form homo- or heterodimers, and translocate to the nucleus, where they regulate expression of their target genes (Darnell, 1996; Ihle, 1996). Among the STATs, STAT3 and STAT5 are involved in a broad spectrum of human hematologic malignancies as well as in solid tumors (Darnell, 2002). Therefore, it is important to understand the signaling pathways of STAT3 and STAT5 for exploring the molecular targets of anticancer drugs.

2. RAC1 AND MGCRACGAP REGULATE NUCLEAR ACCUMULATION OF P-STATS

We took advantage of an internal tandem duplication (ITD) mutant of receptor tyrosine kinase Flt3 (ITD-Flt3) (Yokota *et al.*, 1997), the expression of which could induce clear-cut nuclear import of STAT5A and MgcRacGAP (Fig. 13.1A). Conditional knockout of Rac1 in embryonic fibroblasts impaired the ITD-flt3-induced nuclear translocation of STAT5A (see Fig. 13.1C). This suggested that Rac1 and its GAP, MgcRacGAP, regulate the nuclear accumulation of p-STAT5A. However, the mechanism

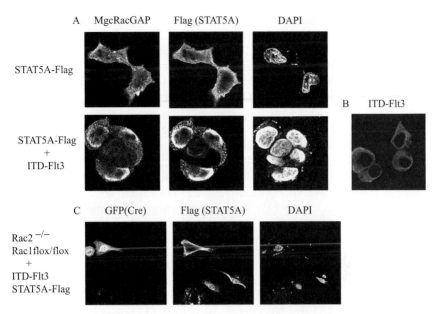

Figure 13.1 Translocation of STATs and MgcRacGAP to the nucleus. (A) Translocation of STAT5A and MgcRacGAP into the nucleus upon ITD-Flt3 stimulation in 293T
cells. Cells were transfected with pME/STAT5A-Flag together with pMKIT (mock)
(top) or pMKIT/ITD-Flt3 (bottom). After 36 h, cells were immunostained with the
anti-MgcRacGAP and anti-Flag Abs and viewed using a FLUOVIEW FV300 confocal
microscope (Olympus). (B) Cytoplasmic localization of ITD-Flt3 in 293T cells. Cells
transfected with pME/STAT5A-Flag together with pMKIT/ITD-Flt3 were immunostained with the anti-Flt3 Ab and viewed using a confocal microscope. (C) Impaired
nuclear translocation of ITD-Flt3-induced p-STAT5A in Rac2$^{-/-}$Rac1flox/flox fibroblasts. Rac2$^{-/-}$Rac1flox/flox fibroblasts were transduced with the pMX-IRES-GFP-Cre
retrovirus vector. After 3 days, cells were transiently cotransfected with pME/STAT5A-
Flag and pMKIT/ITD-Flt3. Cells were fixed and immunostained with anti-Flag Ab.
Note that GFP-positive cells show cytoplasmic localization of STAT5A-Flag, whereas
GFP-negative cells show nuclear localization of STAT5A-Flag.

was still ambiguous. One possibility was that the absence of Rac1 inhibited
nuclear import of p-STAT5A or enhanced its export; alternatively, it might
induce dephosphorylation and/or degradation of p-STAT5A in the nucleus.
To directly examine the role of Rac1/MgcRacGAP in the nuclear translocation of p-STAT5A, we produced and purified the recombinant proteins
and performed an *in vitro* transport assay using digitonin-permeabilized
cells and purified proteins according to the method of Adam *et al.* (1990).

Instead of previously reported protocols using untagged protein and
DEAE-Sepharose columns or ammonium sulfate precipitation (Bromberg
and Chen, 2001; Quelle *et al.*, 1995; Vinkemeier *et al.*, 1996), we selected
a baculovirus expression system using SF-9 cells and Flag-tag-mediated
protein purification using the 3×Flag peptide because we expected that

Flag-mediated purification would allow us to purify proteins quickly and gently, protecting them from losing a native fold and/or aggregating.

3. PROTOCOL FOR PURIFICATION OF FLAG-TAGGED RECOMBINANT PROTEINS USING SF-9 CELLS

To construct baculovirus vectors, Flag-tagged cDNAs encoding STAT5A, MgcRacGAP, V12Rac1, L61Rac1, N17Rac1, importin αs, importin β1, Ran, and NTF2 with the C-terminal Flag epitope tag, and a kinase domain of JAK2 (JH1) (Saharinen et al., 2000) are subcloned into pBacPAK (BD Biosciences). The resulting constructs are used to obtain recombinant baculoviruses by cotransfection with Bsu36 I-digested BacPAK viral DNA (BD Biosciences) into Sf-9 cells using bacfectin (BD Biosciences), according to the manufacturer's protocol.

1. For protein expression, Sf-9 cells (1×10^7) are infected with high-titer viral stocks for 96 h.
2. Cells are lysed in 3 ml of lysis buffer [50 mM Tris-HCl, pH 7.5, 150 mM NaCl, 1.0% Nonidet P-40, 1 mM EDTA, 0.2 mM Na$_3$VO$_4$, 2 mM phenylmethylsulfonyl fluoride (PMSF), 2 μg/ml leupeptin, and 10 μg/ml aprotinin].
3. The lysate is clarified by centrifugation for 20 min at 12000 g, and the supernatant is immunoprecipitated with 200 μl of the anti-Flag M2-agarose-affinity gel (Sigma) prepared as a 50% suspension in phosphate-buffered saline for 2 h at 4°.
4. Agarose beads are washed three times with lysis buffer, and recombinant Flag-tagged proteins are eluted with a final 20 μg/ml of 3×Flag peptide (Sigma).
5. Recombinant proteins are dialyzed extensively against the dialysis buffer [20 mM HEPES, pH 7.3, 110 mM KOAC, 2 mM Mg(OAC)$_2$, 1 mM EGTA, 2 mM dithiothreitol (DTT), and 0.4 mM PMSF].
6. To confirm the purities, eluted proteins are subjected to SDS-PAGE, followed by Coomassie blue staining.
7. The concentration of the purified proteins is determined by a commercial dye-binding assay (BCA★ protein assay reagent; Pierce).

4. RAC1 AND MGCRACGAP SERVE AS A NUCLEAR CHAPERONE FOR P-STATS

Our nuclear import assay is performed according to the method of Adam et al. (1990) with some minor modifications. Treatment of cells with low concentrations of digitonin selectively permeabilizes the plasma membrane because of its relatively high cholesterol content, while most other intercellular membranes, including the nuclear envelope, remain intact. After permeabilization, soluble proteins of the cell can be washed out through the digitonin-induced holes in the plasma membrane. When the permeabilized cells are supplemented with exogenous cytosol extract or purified transport factors, and energies including ATP and GTP, nuclei of these cells rapidly accumulate a tested cargo protein containing the NLS. This *in vitro* assay system, especially together with confocal microscopy, is a powerful tool for directly studying the nuclear import of proteins. We successfully established a STAT nuclear import assay using purified Flag-tagged proteins, including Rac1, MgcRacGAP, and other importin-related transporters (Kawashima *et al.*, 2006). As shown in Fig. 13.2A, in this nuclear transport assay, p-STAT5A did not enter the nucleus, even in the presence of the nuclear transporters, including importin α1, importin β1, Ran, and NTF2. Most p-STAT5A accumulated to the nuclear envelope in the presence of V12Rac1 and MgcRacGAP (see Fig. 13.2C), and further addition of the purified nuclear transporters facilitated the nuclear translocation of p-STAT5A (see Fig. 13.2D). These findings strongly suggest that the complex of p-STAT5A, GTP-bound Rac1, and MgcRacGAP translocates to the nuclear envelope, where it recruits other factors, such as importin $\alpha\beta$ to pass through the nuclear pore complex into the nucleus. Indeed, it was found that direct interaction of both GTP-bound Rac1 and MgcRacGAP facilitated the interaction of p-STAT5A with importin αs (Kawashima *et al.*, 2006). We also found that nuclear translocation of p-STAT3 as well as p-STAT5A was induced in the presence of a combination of purified proteins, including V12Rac1, MgcRacGAP, importin α1, importin β1, Ran, and NTF2 (Kawashima and Kitamura, unpublished data). These results suggested that Rac1/MgcRacGAP functions in the nuclear transport of p-STAT3 as well as p-STAT5A. How activated STATs are transported to the nucleus was previously investigated by other researchers; activated STAT1 and STAT3 were reported to bind importin α5 and several importin αs, respectively, which mediated the nuclear transport of STATs (Liu *et al.*, 2005; Ma *et al.*, 2006; McBride *et al.*, 2002; Sekimoto *et al.*, 1997; Ushijima *et al.*, 2005). However, the NLS sequence of neither STAT3 nor STAT5 has been identified; it also remains ambiguous whether the tyrosine-phosphorylated dimer of STAT3 or STAT5 has a functional NLS sequence that can be directly bound with importin α. It was reported that the ERBB4/HER4 receptor tyrosine kinase, which harbors the NLS sequence, functions as a STAT5A nuclear chaperone, implicating NLS of STAT5A-associated molecules in the nuclear translocation of STAT5A (Williams *et al.*, 2004). Unlike ERBB4/HER4, ITD-Flt3, which does not harbor an NLS, did not enter the nucleus (see Fig. 13.1B). However, we have found that MgcRacGAP harbors a bipartite NLS and binds importin αs (Kawashima and Kitamura, unpublished results). Interestingly, a mutant

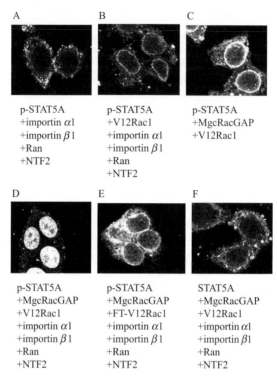

Figure 13.2 Purified p-STAT5A accumulated in the nucleus in the presence of V12Rac1, MgcRacGAP, and transporter proteins of the importin pathway in the nuclear transport assay. HeLa cells were permeabilized and incubated with import mix (IM) at 37° for 60 min. The IM contained TB, an ERS, and a single or combinations of the following purified proteins as indicated: 1 μM of STAT5A, p-STAT5A, V12Rac1, V12Rac1, which had been frozen once for stock and thawed (termed FT-V12Rac1), MgcRacGAP, importin α1, importin β1, Ran, or NTF2. After the import reaction, cells were fixed. STAT5A protein was detected using the anti-STAT5A Ab. Cells were examined using a FLUOVIEW FV300 confocal microscope (Olympus).

of MgcRacGAP lacking NLS strongly blocked the nuclear translocation of p-STATs in the nuclear transport assay, even in the presence of V12Rac1, importin α1, importin β1, Ran, and NTF2, suggesting a role for MgcRac-GAP as an NLS-containing nuclear chaperone of p-STAT3 and p-STAT5A. Because both GTP-bound Rac1 and MgcRacGAP were required for the complex formation of p-STAT5A with importin αs (Kawashima *et al.*, 2006), and a mutant of MgcRacGAP lacking NLS disrupted the complex formation of p-STAT5A with importin αs (Kawashima and Kitamura, unpublished results), it is likely that MgcRacGAP, together with GTP-bound Rac1, functions as a STAT5A nuclear chaperone Therefore, establishment of the nuclear transport assay for p-STATs has enabled us to clearly demonstrate the requirement of Rac1 and MgcRacGAP for the nuclear translocation of p-STATs.

It should be noted that some of the purified proteins, which had been frozen once for stock and thawed for use, did not work in the p-STAT nuclear import assay. Thus, purified proteins to use in this assay must be freshly prepared. As shown in Fig. 13.2E, the previously prepared V12Rac1, which had been frozen and thawed, induced accumulation of p-STAT5A to the nuclear envelope but did not facilitate the nuclear translocation of p-STAT5A, even in the presence of MgcRacGAP, importin α1, importin β1, Ran, and NTF2. Although further studies are required to understand how once-frozen V12Rac1 affects the nuclear import of p-STAT5A, it is possible that the freeze/thaw process ruins the protein conformation of Rac. It should also be noted that V12Rac1, MgcRacGAP, and p-STAT5A tend to aggregate and/or be degraded. Therefore, when using these proteins left overnight at 4°, nuclear import of p-STAT5A was considerably diminished even in the presence of MgcRacGAP, V12Rac1, importin α1, importin β1, Ran, and NTF2 (Kawashima and Kitamura, unpublished results). Thus, again, purification of proteins using the Flag tag and Flag peptide, which make it possible for one to purify proteins quickly and gently, helps in freshly preparing many kinds of recombinant proteins required for this p-STAT import assay.

5. PROTOCOL FOR NUCLEAR IMPORT ASSAYS WITH DIGITONIN-PERMEABILIZED CELLS USING PURIFIED FLAG-TAGGED PROTEINS

1. Forty-eight hours before use in a transport assay, adherent HeLa cells are prepared to culture on 13 × 13-mm poly-L-lysine-coated glass coverslips in 24-well plates.
2. When cells reach subconfluency, cells are rinsed in ice-cold transport buffer [TB; 20 mM HEPES, pH 7.3, 110 mM KOAC, 2 mM Mg (OAC)$_2$, 1 mM EGTA, 2 mM DTT, 0.4 mM PMSF, 3 μg/ml of aprotinin, 2 μg/ml of pepstatin A, 1 μg/ml of leupeptin, and 20 mg/ml of bovine serum albumin (BSA)] and immersed in ice-cold TB containing 40 μg/ml digitonin (Roche) for 10 min at room temperature to permeabilize the cells.
3. Subsequently, cells are washed twice with 1 ml of TB.
4. Incubation with 50 μl import mix (IM) is performed at 37° for 30 min. The IM contains TB, an energy-regenerating system (ERS; 0.5 mM ATP, 0.5 mM GTP, 10 mM creatine phosphate, and 30 U/ml creatine phosphokinase), and 1 μM of purified unphosphorylated or phosphorylated

STAT5A alone, or either STAT5A plus the 1 μM of various combinations of other purified cofactor proteins, including V12Rac1, MgcRacGAP, importin $\alpha 1$, importin $\beta 1$, Ran, and NTF2.

5. Following the import reaction, cells are washed twice with 1 ml of ice-cold TB and fixed with 4% (w/v) paraformaldehyde, followed by immunostaining with the anti-STAT5A Ab as described (Hirose *et al.*, 2001).

6. Cells are examined using a FLUOVIEW FV300 confocal microscope (Olympus).

Note: It must be stressed that an appropriate permeabilization of HeLa cells is critical to achieve successful nuclear translocation of p-STAT5A in this assay. We found that excessive treatment of digitonin, accompanied with destruction of the nuclear envelope, resulted in the artificial nuclear translocation not only of p-STATs but also of unphosphorylated STATs, even in the absence of other cofactors. In contrast, it was also observed that insufficient treatments of digitonin (in either concentration or exposure), accompanied by residual nuclear transporters in the cytoplasm of permeabilized cells, led to a considerable amount of nuclear translocation of p-STATs, but not of unphosphorylated STATs in the absence of other cofactors. Adam *et al.* (1992) also stressed the importance of avoiding these kinds of false results in the nuclear import assay. They recommended that, when one uses another type of cells for the *in vitro* import assay, one should conduct a titration for an optimal concentration and exposure of digitonin using proper control proteins such as FITC-labeled BSA conjugated with a synthetic peptide containing the SV40 large T antigen (CGGGPKKKRKVED) (NLS-conjugated FITC-BSA) and fractioned cytosol (Adam *et al.*, 1990). To establish the nuclear import assay for p-STATs, it is critical to set up an optimal condition in which STATs (both unphosphorylated and phosphorylated) do not enter the nucleus but phosphorylated STATs enter the nucleus in the presence of fractioned cytosol. Purified proteins used in this assay must be freshly prepared.

REFERENCES

Adam, S. A., Marr, R. S., and Gerace, L. (1990). Nuclear protein import in permeabilized mammalian cells requires soluble cytoplasmic factors. *J. Cell Biol.* **111,** 807–816.

Adam, S. A., Marr, R. S., and Gerace, L. (1992). Nuclear protein import using digitonin-permeabilized cells. *Methods Enzymol.* **219,** 97–110.

Bromberg, J., and Chen, X. M. (2001). STAT proteins: Signal transducers and activators of transcription. *Methods Enzymol.* **333,** 138–151.

Darnell, J. E., Jr. (1996). The JAK-STAT pathway: Summary of initial studies and recent advances. *Recent Prog. Horm. Res.* **51,** 391–403.

Darnell, J. E., Jr. (2002). Transcription factors as targets for cancer therapy. *Nat. Rev. Cancer* **2,** 740–749.

Ghoda, L., Lin, X., and Greene, W. C. (1997). The 90-kDa ribosomal S6 kinase (pp90rsk) phosphorylates the N-terminal regulatory domain of IkappaBalpha and stimulates its degradation *in vitro. J. Biol. Chem.* **272,** 21281–21288.

Gorlich, D., and Kutay, U. (1999). Transport between the cell nucleus and the cytoplasm. *Annu. Rev. Cell Dev. Biol.* **15,** 607–660.

Heist, E. K., Srinivasan, M., and Schulman, H. (1998). Phosphorylation at the nuclear localization signal of Ca2+/calmodulin-dependent protein kinase II blocks its nuclear targeting. *J. Biol. Chem.* **273,** 19763–19771.

Hirose, K., Kawashima, T., Iwamoto, I., Nosaka, T., and Kitamura, T. (2001). MgcRac-GAP is involved in cytokinesis through associating with mitotic spindle and midbody. *J. Biol. Chem.* **276,** 5821–5828.

Ihle, J. N. (1996). Janus kinases in cytokine signalling. *Philos. Trans. R. Soc. Lond. B Biol. Sci.* **351,** 159–166.

Kawashima, T., Bao, Y. C., Nomura, Y., Moon, Y., Tonozuka, Y., Minoshima, Y., Hatori, T., Tsuchiya, A., Kiyono, M., Nosaka, T., Nakajima, H., Williams, D. A., *et al.* (2006). Rac1 and a GTPase activating protein MgcRacGAP are required for nuclear translocation of STAT transcription factors. *J. Cell Biol.* **175,** 937–946.

Leslie, D. M., Zhang, W., Timney, B. L., Chait, B. T., Rout, M. P., Wozniak, R. W., and Aitchison, J. D. (2004). Characterization of karyopherin cargoes reveals unique mechanisms of Kap121p-mediated nuclear import. *Mol. Cell. Biol.* **24,** 8487–8503.

Liu, L., McBride, K. M., and Reich, N. C. (2005). STAT3 nuclear import is independent of tyrosine phosphorylation and mediated by importin-alpha3. *Proc. Natl. Acad. Sci. USA* **102,** 8150–8155.

Ma, J., and Cao, X. (2006). Regulation of Stat3 nuclear import by importin alpha5 and importin alpha7 via two different functional sequence elements. *Cell Signal.* **18,** 1117–1126.

Mattaj, I., and Englmeier, L. (1998). Nucleocytoplasmic transport: The soluble phase. *Annu. Rev. Biochem.* **67,** 265–306.

Mattison, C. P., and Ota, I. M. (2000). Two protein tyrosine phosphatases, Ptp2 and Ptp3, modulate the subcellular localization of the Hog1 MAP kinase in yeast. *Genes Dev.* **14,** 1229–1235.

McBride, K. M., Banninger, G., McDonald, C., and Reich, N. C. (2002). Regulated nuclear import of the STAT1 transcription factor by direct binding of importin-α. *EMBO J.* **21,** 1754–1763.

Quelle, F. W., Thierfelder, W., Witthuhn, B. A., Tang, B., Cohen, S., and Ihle, J. N. (1995). Phosphorylation and activation of the DNA binding activity of purified Stat1 by the Janus protein-tyrosine kinases and the epidermal growth factor receptor. *J. Biol. Chem.* **270,** 20775–20780.

Saharinen, P., Takaluoma, K., and Silvennoinen, O. (2000). Regulation of the Jak2 tyrosine kinase by its pseudokinase domain. *Mol. Cell. Biol.* **20,** 3387–3395.

Sekimoto, T., Imamoto, N., Nakajima, K., Hirano, T., and Yoneda, Y. (1997). Extracellular signal-dependent nuclear import of Stat1 is mediated by nuclear pore-targeting complex formation with NPI-1, but not Rch1. *EMBO J.* **16,** 7067–7077.

Tagawa, T., Kuroki, T., Vogt, P. K., and Chida, K. (1995). The cell cycle-dependent nuclear import of v-Jun is regulated by phosphorylation of a serine adjacent to the nuclear localization signal. *J. Cell Biol.* **130,** 255–263.

Ushijima, R., Sakaguchi, N., Kano, A., Maruyama, A., Miyamoto, Y., Sekimoto, T., Yoneda, Y., Ogino, K., and Tachibana, T. (2005). Extracellular signal-dependent nuclear import of STAT3 is mediated by various importin alphas. *Biochem. Biophys. Res. Commun.* **330,** 880–886.

Vinkemeier, U., Cohen, S. L., Moarefi, I., Chait, B. T., Kuriyan, J., and Darnell, J. E., Jr. (1996). DNA binding of *in vitro* activated Stat1 alpha, Stat1 beta and truncated Stat1:

Interaction between NH2-terminal domains stabilizes binding of two dimers to tandem DNA sites. *EMBO J.* **15,** 5616–5626.

Weil, R., Sirma, H., Giannini, C., Kremsdorf, D., Bessia, C., Dargemont, C., Brechot, C., and Israel, A. (1999). Direct association and nuclear import of the hepatitis B virus X protein with the NF-kappaB inhibitor IkappaBalpha. *Mol. Cell. Biol.* **19,** 6345–6354.

Williams, C. C., Allison, J. G., Vidal, G. A., Burow, M. E., Beckman, B. S., Marrero, L., and Jones, F. E. (2004). The ERBB4/HER4 receptor tyrosine kinase regulates gene expression by functioning as a STAT5A nuclear chaperone. *J. Cell Biol.* **167,** 469–478.

Yokota, S., Kiyoi, H., Nakao, M., Iwai, T., Misawa, S., Okuda, T., Sonoda, Y., Abe, T., Kahsima, K., Matsuo, Y., and Naoe, T. (1997). Internal tandem duplication of the FLT3 gene is preferentially seen in acute myeloid leukemia and myelodysplastic syndrome among various hematological malignancies: A study on a large series of patients and cell lines. *Leukemia* **11,** 1605–1609.

CHAPTER FOURTEEN

A Method for Measuring Rho Kinase Activity in Tissues and Cells

Ping-Yen Liu *and* James K. Liao

Contents

Abstract

The Rho-associated kinases (ROCKs) can regulate cell shape and function by modulating the actin cytoskeleton. ROCKs are serine-threonine protein kinases that can phosphorylate adducin, ezrin-radixin-moesin proteins, LIM kinase, and myosin light chain phosphatase. In the cardiovascular system, the RhoA/ROCK pathway has been implicated in angiogenesis, atherosclerosis, cerebral and coronary vasospasm, cerebral ischemia, hypertension, myocardial hypertrophy, and neointima formation after vascular injury. ROCKs consist of two isoforms: ROCK1 and ROCK2. They share overall 65% homology in their amino acid sequence and 92% homology in their amino kinase domains. However, these two isoforms have different subcellular localizations and exert biologically different functions. In particular, ROCK1 appears to be more important for immunological functions, whereas ROCK2 is more important for endothelial and vascular smooth muscle function. Thus, the ability to measure ROCK activity in tissues and cells would be important for understanding mechanisms underlying cardiovascular disease. This chapter describes a method for measuring ROCK activity in peripheral blood, tissues, and cells.

Vascular Medicine Research Unit, Brigham and Women's Hospital and Harvard, Medical School, Boston, Massachusetts

Methods in Enzymology, Volume 439
ISSN 0076-6879, DOI: 10.1016/S0076-6879(07)00414-4

1. INTRODUCTION

The small GTP-binding proteins act as molecular "on–off" switches to control multiple biological signaling pathways (Etienne-Manneville and Hall, 2002; Hall, 1998). The Rho-associated kinases (ROCKs) are found to be one of the downstream targets of RhoA (Ishizaki *et al.*, 1996; Matsui *et al.*, 1996). They not only regulate cell growth, migration, and apoptosis via control of the actin cytoskeletal assembly, but also the contraction of different cells through serine-threonine phosphorylation of adducin, ezrin-radixin-moesin (ERM) proteins, LIM kinase, and myosin light chain phosphatase (MLCP) (Riento and Ridley, 2003). Vascular smooth muscle cells (VSMCs) play an especially important role in the modulation of vascular tone and in the pathogenesis of atherosclerosis and vascular proliferative disease. ROCKs increase myosin light chain (MLC) phosphorylation through phosphorylating and inhibiting the myosin–binding subunit (MBS) of MLC phosphatase, thereby increasing MLC phosphorylation and contraction (Kimura *et al.*, 1996; Somlyo and Somlyo, 1994). They can also phosphorylate and inhibit LIM kinase, which phosphorylates the cofilin/actin-depolymerizing factor complex involved in the depolymerization and severing of actin filaments (Maekawa *et al.*, 1999). Overall, the physiological effects of ROCKs are to enhance actin–myosin association through increasing MLC phosphorylation and preventing actin depolymerization.

The Rho-associated kinases are also important regulators of cellular apoptosis, growth, metabolism, and migration via control of actin cytoskeletal assembly and cell contraction (Riento and Ridley, 2003). In the cardiovascular system, the RhoA/ROCK pathway has been shown to be involved in angiogenesis (Hyvelin *et al.*, 2005), atherosclerosis (Mallat *et al.*, 2003), cerebral and coronary vasospasm (Sato *et al.*, 2000), cerebral ischemia (Toshima *et al.*, 2000), hypertension (Uehata *et al.*, 1997), myocardial hypertrophy (Higashi *et al.*, 2003), and neointima formation after vascular injury (Sawada *et al.*, 2000). Increased leukocyte ROCK activity has been found to be associated with cardiovascular risk factors (Liu *et al.*, 2007), and inhibition of ROCK leads to an improvement in endothelial function (Nohria *et al.*, 2006).

In the mammalian system, ROCKs consist of two isoforms: ROCK1 and ROCK2. ROCK1, which is also known as ROCK-β or p160ROCK, is located on chromosome 18 and encodes a 1354 amino acid protein (Ishizaki *et al.*, 1996). ROCK2, which is also known as ROCK-α, is located on chromosome 12 and contains 1388 amino acids (Riento and Ridley, 2003). ROCK1 and ROCK2 share overall 65% homology in their amino acid sequence and 92% homology in their kinase domains. ROCK1 and ROCK2 are expressed ubiquitously in mouse tissues from early embryonic development to adulthood. However, these two isoforms have different tissue distribution and exert disparate biological functions (Noma *et al.*, 2007).

For example, ROCK1 is expressed preferentially in the lung, liver, spleen, kidney, and testis, whereas ROCK2 is highly expressed in the heart and brain (Nakagawa *et al.*, 1996; Wei *et al.*, 2001). In addition, immunolocalization and cell fractionation studies have shown that inactive ROCK2 is distributed mainly in the cytoplasm but could translocate from the cytoplasm to membranes when activated by GTP-bound RhoA (Matsui *et al.*, 1996). In contrast, ROCK1 colocalizes within or near the centrosomes (Chevrier *et al.*, 2002). Functionally, ROCK2 phosphorylates Ser^{19} of MLC, the same residue that is phosphorylated by MLC kinase and thus increases cellular contractility via dual effects on MLC kinase and MLCP. Indeed, ROCK2 can alter the sensitivity of VSMC contraction in response to changes in Ca^{2+} concentration (Amano *et al.*, 1996). Furthermore, ROCKs could also phosphorylate ERM proteins, which serve as cross-linkers between actin filaments and membrane proteins at the cell surface. ROCK-mediated phosphorylation of ERM proteins, namely Thr^{567} of ezrin, Thr^{564} of radixin, and Thr^{558} of moesin, leads to the disruption of the head-to-tail association of ERM proteins and actin cytoskeletal reorganization (Matsui *et al.*, 1998).

2. DOWNSTREAM TARGETS OF ROCK

The Rho-associated kinases phosphorylate various targets and mediate a broad range of cellular responses that involve the actin cytoskeleton in response to GTP-bound RhoA by activators of RhoA such as lysophosphatidic acid or sphingosine-1 phosphate, which stimulate Rho GEFs. The consensus amino acid sequences for phosphorylation are R/KXS/T or R/KXXS/T (R, arginine; K, lysine; X, any amino acid; S, serine; T, threonine) (Kawano *et al.*, 1999; Sumi *et al.*, 2001). ROCKs can also be autophosphorylated, suggesting that the function of ROCKs may be dependent in part on autoregulation (Maekawa *et al.*, 1999). MBS on MLCP is an important downstream target protein of ROCKs. Phosphorylation of MBS on MLCP by ROCKs leads to the phosphorylation of MLC and subsequent contraction of VSMCs (Somlyo and Somlyo, 2000). The MLCP holoenzyme is composed of three subunits: a catalytic subunit (PP1), a MBS composed of a 58-kDa head and a 32-kDa tail region, and a small noncatalytic subunit, M21.

As mentioned previously, ROCK2 can phosphorylate MBS at Thr^{697}, Ser^{854}, and Thr^{855} (Kawano *et al.*, 1999). The functional significance of MBS phosphorylation at Ser^{854}, however, is not known. Phosphorylation of Thr^{697} or Thr^{855} attenuates MLCP activity and, in some instances, the dissociation of MLCP from myosin (Feng *et al.*, 1999; Velasco *et al.*, 2002). In addition, MLC is one of the major downstream target proteins of ROCKs. However, it is still not known whether phosphorylation of MBS on MLCP or ERM proteins is specific to ROCK isoforms.

Nevertheless, ROCK1 phosphorylates LIM kinase-1 at Thr^{508} and LIM kinase-2 at Thr^{505}, which inhibits cofilin-mediated actin filament disassembly by phosphorylating cofilin (Maekawa *et al.*, 1999; Ohashi *et al.*, 2000; Sumi *et al.*, 2001). Adducin, which is a membrane skeletal protein that associates with and promotes the association of spectrin with F-actin, is also a downstream target of ROCK2 (Fukata *et al.*, 1999; Kimura *et al.*, 1998). Adducin is localized at cell–cell contact sites and is thought to participate in the assembly of the spectrin–actin network by capping the fast-growing ends of actin filaments and recruiting spectrin to the filament ends. The phosphorylation of α-adducin by ROCK2 enhances the binding activity of α-adducin to F-actin, thereby increasing the contractile response.

In order to assess ROCK function *in vitro* and *in vivo*, it is important to accurately measure ROCK activity in tissues and cells. This chapter provides a method for the preparation and analysis of total ROCK activity in cultured cells and tissues, as well as leukocytes isolated from peripheral blood of humans or mice.

3. MEASUREMENT OF ROCK ACTIVITY

For accurate and reproducible measurements of ROCK activity, careful attention should be paid to tissue preparation in order to avoid cell lysis and thus prevent overphosphorylation of MBS. ROCK inhibitors could be added during preparation to stop further phosphorylation in some tissues that are sensitive to ongoing activators of ROCK signaling. For human studies, a protocol is provided showing the proper method for collecting tissues or leukocytes from peripheral blood.

4. CULTURED CELLS

To avoid degradation of target proteins by cellular enzymes, cells must be prepared rapidly in the presence of various protease inhibitors. Thus, we suggest that the number of cell culture dishes be kept at a minimum in order to conserve time. At the appropriate time point, the culture medium is removed rapidly and completely, and the monolayer of adherent cells is washed twice with ice-cold phosphate-buffered saline (PBS) before the addition of cell fixatives and ROCK inhibitor. Approximately 300 μl of fixative solution (see later) is added at room temperature to each 10-mm culture dish, followed by the addition of 1 mM of hydroxyfasudil (ROCK inhibitor) to stop further MBS phosphorylation after cell lysis. A cocktail of proteinase inhibitors (see later) is also added to avoid protein degradation. All culture dishes should be kept on ice during isolation. Scrape the cell contents and transfer them to a microcentrifuge tube on ice. After vortexing and centrifuging the samples at 4° for 5 min at 14,000 rpm, the supernatant is removed and the pellets are used for ROCK assay. Pellets are stored at −80° until use.

4.1. ROCK activity as determined by the level of MBS phosphorylation

Making two 7.5% separating gels with 1.5-mm spacers requires a 20-ml solution consisting of 9.6 ml H_2O, 5 ml 30% acrylamide/bisacrylamide, 5 ml Tris (1.5 M, pH 8.8), 200 μl 10% SDS, 200 μl 10% ammonium persulfate, and 12 μl tetramethylethylenediamine (TEMED). After the separating gels solidify (approximately 30 to 60 min), the stacking gel solution (5% acrylamide) is added to the top of the separating gel. For two gels, prepare a 5-ml solution consisting of 3.44 ml H_2O, 833 μl 30% acrylamide/bisacrylamide, 625 μl Tris (1 M, pH 6.8), 50 μl 10% SDS, 50 μl 10% ammonium persulfate, and 5 μl TEMED.

We use SDS-PAGE buffer as the electrophoresis buffer (for recipe, see later). After centrifugation, pellets are dissolved in 10 μl of 1 mol/liter Tris and mixed with 100 μl of extraction buffer (8 mol/liter urea, 2% SDS, 5% sucrose, 5% 2-mercaptoethanol, 0.02% bromphenol blue). A SDS-PAGE protein standard (i.e., Bio-Rad, Richmond, CA) should be loaded on each gel in a separate lane. To avoid the interference by different exposure durations and variable membrane conditions, we use lipopolysaccharide-pretreated NIH/3T3 cell lysates as a positive control and also to standardize results between different experiments. Standard size gels are electrophoresed at 130 V at 23° for 1.5 h. After complete electrophoresis, the proteins are then transferred to polyvinylidene fluoride membranes (Immobilon P, Millipore Bedford, MA). The membrane is then soaked for 5 s in methanol and washed briefly in H_2O. The gel, transfer membrane, and filter paper are then soaked in transfer buffer (for recipe, see later) for 5 min.

For transferring, mount the following layers in order from bottom (anode of transfer apparatus) to top (cathode of transfer apparatus): one buffer-soaked thick filter paper, the transfer membrane, the gel, and one buffer-soaked thick filter paper. Air bubbles between these layers should be avoided and removed by gently rolling a glass pipette over the transfer membrane. Negatively charged proteins will move downward (from the gel into the membrane). The proteins are transferred at 105 V for about 105 min at 4°. Blocking of unspecific binding sites is achieved by incubating the membrane in PBS with 0.1% Tween and 5% milk for 0.5 h at room temperature or overnight at 4°.

The membranes are then incubated with rabbit antiphospho-specific Thr[853]-MBS polyclonal antibody (Biosource) and rabbit anti-MBS polyclonal antibody (Covance). Bands are visualized using the ECL detection kit (Amersham Corp./New England Nuclear). ROCK activity is expressed as

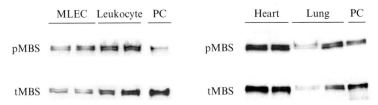

| MLEC Leukocyte PC | Heart   Lung   PC |

pMBS   pMBS

tMBS   tMBS

Figure 14.1 Rho-associated kinase activity as determined by the immunoblotting method in different cells and tissues. (Left) Example of phosphorylation levels of MBS (pMBS) and total MBS (tMBS) in cultured endothelial cells from mouse lung (MLEC) and peripheral blood leukocytes. Positive control (PC) represents cell lysates from 3T3 fibroblasts stimulated with lysophosphatidic acid. ROCK activity is expressed as a relative blot density ratio, which is standardized to the PC blot as (pMBS sample density/tMBS PC density)/(tMBS sample density/tMBS PC density). (Right) Example of ROCK activity in tissues from mouse heart and lung, respectively.

the ratio of pMBS in each sample per pMBS in each positive control divided by MBS in each sample per MBS in each positive control (Fig. 14.1).

4.2. Reagents

Fixative solution: 50% trichloroacetic acid (Sigma), 50 mM dichlorodiphenyltrichloroethane (Sigma), protease inhibitors (Calbiochem, EMD Biosciences, Inc., Darmstadt, Germany), 1 mM phenylmethylsulfonyl fluoride, and 1 mM NaF. The last three substances should be added immediately before use.
SDS-PAGE buffer, 10× (use dilution 1×): 250 mM Tris, 1.92 M glycine, and 1% SDS, pH 8.3
Western blot transfer buffer, 1×: 25 mM Tris, pH 8.3, 190 mM glycine, and 10% methanol
PBST, pH 7.4: 0.1% Tween 20 dissolved in PBS

5. LEUKOCYTES

To isolate human leukocytes, 20 ml of blood sample is mixed with Hanks balanced salt solution (HBSS) in a 50-ml citrate-containing tube. Ten milliliters of Histopaque (Sigma, Histopaque-1077) is layered with 25 ml of diluted blood in two tubes and centrifuged at room temperature for 30 min at 1400–1500 rpm. The supernatant containing the leukocytes is aspirated, mixed with HBSS, diluted with 2% dextran (1:1 ratio), and allowed to sit at room temperature for 30 min. The top layer is then aspirated, mixed with HBSS, and centrifuged at room temperature for 5 min at 1400–1500 rpm. The supernatant is discarded, and the pellet containing the leukocytes is resuspended in 3 ml of cold PBS. After swirling the tubes for 30 s, HBSS is added to stop the lysis. After centrifugation, the supernatant is discarded and the pellet is resuspended with 5 ml of M199. After determining cell yield and viability by using the Trypan blue exclusion test (usually $4-8 \times 10^6$ viable cells with a viability of more than 95%), the suspension is diluted with HBSS to achieve 5×10^6 cells/ml. Then, 400 μl of the leukocyte suspension is transferred to four sterile 1.5-ml tubes. We add 100 μl of fixative solution (see earlier discussion) to each tube. To avoid overphosphorylation, 1 mM of hydroxyfasudil is added to the TCA fixative solution. After vortexing and centrifuging the samples at 4° for 5 min at 12,000 rpm, the supernatant is removed and HBSS is added. The samples are centrifuged again at 4° for 1 min at 12,000 rpm. The supernatant is removed with a micropipette. The remaining leukocyte pellets are stored at $-80°$ until use.

6. Tissues

Similar to cell isolation, extended lysis of cells during preparation is to be avoided. First, HBSS solution is added to the tissue container with proteinase inhibitors and 1 mM hydroxyfasudil in order to avoid protein degradation and continued phosphorylation after lysis, respectively. In addition, a 50% fixative solution (see earlier discussion) is added and the samples are homogenized without sonication. The samples are then centrifuged at 4° for 10 min at 1500 rpm. The supernatant is discarded and the pellet is resuspended in 3 ml of cold PBS. After swirling the tubes for 30 s, HBSS is added to stop the lysis. Then, 400 μl of the suspension is transferred to four sterile 1.5-ml tubes. After vortexing and centrifuging the samples at 4° for 5 min at 12,000 rpm, the supernatant is removed, HBSS is added, and the samples are centrifuged again at 4° for 1 min at 12,000 rpm. Finally, the supernatant is removed with a micropipette. The pellets are stored at $-80°$ until use. ROCK activity is determined by immunoblotting of pellets with the Phospho-Thr[853] antibody as described (see Fig. 14.1, right panel with heart and lung tissue as an example).

7. Summary

By using peripheral blood leukocytes to assess ROCK activity, a less invasive method could be used to monitor ROCK activity *in vivo*. Given that ROCK activity contributes to vascular tone and VSMC contractility, such a method may be useful in assessing the role of ROCK in cardiovascular disease. Further studies are required with antibodies from distal targets of ROCK in order to determine whether methods could be developed that can distinguish ROCK1 and ROCK2 activities from total ROCK activity.

REFERENCES

Amano, M., Ito, M., Kimura, K., Fukata, Y., Chihara, K., Nakano, T., Matsuura, Y., and Kaibuchi, K. (1996). Phosphorylation and activation of myosin by Rho-associated kinase (Rho-kinase). *J. Biol Chem.* **271,** 20246–20249.

Chevrier, V., Piel, M., Collomb, N., Saoudi, Y., Frank, R., Paintrand, M., Narumiya, S., Bornens, M., and Job, D. (2002). The Rho-associated protein kinase p160ROCK is required for centrosome positioning. *J. Cell Biol.* **157,** 807–817.

Etienne-Manneville, S., and Hall, A. (2002). Rho GTPases in cell biology. *Nature* **420,** 629–635.

Feng, J., Ito, M., Ichikawa, K., Isaka, N., Nishikawa, M., Hartshorne, D. J., and Nakano, T. (1999). Inhibitory phosphorylation site for Rho-associated kinase on smooth muscle myosin phosphatase. *J. Biol. Chem.* **274,** 37385–37390.

Fukata, Y., Oshiro, N., Kinoshita, N., Kawano, Y., Matsuoka, Y., Bennett, V., Matsuura, Y., and Kaibuchi, K. (1999). Phosphorylation of adducin by Rho-kinase plays a crucial role in cell motility. *J. Cell Biol.* **145,** 347–361.

Hall, A. (1998). Rho GTPases and the actin cytoskeleton. *Science* **279,** 509–514.

Higashi, M., Shimokawa, H., Hattori, T., Hiroki, J., Mukai, Y., Morikawa, K., Ichiki, T., Takahashi, S., and Takeshita, A. (2003). Long-term inhibition of Rho-kinase suppresses angiotensin II-induced cardiovascular hypertrophy in rats *in vivo*: Effect on endothelial NAD(P)H oxidase system. *Circ. Res.* **93,** 767–775.

Hyvelin, J. M., Howell, K., Nichol, A., Costello, C. M., Preston, R. J., and McLoughlin, P. (2005). Inhibition of Rho-kinase attenuates hypoxia-induced angiogenesis in the pulmonary circulation. *Circ. Res.* **97,** 185–191.

Ishizaki, T., Maekawa, M., Fujisawa, K., Okawa, K., Iwamatsu, A., Fujita, A., Watanabe, N., Saito, Y., Kakizuka, A., Morii, N., and Narumiya, S. (1996). The small GTP-binding protein Rho binds to and activates a 160 kDa Ser/Thr protein kinase homologous to myotonic dystrophy kinase. *EMBO J.* **15,** 1885–1893.

Kawano, Y., Fukata, Y., Oshiro, N., Amano, M., Nakamura, T., Ito, M., Matsumura, F., Inagaki, M., and Kaibuchi, K. (1999). Phosphorylation of myosin-binding subunit (MBS) of myosin phosphatase by Rho-kinase *in vivo*. *J. Cell Biol.* **147,** 1023–1038.

Kimura, K., Fukata, Y., Matsuoka, Y., Bennett, V., Matsuura, Y., Okawa, K., Iwamatsu, A., and Kaibuchi, K. (1998). Regulation of the association of adducin with actin filaments by Rho-associated kinase (Rho-kinase) and myosin phosphatase. *J. Biol. Chem.* **273,** 5542–5548.

Kimura, K., Ito, M., Amano, M., Chihara, K., Fukata, Y., Nakafuku, M., Yamamori, B., Feng, J., Nakano, T., Okawa, K., Iwamatsu, A., and Kaibuchi, K. (1996). Regulation of myosin phosphatase by Rho and Rho-associated kinase (Rho-kinase). *Science* **273,** 245–248.

Liu, P. Y., Chen, J. H., Lin, L. J., and Liao, J. K. (2007). Increased Rho kinase (ROCK) activity in Taiwanese population with metabolic syndrome. *J. Am. Coll. Cardiol.* **49,** 1619–16124.

Maekawa, M., Ishizaki, T., Boku, S., Watanabe, N., Fujita, A., Iwamatsu, A., Obinata, T., Ohashi, K., Mizuno, K., and Narumiya, S. (1999). Signaling from Rho to the actin cytoskeleton through protein kinases ROCK and LIM-kinase. *Science* **285,** 895–898.

Mallat, Z., Gojova, A., Sauzeau, V., Brun, V., Silvestre, J. S., Esposito, B., Merval, R., Groux, H., Loirand, G., and Tedgui, A. (2003). Rho-associated protein kinase contributes to early atherosclerotic lesion formation in mice. *Circ. Res.* **93,** 884–888.

Matsui, T., Amano, M., Yamamoto, T., Chihara, K., Nakafuku, M., Ito, M., Nakano, T., Okawa, K., Iwamatsu, A., and Kaibuchi, K. (1996). Rho-associated kinase, a novel serine/threonine kinase, as a putative target for small GTP binding protein Rho. *EMBO J.* **15**, 2208–2216.

Matsui, T., Maeda, M., Doi, Y., Yonemura, S., Amano, M., Kaibuchi, K., Tsukita, S., and Tsukita, S. (1998). Rho-kinase phosphorylates COOH-terminal threonines of ezrin/radixin/moesin (ERM) proteins and regulates their head-to-tail association. *J. Cell Biol.* **140**, 647–657.

Nakagawa, O., Fujisawa, K., Ishizaki, T., Saito, Y., Nakao, K., and Narumiya, S. (1996). ROCK-I and ROCK-II, two isoforms of Rho-associated coiled-coil forming protein serine/threonine kinase in mice. *FEBS Lett.* **392**, 189–193.

Nohria, A., Grunert, M. E., Rikitake, Y., Noma, K., Prsic, A., Ganz, P., Liao, J. K., and Creager, M. A. (2006). Rho kinase inhibition improves endothelial function in human subjects with coronary artery disease. *Circ. Res.* **99**, 1426–1432.

Noma, K., Goto, C., Nishioka, K., Jitsuiki, D., Umemura, T., Ueda, K., Kimura, M., Nakagawa, K., Oshima, T., Chayama, K., Yoshizumi, M., Liao, J. K., *et al.* (2007). Roles of rho-associated kinase and oxidative stress in the pathogenesis of aortic stiffness. *J. Am. Coll. Cardiol.* **49**, 698–705.

Ohashi, K., Nagata, K., Maekawa, M., Ishizaki, T., Narumiya, S., and Mizuno, K. (2000). Rho-associated kinase ROCK activates LIM-kinase 1 by phosphorylation at threonine 508 within the activation loop. *J. Biol. Chem.* **275**, 3577–3582.

Riento, K., and Ridley, A. J. (2003). Rocks: Multifunctional kinases in cell behaviour. *Nat. Rev. Mol. Cell. Biol.* **4**, 446–456.

Sato, M., Tani, E., Fujikawa, H., and Kaibuchi, K. (2000). Involvement of Rho-kinase-mediated phosphorylation of myosin light chain in enhancement of cerebral vasospasm. *Circ. Res.* **87**, 195–200.

Sawada, N., Itoh, H., Ueyama, K., Yamashita, J., Doi, K., Chun, T. H., Inoue, M., Masatsugu, K., Saito, T., Fukunaga, Y., Sakaguchi, S., Arai, H., *et al.* (2000). Inhibition of rho-associated kinase results in suppression of neointimal formation of balloon-injured arteries. *Circulation* **101**, 2030–2033.

Somlyo, A. P., and Somlyo, A. V. (1994). Signal transduction and regulation in smooth muscle. *Nature* **372**, 231–236.

Somlyo, A. P., and Somlyo, A. V. (2000). Signal transduction by G-proteins, rho-kinase and protein phosphatase to smooth muscle and non-muscle myosin II. *J. Physiol.* **522**(Pt 2), 177–185.

Sumi, T., Matsumoto, K., and Nakamura, T. (2001). Specific activation of LIM kinase 2 via phosphorylation of threonine 505 by ROCK, a Rho-dependent protein kinase. *J. Biol. Chem.* **276**, 670–676.

Toshima, Y., Satoh, S., Ikegaki, I., and Asano, T. (2000). A new model of cerebral microthrombosis in rats and the neuroprotective effect of a Rho-kinase inhibitor. *Stroke* **31**, 2245–2250.

Uehata, M., Ishizaki, T., Satoh, H., Ono, T., Kawahara, T., Morishita, T., Tamakawa, H., Yamagami, K., Inui, J., Maekawa, M., and Narumiya, S. (1997). Calcium sensitization of smooth muscle mediated by a Rho-associated protein kinase in hypertension. *Nature* **389**, 990–994.

Velasco, G., Armstrong, C., Morrice, N., Frame, S., and Cohen, P. (2002). Phosphorylation of the regulatory subunit of smooth muscle protein phosphatase 1M at Thr850 induces its dissociation from myosin. *FEBS Lett.* **527**, 101–104.

Wei, L., Roberts, W., Wang, L., Yamada, M., Zhang, S., Zhao, Z., Rivkees, S. A., Schwartz, R. J., and Imanaka-Yoshida, K. (2001). Rho kinases play an obligatory role in vertebrate embryonic organogenesis. *Development* **128**, 2953–2962.

CHAPTER FIFTEEN

RHO KINASE-MEDIATED VASOCONSTRICTION IN RAT MODELS OF PULMONARY HYPERTENSION

Masahiko Oka,* Noriyuki Homma,* *and* Ivan F. McMurtry[†]

Contents

Abstract

There is current controversy regarding whether vasoconstriction plays a significant role in the elevated pressure of severe, advanced stages of pulmonary hypertension. Results of acute vasodilator testing using conventional vasodilators in such patients suggest there is only a minor contribution of vasoconstriction. However, there is a possibility that these results may underestimate the contribution of vasoconstriction because the most effective vasodilators have not yet been tested. This issue has not been addressed even experimentally, due mainly to a lack of appropriate animal models. A few animal models that mimic the pathology of human severe pulmonary hypertension more closely (i.e., development of occlusive neointimal lesions in small pulmonary

* Cardiovascular Pulmonary Research Laboratory, University of Colorado at Denver and Health Sciences Center, Denver, Colorado
† Department of Pharmacology and Center for Lung Biology, University of South Alabama, Mobile, Alabama

Methods in Enzymology, Volume 439
ISSN 0076-6879, DOI: 10.1016/S0076-6879(07)00415-6

arteries/arterioles) have been introduced, including rat models of left lung pneumonectomy plus monocrotaline injection and vascular endothelial growth factor inhibition plus exposure to chronic hypoxia. We have observed that Rho kinase inhibitors, a novel class of potent vasodilators, reduce the high pulmonary artery pressure of these models acutely and markedly, suggesting that vasoconstriction can significantly be involved in pulmonary hypertension with severely remodeled (occluded) pulmonary vessels. This chapter describes methods used for evaluation of the involvement of Rho kinase-mediated vasoconstriction in rat models of pulmonary hypertension.

1. INTRODUCTION

1.1. Pulmonary hypertension

Pulmonary hypertension is defined by a sustained elevation in pulmonary artery (PA) pressure (mean PA pressure >25 mm Hg at rest or >30 mm Hg with exercise) (Rich *et al.*, 1987). In severe pulmonary hypertension, the increase in pressure is generally progressive and often leads to right heart failure and ultimately to death. Pulmonary hypertension is associated with a variety of adult and pediatric diseases, but is characterized by common features, such as sustained pulmonary vasoconstriction, progressive fixed structural remodeling of the PAs, and *in situ* thrombosis (Humbert *et al.*, 2004; Mandegar *et al.*, 2004; Newman *et al.*, 2004). Patients with severe pulmonary hypertension have combinations of small PA adventitial and medial thickening, occlusive intimal lesions, and obliterating thrombotic and plexiform lesions (Pietra *et al.*, 2004). Although it is widely accepted that the hypertensive component due to vasoconstriction decreases while that due to fixed obstruction increases over time in the development of pulmonary hypertension (Reeves *et al.*, 1986), the exact contribution of vasoconstriction in advanced stages of pulmonary hypertension is unclear. A large clinical study showed that only about 13% of adult pulmonary hypertension patients have a significant acute decrease in PA pressure in response to inhaled nitric oxide (NO), intravenous prostacyclin, or adenosine at the time of diagnosis (Sitbon *et al.*, 2005), suggesting a major fixed structural but minor reversible vasoconstrictor component in this group of pulmonary vascular diseases. However, there is a possibility that the absence of a vasodilator response in these patients may be more apparent than real because the most effective vasodilators have not yet been identified and tested. In fact, we have found that acute administration of Rho kinase inhibitors, a new class of potent vasodilators (Hu and Lee, 2003), reduces the elevated pulmonary artery pressure dramatically in five different rat models of pulmonary hypertension (Nagaoka *et al.*, 2004, 2005, 2006; Oka *et al.*, 2007; M. Oka, unpublished data). Particularly in the vascular endothelial growth factor inhibition plus chronic hypoxia-exposed model, which resembles the advanced stage of human severe pulmonary hypertension (high PA pressure, low cardiac output, and severe occlusive neointimal lesions in small PAs), acute intravenous fasudil, a Rho kinase inhibitor, lowered high PA pressure more effectively and markedly than either inhaled NO or an intravenous infusion of iloprost, a prostacyclin analogue (Oka *et al.*, 2007).

1.2. RhoA/Rho kinase signaling in sustained vasoconstriction

The degree of phosphorylation of the 20-kDa regulatory myosin light chain (MLC) generally determines the degree of vascular smooth muscle cell contraction. The cytosolic Ca^{2+} concentration ($[Ca^{2+}]$) essentially regulates the activity of Ca^{2+}/calmodulin-dependent MLC kinase (MLCK) that phosphorylates MLC. At the same time, phosphorylated MLC is dephosphorylated by Ca^{2+}-independent MLC phosphatase (MLCP). Thus, the balance in activities of MLCK (contraction) and MLCP (relaxation) regulates vascular smooth muscle tone (Pfitzer, 2001; Somlyo and Somlyo, 1994). Inhibition of MLCP promotes MLC phosphorylation and contraction at a constant or decreasing cytosolic $[Ca^{2+}]$. This is referred to as Ca^{2+} sensitization (Fukata *et al.*, 2001; Somlyo and Somlyo, 2003). Evidence indicates that while Ca^{2+}/calmodulin-dependent MLCK-mediated MLC phosphorylation is more important for triggering vascular smooth muscle cell contraction, Ca^{2+} sensitization is important for the sustained phase of contraction (Somlyo and Somlyo, 2003) (Fig. 15.1).

The small GTPase RhoA, a member of the Rho family of small GTP-binding proteins, and its downstream effector, Rho kinase (RhoA/Rho kinase signaling), play a major role in the regulation of MLCP activity. Rho kinase inhibits MLCP by phosphorylating its regulatory subunit MYPT1, thereby inducing Ca^{2+} sensitization (Somlyo and Somlyo, 2003). RhoA is activated by various G protein-coupled receptor (GPCR) agonists (vasoconstrictors), including thromboxane A_2 (TxA_2), endothelin-1 (ET-1), and serotonin (5-HT). Thus, RhoA/Rho kinase-mediated Ca^{2+} sensitization is thought to be a major component in the sustained vasoconstriction induced by GPCR agonists (see Fig. 15.1). In fact, selective Rho kinase inhibitors, such as Y-27632 (Uehata *et al.*, 1997) and fasudil (HA-1077) (Amano *et al.*, 1999; Shimokawa *et al.*, 2002; Uehata *et al.*, 1997), effectively reverse the sustained vasoconstriction induced by many agonists (Batchelor *et al.*, 2001; Nobe and Paul, 2001; Sakurada *et al.*, 2001) and are now regarded as a novel class of potent vasodilators with multiple other actions (Budzyn *et al.*, 2006; Shimokawa and Takeshita, 2005). Several clinical trials with fasudil are ongoing for vasospastic diseases, such as angina and cerebral vasospasm after subarachnoid hemorrhage (Budzyn *et al.*, 2006; Shimokawa and Takeshita, 2005). Low doses of intravenous fasudil have

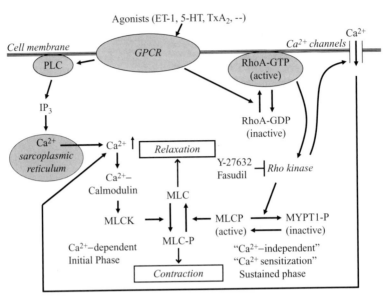

Figure 15.1 Mechanisms of G protein-coupled receptor (GPCR) agonist-induced smooth muscle contraction. ET-1, endothelin-1; 5-HT, 5-hydroxytryptamine (serotonin); TxA₂, thromboxane A₂; PLC, phospholipase C; IP₃, inositol 1,4,5-triphosphate; MLC, myosin light chain; MLCK, MLC kinase; MLCP, MLC phosphatase. See text for details.

been administered to two small groups of patients with moderate pulmonary hypertension and found to cause significant decreases in pulmonary vascular resistance (Fukumoto *et al.*, 2005; Ishikura *et al.*, 2006).

2. RAT MODELS OF SEVERE PULMONARY HYPERTENSION

Among numerous animal models of pulmonary hypertension, the two most frequently studied are chronic hypoxia- and monocrotaline injection-induced models in rats. These models have provided important groundwork for the current use of Ca^{2+} channel blockers, prostanoids, ET-1 receptor antagonists, inhaled NO, and phosphodiesterase inhibitors in the clinical management of pulmonary hypertension and continue to provide new insights into the myriad cellular and molecular mechanisms involved in the regulation of pulmonary vascular tone and structure. However, neither model develops the occlusive neointimal lesions, nor the phenotypically abnormal endothelial cells, found in human severe pulmonary hypertension (Heath, 1992; Voelkel and Tuder, 2000). In fact, if the hypertensive pulmonary vasculature of chronically hypoxic or monocrotaline-injected rats is maximally vasodilated and fixed under positive transmural pressure, there is little structural narrowing of the pulmonary artery lumen (Howell *et al.*, 2004; Hyvelin *et al.*, 2005; van Suylen *et al.*, 1998). This characteristic of pulmonary arterial/arteriolar muscularization without neointimal obstruction of the lumen likely accounts for the fact that these two forms of pulmonary hypertension can be "treated" readily with several different vasodilators and receptor blockers. Notably, this sets these models apart from the situation in human severe pulmonary hypertension, where while the conventionally available drugs improve symptoms and survival, they rarely reverse the hypertension. Thus, to mimic the arteriopathy of human pulmonary hypertension more closely, Botney (1999) and Taraseviciene-Stewart *et al.* (2001) have developed rat models that demonstrate small PA neointimal cell proliferation to the point of lumen occlusion and progressively more severe hypertension ending in right heart failure.

Reasoning that the combined effects of increased blood flow and toxin-induced endothelial injury would cause more severe arteriopathy and pulmonary hypertension than either insult alone, Botney (1999) developed the model of left lung pneumonectomy plus monocrotaline injection in rats. This model develops smooth muscle α-actin-positive and endothelial CD31-negative neointimal lesions in small pulmonary arteries/arterioles of the right lung (Nishimura *et al.*, 2002, 2003). Histological examination suggests more than 50% occlusion of some arteries by medial hypertrophy and neointimal hyperplasia.

In comparison to Botney's approach, Voelkel and colleagues reasoned that physiological vascular endothelial growth factor signaling was required to maintain pulmonary vascular endothelial homeostasis (Voelkel *et al.*, 2006) and developed a rat model of severe pulmonary hypertension based on the single subcutaneous implantation of the highly lipophilic vascular endothelial growth factor receptor blocker, SU5416 (semaxinib), combined with exposure of the rats to 2 to 3 weeks of chronic hypoxia (Taraseviciene-Stewart *et al.*, 2001). This model develops severe progressive pulmonary hypertension associated with the formation of occlusive neointimal lesions in small pulmonary arteries/arterioles. Interestingly, the hypertension and density of occlusive lesions become progressively more severe after the hypoxic animals are returned to the relative normoxia of Denver's altitude.

2.1. Methods

2.1.1. Left lung pneumonectomy plus monocrotaline model

The left lung pneumonectomized rat is available commercially (Zivic Inc.).

Two weeks after left lung pneumonectomy, adult male Sprague–Dawley rats (350 to 400g) are injected subcutaneously with monocrotaline (60 mg/kg, Sigma). They usually develop severe pulmonary hypertension with occlusive neointimal lesions in small pulmonary arteries/arterioles 3 to 4 weeks after monocrotaline injection and start dying after this time point. Several factors influence the severity of pulmonary hypertension in this model, including (1) monocrotaline dose, (2) body weight/age of rats at the time of monocrotaline injection (more severe in smaller/younger rats), and (3) duration between pneumonectomy and monocrotaline injection (more severe in shorter duration).

2.1.2. SU5416 plus hypoxia-exposed model

Adult male Sprague–Dawley rats weighing $\approx$200 g are injected subcutaneously with SU5416 (20 mg/kg, Sugen Inc.), which is suspended in CMC [0.5% (w/v) carboxymethylcellulose sodium, 0.9% (w/v) NaCl, 0.4% (v/v) polysorbate, 0.9% (v/v) benzyl alcohol in deionized water]. The rats are then exposed to chronic hypoxia in a hypobaric chamber (barometric pressure $\approx$410 mm Hg, inspired O_2 tension $\approx$76 mm Hg) for up to 3 weeks. The rats consistently develop severe occlusive pulmonary hypertension 2 to 3 weeks after SU5416 injection and exposure to chronic hypoxia, and the hypertension becomes progressively more severe after the animals are returned to a normoxic environment.

3. ACUTE HEMODYNAMIC EFFECTS OF SELECTIVE RHO KINASE INHIBITORS

To evaluate the contribution of Rho kinase-mediated vasoconstriction to pulmonary hypertension of an experimental model, the first and essential step is to obtain hemodynamic data before and after acute administration of a selective Rho kinase inhibitor. If the Rho kinase inhibitor reduces the high PA pressure significantly without decreasing the cardiac output of the animal, it can be considered that there is a significant contribution of Rho kinase-mediated vasoconstriction. There are currently two selective Rho kinase inhibitors available commercially, i.e., Y-27632 and HA1077 (or fasudil). Although they are relatively selective for Rho kinase up to 10 μM (IC$_{50}$ values of Y-27632 and HA1077 for Rho kinase II are 0.162 and 0.158 μM, respectively) (Tamura *et al.*, 2005), they may also inhibit other kinases, such as protein kinase C and protein kinase N1 (Tamura *et al.*, 2005; Torbett *et al.*, 2003).

3.1. Methods

3.1.1. Catheterized rat

The adult rat is injected with a combination of ketamine (75 mg/kg) and xylaxine (6 mg/kg) intramuscularly as the anesthetic. The ventral neck area and the area dorsally between the scapulae are shaved and the areas are scrubbed with a betadine solution. The animal is placed on a heated sterile surface and is then draped. Lidocaine (2%) is injected into the ventral neck area, and the area is tested for pain response prior to making a 2-cm incision to expose the external jugular vein. A polyvinyl (PV-1, Tygon) catheter is threaded into the pulmonary artery via the external jugular vein (the tip of the catheter is inserted into the jugular vein and then threaded into the right atrium, right ventricle, and lumen of the main PA). The pressure tracing through this catheter is monitored on an oscilloscope, and the location of the tip is identified by the characteristic shapes of the pressure waveforms. The catheter is tied in place. Two additional polyethylene catheters (PE50, Becton Dickinson) are inserted into the jugular vein and tied into place for drug infusion, for venous return for cardiac outputs, and for injecting green dye for cardiac outputs. The right carotid artery is isolated and a catheter (PE50) is inserted 4 cm into the artery and tied in place for blood gases and cardiac outputs (Fig. 15.2). Hemodynamic measurements are made under anesthetic (right after catheterization) or conscious (after 2 days of recovery from anesthesia for catheterization) conditions. Pulmonary and systemic artery pressures are input into a physiograph and microcomputer. The heart rate is determined from the systemic artery pressure trace. Cardiac output is measured using the standard indocyanine green dye dilution technique (Stevens et al., 1993). Briefly, 5 μg of green dye is injected into a jugular catheter while carotid arterial blood is simultaneously withdrawn through a densitometer cuvette. This signal is input to a computer, physiograph, and oscilloscope, generating the dye curve. Total pulmonary and systemic resistances are calculated by mean arterial pressures/cardiac output.

4. DEPHOSPHOLYRATION OF MYPT1

To support that acute *in vivo* administration of the Rho kinase inhibitor reduces the high PA pressure through inhibition of Rho kinase, and thereby activation of MLCP (dephosphorylation of its regulatory subunit MYPT1),

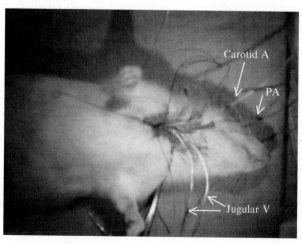

Figure 15.2   A catheterized rat (anesthetized). PA, pulmonary artery.

we compared phospho-MYPT levels in lungs from pulmonary hypertensive rats treated and untreated with a Rho kinase inhibitor. For this measurement, we first immunoprecipitate total MYPT protein and then perform a standard Western blot analysis.

4.1. Methods

Right lung lobes are isolated from anesthetized rats (pentobarbital sodium 30 mg intraperitoneally), snap frozen, and stored at $-80°$ for subsequent measurements. Frozen rat lung tissue is homogenized in lysis buffer [20 mM HEPES, pH 7.4, 1 mM dithiothreitol, 10% glycerol, 0.1% Triton X-100]. The tissue homogenate is centrifuged at 10,000 rpm at $4°$ for 10 min, and the supernatant is collected. Protein concentration in the supernatant is determined by the Bradford assay using Bradford reagent from Sigma. Whole lung protein extracts (500 μl, protein concentration 10 μg/μl) are incubated with 7 μl of anti-MYPT1 antibody (Upstate) for 4 h at $4°$ to allow antibody–antigen complexes to form. Washed and equilibrated EZview Red protein A affinity gel beads (50 μl) (Sigma) are added to the antibody–antigen complex and incubated overnight at $4°$ with gentle mixing. Beads are pelleted by centrifugation and washed, and the antibody–antigen complex is eluted following the manufacturer's protocol (Sigma). Samples are boiled for 5 min and subjected to electrophoresis on 4 to 12% gradient NuPAGE Bis-Tris gels (Invitrogen) and transferred to a Poly-Screen PVDF transfer membrane (NEN Life Science Products) in NuPAGE transfer buffer containing 10% methanol. Prestained molecular mass marker proteins (Bio-Rad) are used as standards for SDS-PAGE. Western blots are carried out for phosphorylated MYPT1 using antipho-spho MYPT1 (pMYPT1[Thr696], Upstate) and MYPT1, and the blots are visualized using Renaissance Western blot chemiluminescence reagent (NEN Life Science Products) and estimated by densitometry. Figure 15.3 shows an example of immunoblots for pMYPT1.

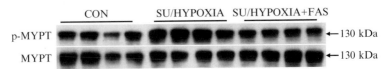

Figure 15.3 Immunoblots for phosphorylated (pMYPT1) and total MYPT1 (MYPT1) of lungs from vehicle control (CON) and a single SU 5416 injection (20 mg/kg, subcutaneous) followed by 3-week hypoxia and additional 2-week normoxia-exposed rats with (SU/HYPOXIA + FAS) and without (SU/HYPOXIA) acute intravenous fasudil treatment (10 mg/kg).

5. Ca²⁺ Sensitization

To assess whether Rho kinase-mediated Ca^{2+} sensitization is present in isolated hypertensive PAs, concentration–response curves for Ca^{2+} are determined in α-toxin permeabilized extralobar pulmonary arteries. We have observed that the concentration–response curve for Ca^{2+} in PAs from mild hypoxia-exposed pulmonary hypertensive fawn-hooded rats is left shifted compared to those from sea level-raised normotensive fawn-hooded rats (Fig. 15.4), indicating that there is increased Ca^{2+} sensitivity in the hypertensive vessels.

5.1. Methods

5.1.1. α-toxin permeabilized PA rings

The left and right first branches of the extralobar PA rings are removed from adult rats after intraperitoneal administration of 30 mg pentobarbital and intracardiac injection of 100 IU heparin. After the arteries are cleaned of adherent fat and connective tissue, endothelium intact rings 3 mm in width are placed on steel wires attached to Grass force displacement transducers,

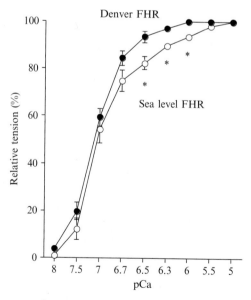

Figure 15.4 Concentration–response curves of calcium in α-toxin-permeabilized pulmonary arteries from sea level-raised (normotensive) and Denver altitude-raised (hypertensive) fawn-hooded rats (FHR). Values are means ± SE of $n = 4$ each. *$P < 0.05$ vs Denver FHR. Reproduced from Nagaoka *et al.* (2006) with permission.

suspended in baths containing 10 ml of physiological salt solution (Earle's balanced salt solution, Sigma), and gassed with 21% O_2, 5% CO_2, and 74% N_2. Resting passive force is adjusted to an optimal tension (determined by maximal response to 80 mM KCl). After 60 min of equilibration, rings are incubated in relaxing solution (described later) for 15 min, gassed with 21% O_2 and 79% N_2, and then permeabilized as described previously (Kitazawa *et al.*, 1989; Shaw *et al.*, 1997) with minor modifications. Briefly, a 10-μl droplet of pCa 6.3 solution containing 750 units of *Staphylococcus aureus* α-toxin (Sigma) and 10 μM A23187, to deplete the sarcoplasmic reticulum of Ca^{2+}, is placed onto each ring segment. After tension development reaches a plateau, the rings are equilibrated in relaxing solution and then exposed to activating solution (described later). All experiments are done at room temperature. Relaxing solution consists of (in mM) PIPES 30, sodium creatine phosphate 10, Na_2ATP 5.16, magnesium chloride 7.31, potassium methane sulfonate 74.1, K_2EGTA 1; pH is adjusted to 7.1 with KOH. In the α-toxin or activating solution, 10 mM EGTA is used, and a specified amount of $CaCl_2$, calculated using the Maxcheletor program (WINMAX Cv 2.00 by C Patton, http//www.Stanford.edu/~cpatton/maxc.html), is added to give the desired concentration of free Ca^{2+} ions (pCa).

6. RhoA Activity

It is well established that various GPCR agonists, such as ET-1, 5-HT, and TxA_2, are involved in the pathogenesis of most forms of human as well as experimental pulmonary hypertension (Farber and Loscalzo, 2004; Said, 2006). These agonists (vasoconstrictors) can activate RhoA, which leads to Rho kinase activation and thereby vascular smooth muscle cell Ca^{2+} sensitization and contraction (see Fig. 15.1). Thus, it is important to know if RhoA is activated in the hypertensive pulmonary circulation to evaluate the involvement of Rho kinase-mediated vasoconstriction. Stimulation of GPCR leads to the exchange of GTP for GDP on RhoA, and GTP-bound RhoA translocates to the membrane where it interacts with Rho kinase. Therefore, there are two ways to estimate RhoA activity, i.e., to measure GTP-bound RhoA and membrane-bound RhoA. We often use the membrane-bound RhoA measurement because under certain circumstances, such as increased protein kinase G activity by NO (Murthy *et al.*, 2003) and decreased 3-hydroxy-3-methylglutaryl-CoA reductase activity by statins (Nakagami *et al.*, 2003), the translocation of GTP-bound RhoA to the membrane can be inhibited, and thus the GTP-bound RhoA measurement may overestimate real RhoA activation.

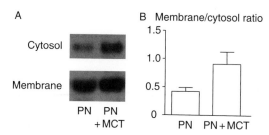

Figure 15.5 (A) RhoA protein levels of lungs from left pneumonectomized rats with a single injection of monocrotaline (60 mg/kg, subcutaneous, PN + MCT) and saline (PN) in cytosolic and membrane fractions. (B) Membrane-to-cytosolic ratio (densitometric values) of RhoA protein expression. Values are means ± SE of $n = 4$.

6.1. Methods

6.1.1. RhoA translocation to the membrane

To assess membrane translocation of RhoA, the frozen lung tissue is homogenized in lysis buffer (10 mM HEPES, 2 mM EDTA, 1 mM MgCl$_2$) containing protease inhibitors after equilibration. The homogenate is centrifuged at 40,000 g for 30 min, and the supernatant is collected as the cytosolic fraction. The pellet is resuspended in lysis buffer containing 0.1% SDS and centrifuged again (40,000 g for 15 min) to generate the membrane fraction. Equal amounts of protein (15 μg) are loaded in the gel. The RhoA protein in membrane and cytosolic fractions is determined by standard Western blot analysis using the mouse monoclonal anti-RhoA antibody (1:250 dilution, Santa Cruz Biotechnology) and a peroxidase-labeled antimouse IgG antibody (3:10,000 dilution, Vector). The relative density of membrane to cytosolic RhoA is determined using NIH image software. An example of measurement of membrane-bound RhoA is shown in Fig. 15.5.

REFERENCES

Amano, M., Chihara, K., Nakamura, N., Kaneko, T., Matsuura, Y., and Kaibuchi, K. (1999). The COOH terminus of Rho-kinase negatively regulates rho-kinase activity. *J. Biol. Chem.* **274,** 32418–32424.

Batchelor, T. J., Sadaba, J. R., Ishola, A., Pacaud, P., Munsch, C. M., and Beech, D. J. (2001). Rho-kinase inhibitors prevent agonist-induced vasospasm in human internal mammary artery. *Br. J. Pharmacol.* **132,** 302–308.

Botney, M. D. (1999). Role of hemodynamics in pulmonary vascular remodeling: Implications for primary pulmonary hypertension. *Am. J. Respir. Crit. Care Med.* **159,** 361–364.

Budzyn, K., Marley, P. D., and Sobey, C. G. (2006). Targeting Rho and Rho-kinase in the treatment of cardiovascular disease. *Trends Pharmacol. Sci.* **27,** 97–104.

Farber, H. W., and Loscalzo, J. (2004). Pulmonary arterial hypertension. *N. Engl. J. Med.* **351**, 1655–1665.

Fukata, Y., Amano, M., and Kaibuchi, K. (2001). Rho-Rho-kinase pathway in smooth muscle contraction and cytoskeletal reorganization of non-muscle cells. *Trends Pharmacol. Sci.* **22**, 32–39.

Fukumoto, Y., Matoba, T., Ito, A., Tanaka, H., Kishi, T., Hayashidani, S., Abe, K., Takeshita, A., and Shimokawa, H. (2005). Acute vasodilator effects of a Rho-kinase inhibitor, fasudil, in patients with severe pulmonary hypertension. *Heart* **91**, 391–392.

Heath, D. (1992). The rat is a poor animal model for the study of human pulmonary hypertension. *Cardioscience* **3**, 1–6.

Howell, K., Ooi, H., Preston, R., and McLoughlin, P. (2004). Structural basis of hypoxic pulmonary hypertension: The modifying effect of chronic hypercapnia. *Exp. Physiol.* **89**, 66–72.

Hu, E., and Lee, D. (2003). Rho kinase inhibitors as potential therapeutic agents for cardiovascular diseases. *Curr. Opin. Invest. Drugs* **4**, 1065–1075.

Humbert, M., Morrell, N. W., Archer, S. L., Stenmark, K. R., MacLean, M. R., Lang, I. M., Christman, B. W., Weir, E. K., Eickelberg, O., Voelkel, N. F., and Rabinovitch, M. (2004). Cellular and molecular pathobiology of pulmonary arterial hypertension. *J. Am. Coll. Cardiol.* **43**, 13S–24S.

Hyvelin, J. M., Howell, K., Nichol, A., Costello, C. M., Preston, R. J., and McLoughlin, P. (2005). Inhibition of Rho-kinase attenuates hypoxia-induced angiogenesis in the pulmonary circulation. *Circ. Res.* **97**, 185–191.

Ishikura, K., Yamada, N., Ito, M., Ota, S., Nakamura, M., Isaka, N., and Nakano, T. (2006). Beneficial acute effects of rho-kinase inhibitor in patients with pulmonary arterial hypertension. *Circ. J.* **70**, 174–178.

Kitazawa, T., Kobayashi, S., Horiuti, K., Somlyo, A. V., and Somlyo, A. P. (1989). Receptor-coupled, permeabilized smooth muscle: Role of the phosphatidylinositol cascade, G-proteins, and modulation of the contractile response to Ca^{2+}. *J. Biol. Chem.* **264**, 5339–5342.

Mandegar, M., Thistlethwaite, P. A., and Yuan, J. X. (2004). Molecular biology of primary pulmonary hypertension. *Cardiol. Clin.* **22**, 417–429 vi.

Murthy, K. S., Zhou, H., Grider, J. R., and Makhlouf, G. M. (2003). Inhibition of sustained smooth muscle contraction by PKA and PKG preferentially mediated by phosphorylation of RhoA. *Am. J. Physiol. Gastrointest. Liver Physiol.* **284**, G1006–G1016.

Nagaoka, T., Fagan, K. A., Gebb, S. A., Morris, K. G., Suzuki, T., Shimokawa, H., McMurtry, I. F., and Oka, M. (2005). Inhaled Rho kinase inhibitors are potent and selective vasodilators in rat pulmonary hypertension. *Am. J. Respir. Crit. Care Med.* **171**, 494–499.

Nagaoka, T., Gebb, S. A., Karoor, V., Homma, N., Morris, K. G., McMurtry, I. F., and Oka, M. (2006). Involvement of RhoA/Rho kinase signaling in pulmonary hypertension of the fawn-hooded rat. *J. Appl. Physiol.* **100**, 996–1002.

Nagaoka, T., Morio, Y., Casanova, N., Bauer, N., Gebb, S., McMurtry, I., and Oka, M. (2004). Rho/Rho kinase signaling mediates increased basal pulmonary vascular tone in chronically hypoxic rats. *Am. J. Physiol. Lung Cell. Mol. Physiol.* **287**, L665–L672.

Nakagami, H., Jensen, K. S., and Liao, J. K. (2003). A novel pleiotropic effect of statins: Prevention of cardiac hypertrophy by cholesterol-independent mechanisms. *Ann. Med.* **35**, 398–403.

Newman, J. H., Fanburg, B. L., Archer, S. L., Badesch, D. B., Barst, R. J., Garcia, J. G., Kao, P. N., Knowles, J. A., Loyd, J. E., McGoon, M. D., Morse, J. H., Nichols, W. C., *et al.* (2004). Pulmonary arterial hypertension. Future directions: Report of a National Heart, Lung and Blood Institute/Office of Rare Diseases workshop. *Circulation* **109**, 2947–2952.

Nishimura, T., Faul, J. L., Berry, G. J., Vaszar, L. T., Qiu, D., Pearl, R. G., and Kao, P. N. (2002). Simvastatin attenuates smooth muscle neointimal proliferation and pulmonary hypertension in rats. *Am. J. Respir. Crit. Care Med.* **166,** 1403–1408.

Nishimura, T., Vaszar, L. T., Faul, J. L., Zhao, G., Berry, G. J., Shi, L., Qiu, D., Benson, G., Pearl, R. G., and Kao, P. N. (2003). Simvastatin rescues rats from fatal pulmonary hypertension by inducing apoptosis of neointimal smooth muscle cells. *Circulation* **108,** 1640–1645.

Nobe, K., and Paul, R. J. (2001). Distinct pathways of Ca(2+) sensitization in porcine coronary artery: Effects of Rho-related kinase and protein kinase C inhibition on force and intracellular Ca(2+). *Circ. Res.* **88,** 1283–1290.

Oka, M., Homma, N., Taraseviciene-Stewart, L., Morris, K. G., Kraskauskas, D., Burns, N., Voelkel, N. F., and McMurtry, I. F. (2007). Rho kinase-mediated vasoconstriction is important in severe occlusive pulmonary arterial hypertension in rats. *Circ. Res.* **100,** 923–929.

Pfitzer, G. (2001). Invited review: Regulation of myosin phosphorylation in smooth muscle. *J. Appl. Physiol.* **91,** 497–503.

Pietra, G. G., Capron, F., Stewart, S., Leone, O., Humbert, M., Robbins, I. M., Reid, L. M., and Tuder, R. M. (2004). Pathologic assessment of vasculopathies in pulmonary hypertension. *J. Am. Coll. Cardiol.* **43,** 25S–32S.

Reeves, J. T., Groves, B. M., and Turkevich, D. (1986). The case for treatment of selected patients with primary pulmonary hypertension. *Am. Rev. Respir. Dis.* **134,** 342–346.

Rich, S., Dantzker, D. R., Ayres, S. M., Bergofsky, E. H., Brundage, B. H., Detre, K. M., Fishman, A. P., Goldring, R. M., Groves, B. M., Koerner, S. K., *et al.* (1987). Primary pulmonary hypertension: A national prospective study. *Ann. Intern. Med.* **107,** 216–223.

Said, S. I. (2006). Mediators and modulators of pulmonary arterial hypertension. *Am. J. Physiol. Lung Cell. Mol. Physiol.* **291,** L547–L558.

Sakurada, S., Okamoto, H., Takuwa, N., Sugimoto, N., and Takuwa, Y. (2001). Rho activation in excitatory agonist-stimulated vascular smooth muscle. *Am. J. Physiol. Cell. Physiol.* **281,** C571–C578.

Shaw, L. M., Ohanian, J., and Heagerty, A. M. (1997). Calcium sensitivity and agonist-induced calcium sensitization in small arteries of young and adult spontaneously hypertensive rats. *Hypertension* **30,** 442–448.

Shimokawa, H., Hiramori, K., Iinuma, H., Hosoda, S., Kishida, H., Osada, H., Katagiri, T., Yamauchi, K., Yui, Y., Minamino, T., Nakashima, M., and Kato, K. (2002). Anti-anginal effect of fasudil, a Rho-kinase inhibitor, in patients with stable effort angina: A multicenter study. *J. Cardiovasc. Pharmacol.* **40,** 751–761.

Shimokawa, H., and Takeshita, A. (2005). Rho-kinase is an important therapeutic target in cardiovascular medicine. *Arterioscler. Thromb. Vasc. Biol.* **25,** 1767–1775.

Sitbon, O., Humbert, M., Jais, X., Ioos, V., Hamid, A. M., Provencher, S., Garcia, G., Parent, F., Herve, P., and Simonneau, G. (2005). Long-term response to calcium channel blockers in idiopathic pulmonary arterial hypertension. *Circulation* **111,** 3105–3111.

Somlyo, A. P., and Somlyo, A. V. (1994). Signal transduction and regulation in smooth muscle. *Nature* **372,** 231–236.

Somlyo, A. P., and Somlyo, A. V. (2003). Ca^{2+} sensitivity of smooth muscle and nonmuscle myosin II: Modulated by G proteins, kinases, and myosin phosphatase. *Physiol. Rev.* **83,** 1325–1358.

Stevens, T., Morris, K., McMurtry, I. F., Zamora, M., and Tucker, A. (1993). Pulmonary and systemic vascular responsiveness to TNF-alpha in conscious rats. *J. Appl. Physiol.* **74,** 1905–1910.

Tamura, M., Nakao, H., Yoshizaki, H., Shiratsuchi, M., Shigyo, H., Yamada, H., Ozawa, T., Totsuka, J., and Hidaka, H. (2005). Development of specific Rho-kinase inhibitors and their clinical application. *Biochim. Biophys. Acta* **1754,** 245–252.

Taraseviciene-Stewart, L., Kasahara, Y., Alger, L., Hirth, P., Mc Mahon, G., Waltenberger, J., Voelkel, N. F., and Tuder, R. M. (2001). Inhibition of the VEGF receptor 2 combined with chronic hypoxia causes cell death-dependent pulmonary endothelial cell proliferation and severe pulmonary hypertension. *FASEB J.* **15,** 427–438.

Torbett, N. E., Casamassima, A., and Parker, P. J. (2003). Hyperosmotic-induced protein kinase N 1 activation in a vesicular compartment is dependent upon Rac1 and 3-phosphoinositide-dependent kinase 1. *J. Biol. Chem.* **278,** 32344–32351.

Uehata, M., Ishizaki, T., Satoh, H., Ono, T., Kawahara, T., Morishita, T., Tamakawa, H., Yamagami, K., Inui, J., Maekawa, M., and Narumiya, S. (1997). Calcium sensitization of smooth muscle mediated by a Rho-associated protein kinase in hypertension. *Nature* **389,** 990–994.

van Suylen, R. J., Smits, J. F., and Daemen, M. J. (1998). Pulmonary artery remodeling differs in hypoxia- and monocrotaline-induced pulmonary hypertension. *Am. J. Respir. Crit. Care Med.* **157,** 1423–1428.

Voelkel, N. F., and Tuder, R. M. (2000). Hypoxia-induced pulmonary vascular remodeling: A model for what human disease? *J. Clin. Invest.* **106,** 733–738.

Voelkel, N. F., Vandivier, R. W., and Tuder, R. M. (2006). Vascular endothelial growth factor in the lung. *Am. J. Physiol. Lung Cell. Mol. Physiol.* **290,** L209–L221.

CHAPTER SIXTEEN

EPITHELIAL RHO GTPASES AND THE TRANSEPITHELIAL MIGRATION OF LYMPHOCYTES

Joanna C. Porter

Contents

Abstract

Tissue injury and inflammation lead to leukocyte recruitment from the bloodstream into the inflamed organ. Because leukocytes in excessive numbers and over prolonged periods can cause tissue damage, it is important that the trafficking of leukocytes is regulated. Although much attention has been focused on leukocyte recruitment, much less is known about the resolution of inflammation. Hollow organs, such as the lung and the gut, are unique in that

Medical Research Council Laboratory of Molecular Cell Biology, University College London, and Department of Respiratory Medicine, University College London Hospitals NHS Trust, University College London Hospital, London, United Kingdom

Methods in Enzymology, Volume 439
ISSN 0076-6879, DOI: 10.1016/S0076-6879(07)00416-8

tissue accumulation of leukocytes is determined by the recruitment of leukocytes from the blood; survival of tissue leukocytes; and migration of leukocytes from the interstitial space, either to the lymphatics or into the lumen of the organ, so-called egression. It has been shown that preventing egression of peribronchial leukocytes in a murine model of bronchial inflammation was fatal. This has led to an interest in the molecular mechanisms underlying egression from the lung. We have used a human bronchial cell line, 16HBE14^{0-}, *in vitro* to analyze transepithelial migration and to investigate the role of Rho GTPases in this process. This chapter describes methods used to establish monolayers of bronchial epithelial cells either the correct way up or inverted on Transwell filters and describes an assay of transepithelial migration of primary human T lymphocytes across this monolayer. This chapter shows how this system can be used to dissect out the molecular events that are required for successful egression. In particular, pretreatment of either the lymphocytes or the epithelium with blocking antibodies against cell surface receptors or with cell-permeable inhibitors directed against signaling molecules allows an analysis of the individual roles played by the T lymphocytes and the epithelial monolayer.

1. INTRODUCTION

1.1. Transepithelial migration in the lung, gut, and other hollow organs

The lung epithelium provides an extensive surface area, in direct contact with the outside environment. This is essential for effective gas exchange but leaves the lung uniquely susceptible to damage or infection by inhaled allergens and pathogens and may explain why lung disease is the single greatest cause of death worldwide (WHO, 2003). Because of this threat there is a need for constant immune surveillance, and lymphocytes traffic through the lung continuously, with rapid recruitment of T lymphocytes when foreign antigens are recognized. Th1/Tc1 effector T lymphocytes play an essential role in the immune response against infectious diseases and may be distinguished from Th2/Tc2 cells by a higher expression of CCR5 and CXCR3 and an enhanced response to the ligands for these receptors (Bonecchi *et al.*, 1998; D'Ambrosio *et al.*, 1998; Moser and Loetscher, 2001). Although essential in fighting infection, excessive or prolonged infiltration of the lung by effector Th1/Tc1 cells may underlie the pathology of diseases as diverse as influenza (Humphreys *et al.*, 2003; Hussell *et al.*, 2004) and tuberculosis (Guyot-Revol *et al.*, 2006), as well as noninfectious lung diseases, such as chronic obstructive pulmonary disease (COPD) (Grumelli *et al.*, 2004), a common debilitating inflammatory disease of the lung caused by tobacco smoke and other inhaled pollutants. An understanding of the pathophysiology behind these disease underlies much of respiratory medicine; tuberculosis is responsible for >1.5 million deaths a year, influenza may occur in devastating epidemics, and COPD is predicted to become the third most common cause of death worldwide by 2020. Much of the damage and death caused by these diseases has been shown to be because of tissue destruction in association with excessive leukocyte recruitment. It is therefore essential that, during the immune response to these diseases, effector T-cell movement into the lung is regulated and that accumulated T cells are promptly cleared when the immediate threat is over.

Although much is known about how effector T lymphocytes enter the lung, the clearance of these cells from inflamed tissue during the resolution phase has been less well studied. The chemokine receptor CCR7 directs the migration of CCR7$^+$ effector and memory lymphocytes from peripheral tissues (via afferent lymphatics) to the lymph nodes (Bromley et al., 2005; Debes et al., 2005), but the factors, if any, that determine the exit of CCR7$^-$ effector and memory T cells from peripheral nonlymphoid tissues such as the lung remain unknown. It has been assumed that most infiltrating leukocytes undergo either necrosis or apoptosis at the site of inflammation, ignoring a potentially very important exit pathway across the bronchial epithelial barrier. Migrating leukocytes, having crossed the epithelial barrier, into the airway would be carried on the mucociliary escalator to the pharynx for removal. Such egression or luminal clearance from the lung has been described for eosinophils and neutrophils (Erjefalt et al., 2004; Uller et al., 2001), and compelling evidence shows that the transepithelial migration of all classes of leukocyte plays a key protective role in experimental murine models of bronchial inflammation (Corry et al., 2002, 2004).

Despite the potential importance of lymphocyte egression from the lung across the bronchial epithelium, this process has not been well studied, in stark contrast to the extensive research into transendothelial migration. There are critical differences in migration across the endothelial and epithelial barriers, the most obvious being that the epithelium forms a much "tighter" barrier than that formed by the endothelium, that maintenance is essential, and that leukocyte recruitment across an epithelium occurs in a basal-to-apical direction, opposite to that across the endothelium.

1.2. Role of Rho GTPases in TEpM

Epithelia form a barrier between the organism and the outside world. All epithelia are polarized into an apical (luminal) compartment and a basolateral compartment, which are separated by the apical junctional complex (AJC). The AJC consists of the apical tight junction and the more basal adherens junction and desmosomes. Rho GTPases have been shown to be involved in the formation and stabilization of intercellular junctions (Etienne-Manneville and Hall, 2002; Van Aelst and Symons, 2002). It is assumed that the majority of leukocytes that cross a tight epithelial barrier will move between adjacent epithelial cells and across the tight interepithelial junctions. It is thought that the leukocytes will have to transiently disrupt the junctions to allow their passage across. It therefore seems highly likely that the Rho family of GTPases and the actin cytoskeleton are involved in tight junction regulation during leukocyte transepithelial migration.

2. Culture of Bronchial Epithelial Cells

We screened several bronchial epithelial cell lines to find one that was suitable for this assay. The 16HBE14^{0-} (16HBE) cell line is used frequently in studies that are looking at tight junctions and has been shown by others to make structural and functional tight junctions (Wan *et al.*, 1999, 2000).

2.1. Materials

The SV40-transformed human bronchial epithelial cell line 16HBE (Gruenert *et al.*, 1988) was a kind gift from Dieter Gruenert. Bovine serum albumin (BSA), EDTA, EGTA, polyvinylpyrrolidone, and human fibronectin are from Sigma–Aldrich (Gillingham, UK); modified Eagle medium (MEM) is from Gibco, Invitrogen Corporation (Paisley, UK). PureCol is from Nutacon BV (Leimuiden, The Netherlands). Fetal bovine serum (FBS; Mycoplex) is from PAA Laboratories (Yeovil, UK). Cotton wool tips are purchased at a local drug store and autoclaved before use (Boots, Nottingham, UK).

2.2. Methods

16HBE cells are best cultured on flasks or membranes that have been precoated overnight at 4° with 30 μg/ml PureCol, 100 μg/ml BSA, and 10 μg/ml fibronectin in PBS-A (we call this coating mix). To coat the flasks, add the coating mix at 100 μl/cm^2, and for filter membranes, coat only the side of the filter on which cells should grow. Add 300 μl/cm^2 to coat the inner membrane or 1200 μl/cm^2 in the outer well to coat the underside. After incubation overnight, the coating mix is removed and the flasks or membranes are rinsed once with PBS-A. The 16HBE cells grow well in MEM with 10% FBS at 37° with 5% CO$_2$. Flasks of cells should be passaged (see later) when the cells are 80% confluent. If monolayers are left for a long period, they will overgrow and cells may pile up on each other.

2.2.1. Culture of 16HBE cells as monolayers on filters

For the transepithelial migration assay, 16HBE cells are passaged under gentle trypsinization. The monolayer of cells is washed in PBS-A to remove inhibitors of trypsin and the cells are then released from the plastic using 0.02% trypsin/ 0.0016% EDTA with 0.02% EGTA and 1% polyvinylpyrrolidine solution in HEPES-buffered saline. The cells take several minutes to be released under these conditions. After release, the cells are centrifuged for 5 min at 1000 rpm and are resuspended in fresh culture medium. Then 100 μl of cells at 1 × 10^6/ml are plated on 8-μm pore size, 6.5-mm-diameter polycarbonate filters in 24-well Transwell chambers (Costar, Corning) and grown for 10 to 11 days. During this time, epithelial cells move across the filter to grow on both sides. The monolayers are checked for impermeability to fluid by the ability to maintain a fluid level difference between inner and outer wells. The inner well is filled with medium and the outer well has 600 μl of medium so that there is a fluid level difference between the two. The monolayers are then returned to the incubator for 24 h. After this time the fluid level in the inner wells should be unchanged. If the inner well fluid level falls, the monolayers are not yet ready and should be left for another 24 h before re-testing. We find that after 10 to 12 days most of the monolayers are tight. Cells are then removed from one side of the monolayer with a cotton wool tip to leave a monolayer on either the top or the bottom of the filter. To do this, wells are emptied and washed with PBS-A, and 100 μl of fresh PBS-A is pipetted into the inner well. Then, one end of a cotton wool tip is gently pushed against the filter and moved in two full circular motions to remove adherent cells from that side of the filter. Any nonadherent cells are removed by a further rinse with PBS-A and the monolayers are put back in the incubator with 300 μl of medium in the inner well and 400 μl of medium in the outer well. If a fluid level difference is maintained after 4 h, then an air–liquid interface is introduced for the bronchial epithelial cells. This involves having medium only on the basolateral side of the monolayer and leaving the apical surface with no medium. This air–liquid interface is thought to help the tight junctions to reform more quickly. The monolayers are left in culture for a further 24 h before recording their transepithelial electrical resistance (TER) and permeability to FITC-dextran 40 kDa (see next section).

3. MEASUREMENT OF TER AND PERMEABILITY TO FITC-DEXTRAN 40 KDA

3.1. Materials

The EVOM epithelial volt ohmmeter is from World Precision Instruments Ltd, (Stevenage, UK). FITC-dextran 40 kDa is from Sigma-Aldrich, and the SPECTRAmax GEMINI XS microplate spectrofluorometer is from

Molecular Devices Ltd. (Wokingham, UK). Hanks balanced salt solution (HBSS) is from Gibco Invitrogen Corporation.

3.2. Methods

Transepithelial electrical resistance is measured using a volt ohmmeter. Confluent monolayers are washed and placed in HBSS to equilibrate for 30 min at 37°. The voltage across the monolayer is recorded and expressed as ohm·cm^2. The voltage across a filter with no epithelial cells is taken as the background reading and may be subtracted from other readings. To measure FITC-dextran permeability, the FITC-dextran is dissolved in water at 50 mg/ml, diluted 1:200 in RPMI, and sonicated briefly prior to use. Permeability to FITC-dextran is measured by washing the monolayers with RPMI/0.5% BSA and taking a filter without epithelial cells as a control. After washing, the monlayers and filters are placed in an empty 24-well plate with 600 μl of RPMI/0.5% BSA in each well. Then, 100 μl of FITC-dextran, molecular mass 40 kDa, is added to the upper chamber. The plate is incubated at 37°. After 30 min, 100 μl of medium is withdrawn from the lower chamber and put in an empty well of a 96-well plate, and 100 μl of medium alone is added to an empty well of the 96-well plate to assess background fluorescence. Fluorescence of the medium in the 96-well plate is measured on a SPECTRAmax GEMINI XS microplate spectrofluorometer (Molecular Devices; excitation: 485 λ; emission: 538 λ). Fluorescence is linear for the concentration of FITC-dextran used, and the permeability of the monolayer is expressed as the percentage permeability of the filter alone.

4. CULTURE OF PRIMARY HUMAN T LYMPHOCYTES

4.1. Materials

T cells are obtained as a buffy coat preparation from UK National Blood Services (Tooting, London, UK). Lymphoprep is from Pharmacia Diagnostics AB (Uppsala, Sweden); RPMI 1640 is obtained from Gibco, Invitrogen Corporation. Phytohemagglutinin (PHA) is from Murex Diagnostics (Dartford, UK). Interleukin (IL)-2 is from Euro Cetus UK Ltd. (Harefield, UK).

4.2. Methods

Peripheral blood mononuclear cells are prepared from single donor leukocyte buffy coats by centrifugation through Lymphoprep. T cells are expanded from this population by culturing in RPMI 1640 plus 10% FCS in the presence of PHA at 1 μg/ml for 72 h as described previously

(Dransfield *et al.*, 1992) . Mononuclear cells attach to the plastic flask during culture and can be removed after 72 h. Cells are washed and maintained for 1 to 2 weeks in medium supplemented with 20 ng/ml recombinant IL-2. The cells are used between days 10 and 14 and are a 99% CD3$^+$ population, containing 65% CD8$^+$ and 35% CD4$^+$ cells. The population is negative for the natural killer cell marker CD56. The lymphocytes are of a Th1/Tc1 effector phenotype expressing CXCR3, with varying levels of CCR5 and CCR7. These T cells will undergo chemotaxis in response to various chemokines, including CCL5, CXCL9, CXCL10, and CXCL11 (data not shown).

5. DEVELOPMENT OF AN ASSAY FOR LYMPHOCYTE TRANSEPITHELIAL MIGRATION

5.1. Materials

Transwell polycarbonate filters are from Corning/Costar (Corning, NY). SonicSeal four-well slides (permanox pastic) are from Nalge Nunc (Rochester, NY). Cell tracker green (CMFDA) is obtained from Molecular Probes, Invitrogen Corporation. Chemokines are from Peptrotech EC (London, UK). The rabbit antihuman ZO-1 antibody (Z-R1) is from Zymed Laboratories (San Francisco, CA). Chemokines and cytokines are from Peprotech. Formaldehyde 16% (w/v) is from TAAB Laboratories (Reading, UK); Triton X-100 is from Sigma-Aldrich. DakoCytomation fluorescent mounting medium is from Dako North America, Inc. (Carpinteria, CA).

5.2. Methods

5.2.1. Chemotaxis assays

Chemotaxis of T cells in response to various chemokines is investigated in 24-well Transwell chambers using 5-μm pore size, 6.5-mm-diameter polycarbonate filters. First, T cells are labeled with 50 nM CMFDA. T cells are washed twice into RPMI alone and are then resuspended at 2×10^6/ml and 50 nM CMDA. The T cells are incubated at 37° and 5% CO_2 for 30 min. Labeled T cells are then washed in RPMI and resuspended in RPMI/0.5% BSA at 2×10^6/ml. The chemokine is placed in the lower chamber in 600 μl of RPMI/0.5% BSA and then 100 μl of T cells is added to the upper chamber. After 90 min, migrating T cells are retrieved from the lower chamber and counted as FL1-positive cells recorded in a set time (30 s) by fluorescence-activated cell scanning (FACS). A sample well, in which 100 μl of T cells is added directly to an outer well containing 600 μl of medium, is also counted and taken as the "total cells added." The chemotactic index is expressed as the number of T cells migrating as percentage of total cells added.

5.2.2. Transepithelial migration assays

To establish an assay of transepithelial migration, we first established a polarized epithelial monolayer of the bronchial epithelial cell line 16HBE, cultured as either standard or inverted monolayers on Transwell filters to allow access to both apical and basal surfaces (Fig. 16.1). The presence of tight junctions was confirmed using a combination of fluorescence microscopy for the junctional component, zona occludens-1 (ZO-1; see Fig. 16.1), combined with measurements of TER using a volt ohmmeter and quantification of permeability to 40 kDa FITC-dextran. We find that the 16HBE cell line provides an accurate representation of human bronchial epithelium as seen *in vivo*; the cells polarize and form tight junctions. The 16HBE monolayers are fluid impermeable with TER >250 $\Omega \cdot cm^2$ and with permeability to 40 kDa F-D of $<0.5\%$ of the filter alone. There is no difference in tight junctions between standard or inverted monolayers.

Only monolayers having a TER of 200 $\Omega \cdot cm^2$ and a permeability of $<0.5\%$ of filter alone are used in transepithelial migration assays. Epithelial monolayers are washed extensively to remove dextran, rinsed with RPMI/

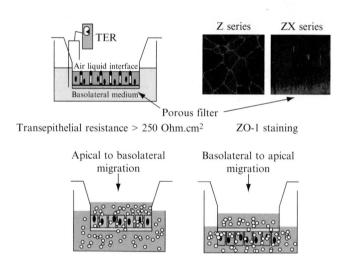

Figure 16.1 Assay for transepithelial migration of human T cells across a confluent monolayer of human bronchial epithelial cells. (Top) Human bronchial epithelial cells (HBE) can be grown on either side of a Costar Transwell filter and polarized with the TJ marker ZO-1 positioned correctly; ZO-1 is in red on the Z series and ZX series, and the porous filter is stained green. (Bottom) T-cell migration can be measured in the apical-to-basal direction (epithelial cells grown on the topside of the filter) and in the more physiological basal-to-apical direction (epithelial cells are grown on the underside of the filter). (See color insert.)

0.5% BSA, and placed in an empty 24-well plate. At this point, 100 μl of labeled T cells at 4×10^6/ml is added to the monolayers, and the inserts are transferred to new wells of a 24-well plate containing 0.6 ml of RPMI/0.5% BSA and the indicated chemokine in each outer well. The Transwells are incubated at 37° in 5% CO_2. After 90 min, T cells, which have migrated through the epithelial monolayer into the lower chambers, are recovered. FL1-positive cells are counted on FACS, and the percentage of migrated cells is calculated as described earlier.

Although very few (<5%) T cells added to the inner well move across an uncoated filter or a monolayer of 16HBE cells in the absence of a chemotactic gradient, 10 to 60% of the T cells move across the filter (chemotaxis) or the epithelial monolayer (transepithelial migration) in response to a chemotactic gradient. The variation occurs because T cells for each experiment are from individual donors. To check that this is true chemotactic movement, the chemotactic gradient can be abolished by placing equal concentrations of chemokine in the upper and lower wells. Under these conditions, both chemotaxis and transepithelial migration should be prevented. Likewise, the specificity of chemotaxis can be examined by the ability of blocking monoclonal antibody (mAb) against the appropriate chemokine receptor to prevent both chemotaxis and transepithelial migration. In this case, the T cells are washed and preincubated for 30 min on ice with the appropriate blocking monoclonal antibody. The T cells are then washed and resuspended in RPMI/0.5% BSA before being used in a chemotaxis or transepithelial migration assay.

5.2.3. Time-lapse digital microscopy of transepithelial migration

Leukocyte/epithelial cell interactions can be visualized in real time to allow further insight into the route taken by migrating T cells. The transepithelial migration assay is set up as described earlier. To look at basal-to-apical migration, a filter with T cells in the inner well and epithelial cells growing on the underside of the filter is moved into a SonicSeal four-well slide. Medium containing a chemoattractant is placed in the outer well. Microscopy is performed with time-lapse photography, and Nomarski differential interference contrast images are captured every 30 s and analyzed on QuickTime movies using ImageJ software. The time course for transepithelial migration is slower than that for chemotaxis, and cells continue to migrate for several hours after the beginning of the assay while the majority of chemotaxis has occurred within 90 min. The video microscope is focused on the basal surface of the monolayer with the filter above the plane of focus so that holes in the filters appear out of focus (Fig. 16.2; open black arrows). This video tracking allows one to see where the T cells (see Fig. 16.2; white arrows) interact with the epithelium and how they move along the epithelium and preferential areas of transepithelial migration (see Fig. 16.2; closed black arrows show migrated T cells).

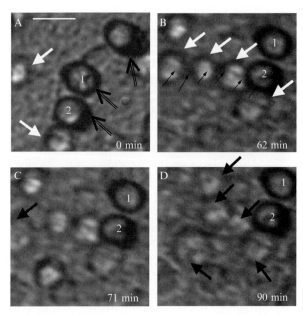

Figure 16.2 Live cell imaging and fixed cell analysis of the different stages of transepithelial migration. (A–D) Time course of video imaging of T lymphocytes during transepithelial migration. (A) White arrows mark T cells on the basal surface of the epithelium. Open black arrows mark the 8-μm pores in the filter that are out of focus and through which the T cells have already passed. Two of these pores have been labeled "1" and "2" in white. (B) The epithelial junctions can be seen, and one section is marked with small black arrows. (C and D) T cells that have successfully undergone transepithelial migration move out of the plane of focus and are labeled with closed black arrows. Scale bar: 10 μm. Reproduced and modified with permission (Porter *et al.*, 2008).

5.2.4. Effect of inhibition of Rho GTPase effectors on transepithelial migration

This model for transepithelial migration is very amenable to manipulation using antibodies against surface molecules on the T cell or the epithelial monolayer, or the use of inhibitors on the T cell or the epithelial monolayer. When using antibodies, we find it best to titrate the antibodies on the T cells or the epithelium, using FACS, to find the optimal concentration of antibody to use to block all the cell surface receptors. T cells or the epithelium can be preincubated with the appropriate antibody for 30 min at 37° before washing twice in RPMI/0.5% BSA and performing the assay as normal. One advantage of the Transwell system is that antibodies can be added to either the apical or the basal surface of the epithelium to look for differential effects. The use of blocking antibodies allows a dissection of the cell surface molecules involved in the different stages of transepithelial migration.

As well as manipulation with blocking antibodies, the system can be used to examine the effects of inhibitors on either T cells or the epithelium. We have used this system to examine the role of epithelial Rho GTPases during the transepithelial migration of T cells. The epithelial monolayer is washed and then incubated in various concentrations of a given inhibitor for various times at 37° and 5% CO_2. The inhibitor is washed out extensively (three times), and the filters are placed into a new 24-well plate. To interpret the results of inhibitor assays, it is very important to document the effect of the inhibitors on the monolayer and epithelial junctions. To do this, we measure the TER and epithelial permeability before treatment, during incubation with the inhibitors, and after washing out the inhibitors. We then fix some of the monolayers and look at the effect on the monolayer and epithelial junctions (see Section 5.2.5 for staining details). Concurrently with these experiments we also carry out transepithelial migration assays with unfixed but treated epithelial monolayers as described earlier. These assays are very good for screening an effect of an inhibitor, but further detailed studies will be required depending on the specific findings. If we identify a blocking antibody or inhibitor that affects transepithelial migration, then we have found it helpful to look in detail at the effect on migrating cells using confocal microscopy.

To perform such confocal experiments, transepithelial migration assays are carried out with and without the relevant blocking monoclonal antibody or inhibitor and at various time points we wash off the nonadherent T cells and fix the epithelial monolayers. The fixed monolayers can then be stained for analysis using confocal microscopy. This approach allows us to identify at which point in the process of transepithelial migration the T cells are inhibited and is described in detail here.

5.2.5. Confocal microscopy of epithelial monolayers with migrating T lymphocytes

For confocal studies, transepithelial migration is performed as described previously, but the T cells are unlabeled. At the indicated time the monolayers are washed briefly in warmed RPMI/0.5% BSA to remove nonadherent T cells and are then fixed in 4% formaldehyde/PBS-A for 20 min at room temperature. Monolayers are permeabilized for the detection of intracellular antigens, such as ZO-1, by incubation with 0.4% Triton-X 100 for 10 min at room temperature. Antibody staining can be performed while the filters are still in the wells. If the antibody is only available in small amounts, then the wells can be inverted and 50 to 100 μl of antibody/PBS-A can be placed on the upturned undersurface. The filters must be kept in a humid chamber during the incubation to prevent them from drying out. Alternatively, the filters can be cut from the well prior to staining using a scalpel blade, taking care to remember which way up the epithelial cells are orientated. Subsequent antibody-binding steps can be carried out by inverting the filters epithelial side down onto 50 to 100 μl of antibody on Parafilm. Depending on the antibody, the primary antibody is usually left on for 30 min at room temperature and then washed three times in PBS-A. An appropriate fluorescent-labeled secondary antibody is then used and incubated for 30 min at room temperature in the dark. We then wash four times with PBS-A before mounting the filters, epithelial cells down, on 12 μl of DakoCytomation fluorescent mounting medium on a clean microscope slide. A further 12 μl of DakoCytomation fluorescent mounting medium is placed on top of the filter, which is then covered with a circular coverslip. The slides are left in the dark at room temperature overnight for the mountant to set. T cells can be well visualized by labeling with mAb against cell surface molecules that allow the lymphocytes to be distinguished from the epithelial cells, for example, anti-CD3, followed with a fluorophore-conjugated goat antimouse secondary. In contrast, epithelial junctions can be visualized with rabbit anti-ZO-1 and a goat antirabbit secondary antibody that has been conjugated to a different fluorophore. We have successfully performed confocal microscopy on such specimens using a Zeiss laser-scanning microscope LSM 510 equipped with a 60× oil immersion objective (Karl Zeiss, Oberkochen, Germany). Images are collected as horizontal sections taken at intervals through whole cell volumes and can be compiled for display from a projection of the complete Z-series, or as a Y-Z display using ImageJ software. This allows a very careful analysis of the spatial distribution of junctional proteins and lymphocyte cell surface molecules during the multistep process of egression.

ACKNOWLEDGMENTS

JP is funded by The Wellcome Trust as an Advanced Clinical Fellow. I thank Nancy Hogg and Dieter Gruenert for generously providing reagents.

REFERENCES

Bonecchi, R., Bianchi, G., Bordignon, P. P., D'Ambrosio, D., Lang, R., Borsatti, A., Sozzani, S., Allavena, P., Gray, P. A., Mantovani, A., and Sinigaglia, F. (1998). Differential expression of chemokine receptors and chemotactic responsiveness of type 1 T helper cells (Th1s) and Th2s. *J. Exp. Med.* **187,** 129–134.

Bromley, S. K., Thomas, S. Y., and Luster, A. D. (2005). Chemokine receptor CCR7 guides T cell exit from peripheral tissues and entry into afferent lymphatics. *Nat. Immunol.* **6,** 895–901.

Corry, D. B., Kiss, A., Song, L. Z., Song, L., Xu, J., Lee, S. H., Werb, Z., and Kheradmand, F. (2004). Overlapping and independent contributions of MMP2 and MMP9 to lung allergic inflammatory cell egression through decreased CC chemokines. *FASEB J.* **18,** 995–997.

Corry, D. B., Rishi, K., Kanellis, J., Kiss, A., Song Lz, L. Z., Xu, J., Feng, L., Werb, Z., and Kheradmand, F. (2002). Decreased allergic lung inflammatory cell egression and increased susceptibility to asphyxiation in MMP2-deficiency. *Nat. Immunol.* **3,** 347–353.

D'Ambrosio, D., Iellem, A., Bonecchi, R., Mazzeo, D., Sozzani, S., Mantovani, A., and Sinigaglia, F. (1998). Selective up-regulation of chemokine receptors CCR4 and CCR8 upon activation of polarized human type 2 Th cells. *J. Immunol.* **161,** 5111–5115.

Debes, G. F., Arnold, C. N., Young, A. J., Krautwald, S., Lipp, M., Hay, J. B., and Butcher, E. C. (2005). Chemokine receptor CCR7 required for T lymphocyte exit from peripheral tissues. *Nat. Immunol.* **6,** 889–894.

Dransfield, I., Cabanas, C., Craig, A., and Hogg, N. (1992). Divalent cation regulation of the function of the leukocyte integrin LFA-1. *J. Cell Biol.* **116,** 219–226.

Erjefalt, J. S., Uller, L., Malm-Erjefalt, M., and Persson, C. G. (2004). Rapid and efficient clearance of airway tissue granulocytes through transepithelial migration. *Thorax* **59,** 136–143.

Etienne-Manneville, S., and Hall, A. (2002). Rho GTPases in cell biology. *Nature* **420,** 629–635.

Gruenert, D. C., Basbaum, C. B., Welsh, M. J., Li, M., Finkbeiner, W. E., and Nadel, J. A. (1988). Characterization of human tracheal epithelial cells transformed by an origin-defective simian virus 40. *Proc. Natl. Acad. Sci. USA* **85,** 5951–5955.

Grumelli, S., Corry, D. B., Song, L. Z., Song, L., Green, L., Huh, J., Hacken, J., Espada, R., Bag, R., Lewis, D. E., and Kheradmand, F. (2004). An immune basis for lung parenchymal destruction in chronic obstructive pulmonary disease and emphysema. *PLoS Med.* **1,** e8.

Guyot-Revol, V., Innes, J. A., Hackforth, S., Hinks, T., and Lalvani, A. (2006). Regulatory T cells are expanded in blood and disease sites in patients with tuberculosis. *Am. J. Respir. Crit. Care Med.* **173,** 803–810.

Humphreys, I. R., Walzl, G., Edwards, L., Rae, A., Hill, S., and Hussell, T. (2003). A critical role for OX40 in T cell-mediated immunopathology during lung viral infection. *J. Exp. Med.* **198,** 1237–1242.

Hussell, T., Snelgrove, R., Humphreys, I. R., and Williams, A. E. (2004). Co-stimulation: Novel methods for preventing viral-induced lung inflammation. *Trends Mol. Med.* **10,** 379–386.

Moser, B., and Loetscher, P. (2001). Lymphocyte traffic control by chemokines. *Nat. Immunol.* **2,** 123–128.

Porter, J., Falzon, M., and Hall, A. (2008). Polarized localization of epithelial CXCL11 in chronic obstructive airways disease and mechanisms of T cells egression. *Journal of Immunology,* vol **180** (in press).

Uller, L., Persson, C. G., Kallstrom, L., and Erjefalt, J. S. (2001). Lung tissue eosinophils may be cleared through luminal entry rather than apoptosis: Effects of steroid treatment. *Am. J. Respir. Crit. Care Med.* **164,** 1948–1956.

Van Aelst, L., and Symons, M. (2002). Role of Rho family GTPases in epithelial morphogenesis. *Genes Dev.* **16,** 1032–1054.

Wan, H., Winton, H. L., Soeller, C., Stewart, G. A., Thompson, P. J., Gruenert, D. C., Cannell, M. B., Garrod, D. R., and Robinson, C. (2000). Tight junction properties of the immortalized human bronchial epithelial cell lines Calu-3 and 16HBE14o. *Eur. Respir. J.* **15,** 1058–1068.

Wan, H., Winton, H. L., Soeller, C., Tovey, E. R., Gruenert, D. C., Thompson, P. J., Stewart, G. A., Taylor, G. W., Garrod, D. R., Cannell, M. B., and Robinson, C. (1999). Der p 1 facilitates transepithelial allergen delivery by disruption of tight junctions. *J. Clin. Invest.* **104,** 123–133.

World Health Organization (2003). "The World Health Report 2003: Shaping the Future." World Health Organization: Geneva, 2003: 3–22 (Chapter 1: Global health: todays' challenges).

CHAPTER SEVENTEEN

INVASION AND METASTASIS MODELS FOR STUDYING RHoGDI2 IN BLADDER CANCER

Matthew D. Nitz, Michael A. Harding, *and* Dan Theodorescu

Contents

Abstract

Invasion and metastasis are the critical steps in cancer progression that lead to death from this disease. Intense investigation into the underlying mechanisms of metastasis has revealed a complex set of signaling pathways that regulate the process. Since the mid-1980s, it has been demonstrated that the Rho family of proteins plays a major role in these pathways. Proteins that regulate Rho, including guanine nucleotide exchange factors, GTPase-activating proteins, and Rho GDP dissociation inhibitors (RhoGDIs), have also been shown to contribute to cancer progression. Among this group of Rho-regulating proteins is RhoGDI2 (RhoGDIbeta/LyGDI/GDID4/RabGDIbeta). Our laboratory initially identified RhoGDI2 as a metastasis suppressor due to its differential expression between metastatically capable and poorly metastatic bladder cancer cell lines. Over the subsequent years, *in vivo* and *in vitro* systems have been used to model steps in the metastatic cascade and to test how the expression of RhoGDI2 affected those processes. This chapter describes several of the more significant methods used to investigate the role of RhoGDI2 in bladder cancer invasion and

Department of Molecular Physiology and Biological Physics, University of Virginia Health Sciences Center, Charlottesville, Virginia

Methods in Enzymology, Volume 439　　　　　　　　　　　　© 2008 Elsevier Inc.
ISSN 0076-6879, DOI: 10.1016/S0076-6879(07)00417-X　　　All rights reserved.

metastasis. These methods include an *in vitro* assay for invasion using bladder organ cultures, lung metastasis assays in immunocompromised murine hosts, polymerase chain reaction-based quantification of metastatic burden, and derivation of increasingly metastatic cell lines.

1. INTRODUCTION

1.1. Rho GTPases in cancer metastasis

Rho GTPases are a family of proteins with a known contribution to invasion and metastasis. Interestingly, the initial discovery in 1985 of this family as Ras-related genes suggested an important role in cancer (Madaule and Axel, 1985). The subsequent finding that Rho and Rac regulate the cytoskeleton helped propel the Rho family into the spotlight of cell migration and later invasion and metastasis studies (Ridley and Hall, 1992; Ridley *et al.*, 1992) . Since that time, approximately 20 Rho family members have been identified, many of which are known to regulate key cellular functions in metastasis, including cytoskeletal rearrangements, cell cycle, vesicle transport, cell polarity, enzyme activation, and gene transcription (Hall, 2005). Various model systems have demonstrated that Rho proteins regulate the metastatic phenotype. In an *in vitro* metastasis model, del Peso (1997) used NIH/3T3 cells transformed with the *Aplysia californica* Rho gene to generate lung metastases from mouse tail vein injections. Clark (2000) used a mouse model of lung metastasis to show that a highly metastatic variant of the melanoma cell line A375P overexpressed RhoC.

In addition to their involvement in these cellular functions and experimental systems, strong clinical evidence shows that Rho GTPase family members are involved in the progression and metastasis of several types of human cancers, including pancreatic, hepatocellular, and inflammatory breast cancer (Suwa *et al.*, 1998; van Golen *et al.*, 1999; Wang *et al.*, 2003). In bladder cancer, high RhoA, RhoC, and ROCK expression was significantly related to muscle invasion, lymph node metastasis, and shortened disease-free survival (Kamai *et al.*, 2003). The associations and mechanisms of Rho family GTPases in tumorigenesis, invasion, and metastasis have been reviewed elsewhere (Gomez del Pulgar *et al.*, 2005; Sahai and Marshall, 2002; Titus *et al.*, 2005).

1.2. RhoGDI2 is a metastasis suppressor

Because the functions of Rho family GTPases are known to be regulated by various guanine nucleotide exchange factors (GEFs), GTPase-activating proteins (GAPs), and GDP dissociation inhibitors (GDIs), it follows that these types of proteins are also involved in acquisition of the metastatic phenotype. Our laboratory has undertaken the study of the role of Rho GDIs in cancer progression, particularly the effects of RhoGDI2 on metastasis. Through the use of various *in vitro* and *in vivo* techniques, we have identified RhoGDI2 as an inhibitor of invasion and metastasis.

Three RhoGDIs have been identified. RhoGDI1 is expressed ubiquitously, whereas RhoGDI2, which is expressed highly in cells of hematopoietic lineage, is expressed at lower, but detectable levels in bladder, amnionic and chorionic cells, transitional epithelium, basal cells of vas, seminal vesicle, epididymis, fallopian tube, basal layer of the skin, renal collecting ducts, acini of breast, sweat glands, Bartholin's gland, lacrimal gland, and salivary glands (Theodorescu et al., 2004). RhoGDI3 is expressed mainly in the brain, pancreas, lung, kidney, and testis (Dovas and Couchman, 2005). RhoGDIs do not bind all Rho GTPases with equal affinity. For example, RhoGDI1 forms a strong interaction with RhoA and RhoC, whereas binding to RhoH is much weaker (Faure and Dagher, 2001). The interactions between other RhoGDIs and Rho proteins are still being investigated. Current data suggests that RhoGDI2 does not bind with the same efficiency to Rho family proteins as does RhoGDI1 (Gorvel et al., 1998). Furthermore, little is known whether RhoGDI2 and RhoGDI3 have binding partners in addition to Rho family members. Through binding of Rho GTPases, RhoGDIs have been proposed to (a) inhibit the dissociation of GDP from Rho proteins, (b) bind GTP-bound Rho and inhibit GTP hydrolysis and block interaction with downstream effectors, and (c) regulate the cycling of Rho GTPases between cytosol and membranes (DerMardirossian and Bokoch, 2005). The RhoGDI and Rho protein interactions have been discussed elsewhere (Dovas and Couchman, 2005; DerMardirossian and Bokoch, 2005; Dransart et al., 2005).

RhoGDI2 became a focus in our laboratory after the initial observation that RhoGDI2 mRNA was downregulated in metastatically competent human bladder cancer cell lines compared to metastatically incompetent cells (Seraj et al., 2000). This observation was explored in a study of 51 bladder tumors with associated normal human urothelium using immunohistochemistry where decreased staining of RhoGDI2 was found in malignant bladder tissue (Theodorescu et al., 2004). Low RhoGDI2 protein levels in tumors also correlated with decreased 5-year disease-free survival and served as an independent predictive factor for disease-specific death (Theodorescu et al., 2004). The association of decreased RhoGDI2 expression with metastasis and poor patient prognosis led us to question whether loss of expression of this gene was causally related to the metastatic process or merely a biomarker of the latter.

Using T24T, a moderately metastatic bladder cancer cell line with decreased expression of RhoGDI2 compared to its isogenic parental line T24, we focused on understanding the connection between RhoGDI2 and metastasis. Data revealed that T24T, transfected with RhoGDI2, produced significantly less lung metastases than the control transfected parent cell line. However, RhoGDI2 did not affect subcutaneous growth. Thus, we concluded that RhoGDI2 is a metastasis suppressor gene (Gildea *et al.*, 2002). To uncover how RhoGDI2 suppressed metastasis, gene expression profiling was used to characterize the RhoGDI2-reconstituted T24T cells and compare these profiles with those of vector control cells. Upon RhoGDI2 expression, endothelin-1 (ET-1), a secreted factor known to induce angiogenesis, was downregulated nearly 10-fold. Using an endothelin receptor A-specific antagonist, atrasentan (Opgenorth *et al.*, 1996), the RhoGDI2 phenotype was recapitulated, decreasing T24T lung metastases in nude mice (Titus *et al.*, 2005). Furthermore, another protein downregulated with RhoGDI2 expression, neuromedin U, has been shown to enhance the ability of T24T cells to metastasize to the lungs of nude mice (Wu *et al.*, 2007).

2. METHODS

2.1. RhoGDI2 blocks tumor invasion in a bladder organ culture (BOC) invasion assay

Metastasis *in vivo* initially begins with the invasion of cancer cells through normal tissue architecture to intravasate blood or lymph vessels. To understand how certain proteins affect this process in bladder cancer, we needed a model that would faithfully replicate the environment at the primary tumor site. To do this, we modified a technique designed by Crook *et al.* (2000) for studying cytotoxic agents on superficial bladder cancer. They noticed cells grown in monolayer were exquisitely sensitive to cytotoxic agents and do not stimulate the resistance seen in multilayered cell populations *in vivo*. To create a more accurate representation of the *in vivo* environment, they cocultured explanted rat bladders with bladder cancer cells. The BOC technique provided the opportunity to study cancer cell invasiveness orthotopically (Fig. 17.1) (Gildea *et al.*, 2004; Theodorescu *et al.*, 1990), yet in an *in vitro* setting, facilitating experimental testing of multiple genetic cell manipulations. This setting would allow tumor cells to interact with transitional epithelium, extracellular matrix, and muscle of bladder, which would be extremely difficult to reproduce with other *in vitro* techniques. The following is a description of the methods used to test the ability of RhoGDI2 to block invasion of the T24T bladder cancer cell line (Gildea *et al.*, 2002).

Discarded breeder, Sprague–Dawley rats are lightly anesthetized with urethane prior to bladder removal. For unknown reasons, attempts to use CO_2 or sodium pentobarbital as the anesthetic render the explanted bladders incapable of surviving in culture. Also, it is highly important for the rats

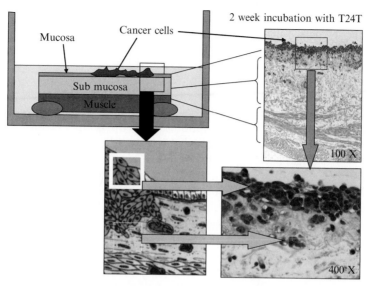

Figure 17.1 Bladder organ culture (BOC) invasion assay. Invasion is a critical step in bladder cancer metastasis. The BOC invasion assay replicates this step on mouse bladders *in vitro*. (Top left) Basic setup of the BOC assay. (Bottom left) Invasion of the submucosa that the BOC assay is designed to test. (Right) Actual H&E sections of a BOC assay using T24T cells. At higher magnification (400×) it is quite clear that these cells have breached the mucosal layer and are invading the submucosa. (See color insert.)

to remain alive prior to bladder removal for the survival of the explanted bladders *in vitro*. After the bladders are removed, the rats are lethally anesthetized with urethane. The rat bladders are removed in a sterile field and immediately transferred to a culture hood for further manipulation. The ureters and urethra are detached immediately, and the bladder is bisected and attached to the base of a single well in a six-well plate with the urothelium facing up. The bladder is opened beginning at the bladder neck and proceeding toward the apex/dome. The bladder is then inverted (turned inside out) so that the urothelium is facing out. Four dots of cyanoacetate glue are placed in the well. When the glue has semisolidified, one corner of the cut bladder is attached and the other three corners are stretched out to meet the glue. We have found that it is very important that tension remains throughout the bladder for the experiment to be successful. Also, it is important for the bladder to be adhered to the plate quickly. The bladder must be removed from the rat and adhered to the plate within 5 min to retain viability.

Fifty microliters of trypsin EDTA in HBSS is added to the bladder and incubated for 15 min at room temperature. Initial difficulty in having cells adhere to the bladder surface was encountered until the trypsin incubation step was added. We believe that this step "loosens" the mucosal layer and better allows the T24T cells to come in contact with the normal urothelial cells, as well as the underlying basement membrane. The trypsin is gently removed by aspirating the side of the bladder. To inactivate any residual trypsin, 100 μl of 100% fetal bovine serum (FBS) is added directly to the bladder and then gently removed similar to trypsin. A solution of 5×10^5 T24T bladder cancer cells in 100 μl of Waymouth medium with 10% FCS containing antibiotic/antimycotic (100 units of penicillin G, 100 μg streptomycin sulfate, and 0.25 μg of amphotericin B/ml) is pipetted directly onto the bladder surface. Waymouth medium is used because normal cultured media of T24T had been reported to transform epithelium. Prior to this step we had transfected T24T cells with RhoGDI2 or the control vector.

The transgene expressing cells are pipetted onto the explanted rat bladder and incubated at 37° and 5% CO_2 for 4 h. The edges of the bladder will dehydrate but the area covered by cells and media will remain viable. Approximately 4 ml of complete media is added to the well. Media are carefully added, allowing it to thinly cover the bladder. In this way the oxygen diffusion distance remains as short as possible. To avoid mechanically removing the cells from the surface of the bladder, we recommend slowly adding media to the edge of the dish. Media are changed twice a week but frequent observation is necessary to check for evaporation. After 3 weeks the bladders are washed twice with phosphate-buffered saline and fixed overnight in 10% buffered formalin. Histologic sections are reviewed by a board-certified anatomical pathologist. After review, it is evident that T24T cells expressing the RhoGDI2 transgene do not invade past the basement membrane. However, T24T vector control cells appear as an aggressive muscle invasive bladder cancer with hematoxylin and eosin (H&E) sections clearly showing invasion of the submucosa (Gildea *et al.*, 2002).

3. EVALUATING LUNG METASTASIS USING THE CLASSIC "EXPERIMENTAL METASTASIS" ASSAY

To successfully form metastases at distant sites, tumor cells must overcome several hurdles. Recent reviews on metastasis have highlighted these steps, which include invasion, intravasation, survival in circulation, arrest in target tissue, extravasation, adaptation to a new environment, and proliferation (Chambers *et al.*, 2002; Gupta and Massague, 2006; Steeg, 2006). In studying metastasis, *in vitro* models of specific steps in the process, such as wound assays to examine migration, are helpful. However, these studies need to be complemented by *in vivo* studies, as none provide a sufficiently complete surrogate of the latter. Hence, *in vivo* animal metastasis models have been developed to address the limitations of *in vitro* studies.

Two forms of metastasis assays are commonly carried out in rodents. The first is the "spontaneous metastasis assay," where cancer cells are injected orthotopically or subcutaneously and metastases occur spontaneously from this primary location. By definition, this model requires tumor cells to complete all steps in the metastatic process. The "experimental metastasis assay" involves direct injection of cells into the venous circulation, eliminating the need for cells to spontaneously escape from a primary tumor (Fidler, 1973). Advantages of the experimental metastasis assay include (1) a reduction of variables, as this assay bypasses some steps of the metastatic cascade (Welch, 1997); (2) a generally decreased time required for metastases to develop (Steeg, 2006); (3) a greater number of metastases formed (Steeg, 2006); and (4) the ability to test whether a specific gene directly affects organ colonization and not indirectly through tumorigenicity (Yang et al., 2001). Limitations of this assay include (1) only the postintravasation stage is modeled (Steeg, 2006; Welch, 1997); (2) a supraphysiologic bolus of cells is injected over a short period of time, which may not parallel the situation in patients (Welch, 1997); and (3) an absence of immune system or lack of prior immune presentation removes the possibility for an immune response (Welch, 1997).

In addition to its role in the study of metastasis, the experimental metastasis assay has also been used to create derivative cell lines with progressively more metastatic phenotypes by reiterative serial injections. For example, Fidler (1973) used the parent melanoma cell line B_{16} to derive cell lines with greater ability to colonize the mouse lung. Lung metastases were isolated, grown in tissue culture, and injected intravenously into a new mouse as depicted in Fig. 17.2. Each successive passage of cells through the

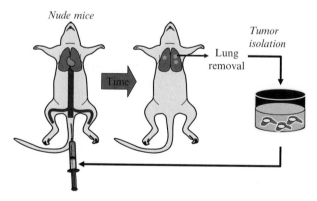

Figure 17.2 Procedure for selecting increasingly metastatic cell lines. Briefly, cells were harvested and injected into the mouse lateral tail vein for lung metastases. After tumors have formed, the lungs were harvested and new cell lines were created from metastatic nodules. The process was repeated until the desired metastatic capacity was reached. (See color insert.)

mouse produced a greater number of lung metastases (Fidler, 1973). Variations of this experiment have been used to study lung metastases of other cancer types (Bandyopadhyay *et al.*, 2005; Chen *et al.*, 2005). Furthermore, injections into other branches of the venous circulation, such as the splenic circulation, where the liver is the host organ with the immediate vascular bed, can generate metastasis models of different end organs (Kusama *et al.*, 2006). We have used the experimental metastasis assay, as described later, to determine if specific genes are mechanistically important in the postintravasation stages of metastasis.

Complete quantification of metastases in mouse lungs by visual inspection is impossible, as nodules less than 1 to 2 mm^3 and those deep to the pleural surface are not visible. Complete sectioning and microscopic evaluation of fixed murine lungs would be difficult and time-consuming and likely suffer from significant inter- and intraobserver variabilities. Furthermore, the mere summation of metastatic nodules does not measure the entire tumor cell burden, as micrometastases are not counted and variation in three-dimensional nodule size is difficult to assess from analysis of tissue sections. In addition, experimental metastasis assays in human tumor models take months to carry out and are associated with significant cost. Finally, because of the reasons just given, a fairly large number of animals may be required for statistically robust conclusions, especially when differences between the metastatic competences of various cell lines are not large. For these reasons, we have developed a method of measuring the amount of human tumor cells within murine tissues by quantitative real-time polymerase chain reaction (PCR). Because this PCR-based method is more sensitive and measures all of the human cells within a tissue sample, it eliminates observer error and allows tumor detection and quantification at earlier time points, reducing both the number of mice required and their housing costs.

This method targets for amplification a sequence of human genomic DNA found on the short arm of chromosome 12 (12p) that is not homologous to any region in the mouse genome. This target sequence does not lie within, or in close proximity to, any known gene. Human-specific primers were designed to amplify a 107-bp product from this locus and were validated by their ability to amplify the product from genomic DNA isolated from various human cell lines but not from mouse tissue DNA. Therefore, amplification with these primers can be used to specifically detect the presence of human cancer cells residing in a mouse organ.

Proper construction of standard curves is critical for accurate quantification by real-time PCR. The standards should closely reflect the composition of the experimental samples to be quantified. Our comparisons of standard curves constructed with mixtures of human and mouse DNA, versus standard curves constructed with serial dilutions of pure human genomic DNA, revealed significant differences in amplification kinetics and thus quantification (data not shown). Therefore, for the purpose of detecting human cancer cells in a mouse tissue background, we used standards consisting of genomic DNA from the human cell line added to mouse genomic DNA at various defined ratios, while keeping the total amount of DNA constant.

Genomic DNA is isolated from *in vitro* cultures of the human cancer cell line to be used in the metastasis assay. We use Purgene reagents (Gentra Systems, Minneapolis, MN); however, any commercial set of reagents that yields high-quality genomic DNA from *in vivo* and *in vitro* samples should work. DNA is dissolved in pure water so that the presence of other chemicals does not interfere with subsequent PCR reactions (EDTA especially can alter the efficiency of the PCR). Genomic DNA is isolated from tissue samples of a control mouse (one not injected with human cancer cells). DNA from lung tissue is ideal because this is the experimental target organ, although DNA from any mouse tissue should be theoretically equivalent. The DNA samples are quantified by absorbance spectrophotometry or by a fluorescence-based method (i.e., PicoGreen, Molecular Probes). Mixtures of human and mouse DNA are made with defined proportions while keeping the total concentration of DNA constant. A typical set of standards is shown in Table 17.1.

Sufficient amounts of these standard mixtures should be made at any one time to run the required number of replicates (typically three per analysis). If results from multiple experiments are to be compared, it is preferable to use the same set of standards for each PCR run; therefore, one large set of standards should be made at the outset of a study and can be stored at $-20°$.

Cells are prepared and injected into the lateral tail vein of an NCr nu/nu mouse as described earlier. At 5 weeks post-tail vein injection, mice are euthanized and lungs harvested. Total DNA is isolated from the lungs. Prior to isolation using the Puregene system, it is necessary to break up the lung

Table 17.1 Dilution table for constructing the standard curve in a molecular metastasis assay

Human DNA (ng)	Mouse DNA (ng)	Total volume (μl)	Total DNA in 10-μl aliquot (ng)
1000	0	10	1000
100	900	10	1000
10	990	10	1000
1	999	10	1000
0.1	999.9	10	1000
0.01	999.99	10	1000
0	1000	10	1000

tissue for efficient homogenization. An easy, inexpensive method that minimizes the chances of sample cross–contamination is to place the lung tissue in a packet made from folded aluminum foil, which is flash frozen in liquid nitrogen. While the tissue is still frozen, the packet is struck with a hammer, reducing the tissue to a fine powder. This powder is scraped into an appropriate tube and DNA isolation proceeds as per the manufacturer's directions. The sample DNA concentration is measured by the same method used to measure the concentration of the standards. The volume of sample DNA is adjusted with water so that equal amounts of DNA are placed in the following PCR reactions.

Polymerase chain reaction is performed in a Bio-Rad iCycler thermo-cycler (Bio-Rad, Hercules, CA) using Bio-Rad iQ SYBR Green Supermix reagents according to directions. As with all qPCR, it is important to make a master mix of primers, iQ SYBR Green Supermix, and water, from which constant volumes are aliquoted into each well of the PCR plate. Typically, each PCR reaction contains a total volume of 25 μl with each sample and standard run in triplicate. The final concentration of each primer is 250 nM with 1000 ng of total DNA in each reaction well for samples and standards. An additional set of wells is run as negative controls for contamination by replacing the DNA template with the correct volume of water. The following primers are from Sigma-Genosys and are desalted and cartridge purified at a minimum:

Forward primer: 5′-ggg aca gac act gag cct tga g -3′
Reverse primer: 5′-tga ccc tga taa agt ttc ttg gaa -3′

Polymerase chain reaction cycling parameters: Initial denaturation at 95° for 1 min 30 s followed 40 cycles of 95° for 15 s, 60° for 30 s, and 72° for 30 s. Fluorescence data are acquired at the end of each cycle (during 72° extension phase). Following PCR amplification, the thermocycler is set up to perform a melt curve analysis from 60 to 95° in 0.5° increments. Melt curve analysis allows for the discrimination of amplification of the human genomic target from spurious PCR products (i.e., primer dimers). Samples and standards that have a significant amount of fluorescence attributable to products other than the authentic target should be excluded from further analysis.

The Bio-Rad iCycler software is used for data analysis. The standard curve appears as in Fig. 17.3. Ideally, all of the points on the standard curve should conform to acceptable PCR amplification statistics. We strive for efficiencies between 95 and 105%, a slope between −3.1 and −3.6, and a correlation coefficient of greater than 0.95. Samples and standards that fall outside of these predetermined parameters are excluded. Only experimental samples that fall between the extreme ends of an acceptable standard curve can be quantified accurately. From this standard curve, the amount of human cancer cells in each sample is interpolated. Units derived from this

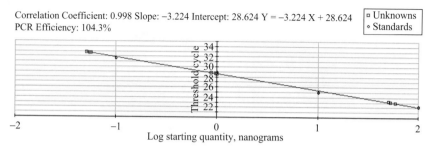

Figure 17.3 Standard curve of human DNA in a murine DNA mixture. In the molecular metastasis assay, human tumors that colonize mouse lungs were quantified by detecting unique regions of human DNA via real-time PCR. Known quantities of human DNA mixed with murine DNA were amplified, and when their threshold cycle was plotted against the logarithmic quantity of human DNA within the sample, a linear regression curve could be fit to data. The slope of that curve could then be used to calculate the quantity of human DNA in an experimental sample.

analysis are expressed as the equivalent in nanograms of human genomic DNA derived from the cell line used to construct the standards. If two or more cell lines are being compared for metastatic ability, one should be selected as the standard and the others compared to it.

We have successfully used the described method to detect and quantify human cancer cells in the lungs of mice with no visible metastasis. On occasion we have also used variations of the method to increase the sensitivity even further. For example, the standards and samples can be subjected to an initial round of PCR amplification with primers flanking the described primer pair. After this initial, abbreviated (less than 10 cycles), amplification step, a 1-μl aliquot from each reaction is used in the subsequent real-time qPCR analysis as described. Alternatively, an amplification primer/probe strategy, such as Molecular Beacons or TaqMan, can be used for qPCR analysis, which may be somewhat more product specific than detection with SYBR green, and obviates the need for melt curve analysis. When employing any of these alternative strategies, it is paramount to simultaneously carry the standards through the same set of manipulations as the experimental samples.

4. CREATING INCREASINGLY LUNG METASTATIC BLADDER CANCER CELL LINES

One of the key tools used to discover genes associated with the metastatic phenotype is the availability of lineage-related human cell lines with various metastatic capabilities in animal models of lung metastasis.

T24T   T24

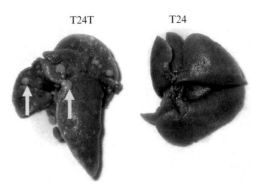

Figure 17.4   Mouse lungs fixed with Bouin's solution. T24T is a moderately metastatic cell line and a variant of the poorly metastatic T24 cell line. Mice were injected with 1×10^6 cells of T24 or T24T into the lateral tail vein. The lungs were fixed in Bouin's solution and neutral-buffered formalin. The greater numbers of tumors formed by T24T are easily seen.

The parental line T24 cells were originally isolated from an invasive grade III human transitional cell bladder cancer in a female patient (Bubenik *et al.*, 1973). The T24T cell line was characterized as a variant of T24 that shows higher tumorigenicity and metastatic potential (see Fig. 17.4) (Gildea *et al.*, 2000). We then sought to further develop lineage-related and progressively metastatic cell lines by adapting the methods of Fidler (1973) described earlier. T24T cells are passaged a minimum of three times at 80%, subconfluent density. Cells are counted using a Coulter counter to ensure reproducible results. T24T cells are injected at a concentration of 1×10^6 per 0.1 ml of serum-free Dulbecco's modified Eagle's medium (DMEM)/F-12 medium into the lateral tail vein of 6-week-old NCr nu/nu mice. The mice are examined and weighed weekly. Mice are euthanized at 16 weeks, and lungs are removed under sterile conditions and examined under a dissecting microscope. Metastatic lesions are harvested from the lungs, cut into 1- to 3-mm^3 cubes, and placed in DMEM/F-12 media supplemented with 2× FBS and 1× antibiotic–antimycotic solution (100 units of penicillin G, 100 μg streptomycin sulfate, and 0.25 μg of amphotericin B/ml). Murine stromal cells have the propensity to contaminate epithelial tumor explants. To mitigate this problem, nonimmortalized murine cells are removed by keeping explanted lung tumor cells in culture until the murine stromal cells are no longer visible microscopically (approximately 8 weeks). Other researchers have used immunoperoxidase staining for vimenten, a cytoskeletal protein of fibroblasts, for detecting murine fibroblast contamination of human epithelial cells (Isseroff *et al.*, 1987). When we are confident that no murine cells remain, tumor cells are passaged into media with 1× FBS and no antibiotic–antimycotic solution. In situations where the tumors cells are stably transfected with a selectable marker such as G418 or puromycin, treatment with the drug eliminates any nontransfected cells (i.e., fibroblasts) with greater ease.

After the first *in vivo* passage and pure *in vitro* culture, the resulting cells are called FL1 ("From Lung" 1). FL1 cells are injected into the lateral tail vein of a new cohort of mice. Time to tumor formation for FL1 cells decreases to 8 weeks from 16 weeks for the T24T cells. Reiterative repetitions of these steps are employed to create the increasingly metastatic cell lines FL2 and FL3. FL3 cells took only 3 weeks to form lung tumors (Nicholson *et al.*, 2004). To estimate the number of metastases, murine lungs are dissected and placed in a mixture of one part Bouin's fixative in five parts neutral-buffered formalin. This allows for fixation of tissue without tissue becoming too brittle. With this preparation, tumors appear as pale yellow spots after fixation, making it easier to count and measure the size of the lesions (Welch, 1997). Macroscopic tumors are measured by calipers and classified as "gross," whereas "microscopic" tumors are detected with the aid of a $10\times$ dissection microscope (Wu *et al.*, 2007). A typical example is shown in Fig. 17.4, where lung tumors of T24T cells are easily distinguished from the lack of tumors created by T24 cells.

ACKNOWLEDGMENTS

The authors thank Dr. John Gildea for his insightful comments and suggestions. This work was supported by Grant NIH R01CA075115 to D.T.

REFERENCES

Bandyopadhyay, A., Elkahloun, A., Baysa, S. J., Wang, L., and Sun, L. Z. (2005). Development and gene expression profiling of a metastatic variant of the human breast cancer MDA-MB-435 cells. *Cancer Biol. Ther.* **4,** 168–174.

Bubenik, J., Baresova, M., Viklicky, V., Jakoubkova, J., Sainerova, H., and Donner, J. (1973). Established cell line of urinary bladder carcinoma (T24) containing tumour-specific antigen. *Int. J. Cancer* **11,** 765–773.

Chambers, A. F., Groom, A. C., and MacDonald, I. C. (2002). Dissemination and growth of cancer cells in metastatic sites. *Nat. Rev. Cancer* **2,** 563–572.

Chen, X., Su, Y., Fingleton, B., Acuff, H., Matrisian, L. M., Zent, R., and Pozzi, A. (2005). Increased plasma MMP9 in integrin alpha1-null mice enhances lung metastasis of colon carcinoma cells. *Int. J. Cancer* **116,** 52–61.

Clark, E. A., Golub, T. R., Lander, E. S., and Hynes, R. O. (2000). Genomic analysis of metastasis reveals an essential role for RhoC. *Nature* **406,** 532–535.

Crook, T. J., Hall, I. S., Solomon, L. Z., Birch, B. R., and Cooper, A. J. (2000). A model of superficial bladder cancer using fluorescent tumour cells in an organ-culture system. *BJU Int.* **86,** 886–893.

del Peso, L., Hernandez-Alcoceba, R., Embade, N., Carnero, A., Esteve, P., Paje, C., and Lacal, J. C. (1997). Rho proteins induce metastatic properties *in vivo*. *Oncogene* **15,** 3047–3057.

DerMardirossian, C., and Bokoch, G. M. (2005). GDIs: Central regulatory molecules in Rho GTPase activation. *Trends Cell Biol.* **15,** 356–363.

Dovas, A., and Couchman, J. R. (2005). RhoGDI: Multiple functions in the regulation of Rho family GTPase activities. *Biochem. J.* **390,** 1–9.

Dransart, E., Olofsson, B., and Cherfils, J. (2005). RhoGDIs revisited: Novel roles in Rho regulation. *Traffic* **6,** 957–966.

Faure, J., and Dagher, M. C. (2001). Interactions between Rho GTPases and Rho GDP dissociation inhibitor (Rho-GDI). *Biochimie* **83,** 409–414.

Fidler, I. J. (1973). Selection of successive tumour lines for metastasis. *Nat. New Biol.* **242,** 148–149.

Gildea, J. J., Golden, W. L., Harding, M. A., and Theodorescu, D. (2000). Genetic and phenotypic changes associated with the acquisition of tumorigenicity in human bladder cancer. *Genes Chromosomes Cancer* **27,** 252–263.

Gildea, J. J., Herlevsen, M., Harding, M. A., Gulding, K. M., Moskaluk, C. A., Frierson, H. F., and Theodorescu, D. (2004). PTEN can inhibit *in vitro* organotypic and *in vivo* orthotopic invasion of human bladder cancer cells even in the absence of its lipid phosphatase activity. *Oncogene* **23,** 6788–6797.

Gildea, J. J., Seraj, M. J., Oxford, G., Harding, M. A., Hampton, G. M., Moskaluk, C. A., Frierson, H. F., Conaway, M. R., and Theodorescu, D. (2002). RhoGDI2 is an invasion and metastasis suppressor gene in human cancer. *Cancer Res.* **62,** 6418–6423.

Gomez del Pulgar, T., Benitah, S. A., Valeron, P. F., Espina, C., and Lacal, J. C. (2005). Rho GTPase expression in tumourigenesis: Evidence for a significant link. *Bioessays* **27,** 602–613.

Gorvel, J. P., Chang, T. C., Boretto, J., Azuma, T., and Chavrier, P. (1998). Differential properties of D4/LyGDI versus RhoGDI: Phosphorylation and rho GTPase selectivity. *FEBS Lett.* **422,** 269–273.

Gupta, G. P., and Massague, J. (2006). Cancer metastasis: Building a framework. *Cell* **127,** 679–695.

Hall, A. (2005). Rho GTPases and the control of cell behaviour. *Biochem. Soc. Trans.* **33,** 891–895.

Isseroff, R. R., Ziboh, V. A., Chapkin, R. S., and Martinez, D. T. (1987). Conversion of linoleic acid into arachidonic acid by cultured murine and human keratinocytes. *J. Lipid Res.* **28,** 1342–1349.

Kamai, T., Tsujii, T., Arai, K., Takagi, K., Asami, H., Ito, Y., and Oshima, H. (2003). Significant association of Rho/ROCK pathway with invasion and metastasis of bladder cancer. *Clin. Cancer Res.* **9,** 2632–2641.

Kusama, T., Mukai, M., Endo, H., Ishikawa, O., Tatsuta, M., Nakamura, H., and Inoue, M. (2006). Inactivation of Rho GTPases by p190 RhoGAP reduces human pancreatic cancer cell invasion and metastasis. *Cancer Sci.* **97,** 848–853.

Madaule, P., and Axel, R. (1985). A novel ras-related gene family. *Cell* **41,** 31–40.

Nicholson, B. E., Frierson, H. F., Conaway, M. R., Seraj, J. M., Harding, M. A., Hampton, G. M., and Theodorescu, D. (2004). Profiling the evolution of human metastatic bladder cancer. *Cancer Res.* **64,** 7813–7821.

Opgenorth, T. J., Adler, A. L., Calzadilla, S. V., Chiou, W. J., Dayton, B. D., Dixon, D. B., Gehrke, L. J., Hernandez, L., Magnuson, S. R., Marsh, K. C., Novosad, E. I., Von Geldern, T. W., *et al.* (1996). Pharmacological characterization of A-127722: An orally active and highly potent ETA-selective receptor antagonist. *J. Pharmacol. Exp. Ther.* **276,** 473–481.

Ridley, A. J., and Hall, A. (1992). The small GTP-binding protein rho regulates the assembly of focal adhesions and actin stress fibers in response to growth factors. *Cell* **70,** 389–399.

Ridley, A. J., Paterson, H. F., Johnston, C. L., Diekmann, D., and Hall, A. (1992). The small GTP-binding protein rac regulates growth factor-induced membrane ruffling. *Cell* **70,** 401–410.

Sahai, E., and Marshall, C. J. (2002). RHO-GTPases and cancer. *Nat. Rev. Cancer* **2,** 133–142.

Seraj, M. J., Harding, M. A., Gildea, J. J., Welch, D. R., and Theodorescu, D. (2000). The relationship of BRMS1 and RhoGDI2 gene expression to metastatic potential in lineage related human bladder cancer cell lines. *Clin. Exp. Metastasis* **18,** 519–525.

Steeg, P. S. (2006). Tumor metastasis: Mechanistic insights and clinical challenges. *Nat. Med.* **12,** 895–904.

Suwa, H., Ohshio, G., Imamura, T., Watanabe, G., Arii, S., Imamura, M., Narumiya, S., Hiai, H., and Fukumoto, M. (1998). Overexpression of the rhoC gene correlates with progression of ductal adenocarcinoma of the pancreas. *Br. J. Cancer* **77,** 147–152.

Theodorescu, D., Cornil, I., Fernandez, B. J., and Kerbel, R. S. (1990). Overexpression of normal and mutated forms of HRAS induces orthotopic bladder invasion in a human transitional cell carcinoma. *Proc. Natl. Acad. Sci. USA* **87,** 9047–9051.

Theodorescu, D., Sapinoso, L. M., Conaway, M. R., Oxford, G., Hampton, G. M., and Frierson, H. F., Jr. (2004). Reduced expression of metastasis suppressor RhoGDI2 is associated with decreased survival for patients with bladder cancer. *Clin. Cancer Res.* **10,** 3800–3806.

Titus, B., Frierson, H. F., Jr., Conaway, M., Ching, K., Guise, T., Chirgwin, J., Hampton, G., and Theodorescu, D. (2005). Endothelin axis is a target of the lung metastasis suppressor gene RhoGDI2. *Cancer Res.* **65,** 7320–7327.

Titus, B., Schwartz, M. A., and Theodorescu, D. (2005). Rho proteins in cell migration and metastasis. *Crit. Rev. Eukaryot. Gene Expr.* **15,** 103–114.

van Golen, K. L., Davies, S., Wu, Z. F., Wang, Y., Bucana, C. D., Root, H., Chandrasekharappa, S., Strawderman, M., Ethier, S. P., and Merajver, S. D. (1999). A novel putative low-affinity insulin-like growth factor-binding protein, LIBC (lost in inflammatory breast cancer), and RhoC GTPase correlate with the inflammatory breast cancer phenotype. *Clin. Cancer Res.* **5,** 2511–2519.

Wang, W., Yang, L. Y., Yang, Z. L., Huang, G. W., and Lu, W. Q. (2003). Expression and significance of RhoC gene in hepatocellular carcinoma. *World J. Gastroenterol.* **9,** 1950–1953.

Welch, D. R. (1997). Technical considerations for studying cancer metastasis *in vivo. Clin. Exp. Metastasis* **15,** 272–306.

Wu, Y., McRoberts, K., Berr, S. S., Frierson, H. F., Jr., Conaway, M., and Theodorescu, D. (2007). Neuromedin U is regulated by the metastasis suppressor RhoGDI2 and is a novel promoter of tumor formation, lung metastasis and cancer cachexia. *Oncogene* **26,** 765–773.

Yang, X., Wei, L. L., Tang, C., Slack, R., Mueller, S., and Lippman, M. E. (2001). Overexpression of KAI1 suppresses *in vitro* invasiveness and *in vivo* metastasis in breast cancer cells. *Cancer Res.* **61,** 5284–5288.

CHAPTER EIGHTEEN

CHARACTERIZATION OF THE ROLES OF RAC1 AND RAC2 GTPASES IN LYMPHOCYTE DEVELOPMENT

Celine Dumont, Robert Henderson, *and* Victor L. J. Tybulewicz

Contents

Abstract

This chapter describes methods for the analysis of B and T lymphocyte development in mice deficient in Rac1 and/or Rac2 GTPases. The development of both B and T cells is critically dependent on transition through checkpoints monitoring for correct rearrangement of antigen receptor genes. Progression through these checkpoints depends on signaling from the antigen receptors.

Division of Immune Cell Biology, National Institute for Medical Research, London, United Kingdom

Methods in Enzymology, Volume 439
ISSN 0076-6879, DOI: 10.1016/S0076-6879(07)00418-1

In addition, signals from cytokine, chemokine, Notch, and death receptors play important roles in the survival, proliferation, and migration of developing lymphocytes. Analysis of these processes in mice deficient in these GTPases can illuminate their roles in transducing signals from these different receptors.

1. INTRODUCTION

The Rac GTPases (Rac1, Rac2, and Rac3) are members of the Rho family of GTPases. Like many members of this family, they have been implicated in the regulation of the actin cytoskeleton, as well as other cellular processes, including cell survival, proliferation, gene transcription, and activation of the NADPH oxidase. Primary lymphocytes express Rac1 and Rac2, whereas Rac3 expression has not been reported in these cells. Studies using mice deficient in one or more of these GTPases have started to define the physiological function of the Rac GTPases in lymphocytes.

In the B lymphocyte lineage, a deficiency in Rac2 results in reduced numbers of peritoneal B1 cells and splenic marginal zone (MZ) B cells (Croker *et al.*, 2002b; Walmsley *et al.*, 2003). Furthermore, Rac2-deficient B cells demonstrate reduced B-cell antigen receptor (BCR)-induced proliferation, survival, and calcium flux, as well as defective chemotaxis in response to the chemokines CXCL13 and CXCL12 (Croker *et al.*, 2002b; Walmsley *et al.*, 2003). Analysis of the role of Rac1 in lymphocytes is complicated by the early embryonic lethality of Rac1-deficient mice (Sugihara *et al.*, 1998), requiring the use of a conditional system of gene ablation. Generation of a loxP-flanked *Rac1* gene has allowed the selective inactivation of *Rac1* in specific tissues and at specific times by crossing these mice to strains expressing the Cre recombinase in defined lineages or in an inducible form. For B-cell-specific deletion of *Rac1*, we used a strain carrying a knock-in of Cre into the CD19 locus (Rickert *et al.*, 1997). This showed that loss of Rac1 alone had no detectable effect on the development of B-cell lineages or on B-cell activation (Walmsley *et al.*, 2003). However, combining this B-cell-specific Rac1 deficiency with a constitutive ablation of Rac2 resulted in a dramatic loss of all mature subsets of B cells, including the mature recirculating follicular (MRF) B cells, B1 and MZ B cells (Walmsley *et al.*, 2003). These results show that Rac1 does indeed play a role in B-cell development, but this is normally redundant due to the coincident expression of Rac2. Such overlapping function is not unexpected in view of the high degree of sequence similarity between Rac1 and Rac2.

In the T lymphocyte lineage, loss of Rac2 does not grossly perturb T-cell development. Rac2-deficient T cells show a mild impairment in T-cell antigen receptor (TCR)-induced activation, as well as decreased interferon-γ secretion under Th1 differentiation conditions *in vitro* (Li *et al.*, 2000;

Yu *et al.*, 2001). However, no gross defect in the Th1 response was seen *in vivo* (Croker *et al.*, 2002a). As seen with B cells, Rac2-deficient T cells also show reduced *in vitro* chemotaxis (Croker *et al.*, 2002a). The role of Rac1 in T-cell biology remains less well characterized. Forced expression of a constitutively active Rac1 protein in the T-cell lineage results in differentiation of early thymic CD4$^-$CD8$^-$ double negative (DN) thymocytes into CD4$^+$CD8$^+$ double positive (DP) thymocytes, even in the absence of a pre-TCR, which is normally required for this developmental transition, suggesting that Rac1 may normally transduce signals from the pre-TCR that control this developmental step (Gomez *et al.*, 2000). Furthermore, the same active Rac1 transgene can switch the developmental fate of DP thymocytes from positive to negative selection, consistent with a role for Rac1 in contributing to these developmental processes (Gomez *et al.*, 2001). More recently, we have used conditional ablation of *Rac1* in the T-cell lineage to show that, as in the B-cell lineage, there is considerable redundancy of function between the two GTPases (C. Dumont and V. L. J. Tybulewicz, manuscript in preparation). While the loss of Rac1 has no noticeable effect on T-cell development, simultaneous loss of both Rac1 and Rac2 leads to a complete block in T-cell development at the DN to DP transition, consistent with a critical role for both GTPases in signaling from the pre-TCR.

This chapter describes methods for the detailed analysis of both B- and T-cell development in mice. While we have used these methods to characterize the role of Rac1 and Rac2 in these processes, the same methods could be applied to any genetically modified mice.

2. ANALYSIS OF B-CELL DEVELOPMENT

2.1. Early B-cell development

In the adult mouse, the earliest stages of B-cell development occur in the bone marrow, where hematopoietic precursors acquire expression of markers characteristic of the B-cell lineage (B220 and then CD19), begin to express Igα and Igβ proteins, which are nonpolymorphic signaling subunits of the BCR, and start to rearrange the immunoglobulin heavy chain (IgH) locus. This stage of B-cell development is referred to as the pro-B cell. If rearrangement of the D$_H$ to J$_H$ and subsequently the V$_H$ to DJ$_H$ gene segments is successful, a μ-type IgH molecule is synthesized, associates with the surrogate light chains (SLC) λ5 and VpreB, and with Igα and Igβ to form a receptor known as the pre-BCR. Signaling from the pre-BCR is required for a number of cellular processes: differentiation of the pro-B cells into the next stage of development termed a pre-B cell, proliferation, survival, and inhibition of further IgH gene rearrangement. The effect of these processes is to allow only cells with a functional pre-BCR to progress in development, with cells failing to generate a functional IgH rearrangement destined to die by apoptosis. The inhibition of IgH gene rearrangement in pre-B cells ensures that the cells have only one functional μ chain, a process termed allelic exclusion. The transition from pro-B to pre-B cells constitutes a developmental checkpoint, controlled by signals from the pre-BCR.

Pre-B cells can be distinguished from pro-B cells by a number of cell surface markers. While both cell types express B220 and CD19, only pro-B cells express CD43, whereas only pre-B cells express CD2 and CD25. Pre-B cells turn off expression of the SLC and initiate rearrangement of the immunoglobulin light chains (IgL), first at the Igκ locus and subsequently at the Igλ locus. Successful rearrangement of a light chain gene results in expression of κ or λ light chain, which then associates with the μ heavy chain and with Igα and Igβ to generate a mature BCR in the form of membrane-tethered immunoglobulin M (IgM). These IgM expressing cells are termed immature B cells.

2.2. Late B-cell development

B-cell antigen receptor signaling in immature B cells turns off IgL rearrangement and controls a second developmental checkpoint, which cells must transit to become mature B cells. Immature B cells are released from the bone marrow and migrate through the blood to the spleen and perhaps to other organs. Upon arriving in the spleen, these immature B cells, referred to as "newly formed" or "transitional" B cells, complete their maturation into the two main B-cell lineages: mature recirculating follicular (MRF) and marginal zone (MZ) B cells. These maturational steps are dependent on BCR signaling and have sometimes been described as B-cell positive selection, as cells failing to make a functional BCR do not progress in development and are eliminated by apoptosis. The same immature B cells are also subject to the process of negative selection. If the BCR they express binds strongly to an endogenous ligand, the B cells might become autoreactive. Such cells are killed by apoptosis, anergized, or their receptor specificity is altered by the process of receptor editing (Melchers and Rolink, 2006).

Two main marker expression patterns have been devised to distinguish transitional, MRF and MZ B cells, as well as to subdivide transitional B cells into different fractions. In one scheme, expression of CD21, CD23, HSA, and IgM is used to distinguish transitional type 1 (T1; CD21$^-$CD23$^-$HSA^{hi} IgM^{hi}) and transitional type 2 (T2; CD21$^+$CD23$^+$HSA^{hi}IgM^{hi}) cells from MRF B cells (CD21$^+$CD23$^+$HSA^{lo}IgM^{lo}) and MZ B cells (CD21$^+$ CD23$^-$HSA^{lo}IgM^{hi}) (Loder *et al.*, 1999). MRF B cells constitute the main pool of mature B cells in the adult mouse and are also characterized by the expression of high levels of the IgD form of the BCR. Alternatively, transitional cells have been distinguished from the more mature compartments by expression of C1qRp, which is sometimes termed AA4.1 from the name of the monoclonal antibody used to detect expression (Allman *et al.*, 2001). Using this marker it is possible to subdivide transitional B cells into three fractions: transitional type 1 (T1; C1qRp$^+$CD23$^-$IgM^{hi}), transitional type 2 (T2; C1qRp$^+$CD23$^+$IgM^{hi}), and transitional type 3 (T3; C1qRp$^+$CD23$^+$IgM^{lo}), and to identify MRF (C1qRp$^-$CD23$^+$IgM^{lo}) and MZ (C1qRp$^-$CD23$^-$IgM^{hi}) B cells. While transitional B cells are thought to mature from T1 to T2 stages and then onto the MRF and MZ compartments, recent work has shown that T3 cells are likely to be an anergic population of B cells destined to die, generated from autoreactive transitional B cells engaging self-antigens (Merrell *et al.*, 2006). Measurements of life span show that cells in the transitional compartments have half-lives of 2 to 4 days, whereas MRF B cells have lifetimes of many weeks. For more comprehensive discussions of early B-cell development in the bone marrow, and later developmental steps leading to MRF and MZ B cells in the spleen, the reader is directed to more detailed reviews (Hardy and Hayakawa, 2001; Pillai *et al.*, 2005; Thomas *et al.*, 2006).

2.3. B-1 cells

In addition to MRF and MZ B cells, a third B-cell lineage has been identified. These cells, termed B-1 cells, are present in large numbers in the peritoneal and pleural cavities. Their antibody repertoire seems to encode a high frequency of specificities to bacterial cell wall components, allowing them to respond rapidly to bacterial infection. Thus B-1 cells have been proposed to represent an innate-like response, which can function early in infection before the adaptive immune response is fully active. B-1 cells are characterized by the expression of high levels of IgM, but low levels of IgD, as well as expression of Mac1 (CD11b) and CD43, and have been subdivided according to the expression of CD5 into B-1a (CD5$^+$) and B-1b (CD5$^-$) cells. B-1a cells appear to arise mainly from fetal hematopoietic precursors and are likely to be positively selected by reactivity toward self-antigens. In contrast, B-1b cells arise from both fetal and adult progenitors, are less likely to be autoreactive, and thus seem to resemble MRF B cells. As with MRF and MZ B cells, the development of B-1 cells is also critically dependent on BCR signaling. For a more comprehensive discussion of B-1 cell development, the reader is directed to a more detailed review (Hardy, 2006).

2.4. Signaling during B-cell development

While all three B-cell lineages require BCR signaling for their development and maturation, it has been proposed that the strength of BCR signaling may determine whether the B cell develops into a MRF, MZ, or B-1 cell. The evidence for this comes from studies using transgenic mice expressing monoclonal BCRs of defined specificity, which direct development into particular lineages. These studies have proposed that the strongest BCR signals (caused by binding to self-antigen) may drive B cells into the B-1 lineage with intermediate and weak BCR signals specifying development into the MRF and MZ lineages (Hardy, 2006; Pillai *et al.*, 2005).

The aforementioned discussion emphasized the importance of pre-BCR or BCR signaling during B-cell development, but signals from many other receptors also play key roles. For example, interleukin (IL)-7 receptor signaling is required for pro-B cell survival, BAFF receptor signaling controls the survival of transitional, MRF, and MZ B cells, and Notch2 is required for specification of the MZ lineage. Thus analysis of B-cell development in the absence of Rac GTPases, or other signaling molecules, can be used to identify *in vivo* functions for the GTPases in transducing signals from the antigen receptors, as well as other receptors.

2.5. Mice deficient for Rac GTPases in the B-cell lineage

Mice bearing a deletion of the *Rac2* gene are viable (Roberts *et al.*, 1999). In contrast, ablation of the *Rac1* gene leads to an early embryonic lethality (Sugihara *et al.*, 1998), necessitating the use of a conditional loxP-flanked allele of *Rac1* (Walmsley *et al.*, 2003). To delete *Rac1* throughout the B-cell lineage we have used mice carrying a knock-in of Cre into the *CD19* locus (Rickert *et al.*, 1997). While expression of CD19 itself begins within the pro-B-cell compartment, we find that measuring deletion with a reporter mouse, which induces YFP expression following Cre-mediated excision of a loxP-flanked Stop transcription signal (Srinivas *et al.*, 2001), the $CD19^{Cre}$-induced deletion is only partial in pro-B cells (around 14%), rising to 34% in pre-B cells, 44% in immature B cells, and 95% in MRF B cells (R. Henderson and V. L. J. Tybulewicz, unpublished data). Direct measurement of deletion of *Rac1* showed somewhat more efficient deletion of 80% in immature B cells and >90% in MRF B cells, underlining the different efficiencies of deletion at different loci (Walmsley *et al.*, 2003). We have also used the CD2-Cre transgenic line to delete *Rac1* in the B-cell lineage (de Boer *et al.*, 2003). This comes on earlier than CD19 and shows much more efficient deletion in pro-B and pre-B cells (91 and 99%, respectively) than the $CD19^{Cre}$ allele as measured by the YFP reporter strain (R. Henderson and V. L. J. Tybulewicz, unpublished result). To generate efficient deletion at the very earliest stage of B-cell development, it may be necessary to use Cre drivers that come on within hematopoietic precursors, such as the Vav1-Cre transgenic (de Boer *et al.*, 2003). Deletion of *Rac1* using $CD19^{Cre}$ in the absence of Rac2 leads to a complete block in B-cell development at the T1 stage in the spleen, such that no mature B cells are generated (Walmsley *et al.*, 2003). To study MRF B cells missing both Rac1 and Rac2, it would be necessary to delete *Rac1* after positive selection. This could be achieved using the CD21-Cre transgenic line (Kraus *et al.*, 2004).

2.6. Enumeration of number of cells in B-cell compartments

The most effective method for evaluating whether a genetic mutation is perturbing signaling events operating during B-cell development is to use flow cytometry to enumerate the number of cells in each stage of B-cell development. While flow cytometry will give the percentage of cells with a given phenotype, the absolute number is obtained by combining this percentage with a total cellular count for the organ being analyzed. The organs typically analyzed for B-cell development include bone marrow, spleen, lymph nodes, peritoneal cavity, and Peyer's patches. The number of B cells in the different subsets is altered in mice as they age, for example, transitional B cells are present in low numbers in mice older than 8 weeks, whereas marginal zone B-cell numbers increase with age. Comparisons must therefore always be made between age-matched, sex-matched mice on the same genetic background.

2.6.1. Cell preparation

To isolate bone marrow, prepare femurs and tibias by dissecting away muscle, cutting below the hip joint, immediately above and below the knee, and just above the fibula/tibia junction. Flush bone marrow out of the femurs and tibias using air-buffered Iscove's modified Dulbecco's medium (AB-IMDM; IMDM with 25 mM HEPES and no sodium bicarbonate, Invitrogen) supplemented with 5% fetal calf serum (FCS) introduced into the bone cavity with a syringe and 27-gauge needle. Flush bones until they turn translucent, collecting flushed cells in a small dish. For spleen, lymph nodes, and Peyer's patches, prepare single cell suspensions by passage through a 70-μm cell sieve using the plunger of a 5-ml syringe. To wash resident cells out of the peritoneal cavity, cut open the skin over the abdomen, revealing the outer lining of the peritoneum. Inject 5 ml of cold phosphate-buffered saline (PBS, 155.17 mM NaCl, 1.06 mM KH$_2$PO$_4$, 2.97 mM Na$_2$HPO$_4\cdot$7H$_2$O) and 5 mM EDTA into the left-hand side of the cavity using a syringe fitted with a 25-gauge needle. Gently massage the abdomen to release cells and collect the lavage from the right-hand side. Absolute numbers of peritoneal cells are calculated based on the percentage of input PBS that is recovered. Any phenotype in the peritoneal cavity can be confirmed by performing a lavage of the pleural cavity. Centrifuge all cell preparations at 4°C, resuspend in ACK buffer (155 mM NH$_4$Cl, 10 mM KHCO$_3$, 100 μM EDTA) for 2 min at room temperature to lyse red blood cells, centrifuge again at 4°, wash in AB-IMDM, 5% FCS, resuspend in AB-IMDM, 5% FCS, and count the number of cells using a hemocytometer or an automated cell counter (e.g., CASY TT cell Counter, Schärfe Systems).

2.6.2. Cell staining for flow cytometry

We typically stain 2×10^6 cells in each sample. If rare subsets are being analyzed, cell numbers can be increased, as long as staining volumes are increased to keep cell and antibody concentrations unchanged.

Place each cell sample to be stained in a well of a 96-well V-bottom plate. Centrifuge the plate at 4° and resuspend in 100 μl of FACS buffer (PBS, 0.1% NaN_3, 0.5% bovine serum albumin) containing 5 μg/ml anti-CD16/CD32 (monoclonal antibody 2.4G2 against $Fc\gamma RII/III$) to block nonspecific antibody staining. Leave on ice for 10 min, centrifuge, and resuspend in 100 μl FACS buffer containing fluorophore- or biotin-labeled antibodies. The optimum concentration for each antibody needs to be determined individually by a titration experiment to establish the concentration that gives rise to sufficient discrimination between positive and negative populations, without causing too much nonspecific staining. Typical concentrations used are listed in Table 18.1. Incubate primary staining for 20 min at 4° in the dark. If a biotin-labeled antibody has been used in primary staining, centrifuge and resuspend the samples in 100 μl FACS buffer containing fluorophore-labeled streptavidin and incubate for a further 20 min at 4° in the dark. Finally, centrifuge samples, resuspend in 100 μl FACS buffer at 4°, and analyze within a few hours using a multicolor flow cytometer.

2.6.3. Identification of B-cell subsets

Within the bone marrow pro-B-cell compartment, the earliest subset, sometimes termed fraction A, is identified as being $B220^+CD19^-$ (Fig. 18.1A) (Hardy and Hayakawa, 2001). Subsequently, all further subsets are $B220^+CD19^+$. Among these cells we use CD2, IgM, and IgD to separate pro-B ($CD2^-IgM^-IgD^-$), pre-B ($CD2^+IgM^-IgD^-$), immature ($CD2^+IgM^+IgD^-$), and MRF B cells ($CD2^+IgM^+IgD^+$) (see Fig. 18.1A). A number of other markers can be useful. CD43 is expressed on pro-B, but not on later developmental stages, whereas CD25 is expressed on pre-B but not on pro-B cells. Normal pre-BCR signaling leads to extensive pre-B-cell proliferation, whereas aberrant selection can cause increased rates of cell death. Thus measurements of cell division and death at these developmental stages can be informative. Methods to measure these are described in the T-cell development section.

To separate splenic transitional subsets, we usually use the Allman *et al.* (2001) staining scheme. Transitional cells are identified as $B220^+C1qRp^+$ and subdivided into T1 ($CD23^-IgM^{hi}$), T2 ($CD23^+IgM^{hi}$), and T3 ($CD23^+IgM^{lo}$) by virtue of CD23 and IgM expression (see Fig. 18.1B). The same staining can also be used to identify MRF ($B220^+$ $C1qRp^-CD23^+IgM^{lo}$) and MZ ($B220^+C1qRp^-CD23^-IgM^{hi}$) B cells. However, for MZ B cells, we prefer to use the Loder *et al.* (1999) staining scheme, as it gives better resolution of this subset (see Fig. 18.1C).

Mature recirculating follicular B cells in lymph nodes can be identified using the same stains as in spleen. Alternatively, because there are no MZ cells and few transitional B cells in these organs, MRF B cells can be identified as $CD19^+IgM^{lo/hi}IgD^{hi}$ cells (see Fig. 18.1E). Finally, B-cell

Table 18.1   Antibodies used to generate the flow cytometry plots in Figs. 18.1 and 18.2[a]

Antibody specificity	Fluorophore/ biotin	Supplier	Concentration
B220	APC-Cy7	BD Bioscience	0.5 μg/ml
B220	Biotin	eBiosciences	1.25 μg/ml
C1qRp (AA4.1)	APC	eBiosciences	1 μg/ml
CD11b (Mac1)	Biotin	eBiosciences	1.25 μg/ml
CD19	APC	Caltag Laboratories	1 μg/ml
CD19	APC-Cy7	BD Bioscience	1 μg/ml
CD2	PE	BD Bioscience	0.25 μg/ml
CD21	FITC	BD Bioscience	2.5 μg/ml
CD23	PE	eBiosciences	0.25 μg/ml
CD25	PE	BD Bioscience	0.3 μg/ml
CD3ϵ	Biotin	eBiosciences	1.25 μg/ml
CD4	Biotin	BD Bioscience	1.25 μg/ml
CD4	PerCP	BD Bioscience	0.7 μg/ml
CD44	Biotin	BD Bioscience	2.5 μg/ml
CD44	FITC	eBiosciences	1.25 μg/ml
CD5	APC	BD Bioscience	0.5 μg/ml
CD8α	APC	eBiosciences	0.5 μg/ml
CD8α	Biotin	Caltag	0.25 μg/ml
CD8α	FITC	eBiosciences	1.25 μg/ml
Dx5	Biotin	eBiosciences	1.25 μg/ml
Gr-1	Biotin	eBiosciences	1.25 μg/ml
IgD	Biotin	Southern Biotech	2.5 μg/ml
IgM[b]	FITC	Jackson ImmunoResearch	2.5 μg/ml
TCRβ	Biotin	BD Bioscience	2.5 μg/ml
TCRβ	FITC	BD Bioscience	2.5 μg/ml
TCRβ	PE	eBiosciences	1 μg/ml
TCR$\gamma\delta$	Biotin	eBiosciences	0.5 μg/ml
Thy1.2	APC	BD Bioscience	0.2 μg/ml
Secondary reagent			
Streptavidin	PerCP	BD Bioscience	1 μg/ml

[a] The specificity of the antibody, which fluorophore the antibody was conjugated to, or whether conjugated to biotin, the supplier, and the concentration at which it was used are shown. While the optimal concentration for each antibody will vary from batch to batch and must be individually optimized by titration, the listed concentration will give an indication of typical concentrations to use. Finally, a secondary reagent is used to reveal biotin-labeled antibodies. FITC, fluorescein isothiocyanate; PE, phycoerythrin; APC, allophycocyanin; APC-Cy7, allophycocyanin-cyanin 7; PerCP, peridinin chlorophyll protein.
[b] This is a Fab antibody fragment. All other antibodies are whole.

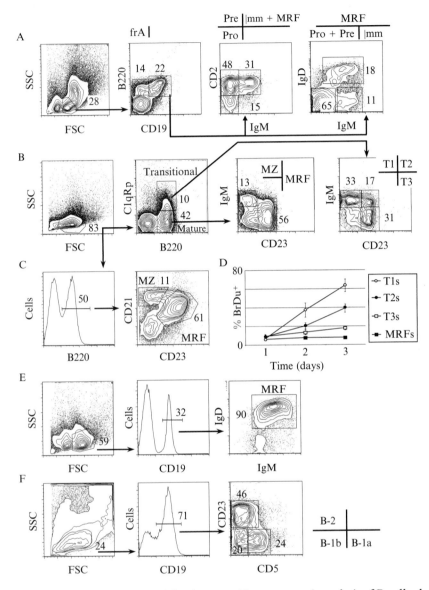

Figure 18.1 Analysis of B-cell development. Flow cytometric analysis of B-cell subsets from a 6-week-old C57BL/6 mouse. Numbers indicate percentages of cells falling into marked gates. (A) Bone marrow cells were first gated on live lymphoid-sized cells using forward scatter (FSC) and side scatter (SSC). These scatter-gated cells were analyzed for expression of B220 and CD19. B220$^+$CD19$^-$ cells are the earliest cells within the pro-B-cell subset and have been referred to as "fraction A" (frA) (Hardy and Hayakawa, 2001). The remaining B-cell subsets fall into the B220$^+$CD19$^+$ gate. This is broken down further by expression of CD2, IgM, and IgD, allowing the identification of pro-B (CD2$^-$IgM$^-$IgD$^-$), pre-B (CD2$^+$IgM$^-$IgD$^-$), immature (Imm,

subsets in the peritoneal cavity can be analyzed using CD19 (or B220) to identify all B cells, and CD5 and CD23 to separate the subsets: B-1a ($CD5^+CD23^-$), B-1b ($CD5^-CD23^-$), and MRF ($CD5^-CD23^+$) (see Fig. 18.1F). This MRF subset is sometimes referred to as B-2 cells, to distinguish it from the B-1 lineage. Other useful markers for B-1 cells (both B-1a and B-1b) are Mac1 (CD11b), CD43, and IgM; B-1 cells are $Mac1^+CD43^+IgM^{hi}$.

2.7. Turnover of cells in different B-cell compartments

Turnover of transitional and MRF B cells in the spleen can be a useful measure of changes in the rate of differentiation and death of a given population. Signals from the pre-BCR drive extensive pre-B-cell proliferation, but after this point, developing B cells become quiescent and do not divide again until they are stimulated by antigen or another B-cell agonist. To measure turnover, we treat mice with 5-bromo-2-deoxyuridine (BrdU), which is incorporated into dividing pre-B cells, which in turn begin to contribute to downstream B-cell subsets: immature B cells, T1, T2, T3, MRF, and MZ B cells. The rate at which cells in these compartments become $BrdU^+$ is a measure of the turnover of the population.

2.7.1. Method

Inject mice intraperitoneally with 1 mg of BrdU in 200 μl PBS. Then give mice BrdU in their drinking water (1 mg/ml) continuously. Harvest organs for analysis at suitable time points. In the case of transitional cells in the

$CD2^+IgM^+IgD^-$), and MRF B cells ($CD2^+IgM^+IgD^+$). (B) Splenic cells are first gated on the scatter profile for live lymphoid-sized cells and then the expression of B220 and C1qRp (AA4.1) is used to distinguish transitional ($B220^+C1qRp^+$) and mature B cells ($B220^+C1qRp^-$). The expression of CD23 and IgM is used to further subdivide transitional cells into T1 ($CD23^-IgM^{hi}$), T2 ($CD23^+IgM^{hi}$), and T3 ($CD23^+IgM^{lo}$) B cells and to subdivide mature cells into MZ ($CD23^-IgM^{hi}$) and MRF ($CD23^+IgM^{lo}$) B cells. (C) Splenocytes scatter gated as in B were gated on $B220^+$ cells and subdivided into MRF ($CD21^+CD23^+$) and MZ ($CD21^+CD23^-$) B cells according to the expression of CD21 and CD23. Note that this is a good way to resolve the MZ subset, but that, in our view, MRF B cells are best identified using the stain shown in B. (D) Graph showing percentage of T1, T2, T3, and MRF B cells in the spleen positive for BrdU staining as a function of time following the start of continuous administration of BrdU to the mice. Note that the transitional subsets show significant labeling within 2 to 3 days and that T1 cells label faster than T2 cells, which in turn label more rapidly than T3 cells. In the time course shown, MRF B cells do not show significant labeling—the half-life of this subset is typically several weeks. (E) Lymph node cells were scatter gated as before, and $CD19^+$ cells were analyzed for the expression of IgM and IgD. MRF cells are $CD19^+IgM^{lo/hi}IgD^{hi}$. (F) Cells from a peritoneal lavage were scatter gated as before, and $CD19^+$ cells were analyzed for the expression of CD5 and CD23, allowing for the separation of B-1a ($CD5^+CD23^-$), B-1b ($CD5^-CD23^-$), and MRF (B-2, $CD5^-CD23^+$) B cells.

spleen, prepare them and stain them as described earlier for B220, C1qRp, CD23, and IgM. Fix in 3% paraformaldehyde at 4° for at least 20 min. Wash with PBS, quench with PBS, 50 mM NH$_4$Cl for 10 min at room temperature, and permeabilize with PBS, 0.1% Nonidet P-40 (NP-40) for 3 min at room temperature. Wash again, spin down cells, resuspend in FITC-labeled anti-BrdU plus DNase (FastImmune anti-BrdU-DNase, Becton Dickinson), incubate for 1 h in the dark at room temperature, wash, and analyze by flow cytometry.

2.7.2. Expected results

Transitional B-cell subsets (T1, T2, T3) label up with BrdU with a typical half-time of 2 to 4 days, whereas the half-life for MRF and MZ subsets is several weeks, reflecting their much slower turnover (see Fig. 18.1D).

3. Analysis of T-Cell Development

3.1. Early T-cell development

The first stages of T-cell development occur in the thymus. Hematopoietic progenitors enter the thymus at the cortico-medullary junction and then migrate to the subcapsular zone of the thymic cortex. These earliest T-cell progenitors are typified by the absence of expression of both CD4 and CD8 and have thus been termed DN thymocytes. These have been further subdivided according to the expression of CD25 and CD44 to give rise to the following subsets in chronological order: DN1 (CD25$^-$CD44$^+$), DN2 (CD25$^+$CD44$^+$), DN3 (CD25$^+$CD44$^-$), and DN4 (CD25$^-$CD44$^-$) cells. Within the DN2 and DN3 compartments, rearrangement of the TCRβ genes begins, first with joining of Dβ to Jβ and then Vβ to DJβ. Successful rearrangement results in synthesis of a TCRβ protein, which associates with pTα (a surrogate TCRα), and the signaling CD3$\gamma\delta\zeta$ complex to form the pre-TCR. Signaling from the pre-TCR in DN3 cells allows them to survive, differentiate into DN4 cells, proliferate, and eventually become CD4$^+$CD8$^+$ DP thymocytes. As in B-cell development, signals from the pre-TCR cause TCRβ rearrangement to be switched off, resulting in allelic exclusion of the TCRβ locus, such that developing thymocytes express only one TCRβ protein. For a more comprehensive discussion of this developmental transition, the reader is directed to a more detailed review (Aifantis *et al.*, 2006). Also differentiating at the DN2/3 stages is the alternative TCR$\gamma\delta$ lineage, in which the TCRγ and then TCRδ genes are rearranged in preference to TCRβ. For more details on this, the reader is directed to a review (Xiong and Raulet, 2007).

3.2. Positive and negative selection

The extensive pre-TCR-driven proliferation at the DN4 stage results in a large cellular expansion, such that DP thymocytes are typically the most abundant cell type in the thymus. DP cells initiate rearrangement of the TCRα genes. Successful rearrangement results in association of TCRα and TCRβ with the CD3γδζ signaling complex to form a mature TCR. At this stage, the immature T cells are subject to the processes of positive and negative selection. Cells bearing a TCR with an ability to interact weakly with endogenous peptides presented by MHC molecules undergo positive selection—they survive, upregulate expression of TCR, and differentiate into either CD4 single positive (CD4SP, $CD4^+CD8^-$) or CD8 single positive (CD8SP, $CD4^-CD8^+$) cells by turning off expression of either CD8 or CD4, respectively. The choice of differentiation into the CD4 or CD8 lineage is determined by the specificity of the TCR: if the receptor binds peptides presented by MHC class I, the cells will become CD8SP, whereas if it binds MHC class II, the cells differentiate into CD4SP cells. This lineage choice determines the function of the T cell, with CD4SP cells emigrating from the thymus and becoming $CD4^+$ helper T cells and CD8SP thymocytes becoming $CD8^+$ cytotoxic T cells. While weak binding of peptide/MHC ligands to the TCR results in positive selection, strong binding results in negative selection whereby the cells die by apoptosis. This ensures that potentially autoreactive T cells are eliminated. For a comprehensive discussion of positive and negative selection, see Starr et al. (2003).

3.3. Peripheral T-cell subsets

Positive selection of DP thymocytes into CD4SP and CD8SP cells is accompanied by migration of the cells into the medulla of the thymus and subsequently emigration from the thymus to seed the peripheral lymphoid organs (spleen, lymph nodes, etc.) as $CD4^+$ and $CD8^+$ T cells. These quiescent T cells are termed naïve until they have encountered antigen. Stimulation of T cells by antigen, for example, a viral peptide presented by MHC, results in activation of the cells, proliferation, and differentiation first into effector cells and then into memory cells. A useful marker of activation on T cells is CD44, which identifies recently activated cells, as well as the persisting memory cells. An additional lineage of $CD4^+$ T cells is regulatory T cells (T_{reg}), whose function is to suppress immune responses and maintain immune tolerance. These develop in the thymus, probably as a result of expressing self-reactive TCRs with an affinity somewhere between that required for positive and negative selection. They are characterized by the expression of CD25 and the transcription factor Foxp3 (Fontenot and Rudensky, 2005).

3.4. Signaling during T-cell development

Signaling from the pre-TCR and the TCR is critical during T development. Signals from the pre-TCR allow only cells with a functional TCRβ to differentiate into DP cells, whereas signals from the TCR at the DP stage determine whether the cells undergo positive selection into CD4$^+$, CD8$^+$, or T$_{reg}$ lineages or whether they are eliminated by negative selection, if potentially autoreactive. In addition, signals from many other receptors play a critical role in these processes. IL-7 receptor and Notch signaling is crucial for cell survival during the DN stage, signals from death receptors may be required to eliminate cells that have failed to rearrange TCR genes, and chemokine receptors control the movement of thymocytes within the thymus, and their subsequent emigration and localization in peripheral lymphoid organs. Thus, as with B cells, analysis of T-cell development in the absence of Rac GTPases, or other signaling molecules, can be used to identify *in vivo* functions for the GTPases in transducing signals from the antigen receptors, as well as other receptors.

3.5. Mice deficient for Rac GTPases in the T-cell lineage

To delete *Rac1* in the T-cell lineage, we initially used a transgenic mouse strain with Cre under the control of the Lck promoter (Orban *et al.*, 1992). In our hands this gave very variable deletion from mouse to mouse (M. Walmsley and V. L. J. Tybulewicz, unpublished results). Far more consistent and efficient deletion has been achieved using the CD2-Cre transgenic, which starts to delete loxP-flanked genes already at the DN1 stage, and is almost complete by the DN4 stage (de Boer *et al.*, 2003). This is very useful for studying signaling events at the pre-TCR checkpoint. To delete *Rac1* at later stages, it would be possible to use the CD4-Cre transgene, which starts to delete at the DP stage (Lee *et al.*, 2001), and the OX40-Cre, which deletes in activated T cells (Zhu *et al.*, 2004).

3.6. Enumeration of number of cells in T-cell compartments

As with the B-cell lineage, enumeration of different T-cell subsets by flow cytometry is an effective way to establish whether a specific genetic mutation has affected signaling pathways operating during T-cell development. Organs typically analyzed for T-cell development are the thymus, spleen, and lymph nodes. More specialized T-cell subsets can also be analyzed from the skin and the epithelium of the gut. The number of cells in T-cell subsets changes with the age of mouse. The thymus reaches its maximum size around 6 to 8 weeks of age, after which time it begins to shrink. Peripheral T-cell numbers rise from birth, reaching a plateau around 8 weeks of age.

In view of this, comparisons must always be carried out on age-matched mice, which should also be matched for gender and genetic background.

3.6.1. Cell preparation and staining for flow cytometry

Single cell suspensions from thymus, spleen, and lymph nodes are prepared and counted as described previously. It is not usually necessary to remove red cells from the thymus using ACK buffer lysis. Cell staining procedures are again as described previously. To stain for intracellular proteins, such as TCRβ, first stain cells with antibodies to cell surface markers, wash once in FACS buffer, then in PBS, fix in 1% paraformaldehyde for 15 min at room temperature, wash again, and quench autofluorescence from the paraformaldehyde by incubating the cells in PBS, 10 mM glycine for 10 min at room temperature in the dark. Wash the cells, permeabilize using PBS, 10 mM HEPES, pH 7.4, 0.5% saponin, 5% FCS for 10 min at room temperature, and stain with antibodies to TCRβ diluted in the permeabilization buffer for 45 min at room temperature in the dark. Wash once with saponin-containing buffer, once with PBS, and analyze on a flow cytometer.

3.6.2. Identification of T-cell subsets

The thymus is easily divided into DN, DP, CD4SP, and CD8SP cells using the expression of CD4 and CD8 (Fig. 18.2A). Because true mature CD4SP and CD8SP cells express high levels of TCRβ, it is useful to include expression of TCRβ. Usually this shows that all cells in the CD4SP gate are uniformly high for TCRβ, but some cells in the CD8SP gate have little or no TCRβ expression (see Fig. 18.2A). These are intermediate single positive (ISP) cells, which represent cells in transit from the DN4 to DP compartments that have started to express CD8, but do not yet express CD4.

To separate DN cells into DN1–4 subsets on the basis of CD25 and CD44 expression, it is necessary to carefully exclude all non-T lineage cells from this compartment and to positively identify the cells as being from the T lineage. We do this by staining for expression of Thy1.2 and for a mixture of markers termed "Lineage" (see Fig. 18.2B). This mix includes antibodies to CD4, CD8, TCRβ, TCR$\gamma\delta$, B220, Mac1, Gr1, and DX5. Gating on cells that are Thy1.2$^+$ lineage gives a population of T lineage DN cells, which do not include DP, CD4SP, CD8SP, $\gamma\delta$T, B, natural killer cells, macrophages, and neutrophils. A useful measure of successful transition through the pre-TCR checkpoint is the expression of intracellular TCRβ in DN3 and DN4 cells (see Fig. 18.2C). While only a small fraction of DN3 cells express intracellular TCRβ (typically around 20%), almost all DN4 cells should express the protein if the checkpoint is operating correctly.

In peripheral T-cell subsets, separation of CD4 and CD8 T cells can be usefully combined with the measurement of expression of TCRβ and CD5,

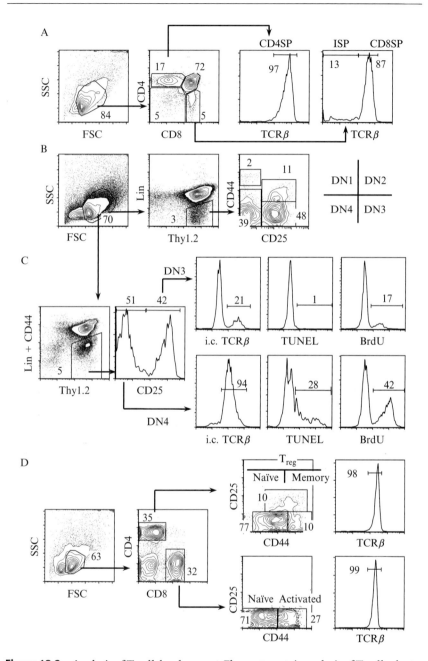

Figure 18.2 Analysis of T-cell development. Flow cytometric analysis of T-cell subsets from a 6-week-old mouse containing the CD2-Cre transgene, but no loxP-flanked target. Numbers indicate percentages of cells falling into marked gates. (A) Thymocytes were first gated on live lymphoid-sized cells using forward scatter (FSC) and side scatter (SSC). These scatter-gated cells were analyzed for the expression of CD4 and CD8,

which should both be uniformly expressed at a high level (see Fig. 18.2D and data not shown). Alterations in signaling can be manifested by changes in the level of expression of CD4, CD8, TCRβ, or CD5 molecules. Activated T cells can be identified by the expression of CD44, which also marks memory T cells. T$_{reg}$ cells are characterized by expression of CD25 and Foxp3; the latter can be measured by intracellular antibody staining (not shown).

3.7. Measurement of cell proliferation

Successful TCRβ rearrangement and subsequent signaling from the pre-TCR result in extensive proliferation of DN4 cells. To measure this, inject mice intraperitoneally with 1 mg of BrdU in 200 μl PBS. After 4 h, sacrifice the mouse and prepare thymocytes. Stain cells with antibodies to Thy1.2, CD25, and the lineage mix as before, but include a biotinylated antibody to CD44 in the mix (Lin+CD44) in order to exclude DN1 and DN2 cells. Wash cells once in FACS buffer and then in PBS and fix in 3% paraformaldehyde at 4° for at least 1 h. Wash with PBS, quench with PBS, 10 mM glycine for 10 min at room temperature, and permeabilize with PBS, 0.1% NP-40 for 3 min at room temperature. Wash in PBS, spin cells, resuspend

---

identifying double negative (DN, CD4$^-$CD8$^-$), double positive (DP, CD4$^+$CD8$^+$), CD4 single positive (CD4SP, CD4$^+$CD8$^-$), and CD8 single positive (CD8SP, CD4$^-$CD8$^+$) subsets. While all the cells in the CD4$^+$CD8$^-$ gate are uniformly high for TCRβ expression, and thus are all mature CD4SP cells, some of the cells in the CD4$^-$CD8$^+$ gate express low levels of TCRβ and are thus identified as ISP cells. True mature CD8SP cells express high levels of TCRβ. (B) Thymocytes scatter gated as in A are analyzed for the expression of Thy1.2, a T lineage marker, and a lineage (Lin) mix of antibodies against CD4, CD8, TCRβ, TCR$\gamma\delta$, B220, Mac1, Gr1, and DX5. All lineage mix antibodies were biotin labeled and detected with a streptavidin-PerCP secondary reagent. Cells that are Thy1.2$^+$Lin$^-$ are true T lineage DN cells. These are subdivided further by the expression of CD25 and CD44 into DN1 (CD25$^-$CD44$^+$), DN2 (CD25$^+$CD44$^+$), DN3 (CD25$^+$CD44$^-$), and DN4 (CD25$^-$CD44$^-$) cells. (C) Thymocytes scatter gated as in A were stained for a lineage mix, including antibody to CD44 to exclude DN1 and DN2 cells. The resultant Thy1.2$^+$(Lin+CD44)$^-$ cells are a mixture of DN3 and DN4 cells, which can be separated by the expression of CD25. In each of these populations, staining of permeabilized cells is used to quantify the expression of intracellular TCRβ (i.e. TCRβ), TUNEL staining as a measure of cell death, and BrdU staining as a measure of cell division. BrdU staining is carried out on cells from mice that had been given BrdU 4 h earlier. In the DN3 compartment, only a small fraction of cells express intracellular TCRβ, most cells are alive, and few are dividing. In contrast, almost all DN4 cells express intracellular TCRβ, some are dying, and many are dividing. (D) Lymph node cells were scatter gated as before and separated into CD4 and CD8 T cells, which in turn were analyzed for expression of CD25, CD44, and surface TCRβ. While both T-cell lineages uniformly express high levels of TCRβ, CD25 and CD44 expression can be used to divide CD4 cells into naïve (CD25$^-$CD44$^-$), memory/activated (CD25$^-$CD44$^+$), and regulatory (T$_{reg}$, CD25$^+$CD44^{lo}) cells. Similarly, CD44 identifies activated CD8 T cells.

in FITC-labeled anti–BrdU plus DNase solution (FastImmune anti–BrdU–DNase, Becton Dickinson), incubate for 1 h in the dark at room temperature, wash, and analyze by flow cytometry. Identify DN3 [Thy1.2$^+$ (Lin+CD44)$^-$CD25$^+$] and DN4 [Thy1.2$^+$(Lin+CD44)$^-$CD25$^-$] cells as shown in Fig. 18.2C. Typically, DN3 cells show some level of cell division, which is much higher in DN4 cells. In contrast, later stages of T-cell development (DP, CD4SP, CD8SP, and naïve peripheral T cells) are quiescent, showing little or no cell division (not shown).

3.8. Measurement of cell death

Both the pre-TCR and the TCR checkpoints in thymic development are associated with extensive apoptosis of cells that either have failed to rearrange TCRβ or TCRα genes or have generated an autoreactive TCR. Thus measurement of cell death can be a useful indicator of correct signaling events. One method to measure apoptotic cell death by flow cytometry is TdT-mediated dUTP nick end labeling (TUNEL), which measures the frequency of DNA breaks using terminal deoxynucleotidyl transferase to label free 3'-OH ends in the DNA molecule with fluorescein-dUTP. To measure cell death at the DN3/DN4 checkpoint, stain thymocytes as before with antibodies to Thy1.2, CD25, and a lineage mix, including a biotinylated antibody to CD44 (Lin+CD44) as before. Wash in PBS, fix in 3% paraformaldehyde for 20 min at room temperature, wash with PBS, quench with PBS, 10 mM glycine for 10 min at room temperature, and permeabilize with PBS, 1% Triton X-100, 1% sodium citrate for 2 min on ice. Wash in PBS and stain using the *in situ* cell death detection kit (Roche), following the manufacturer's instructions, and analyze by flow cytometry. Usually very little cell death is detectable in DN3 cells, whereas more is visible in DN4 cells (see Fig. 18.2C).

ACKNOWLEDGMENT

This work was supported by the Medical Research Council, UK.

REFERENCES

Aifantis, I., Mandal, M., Sawai, K., Ferrando, A., and Vilimas, T. (2006). Regulation of T-cell progenitor survival and cell-cycle entry by the pre-T-cell receptor. *Immunol. Rev.* **209,** 159–169.

Allman, D., Lindsley, R. C., DeMuth, W., Rudd, K., Shinton, S. A., and Hardy, R. R. (2001). Resolution of three nonproliferative immature splenic B cell subsets reveals multiple selection points during peripheral B cell maturation. *J. Immunol.* **167,** 6834–6840.

Croker, B. A., Handman, E., Hayball, J. D., Baldwin, T. M., Voigt, V., Cluse, L. A., Yang, F. C., Williams, D. A., and Roberts, A. W. (2002a). Rac2-deficient mice display perturbed T-cell distribution and chemotaxis, but only minor abnormalities in T(H)1 responses. *Immunol. Cell Biol.* **80,** 231–240.

Croker, B. A., Tarlinton, D. M., Cluse, L. A., Tuxen, A. J., Light, A., Yang, F. C., Williams, D. A., and Roberts, A. W. (2002b). The Rac2 guanosine triphosphatase regulates B lymphocyte antigen receptor responses and chemotaxis and is required for establishment of B-1a and marginal zone B lymphocytes. *J. Immunol.* **168,** 3376–3386.

de Boer, J., Williams, A., Skavdis, G., Harker, N., Coles, M., Tolaini, M., Norton, T., Williams, K., Roderick, K., Potocnik, A. J., and Kioussis, D. (2003). Transgenic mice with hematopoietic and lymphoid specific expression of Cre. *Eur. J. Immunol.* **33,** 314–325.

Fontenot, J. D., and Rudensky, A. Y. (2005). A well adapted regulatory contrivance: Regulatory T cell development and the forkhead family transcription factor Foxp3. *Nat. Immunol.* **6,** 331–337.

Gomez, M., Kioussis, D., and Cantrell, D. A. (2001). The GTPase Rac-1 controls cell fate in the thymus by diverting thymocytes from positive to negative selection. *Immunity* **15,** 703–713.

Gomez, M., Tybulewicz, V., and Cantrell, D. A. (2000). Control of pre-T cell proliferation and differentiation by the GTPase Rac-1. *Nat. Immunol.* **1,** 348–352.

Hardy, R. R. (2006). B-1 B cell development. *J. Immunol.* **177,** 2749–2754.

Hardy, R. R., and Hayakawa, K. (2001). B cell development pathways. *Annu. Rev. Immunol.* **19,** 595–621.

Kraus, M., Alimzhanov, M. B., Rajewsky, N., and Rajewsky, K. (2004). Survival of resting mature B lymphocytes depends on BCR signaling via the Igalpha/beta heterodimer. *Cell* **117,** 787–800.

Lee, P. P., Fitzpatrick, D. R., Beard, C., Jessup, H. K., Lehar, S., Makar, K. W., Perez-Melgosa, M., Sweetser, M. T., Schlissel, M. S., Nguyen, S., Cherry, S. R., Tsai, J. H., *et al.* (2001). A critical role for Dnmt1 and DNA methylation in T cell development, function, and survival. *Immunity* **15,** 763–774.

Li, B., Yu, H., Zheng, W., Voll, R., Na, S., Roberts, A. W., Williams, D. A., Davis, R. J., Ghosh, S., and Flavell, R. A. (2000). Role of the guanosine triphosphatase Rac2 in T helper 1 cell differentiation. *Science* **288,** 2219–2222.

Loder, F., Mutschler, B., Ray, R. J., Paige, C. J., Sideras, P., Torres, R., Lamers, M. C., and Carsetti, R. (1999). B cell development in the spleen takes place in discrete steps and is determined by the quality of B cell receptor-derived signals. *J. Exp. Med.* **190,** 75–89.

Melchers, F., and Rolink, A. (2006). B cell tolerance: How to make it and how to break it. *Curr. Top. Microbiol. Immunol.* **305,** 1–23.

Merrell, K. T., Benschop, R. J., Gauld, S. B., Aviszus, K., Decote-Ricardo, D., Wysocki, L. J., and Cambier, J. C. (2006). Identification of anergic B cells within a wild-type repertoire. *Immunity* **25,** 953–962.

Orban, P. C., Chui, D., and Marth, J. D. (1992). Tissue- and site-specific DNA recombination in transgenic mice. *Proc. Natl. Acad. Sci. USA* **89,** 6861–6865.

Pillai, S., Cariappa, A., and Moran, S. T. (2005). Marginal zone B cells. *Annu. Rev. Immunol.* **23,** 161–196.

Rickert, R. C., Roes, J., and Rajewsky, K. (1997). B lymphocyte-specific, Cre-mediated mutagenesis in mice. *Nucleic Acids Res.* **25,** 1317–1318.

Roberts, A. W., Kim, C., Zhen, L., Lowe, J. B., Kapur, R., Petryniak, B., Spaetti, A., Pollock, J. D., Borneo, J. B., Bradford, G. B., Atkinson, S. J., Dinauer, M. C., *et al.* (1999). Deficiency of the hematopoietic cell-specific Rho family GTPase Rac2 is characterized by abnormalities in neutrophil function and host defense. *Immunity* **10,** 183–196.

Srinivas, S., Watanabe, T., Lin, C. S., William, C. M., Tanabe, Y., Jessell, T. M., and Costantini, F. (2001). Cre reporter strains produced by targeted insertion of EYFP and ECFP into the ROSA26 locus. *BMC Dev. Biol.* **1,** 4.

Starr, T. K., Jameson, S. C., and Hogquist, K. A. (2003). Positive and negative selection of T cells. *Annu. Rev. Immunol.* **21,** 139–176.

Sugihara, K., Nakatsuji, N., Nakamura, K., Nakao, K., Hashimoto, R., Otani, H., Sakagami, H., Kondo, H., Nozawa, S., Aiba, A., and Katsuki, M. (1998). Rac1 is required for the formation of three germ layers during gastrulation. *Oncogene* **17,** 3427–3433.

Thomas, M. D., Srivastava, B., and Allman, D. (2006). Regulation of peripheral B cell maturation. *Cell. Immunol.* **239,** 92–102.

Walmsley, M. J., Ooi, S. K., Reynolds, L. F., Smith, S. H., Ruf, S., Mathiot, A., Vanes, L., Williams, D. A., Cancro, M. P., and Tybulewicz, V. L. (2003). Critical roles for Rac1 and Rac2 GTPases in B cell development and signaling. *Science* **302,** 459–462.

Xiong, N., and Raulet, D. H. (2007). Development and selection of $\gamma\delta$ T cells. *Immunol. Rev.* **215,** 15–31.

Yu, H., Leitenberg, D., Li, B., and Flavell, R. A. (2001). Deficiency of small GTPase Rac2 affects T cell activation. *J. Exp. Med.* **194,** 915–926.

Zhu, J., Min, B., Hu-Li, J., Watson, C. J., Grinberg, A., Wang, Q., Killeen, N., Urban, J. F., Jr., Guo, L., and Paul, W. E. (2004). Conditional deletion of Gata3 shows its essential function in TH1-TH2 responses. *Nat. Immunol.* **5,** 1157–1165.

CHAPTER NINETEEN

CHARACTERIZATION OF OLIGOPHRENIN-1, A RHOGAP LOST IN PATIENTS AFFECTED WITH MENTAL RETARDATION: LENTIVIRAL INJECTION IN ORGANOTYPIC BRAIN SLICE CULTURES

Nael Nadif Kasri,* Eve-Ellen Govek,[†] *and* Linda Van Aelst*

Contents

Abstract

Mutations in regulators and effectors of the Rho GTPases underlie various forms of mental retardation (MR). Among them, *oligophrenin*-1 (*OPHN*1), which encodes a Rho-GTPase activating protein, was one of the first Rho-linked MR genes identified. Upon characterization of OPHN1 in hippocampal brain slices, we obtained evidence for the requirement of OPHN1 in dendritic spine morphogenesis and neuronal function of CA1 pyramidal neurons. Organotypic hippocampal brain slice cultures are commonly used as a model system to investigate the morphology and synaptic function of neurons, mainly because they allow for the long-term examination of neurons in a preparation where the

* Cold Spring Harbor Laboratory, Cold Spring Harbor, New York
[†] The Rockefeller University, Laboratory of Developmental Neurobiology, New York

Methods in Enzymology, Volume 439
ISSN 0076-6879, DOI: 10.1016/S0076-6879(07)00419-3

gross cellular architecture of the hippocampus is retained. In addition, maintenance of the trisynaptic circuitry in hippocampal slices enables the study of
synaptic connections. Today, a multitude of gene transfer methods for postmitotic neurons in brain slices are available to easily manipulate and scrutinize
the involvement of signaling molecules, such as Rho GTPases, in specific
cellular processes in this system. This chapter covers techniques detailing the
preparation and culturing of organotypic hippocampal brain slices, as well as
the production and injection of lentivirus into brain slices.

1. INTRODUCTION: RHO GTPASES, SYNAPTIC STRUCTURE AND FUNCTION

Dendritic spines are highly specialized, actin-rich structures that pro-
trude from dendrites and serve as the postsynaptic compartment for the
majority of excitatory synapses (Hering and Sheng, 2001). Spine morphology is ultimately linked to synaptic function, which underlies cognitive
functions, such as learning and memory. In accordance with this idea,
pathological studies have revealed dendritic spine abnormalities in patients
with mental retardation (MR) (Fiala *et al.*, 2002). Recent progress in the
field of MR suggests that defects in such crucial cellular processes as spine
morphogenesis, synaptogenesis, and synaptic plasticity contribute to cognitive impairment resulting from mutations in MR-related genes (Newey
et al., 2005; van Galen and Ramakers, 2005).

Small GTP-binding proteins of the Rho subfamily, and in particular
Rac1, Cdc42, and RhoA GTPases, have emerged as key modulators of
dendritic spine morphogenesis through their regulation of the actin
cytoskeleton (Govek *et al.*, 2005). Hence, it is not surprising that mutations
in genes encoding regulators and effectors of the Rho GTPases have been
found to underlie human neurological diseases. To date, several Rho
GTPase-linked genes associated with MR have been identified (Newey
et al., 2005), including *OPHN1* (Bienvenu *et al.*, 1997; Billuart *et al.*, 1998;
Govek *et al.*, 2004), *PAK3* (Allen *et al.*, 1998; Bienvenu *et al.*, 2000; Boda
et al., 2004; Gedeon *et al.*, 2003; Meng *et al.*, 2005), and *ARHGEF6*
(Kutsche *et al.*, 2000; Node-Langlois *et al.*, 2006). We embarked on the
functional characterization of OPHN1. *OPHN1* was first found mutated in
patients with nonsyndromic MR (Bienvenu *et al.*, 1997; Billuart *et al.*,
1998), but the presence of *OPHN1* mutations has been documented
more recently in families with syndromic forms of MR (Bergmann *et al.*,
2003; des Portes *et al.*, 2004; Philip *et al.*, 2003; Zanni *et al.*, 2005).
To better understand how mutations in *OPHN1* result in defects in neuronal signaling, we took advantage of the organotypic hippocampal slice
culture system.

The hippocampus is a structure central to learning and memory. Previous studies have demonstrated experience- and learning-dependent structural changes in dendrites and spines of hippocampal neurons in mammalian brain (Yuste and Bonhoeffer, 2001). The highly organized and laminar arrangement of synaptic pathways, from the entorhinal cortex to the dentate gyrus (DG), DG to CA3, and CA3 to CA1, makes the hippocampus a convenient model for studying synaptic connections. Identifiable neurons (e.g., CA1 pyramidal cells) in hippocampal slices are amenable to specific genetic manipulations, including RNAi and overexpression, and neighboring, untransfected neurons with comparable connectivity, developmental history, and exposure to culture conditions in the same slice can serve as ideal control neurons.

2. PREPARATION OF ORGANOTYPIC BRAIN SLICE CULTURES

Several methods have been developed for the preparation of organotypic brain slice cultures. This section describes the interface culturing technique introduced originally by Stoppini *et al.* (1991), in which the slices are placed on Millicell inserts so that the top of the slice is exposed to the incubator atmosphere (35° and 5% CO_2), while the bottom of the slice contacts the culture medium.

2.1. Preparation of media

1. Place an insert (Millicell-CM 0.4-μm culture plate insert, 30 mm diameter, Millipore) in a culture dish (Corning, 35 mm diameter).
2. Add 750 μl of filter-sterilized (using 0.22-μm disposable filter, Millipore) long-term slice culture medium (MEM Hanks medium, 1 mM L-glutamine, 1 mM $CaCl_2$, 2 mM $MgSO_4$, 1 mg/liter insulin, 1 mM $NaHCO_3$, 20% heat-inactivated horse serum, 0.5 mM L-ascorbate, 30 mM HEPES) under the insert. Make sure that the insert is completely wet on the bottom, avoiding any air bubbles.
3. Place 2 ml of long-term slice medium in a second cell culture dish. This will be used for the transfer of sliced hippocampi.
4. Place both cell culture dishes in a cell culture incubator at 35°/5% CO_2 until hippocampi are ready.

2.2. Preparation for dissection and slicing

1. Prepare the tissue chopper (Stoelting). Tape a square piece of Teflon to the chopping platform. Insert a clean razor blade and align it with

the chopping platform. Clean the chopping platform and blade with 70% ethanol.
2. Clean the dissection tools, including large scissors, small 3 1/2-in. straight microdissecting scissors (such as Biomedical Research Instruments), straight forceps, a straight, tapered spatula for removing the brain from the skull, a flat, spoon-shaped spatula for handling the whole brain, curved spatulae (such as Fine Science Tools), scalpels or dissecting chisels (such as Fine Science Tools), and wide-bore pipettes.
3. Prepare dissection buffer (low Na^+ artificial cerebral spinal fluid: 1 mM $CaCl_2$, 5 mM $MgCl_2$, 4 mM KCl, 26 mM $NaHCO_3$, 8% sucrose, 0.5% phenol red). Filter sterilize and oxygenate with 95% O_2/5% CO_2 gas mixture in a beaker chilled on salted ice. To make salted ice, simply add a large scoop of NaCl to a bucket of ice and mix the salt into the ice. The buffer will be ready when the color changes from purple to orange.
4. Prepare a cell culture dish (Corning, 60-mm-diameter well) with a Whatman filter (Whatman 42.5 mm) in the lid and cover it with cold dissection buffer. Keep it on ice until ready to use.

2.3. Removing and dissecting the hippocampus

1. Wipe the head of a postnatal day (P)6–P8 rat pup with 70% ethanol.
2. Using a large pair of scissors, collect the head of the animal.
3. Cut the skin down the midline, from the neck up to and between the eyes. Insert the scissors into the foramen magnum and cut the skull along the midline up to and between the eyes. Make four small incisions perpendicular to the main cut at the ends of the main incision. Peel the skull back with straight forceps, remove the brain with a straight, tapered spatula, and place it immediately in a small beaker with chilled dissection buffer.
4. Place the brain in the culture dish lid containing the chilled dissection medium with a Whatman filter (from step 4 under Section 2.2). Keep the brain covered with chilled dissection medium during the rest of the dissection.
5. Using a scalpel, remove the cerebellum by making a coronal cut just behind the inferior colliculi. Make a second cut sagittally down the midline, completely through the brain. Separate the hemispheres. Turn the hemibrain so that the medial surface is facing up. Separate the neocortex with the underlying hippocampus from the midbrain and brain stem. The hippocampus will now be exposed. Using a curved spatula, disrupt the connection of the hippocampus on the ventral side (fimbria). Rotate the hippocampus gently with the curved spatula around the longitudinal axis of the hippocampus to separate it from the neocortex.

6. Repeat this procedure to isolate the second hippocampus.

7. Remove hippocampi from the dish with a wide-bore pipette and place them on the chopping block.

8. Position hippocampi, smooth side up, using a pipette filled with dissection medium to push hippocampi into the correct position. Place them so that the long axis of each hippocampus is parallel to the direction of stage movement, but allow the septal end to curve away from the parallel axis.

9. Remove excess liquid. It is important to achieve the right amount of wetness. If the tissue is too wet, hippocampi tend to move each time the blade is lifted, whereas if the tissue is too dry, hippocampi tend to stick to the blade.

10. Position the stage of the tissue chopper to the edge of hippocampi. To make 400-μm slices, raise the blade, adjust the actuator by 400 μm, and drop the blade. Continue to raise the blade and adjust the micrometer by 400 μm before each drop of the blade until you are done cutting.

11. Transfer the slices from the chopping platform into a tissue culture plate with prewarmed medium (from step 3 under Section 2.1) using a wide-bore pipette.

12. Separate the slices under a dissecting microscope, either by agitating the dish gently or by using a curved spatula and forceps. Be careful not to poke or tear the slices.

13. Select slices with an intact structure that display distinct CA1, CA3, and DG cell layers. Typically, 12 to 15 slices are obtained for each preparation.

14. Place three to four slices per Millicell-CM insert in a well with prewarmed long-term slice medium (from steps 1 and 2 under Section 2.1). Keep slices distant from each other, but relatively centered in the middle of the membrane. Gently remove excess medium from the top of the insert, as medium on top of the insert will prevent oxygen exchange.

15. Place the cell culture dishes in a cell culture incubator (35° and 5% CO_2). When medium is changed every 3 days, the slices will continue to survive for several weeks in culture, forming layer-specific connections. It is important to assess the viability of the organotypic slices before proceeding with the experiment. Healthy slices should look transparent and the DG should be visible by eye. Viable slices are obtained most easily from young animals. We usually use P6 to P8 rat pups.

The quality of organotypic brain slice cultures will strongly depend on the isolation and dissection procedure described earlier. Time is an important parameter. Typically, the actual dissection should not take longer than 10 min for obtaining healthy slices. Although time is an important parameter, it is more important to do the dissection well than fast. During the procedure, it is essential to treat the slices gently and to avoid touching them with any sharp objects. Slices that are treated too harshly tend to become epileptic and will die prematurely.

3. Lentivirus Preparation and Injection

A multitude of gene transfer methods for postmitotic neurons are now available, including different viral vectors (Ehrengruber et al., 2001) to make adenovirus, adeno-associated virus, Sindbis virus (Ehrengruber et al., 1999), Semliki Forest virus, measles virus, and lentivirus (Zufferey et al., 1997). Each viral vector has its own specificity and advantages (Ehrengruber et al., 2001). This section describes the use of lentiviral vectors to transduce genes into CA1 pyramidal cells in hippocampal brain slices, primarily for long-term expression studies.

Loss of function by RNAi and overexpression are two convenient methods used to study the involvement of a Rho GTPase, or a related molecule, in a specific cellular process. For the expression of shRNAs in hippocampal neurons in organotypic slices, we have successfully used the TRIPΔU3-EF1α-EGFP (pTRIP) vector, a replication-defective self-inactivating lentiviral vector (Gimeno et al., 2004; Stove et al., 2005; Zennou et al., 2000). Use of this vector resulted in a very efficient knockdown of OPHN1. Details concerning the pTRIP vector and how to clone the shRNA expression cassette are described in Janas et al. (2006). To express OPHN1 ectopically, we successfully used a lentiviral backbone, based on a self-inactivating FUGW vector (described previously) (Lois et al., 2002). The FUGW vector is modified by introducing an internal ribosomal entry site (IRES), followed by the green fluorescent protein (GFP) cDNA, to create the vector FUIGW that contains an IRES-GFP cassette downstream of the ubiquitin-C promoter. The full-length cDNA of OPHN1 is then cloned into the BamHI and EcoRI sites. This lentiviral vector allows us to ectopically express full-length OPHN1, driven by an ubiquitin-C promoter, while coexpressing GFP in order to visualize the infected neurons (Fig. 19.1).

3.1. Lentivirus production and concentration

The pTRIP and FUIGW vectors are pseudotyped with VSV-G encoded by pMD.G (Dull et al., 1998). Gag, Pol, and Tat are expressed from the packaging construct pCMVΔR8.91 (Zufferey et al., 1997).

1. Plate 2 to 3 × 10^6 HEK 293T cells in a 10-cm tissue culture dish with 10 ml of medium 24 h before transfection. Cells should be approximately 60 to 80% confluent at the time of transfection.

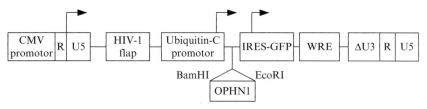

Figure 19.1 Schematic representation of the FUIGW vector. The FUIGW vector was adapted from the previously described FUGW vector (Lois *et al.*, 2002). The woodchuck hepatitis virus post-transcriptional regulatory element (WRE) serves to increase the level of transcription, and the human immunodeficiency virus-1 (HIV-1) flap element, between the 59 long terminal repeat and the human ubiquitin-C promoter, serves to increase the titer of the virus. For visualization of infected cells, an IRES-GFP cassette was introduced downstream of the ubiquitin-C promoter. Full-length cDNA of OPHN1 was inserted between the ubiquitin-C promoter and the IRES-GFP cassette into the unique restriction sites *Bam*HI and *Eco*RI.

2. Prepare 1 ml of calcium phosphate–DNA precipitate for the transfection of one 10-cm plate of HEK 293T cells as follows.

   a. Combine 7 μg of pTRIP or FUIGW vector, 10 μg of pCMVΔR8.91, and 6 μg of pMD.G (VSVG) in a sterile microcentrifuge tube. *Note*: For optimal results, the relative amounts of plasmids used for cotransfection should be determined empirically.

   b. Adjust the volume to 250 μl with 0.1$\times$ TE (1 mM Tris-HCl, pH 8.0, 0.1 mM EDTA).

   c. Add 250 μl of 2$\times$ CaCl$_2$ solution (0.5 M CaCl$_2$, 1 mM Tris-base, 0.1 mM EDTA) and mix by vortexing.

   d. Place 500 μl of 2$\times$ HEPES-buffered saline (HBS) (0.05 M HEPES-NaOH, pH 7.12, 0.28 M NaCl, 1.5 mM sodium phosphate buffer, pH 7.0) in a sterile tube (e.g., 2065 Falcon tube).

   e. Add the Ca^{2+}–DNA solution drop wise to 500 μl of 2$\times$ HBS while vortexing slowly. Let the suspension stand for 10 min to allow precipitate formation.

Note: 2$\times$ HBS should be stored at 4° and can be kept for up to 2 weeks. Aliquots of 2$\times$ CaCl$_2$ solution and 0.5 M HEPES-NaOH stock should be stored frozen at $-20°$. All solutions should be thawed at room temperature before use.

3. Distribute the precipitate evenly over the cells in the culture dish. Mix gently by rocking the dish back and forth, and return the plates to the incubator.

4. Replace medium 5 to 8 h after transfection with 10 ml of fresh culture medium.

5. Collect the medium containing viral particles 48 to 60 h after transfection.

6. Centrifuge freshly collected supernatant at 2000 *g* to remove cell debris.
7. Filter the supernatant through a 0.45-μm filtration unit. The filter should be prerinsed with cell culture medium before use.
8. Aliquot and store at $-80°$ or proceed to the concentration step.

   For infecting neurons in brain slices, further concentration of the supernatant is required.

3.2. Concentration

1. Transfer the filtered supernatant containing the virus in an Ultra-Clear tube for an SW28 Rotor (Beckman).
2. Centrifuge at 120,000 *g* for 3 h at $4°$.
3. Remove the supernatant and resuspend the viral pellet in an appropriate volume of Hanks balanced salt solution without calcium (Ca^{2+}) and magnesium (Mg^{2+}) (Gibco/Invitrogen) or suitable culture medium.
4. Store the tube containing the concentrated viral suspension at $4°$ for 12 h to ensure complete pellet resuspension. Aliquot and store at $-80°$.

3.3. Injection

Infection with lentivirus is performed as soon as possible after the preparation of the brain slices, as the efficiency of infection will decrease with time.

1. Prepare a glass pipette pulled to an outer diameter of 5 to 15 μm (Premium standard wall borosilicate capillaries with filament, 2.00 mm outside diameter, Warner Instruments).
2. Add 0.3 μl of a 1% fast green solution (Sigma) to 2.5 μl of concentrated virus for visualization during the injection.
3. Fill the glass pipette with the viral solution.
4. Position the glass pipette through a microelectrode holder (World Precision Instruments) on a micromanipulator (World Precision Instruments).
5. Connect the microelectrode holder to a controlled air pressure system, a Picospritzer (General Valve, Fairfield, NJ).
6. Under a stereomicroscope (Zeiss, Stemi 2000), lower the glass pipette into the desired cell layer (e.g., CA1 pyramidal cells), just under the surface membrane. The longitudinal axis of the glass pipette should be almost parallel with the dendritic projections of the pyramidal cells.
7. Inject virus using the Picospritzer. For each position, one or multiple injections of short duration (20 ms) can be performed, using controlled air pressure (10–20 psi). By varying the pulse duration and pressure on the Picospritzer, the multiplicity of infection can be adjusted from very few to 10–50 infected cells. Alternatively, a sharpened polymerase chain

reaction micropipette with plunger (Drummond) can be used to manually inject the virus into the slice.

Typically, two to three injection sites are selected per slice, and a total of 0.5 to 1 μl of concentrated virus solution is applied. When injecting the virus, a rapid diffusion of the fast green dye should be observed between the cells.

8. A few hours (5–6 h) after the injection, change the culture medium to prevent toxicity from the fast green dye.

Using the lentiviral-based system described, one can expect a very high efficiency of infection that is restricted to the area of injection (Fig. 19.2). Typically, transgene expression increases slowly with lentivirus. We observed the initial onset of GFP expression 4 days after infection. At this time point, the fluorescence intensity is still low, but increases rapidly at 6 to 8 days postinfection and lasts for many days. Lentiviral-based infection is therefore ideal for long-term gene expression.

Note that the amount of infection (dependent on injection duration, pressure, and number of injection sites) mainly depends on the purpose of the experiment. For electrophysiological experiments, one or two injection sites per slice are sufficient. In contrast, for biochemistry experiments, slices should receive multiple viral injections, spaced 100 to 150 μm apart, spanning the area of interest (e.g., CA1 area). These multiple injections and spacing ensure high infection efficiency. The infected regions can then be dissected under a fluorescence magnifying scope.

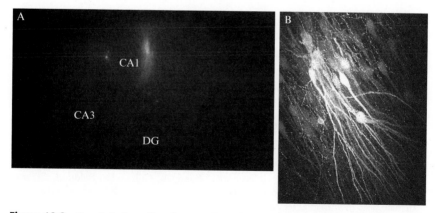

Figure 19.2 Lentiviral-mediated expression of GFP using the pTRIP lentiviral vector. (A) An organotypic brain slice prepared from a postnatal day 6 rat was infected in the CA1 region with a GFP expressing lentivirus (pTRIP). The image was taken 7 days post-infection. (B) A higher magnification two-photon microscope image of CA1 pyramidal cells shows a clear fluorescent signal in the cell body and apical dendrites. CA1–CA3, cornu ammonis; DG, dentate gyrus.

4. CONCLUSIONS

Lentiviral injection can be used reliably to infect postmitotic neurons in intact brain tissue. Once the specific parameters of the infection (titer, volume, pulse duration) are determined, this method is efficient, reproducible, and does not require advanced molecular biological facilities for its application. Significantly, the use of lentiviral vectors to express hairpins or recombinant proteins allows for gene manipulation in specific neurons (e.g., CA1 pyramidal cells) in an environment of normal background tissue at a particular time point during development, thus circumventing many of the problems of interpreting data from transgenic and knockout animals. The preserved tissue architecture in slices also facilitates the analysis of defined hippocampal cells, synapses, and connections. Applications are numerous and include electrophysiology, morphology, biochemistry, pharmacology, and development.

ACKNOWLEDGMENTS

We thank members of the Van Aelst and Malinow laboratories for sharing their expertise and helpful discussions. L.V.A is supported by the National Institutes of Health, National Science Foundation, and National Alliance for Autism Research. N.N.K is a postdoctoral fellow from the Fund for Scientific Research Flanders and is supported by the Human Frontiers Science Program. E.E.G. is a postdoctoral associate from the Rockefeller University supported by a Women & Science fellowship.

REFERENCES

Allen, K. M., Gleeson, J. G., Bagrodia, S., Partington, M. W., MacMillan, J. C., Cerione, R. A., Mulley, J. C., and Walsh, C. A. (1998). PAK3 mutation in nonsyndromic X-linked mental retardation. *Nat. Genet.* **20,** 25–30.

Bergmann, C., Zerres, K., Senderek, J., Rudnik-Schoneborn, S., Eggermann, T., Hausler, M., Mull, M., and Ramaekers, V. T. (2003). Oligophrenin 1 (OPHN1) gene mutation causes syndromic X-linked mental retardation with epilepsy, rostral ventricular enlargement and cerebellar hypoplasia. *Brain* **126,** 1537–1544.

Bienvenu, T., Der-Sarkissian, H., Billuart, P., Tissot, M., Des Portes, V., Bruls, T., Chabrolle, J. P., Chauveau, P., Cherry, M., Kahn, A., Cohen, D., Beldjord, C., *et al.* (1997). Mapping of the X-breakpoint involved in a balanced X;12 translocation in a female with mild mental retardation. *Eur. J. Hum. Genet.* **5,** 105–109.

Bienvenu, T., des Portes, V., McDonell, N., Carrie, A., Zemni, R., Couvert, P., Ropers, H. H., Moraine, C., van Bokhoven, H., Fryns, J. P., Allen, K., Walsh, C. A., *et al.* (2000). Missense mutation in PAK3, R67C, causes X-linked nonspecific mental retardation. *Am. J. Med. Genet.* **93,** 294–298.

Billuart, P., Bienvenu, T., Ronce, N., des Portes, V., Vinet, M. C., Zemni, R., Roest Crollius, H., Carrie, A., Fauchereau, F., Cherry, M., Briault, S., Hamel, B., *et al.* (1998).

Oligophrenin-1 encodes a rhoGAP protein involved in X-linked mental retardation. *Nature* **392**, 923–926.

Boda, B., Alberi, S., Nikonenko, I., Node-Langlois, R., Jourdain, P., Moosmayer, M., Parisi-Jourdain, L., and Muller, D. (2004). The mental retardation protein PAK3 contributes to synapse formation and plasticity in hippocampus. *J. Neurosci.* **24**, 10816–10825.

des Portes, V., Boddaert, N., Sacco, S., Briault, S., Maincent, K., Bahi, N., Gomot, M., Ronce, N., Bursztyn, J., Adamsbaum, C., Zilbovicius, M., Chelly, J., *et al.* (2004). Specific clinical and brain MRI features in mentally retarded patients with mutations in the Oligophrenin-1 gene. *Am. J. Med. Genet. A* **124**, 364–371.

Dull, T., Zufferey, R., Kelly, M., Mandel, R. J., Nguyen, M., Trono, D., and Naldini, L. (1998). A third-generation lentivirus vector with a conditional packaging system. *J. Virol.* **72**, 8463–8471.

Ehrengruber, M. U., Hennou, S., Bueler, H., Naim, H. Y., Deglon, N., and Lundstrom, K. (2001). Gene transfer into neurons from hippocampal slices: Comparison of recombinant Semliki Forest Virus, adenovirus, adeno-associated virus, lentivirus, and measles virus. *Mol. Cell. Neurosci.* **17**, 855–871.

Ehrengruber, M. U., Lundstrom, K., Schweitzer, C., Heuss, C., Schlesinger, S., and Gahwiler, B. H. (1999). Recombinant Semliki Forest virus and Sindbis virus efficiently infect neurons in hippocampal slice cultures. *Proc. Natl. Acad. Sci. USA* **96**, 7041–7046.

Fiala, J. C., Spacek, J., and Harris, K. M. (2002). Dendritic spine pathology: Cause or consequence of neurological disorders? *Brain Res. Brain Res. Rev.* **39**, 29–54.

Gedeon, A. K., Nelson, J., Gecz, J., and Mulley, J. C. (2003). X-linked mild non-syndromic mental retardation with neuropsychiatric problems and the missense mutation A365E in PAK3. *Am. J. Med. Genet. A* **120**, 509–517.

Gimeno, R., Weijer, K., Voordouw, A., Uittenbogaart, C. H., Legrand, N., Alves, N. L., Wijnands, E., Blom, B., and Spits, H. (2004). Monitoring the effect of gene silencing by RNA interference in human CD34+ cells injected into newborn RAG2-/- gammac-/- mice: Functional inactivation of p53 in developing T cells. *Blood* **104**, 3886–3893.

Govek, E. E., Newey, S. E., Akerman, C. J., Cross, J. R., Van der Veken, L., and Van Aelst, L. (2004). The X-linked mental retardation protein oligophrenin-1 is required for dendritic spine morphogenesis. *Nat. Neurosci.* **7**, 364–372.

Govek, E. E., Newey, S. E., and Van Aelst, L. (2005). The role of the Rho GTPases in neuronal development. *Genes Dev.* **19**, 1–49.

Hering, H., and Sheng, M. (2001). Dendritic spines: Structure, dynamics and regulation. *Nat. Rev. Neurosci.* **2**, 880–888.

Janas, J., Skowronski, J., and Van Aelst, L. (2006). Lentiviral delivery of RNAi in hippocampal neurons. *Methods Enzymol.* **406**, 593–605.

Kutsche, K., Yntema, H., Brandt, A., Jantke, I., Nothwang, H. G., Orth, U., Boavida, M. G., David, D., Chelly, J., Fryns, J. P., Moraine, C., Ropers, H. H., *et al.* (2000). Mutations in ARHGEF6, encoding a guanine nucleotide exchange factor for Rho GTPases, in patients with X-linked mental retardation. *Nat. Genet.* **26**, 247–250.

Lois, C., Hong, E. J., Pease, S., Brown, E. J., and Baltimore, D. (2002). Germline transmission and tissue-specific expression of transgenes delivered by lentiviral vectors. *Science* **295**, 868–872.

Meng, J., Meng, Y., Hanna, A., Janus, C., and Jia, Z. (2005). Abnormal long-lasting synaptic plasticity and cognition in mice lacking the mental retardation gene Pak3. *J. Neurosci.* **25**, 6641–6650.

Newey, S. E., Velamoor, V., Govek, E. E., and Van Aelst, L. (2005). Rho GTPases, dendritic structure, and mental retardation. *J. Neurobiol.* **64**, 58–74.

Node-Langlois, R., Muller, D., and Boda, B. (2006). Sequential implication of the mental retardation proteins ARHGEF6 and PAK3 in spine morphogenesis. *J. Cell Sci.* **119,** 4986–4993.

Philip, N., Chabrol, B., Lossi, A. M., Cardoso, C., Guerrini, R., Dobyns, W. B., Raybaud, C., and Villard, L. (2003). Mutations in the oligophrenin-1 gene (OPHN1) cause X linked congenital cerebellar hypoplasia. *J. Med. Genet.* **40,** 441–446.

Stoppini, L., Buchs, P. A., and Muller, D. (1991). A simple method for organotypic cultures of nervous tissue. *J. Neurosci. Methods* **37,** 173–182.

Stove, V., Van de Walle, I., Naessens, E., Coene, E., Stove, C., Plum, J., and Verhasselt, B. (2005). Human immunodeficiency virus Nef induces rapid internalization of the T-cell coreceptor CD8alphabeta. *J. Virol.* **79,** 11422–11433.

van Galen, E. J., and Ramakers, G. J. (2005). Rho proteins, mental retardation and the neurobiological basis of intelligence. *Prog. Brain Res.* **147,** 295–317.

Yuste, R., and Bonhoeffer, T. (2001). Morphological changes in dendritic spines associated with long-term synaptic plasticity. *Annu. Rev. Neurosci.* **24,** 1071–1089.

Zanni, G., Saillour, Y., Nagara, M., Billuart, P., Castelnau, L., Moraine, C., Faivre, L., Bertini, E., Durr, A., Guichet, A., Rodriguez, D., des Portes, V., *et al.* (2005). Oligophrenin 1 mutations frequently cause X-linked mental retardation with cerebellar hypoplasia. *Neurology* **65,** 1364–1369.

Zennou, V., Petit, C., Guetard, D., Nerhbass, U., Montagnier, L., and Charneau, P. (2000). HIV-1 genome nuclear import is mediated by a central DNA flap. *Cell* **101,** 173–185.

Zufferey, R., Nagy, D., Mandel, R. J., Naldini, L., and Trono, D. (1997). Multiply attenuated lentiviral vector achieves efficient gene delivery *in vivo*. *Nat. Biotechnol.* **15,** 871–875.

CHAPTER TWENTY

Rho GTPases and Hypoxia in Pulmonary Vascular Endothelial Cells

Beata Wojciak-Stothard *and* James Leiper

Contents

BHF Laboratories, Department of Medicine, University College London, London, United Kingdom

Methods in Enzymology, Volume 439
ISSN 0076-6879, DOI: 10.1016/S0076-6879(07)00420-X

Abstract

The pulmonary endothelium is a single-cell layer forming the inner lining of a vast network of arteries, veins, and capillaries in the lung. Its main function is to regulate the contractility of underlying vascular smooth muscle cells (SMCs), which determines vascular tone and allows adaptation of blood flow to oxygenative conditions. Low oxygen tension (hypoxia) causes vasoconstriction of pulmonary vasculature and, depending on the duration of hypoxia, this effect may be reversed by reoxygenation. The key role of the pulmonary endothelium in the regulation of vascular tone has focused considerable attention on the effects of hypoxia/reoxygenation on pulmonary endothelial barrier function. Hypoxia increases endothelial permeability, which is believed to promote vasoconstriction by facilitating the leakage of vasoactive agents from the blood to the underlying SMCs. Data show that Rho GTPases RhoA and Rac1 regulate pulmonary endothelial barrier function in response to changes in oxygen tension. This chapter describes methods to isolate and culture primary pulmonary artery endothelial cells, to measure changes in endothelial barrier function and reactive oxygen species production, and to study the role of Rho GTPases in endothelial responses to hypoxia and reoxygenation.

1. INTRODUCTION

The pulmonary endothelium regulates pulmonary vascular tone and thus plays a key role in optimizing blood oxygenation for the whole body. Endothelial cells can sense and respond to oxygen tensions (PO_2) falling below 70 mm Hg (approximately 12% oxygen) (Faller, 1999). The PO_2 of arterial blood is approximately 150 mm Hg, whereas oxygen tensions in tissues are substantially lower (40 mm Hg). Ascent to the summit of Mount Everest defines the limit of hypoxia tolerated by humans ($PO_2 = 43$ mm Hg), where the inspired and arterial PO_2 values are nearly identical (Malconian et al., 1993). Acute (minutes to hours) hypoxia causes constriction of pulmonary arteries and this process is reversed by reoxygenation. Chronic hypoxia is characterized by extended (days to weeks) exposure to low oxygen concentrations and results in the development of chronic pulmonary hypertension, characterized by sustained vasoconstriction that is not immediately reversible on return to normoxic conditions.

The endothelium modulates SMC contractility by releasing vasorelaxants and vasoconstrictors, and in hypoxia the release of vasoconstrictors such as endothelin-1 and tromboxane exceeds the release of vasorelaxants, nitric oxide, and prostacyclin (Budhiraja *et al.*, 2004). In addition, hypoxia compromises endothelial barrier function. Increased pulmonary endothelial permeability is believed to facilitate the leakage of plasma components to the underlying SMCs, causing further SMC contraction and proliferation (Budhiraja *et al.*, 2004).

Rho GTPases have been implicated in oxygen sensing ever since Rac1 was identified as a component of the NADPH oxidase complex in phagocytes (Abo *et al.*, 1992). Subsequent studies demonstrated that Cdc42 activity is required for hypoxia-induced activation of hypoxia-inducible factor (HIF-1α) in renal carcinoma cells (Turcotte *et al.*, 2003), and in vascular SMCs RhoA and Rho kinase are activated by acute and chronic hypoxia (Fagan *et al.*, 2004; Wang *et al.*, 2001). Activation of RhoA and Rho kinase in vascular SMCs leads to calcium sensitization, increased contractility in response to vasoconstrictors, and decreased response to vasorelaxants, the effects important in mediating the sustained phase of hypoxic vasoconstriction. The role of Rho GTPases in endothelial responses to hypoxia has not been studied to the same extent. We have shown for the first time that hypoxia/reoxygenation-induced changes in endothelial permeability result from coordinated actions of the Rho GTPases Rac1 and RhoA. Rac1 and RhoA respond rapidly to changes in oxygen tension, and this response depends on NADPH oxidase- and PI3 kinase-dependent production of reactive oxygen species (ROS) (Wojciak-Stothard *et al.*, 2005). The activities of Rac1 and RhoA have also been shown to change in a sustained manner in porcine pulmonary artery endothelial cells from an animal model of chronic hypoxia-induced pulmonary hypertension (Wojciak-Stothard *et al.*, 2006).

This chapter describes how to (1) culture the cells taken from the pulmonary artery and expose them to hypoxia and reoxygenation, (2) detect changes in endothelial permeability and analyze the structure of endothelial adherens junctions, (3) measure the activities of Rho GTPases in pulmonary endothelial cells, and (4) measure ROS production in cells and study the role of NADPH oxidase in this process.

2. PURIFICATION AND CULTURE OF PRIMARY PULMONARY ARTERY ENDOTHELIAL CELLS

The porcine model of chronic hypoxia-induced pulmonary hypertension resembles the pathophysiology of the disease in the human lung. In response to hypoxia the pig develops changes in vascular smooth muscle cells and small arteries similar to those seen in pulmonary hypertensive humans (Haworth and Hislop, 1981, 1982). In addition, as compared to other animal or human sources, porcine pulmonary artery endothelial cells can be easily isolated and grown without additional growth factors.

2.1. Materials

One butterfly needle, $2\times$ 10-ml syringes, two to three surgical clamps, forceps, 0.1% solution of collagenase 1A (Sigma) in serum-free RPMI medium containing 200 U/ml penicillin and 200 μg/ml streptomycin; 10 μg/ml bovine plasma fibronectin in phosphate-buffered (PBS; Sigma); RPMI medium containing 10% fetal calf serum (FCS), 100 U/ml penicillin, and 100 μg/ml streptomycin

2.2. Method

Porcine pulmonary artery endothelial cells (PPAECs) are isolated by collagenase digestion from conduit pulmonary arteries of adult large White pigs collected in serum-free RPMI medium containing antibiotics. The arteries should be used within 1 h to maximize the efficiency of isolation. Pulmonary arteries are manipulated in a laminar flow cabinet, placed on a 10-cm petri dish, and washed externally with PBS. In order to remove clotted blood from inside of the artery, one end of a butterfly needle, together with its plastic sheath, is inserted into the artery and the artery opening is sealed tightly with surgical clamps (Fig. 20.1). A 10-ml syringe filled with PBS is attached to the other end of tubing of the butterfly needle and PBS is flushed down the vessel. The artery is then squashed gently to remove the air from the inside of the vessel and the other end is clamped tight to prevent any leakage. Four to 5 ml of collagenase solution is gently injected into the vessel and the syringe is left at the end of a flexible tubing of a butterfly needle.

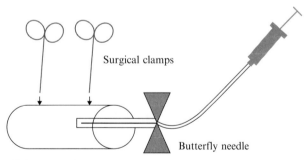

Figure 20.1 During the isolation of PAECs, a butterfly needle, together with its sheath, is inserted into the pulmonary artery. Both ends of the vessel are clamped (arrows) and the artery is filled with a 0.1% solution of collagenase.

At this stage the artery should be inflated only slightly to avoid damaging the arterial wall and contamination with medial SMCs. The outside of the artery with the attached clamps and a butterfly needle is sprayed with 70% ethanol and transferred into a 37° incubator for 20 min in a sterile plastic box. It is important not to exceed digestion time as this will result in contamination with SMCs. Following digestion, the pulmonary artery is massaged gently to facilitate detachment of the endothelium, and a new syringe containing 10 ml of RPMI with 10% FCS is attached to the butterfly needle. The other end of the vessel is placed above the 50-ml centrifuge tube, unclamped carefully, and the contents of the artery are collected in the 50-ml tube. The artery is flushed twice with 10 ml medium with FCS to remove the remaining endothelium, and the medium is collected in the same 50-ml tube. Cell suspension is centrifuged at 1000 rpm for 5 min and the cells are resuspended in 10 ml culture medium and plated in a T75 flask covered with fibronectin. In order to obtain coated flasks, the flasks are incubated with a 10-μg/ml solution of fibronectin for 30 min at 37°.

The medium is changed on the following day. Porcine extrapulmonary arteries are approximately 4 to 5 cm in length, 1 cm in diameter and provide 0.5 to 0.8 × 10^6 cells/isolation. Provided that the arteries are not damaged or overdigested, the isolation provides 98% pure population of endothelial cells and no further purification is required. We use cells between passages 2 and 5, as at higher passage number the cells tend to lose cobblestone morphology and junctional integrity.

In order to estimate the purity of the obtained endothelial cell population, the cells are plated on Thermanox plastic coverslips covered with fibronectin. The following day, the acetylated low-density lipoprotein labeled with 1,1′-dioctadecyl-3,3,3′,3′-tetramethylindocarbo-cyanine (DiL-AcLDL) (Molecular Probes) is added into the culture medium for 4h at a final concentration of 5 μg/ml. AcLDL is taken up by scavenger receptors on cells of endothelial-macrophage lineages and accumulates inside the cells. The fluorescent tag (DiL) enables an easy identification ofthe cells under a fluorescent microscope (excitation 554 nm, emission 571 nm). Alternatively, cells grown on coverslips can be fixed and immunostained with antibodies against endothelial-specific markers: anti-PECAM-1 and anti-VE-cadherin.

3. HYPOXIC EXPOSURE AND REOXYGENATION

In cell culture conditions, the control of pericellular oxygen tension is traditionally achieved by changing the oxygen tension in the gas phase above the medium. However, pericellular oxygen tension also depends on the number of cells, their metabolic rate, and whether the medium was stirred, as well as the diffusion distance from the medium surface to the bottom of the culture dish (Chapman *et al.*, 1970; Pettersen *et al.*, 2005). It is therefore important to consistently apply the same cell density and the same media thickness in cells cultured in different oxygenative conditions.

3.1. Method

Hypoxic conditions are achieved by setting the incubator at 3% O_2, 92% N_2, and 5% CO_2. For short-term hypoxic exposure (15 min to 1 h), it is important to remove excess media, leaving a 3-mm layer of media above the cells, and take the lids off culture dishes to allow easy gas exchange. A 3-mm media thickness has been shown to allow PO_2 at the bottom of the dish to reach 90% of the steady state within 3 to 4 min (Baumgardner and Otto, 2003). As the precise measurement of the PO_2 at the cellular level is extremely difficult, time points shorter than 15 min should be studied with a forced convection apparatus (Baumgardner and Otto, 2003). For a longer term exposure, the lids should be kept on to prevent drying out or contamination. Ideally, each time point has to be established separately, as frequent opening of the door of the incubator increases O_2 concentration (normally a quick opening and closing of the door increases O_2 from 3 to 5%, which then returns to 3% within the next 2 min).

Reoxygenation of cells is carried out by transferring the cells from a hypoxic to a normoxic (20% O_2, 5% CO_2, 75% N_2) incubator. It is important to remove the lid to allow fast gas exchange.

4. STUDYING ENDOTHELIAL INTEGRITY

Endothelial barrier function depends on the integrity of intercellular junctions. Vascular-endothelial (VE)-cadherin (cadherin-5, CD144) is a marker of endothelial adherens junctions and mediates homotypic cell–cell adhesion that requires the association of VE-cadherin with the cortical actin cytoskeleton (Dejana *et al.*, 1999). RhoA and Rac1 generally have opposing effects on endothelial permeability, Rho stimulates actomyosin contractility, which facilitates the breakdown of intercellular junctions, while Rac1 stabilizes endothelial junctions and counteracts the effects of Rho (Wojciak-Stothard and Ridley, 2002). Thus, in a simplified model, improvement of endothelial barrier function may be achieved by inhibiting RhoA and activation of Rac1. The integrity of endothelial junctions can be studied using a permeability assay in the Transwell culture system and fluorescence microscopy.

5. Transendothelial Permeability Assays

5.1. Materials

Transwell-Clear filters (3-μm pore size, 12 mm diameter, Costar Corning; FITC-dextran (molecular weight 42,000, 1 mg/ml; Sigma); PBS, 10 μg/ml solution of bovine fibronectin (Sigma) in PBS; Coomassie blue solution (50% methanol, 10% acetic acid, 0.025% Coomassie blue)

5.2. Method

Cells are plated at a density of 1×10^4 cells/well on polyester, fibronectin-covered Transwell filters and grown until confluence. The cells should be used 3 days postconfluence, as this time is usually required to establish mature adherens junctions. The filters are transparent and the cells can be clearly seen in the center of the filter in phase-contrast microscopy.

FITC-dextran molecular weight 42,000, which has a Stokes radius similar to that of albumin (Draijer et al., 1995), is added to the upper compartment of Transwell-Clear chambers and cells are incubated in normoxic or hypoxic conditions. At various time points, 200-μl samples are taken from the lower compartment, and the amount of FITC-dextran is determined with a microplate reader using an excitation wavelength of 485 nm and emission at 510 nm. A standard curve with serial dilutions of FITC-dextran should be made in order to calculate the amount of FITC-dextran that passed through the monolayer. The mass flux of FITC-dextran is then expressed in micrograms of FITC-dextran per square centimeter per hour. To determine endothelial permeability following reoxygenation, FITC-dextran is added to the upper chamber of Transwell filters after a 1h exposure to hypoxia, and filters are left for 1 h in normoxic conditions before the flux is measured. At the end of the experiment, in order to ensure that the whole filters were covered with cells and that the filter was not damaged, the filters may be taken out, washed in PBS, and fixed in 4% formaldehyde and incubated in Coomassie blue solution for 1 h. After washing off the unbound Coomassie blue in PBS, the filter should be examined under the bright field of a phase-contrast microscope. Coomassie blue stains proteins so the cells appear dark blue and the gaps between them are light.

6. Localization of F-actin and VE-cadherin

Confluent PAECs show F-actin localization around the cell periphery representing junction-bound actin but also show a certain basal level of stress fiber formation, consistent with their appearance *in vivo* in the arterial

F-actin   VE-cadherin

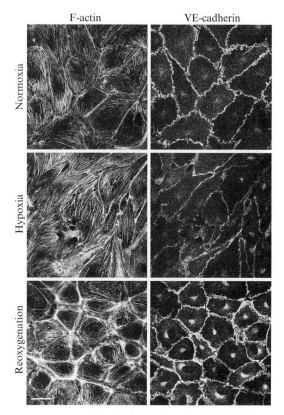

Figure 20.2 Changes in the distribution of F-actin and VE-cadherin in PAECs cultured in normoxic conditions, incubated in hypoxic conditions (3% oxygen, 1 h), or incubated in hypoxic conditions and then subjected to reoxygenation for 1 h. Bar: 25 μm.

system. VE-cadherin is organized around the cell periphery (Fig. 20.2). Upon exposure to hypoxia, the levels of stress fibers increase and VE-cadherin is dispersed from intercellular junctions while reoxygenation restores cortical localization of F-actin and VE-cadherin (see Fig. 20.2).

6.1. Materials

4% formaldehyde in PBS, 1% solution of bovine serum albumin (Sigma) in PBS, 0.1% Triton X-100 (Sigma), mouse monoclonal anti-VE-cadherin antibody (Santa Cruz Biotechnology), FITC-labeled donkey anti-mouse antibody (Jackson ImmunoResearch Laboratories), tetramethylrhodamine isothiocyanate (TRITC)-phalloidin (Sigma)

6.2. Method

To visualize the distributions of VE-cadherin and F-actin, PPAECs grown on Thermanox coverslips are washed in PBS and then fixed with formaldehyde for 15 min. Cells are washed twice in PBS, permeabilized for 3 min in Triton X-100 solution, washed twice in PBS, incubated with 1% BSA in PBS for 1 h to block nonspecific binding, washed once in PBS, and incubated with mouse monoclonal anti–VE-cadherin antibody (diluted 1:200 in PBS) for 1 h. Coverslips are washed three times in PBS allowing 1 to 2 min for each wash, and 50 μl of a secondary antibody solution (FITC-labeled donkey antimouse antibody, 1:100 dilution in PBS), together with 1 μg/ml TRITC-phalloidin (Sigma), is placed on top of each coverslip. Phalloidin binds to actin at the junction between subunits (Faulstich *et al.*, 1993) and because this is not a site at which many actin-binding proteins bind, most of the F-actin in cells is available for phalloidin labeling. The cells are washed three times in PBS, mounted in an antifadent fluorescent mountant solution, and examined under a conventional fluorescent microscope or a confocal laser-scanning microscope.

7. INVESTIGATING THE MECHANISM OF HYPOXIA/ REOXYGENATION-INDUCED CHANGES: THE ROLE OF RHO GTPASES

Hypoxia activates RhoA in pulmonary artery smooth muscle and endothelial cells (Takemoto *et al.*, 2002; Wang *et al.*, 2001). We have shown that RhoA and Cdc42 are activated and that Rac1 is inhibited in pulmonary artery endothelial cells exposed to acute hypoxia, while reoxygenation has an opposing effect (Fig. 20.3). The coordinated actions of RhoA and Rac1 were found to regulate changes in endothelial barrier function in response to hypoxia and reoxygenation, while Cdc42 had no measurable effect on endothelial permeability (Wojciak-Stothard *et al.*, 2005).

8. MEASURING RHO GTPASES ACTIVITY

The activity of Rho, Rac, and Cdc42 can be measured in "pull-down assays." Glutathione *S*-transferase (GST)-Rhotekin is used to precipitate the active, GTP-bound Rho, whereas GST-PAK-CRIB and GST-WASP-CRIB are used for GTP-bound Rac and Cdc42, respectively. The detailed protocol for these assays in other cell types is described elsewhere (Erasmus and Braga, 2005). This section describes variations in the method required in studies on the effects of hypoxia/reoxygenation on Rho GTPases activity in pulmonary endothelial cells.

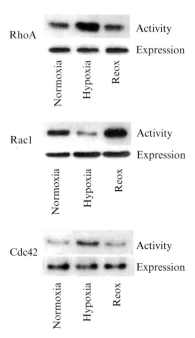

Figure 20.3 Hypoxia increases the activities of RhoA and Cdc42 and decreases the activity of Rac1 while reoxygenation has the opposite effect. Western blots show the levels of activated RhoA, Rac1, and Cdc42 obtained in pull-down assays as well as expression levels of the proteins in PAECs cultured in normoxic conditions, incubated in hypoxic conditions (3% oxygen, 1 h), or incubated in hypoxic conditions and then subjected to reoxygenation for 1 h.

Things to prepare before the assay:

1. Recombinant GST-Rhotekin Rho-binding domain (RBD) and GST-p21–activated kinase (PAK) Rac-binding domain bound to glutathione beads can be purchased from Upstate Biotechnology. Alternatively, GST-Rhotekin RBD, GST-PAK RBD, and GST-Wiscott-Aldrich syndrome protein (WASP)-Cdc42-binding domain bound to glutathione-Sepharose beads can be prepared and stored at −80° as described in Erasmus and Braga (2006).
2. Prepare 5× wash/lysis buffer and store in a refrigerator (detailed protocol provided by Upstate Biotechnology, see Rac and Rho activation assay kits).
3. Cells should be grown in 6-cm petri dishes and used 3 days postconfluence.

A day before the assay:

1. Starve PAECs in media containing 5% serum overnight before the assay. Incubating the cells with a reduced amount of serum reduces the basal

activity of Rho GTPases. PAECs are sensitive to serum deprivation, and reduction from 10% to 5% FCS is sufficient to downregulate basal levels of Rho GTPases without affecting monolayer integrity.

2. Prepare 2× 500-ml bottles of sterile PBS and place one of them in the refrigerator. The other bottle with a loosened cap should be placed overnight in the hypoxic incubator to reduce the oxygen content.

On the day of the assay:

1. Take PBS out of the hypoxic incubator, close the cap tightly, and chill out on ice. Prechill the centrifuge to 4° and prepare four sets of labeled tubes. The first set is for the protein assay, the second set for the determination of Rho expression (the tubes should have the lids pierced and contain 5 μl of 4× sample buffer in each tube), and the third set should contain Sepharose beads. The fourth set is for collecting cell lysates. The tubes should be kept on ice.
2. Place a rotating wheel at 4° and switch on the heat block.
3. Prepare 1× lysis/wash buffer prechilled on ice (a total of 4 ml is required for 1× 6-cm petri dish). Prepare 1-ml, 200-μl and 10-μl pipettes, pipette tips, and a cell scraper.

8.1. Method

1. Place petri dishes with cells on ice. Wash cells once in prechilled PBS and remove any remaining PBS from the dish. For hypoxic cell cultures, use prechilled "hypoxic" PBS. Make sure that cells are lysed immediately after removing from the hypoxic incubator, and avoid lysing several dishes at the same time, as this prolongs the exposure of cells to air.
2. Place 1 ml of cold lysis buffer over the cells and scrape the cells off immediately, place the lysate in prechilled labeled vials, and centrifuge at 13,000 rpm for 3 min. Take 10 μl of the supernatant for protein assay and 20 μl for determination of Rho expression. The remaining supernatant should be added to the GST-Rhotekin beads and rotated for 1 h at 4°.
3. Pulse spin the beads in a benchtop centrifuge. Discard the supernatant and wash the beads three times with 1 ml of 1× lysis/wash buffer. After the last spin, collect the supernatant carefully, leaving 10 to 15 μl of buffer above the pellet. Add the appropriate volume of 4× sample buffer and boil the samples for 5 min on the heat block. Resolve affinity-precipitated RhoA, Rac1, and Cdc42 proteins by SDS-PAGE and detect by Western blotting. For detection, we use primary antibodies: mouse monoclonal anti-RhoA antibody (sc-418, Santa Cruz Biotechnology), mouse monoclonal anti-Rac1 antibody (05–389, Upstate Biotechnology), and rabbit polyclocal anti-Cdc42 (sc-87, Santa Cruz Biotechnology).

Secondary antibodies, horseradish peroxidase-conjugated mouse, and rabbit IgG are from DAKO (Glostrup, Denmark).

9. Analysis of Rho GTPase Function in Hypoxia/ Reoxygenation-Induced Changes in Endothelial Permeability

The role of Rho GTPases in hypoxia/reoxygenation–induced changes in PAECs can be verified by overexpressing mutant Rho GTPases in PAECs or by using chemical activators or inhibitors of Rho GTPases. *Clostridium difficile* toxin B inactivates Rho GTPases by glucosylation, which blocks GTP exchange and GAP activity, as well as the interaction of Rho with effectors (Aktories and Schmidt, 2003). C3 transferase produced by *Clostridium botulinum* types C and D inactivates RhoA, RhoB, and RhoC by ADP-ribosylation on Asn41, inhibiting nucleotide exchange catalyzed by GEFs, and rapidly leading to inactive proteins (Barbieri *et al.*, 2002). Pyridine derivative Y-27632 acts as an inhibitor of Rho kinase but can also inhibit protein kinase C-related protein kinase 2 *in vitro* (Davies *et al.*, 2000). The lipid mediator lysophosphatidic acid (LPA) activates RhoA and increases endothelial permeability through Rho-mediated acto-myosin contractility and possibly through Rho kinase-mediated phosphorylation of the tight junctional component, occludin (Aepfelbacher and Essler, 2001; Hirase *et al.*, 2001).

9.1. Materials

C3 transferase, adenoviral vectors for expression of Rho GTPase mutants: dominant negative RhoA, Rac1, and Cdc42 (N19RhoA, N17Rac1, N17Cdc42), and constitutively active Rac1 (V12Rac1), the Rho kinase inhibitor Y-27632 (5 μM, Yoshitomi Pharmaceutical Industries), LPA (1 μg/ml, Sigma), Rho GTPases inhibitor *C. difficile* toxin B (50 ng/ml, Calbiochem)

9.2. Methods

1. High-titer-purified adenoviruses expressing myc-tagged Rho GTPases mutants are prepared as described (Millan *et al.*, 2006). Adenoviral infection is carried out 18 to 24 h before exposing cells to hypoxia/ reoxygenation. Adenoviruses suspended in prewarmed serum-free RPMI medium are placed over the cells at a multiplicity of infection of 500. For optimal infection, a minimal volume of the medium should be used (2.5 ml/6-cm petri dish, 1 ml/3-cm petri dish, 250 μl/well in a

24-well plate well, and 50 μl/well in a 96-well plate). After 1-h incubation at 37°, the medium is replaced with normal culture medium. In such conditions, approximately 70% cells show overexpression that can be determined by immunostaining with an anti-myc antibody.

2. Inhibitors and activators of Rho GTPases: The Rho kinase inhibitor Y-27632 (5 μM, Yoshitomi Pharmaceutical Industries), LPA (1 μg/ml, Sigma), and Rho GTPase inhibitor *C. difficile* toxin B (50 ng/ml, Calbiochem) are added to cell cultures 1 h before the experiments. The RhoA inhibitor C3 transferase is expressed in *Escherichia coli* from the pGEX-2T vector as GST fusion protein and purified as described previously (Ridley *et al.*, 1992). C3 transferase is added to the cells for overnight incubation at 10 μg/ml.

10. Effects of Changes in NADPH-Mediated ROS Production on the Activity of Rho GTPases and Endothelial Barrier Function

The NOX family NADPH oxidases are proteins that transfer electrons across biological membranes. Oxygen acts as an acceptor of electrons generating superoxide, which starts a cascade of reactions producing other reactive oxygen species (ROS), such as oxygen radicals hydroxyl (OH), peroxyl (RO2), and alkoxyl (RO·) and certain nonradicals that are either oxidizing agents and/or are converted into radicals easily, such as hypochlorous acid (HOCl), ozone (O_3), singlet oxygen ($1O_2$), and hydrogen peroxide (H_2O_2) (Bedard and Krause, 2007). ROS modify cysteine thiols of various proteins and are known to inhibit the activity of protein phosphatase activity (Finkel, 2003; Forman *et al.*, 2004).

Results show that NADPH oxidase is the main source of ROS generation in PAECs (Wojciak-Stothard *et al.*, 2005). Inhibition of NADPH oxidase-mediated ROS generation in hypoxia leads to Rac1-mediated activation of RhoA and a breakdown of endothelial barrier function. In reoxygenation, ROS are required for the restoration of normoxic levels of Rac1 and RhoA activities and recovery of endothelial barrier function. It is becoming clear that an optimal level of ROS is necessary for the maintenance of endothelial barrier function. A modest 2-fold increase in ROS production leads to endothelial recovery and restoration of intercellular junctional integrity (Wojciak-Stothard *et al.*, 2005), whereas growth factors and cytokines known to disrupt endothelial barrier function induce a much larger (up to 10-fold) increase in ROS production (Sundaresen *et al.*, 2002).

In order to verify the role of NADPH in hypoxia/reoxygenation-induced changes in the activity of Rho GTPases, we used two structurally unrelated inhibitors: diphenylene iodonium (DPI) and 4-(2-aminoethyl) benzenesulfonylfluoride (AEBSF). DPI acts by abstracting an electron from an electron transporter and forming a radical, which then inhibits the respective electron transporter through a covalent binding step (O'Donnell *et al.*, 1993). However, apart from inhibiting all NOX isoforms, DPI also inhibits nitric oxide synthase, xanthine oxidase, mitochondrial complex I, and cytochrome P-450 reductase (Bedard and Krause, 2007). AEBSF inhibits NOX enzymes by interfering with the association of the cytoplasmic subunit $p47^{phox}$ with cytochrome b_{559}, therefore preventing the assembly of functionally active NADPH oxidase at the cell membrane (Diatchuk *et al.*, 1997). Activation of NADPH oxidase requires the activities of Rac1 (Hordijk, 2006) and PI3 kinase activity (Seshiah *et al.*, 2002). Therefore, overexpression of dominant negative Rac1, V12Rac1, and the inhibitors of PI3 kinase, LY294002 and wortmannin, can also be used in studies on NADPH oxidase function. LY294002 is a potent, cell-permeable inhibitor that binds to the ATP-binding site of PI3K, whereas wortmannin binds irreversibly to the catalytic subunit of class IA PI3K (Vanhaesebroeck and Waterfield, 1999).

10.1. Materials

LY294002 (10 μM, Calbiochem), wortmannin (50 nM, Calbiochem), DPI (10 μM; Calbiochem), AEBSF (2 mM, Sigma), adenoviral expression vector for dominant negative Rac1, N17Rac1. The stock solution of DPI in dimethyl sulfoxide (DMSO) is protected from light and stored at $-20°$, whereas the stock solution of AEBSF in H_2O, pH 4, is stored at $-20°$ (solutions above pH 7 are less stable). The stock solution of LY294002 in DMSO is stored at $-20°$.

10.2. Method

1. Infection with recombinant adenoviruses to overexpress myc-tagged dominant negative Rac1, N17Rac1 is carried out 18 to 24 h before exposing the cells to hypoxia/reoxygenation. N17Rac1 acts as a dominant inhibitor of Rac1, as a single mutation from Ser to Asn at position 17 results in low affinity for GTP and high affinity for GDP (Ridley *et al.*, 1992).
2. The inhibitors of NADPH oxidase, DPI, AEBSF or the inhibitor of PI3 kinase are added into the culture medium 1 h before the experiment. Control and treated cells are then transferred to the hypoxic incubator for a 1-h incubation. In order to study the effects of reoxygenation, some of the hypoxic cultures are transferred into the normoxic incubator for 1 h. Cells are then examined for changes in endothelial cytoskeleton, junctions, Rho GTPases activity, ROS production, and endothelial permeability.

11. MEASUREMENT OF ROS

The fluorescent probe dichlorofluorescein (DCFH) is widely used for the measurement of intracellular H_2O_2 production. However, it is becoming clear that this probe is not entirely specific for H_2O_2 and therefore should be used as an indicator of general intracellular ROS levels (Thannickal and Fanburg, 2000). The esterified form of DCFH, dichlorofluorescin diacetate (DCFH-DA), is able to cross cell membranes and is hydrolyzed by intracellular esterases to liberate DCFH, which accumulates inside the cell. Upon reaction with oxidizing species, the highly fluorescent compound 2′,7′-dichlorofluorescein is formed, which can be quantified spectrophotometrically by measuring the intensity of fluorescence at the excitation wavelength of 485 nm and emission at 530 nm. Stable intracellular concentrations of DCFH-DA are established within 15 min of the incubation and remain stable for at least 1 h, provided that DCFH-DA remains in the culture medium (Royall and Ischiropoulos, 1993).

11.1. Materials

TC grade transparent 96-well plates (NUNC), DCFH-DA (5 μM, Molecular Probes). The stock solution of DCFH-DA in ethanol should be purged with nitrogen and stored at $-80°$ and protected from light to avoid photooxidation.

11.2. Method

1. Grow PAECs until confluence on 96-well plates.
2. Before placing cells in a hypoxic incubator, the culture medium is replaced with serum-free Krebs saline, pH 7.4 (118 mM NaCl, 4.8 mM KCl, 1.2 mM KH$_2$PO$_4$, 1.2 mM MgSO$_4$, 1 mM CaCl$_2$, 25 mM HEPES), containing 5.6 mM glucose and 5 μM of DCFH-DA. Alternatively, phenol red-free and serum-free medium can also be used instead of Krebs saline. Phenol red interferes with DCFH by reacting with H_2O_2. A working solution of DCFH-DA should be kept in the dark and used soon after preparation.
3. On removal from the hypoxic incubator, the lid is placed onto the 96-well plate together with aluminum foil to protect the solution from light-induced ROS generation, and the plate is quickly (<20 s) placed in a microplate reader to measure fluorescent emission at 530 nm.

REFERENCES

Abo, A., Boyhan, A., West, I., Thrasher, A. J., and Segal, A. W. (1992). Reconstitution of neutrophil NADPH oxidase activity in the cell-free system by four components: p67-phox, p47-phox, p21rac1, and cytochrome b-245. *J. Biol. Chem.* **267,** 16767–16770.

Aepfelbacher, M., and Essler, M. (2001). Disturbance of endothelial barrier function by bacterial toxins and atherogenic mediators: A role for Rho/Rho kinase. *Cell Microbiol.* **3,** 649–658.

Aktories, K., and Schmidt, G. (2003). A new turn in Rho GTPase activation by Escherichia coli cytotoxic necrotizing factors. *Trends Microbiol.* **11,** 152–155.

Barbieri, J. T., Riese, M. J., and Aktories, K. (2002). Bacterial toxins that modify the actin cytoskeleton. *Annu. Rev. Cell. Dev. Biol.* **18,** 315–344.

Baumgardner, J. E., and Otto, C. M. (2003). *In vitro* intermittent hypoxia: Challenges for creating hypoxia in cell culture. *Respir. Physiol. Neurobiol.* **136,** 131–139.

Bedard, K., and Krause, K. H. (2007). The NOX family of ROS-generating NADPH oxidases: Physiology and pathophysiology. *Physiol. Rev.* **87,** 245–313.

Budhiraja, R., Tuder, R. M., and Hassoun, P. M. (2004). Endothelial dysfunction in pulmonary hypertension. *Circulation* **109,** 159–165.

Chapman, J. D., Sturrock, J., Boag, J. W., and Crookall, J. O. (1970). Factors affecting the oxygen tension around cells growing in plastic petri dishes. *Int. J. Radiat. Biol. Relat. Stud. Phys. Chem. Med.* **17,** 305–328.

Davies, S. P., Reddy, H., Caivano, M., and Cohen, P. (2000). Specificity and mechanism of action of some commonly used protein kinase inhibitors. *Biochem. J.* **351,** 95–105.

Dejana, E., Bazzoni, G., and Lampugnani, M. G. (1999). Vascular endothelial (VE)-cadherin: Only an intercellular glue? *Exp. Cell Res.* **252,** 13–19.

Diatchuk, V., Lotan, O., Koshkin, V., Wikstroem, P., and Pick, E. (1997). Inhibition of NADPH oxidase activation by 4-(2-aminoethyl)-benzenesulfonyl fluoride and related compounds. *J. Biol. Chem.* **272,** 13292–13301.

Draijer, R., Atsma, D. E., van der Laarse, A., and van Hinsbergh, V. W. (1995). cGMP and nitric oxide modulate thrombin-induced endothelial permeability: Regulation via different pathways in human aortic and umbilical vein endothelial cells. *Circ. Res.* **76,** 199–208.

Erasmus, J. C., and Braga, V. M. (2006). Rho GTPase activation by cell-cell adhesion. *Methods Enzymol.* **406,** 402–415.

Fagan, K. A., Oka, M., Bauer, N. R., Gebb, S. A., Ivy, D. D., Morris, K. G., and McMurtry, I. F. (2004). Attenuation of acute hypoxic pulmonary vasoconstriction and hypoxic pulmonary hypertension in mice by inhibition of Rho-kinase. *Am. J. Physiol. Lung Cell. Mol. Physiol.* **287,** L656–L664.

Faller, D. V. (1999). Endothelial cell responses to hypoxic stress. *Clin. Exp. Pharmacol. Physiol.* **26,** 74–84.

Faulstich, H., Zobeley, S., Heintz, D., and Drewes, G. (1993). Probing the phalloidin binding site of actin. *FEBS Lett.* **318,** 218–222.

Finkel, T. (2003). Oxidant signals and oxidative stress. *Curr. Opin. Cell Biol.* **15,** 247–254.

Forman, H. J., Fukuto, J. M., and Torres, M. (2004). Redox signaling: Thiol chemistry defines which reactive oxygen and nitrogen species can act as second messengers. *Am. J. Physiol. Cell Physiol.* **287,** C246–C256.

Haworth, S. G., and Hislop, A. A. (1981). Adaptation of the pulmonary circulation to extra-uterine life in the pig and its relevance to the human infant. *Cardiovasc. Res.* **15,** 108–119.

Haworth, S. G., and Hislop, A. A. (1982). Effect of hypoxia on adaptation of the pulmonary circulation to extra-uterine life in the pig. *Cardiovasc. Res.* **16,** 293–303.

Hirase, T., Kawashima, S., Wong, E. Y., Ueyama, T., Rikitake, Y., Tsukita, S., Yokoyama, M., and Staddon, J. M. (2001). Regulation of tight junction permeability and occludin phosphorylation by Rhoa-p160ROCK-dependent and -independent mechanisms. *J. Biol. Chem.* **276,** 10423–10431.

Hordijk, P. L. (2006). Regulation of NADPH oxidases: The role of Rac proteins. *Circ. Res.* **98,** 453–462.

Malconian, M. K., Rock, P. B., Reeves, J. T., Cymerman, A., and Houston, C. S. (1993). Operation Everest II: Gas tensions in expired air and arterial blood at extreme altitude. *Aviat. Space Environ. Med.* **64,** 37–42.

Millan, J., Williams, L., and Ridley, A. J. (2006). An *in vitro* model to study the role of endothelial rho GTPases during leukocyte transendothelial migration. *Methods Enzymol.* **406,** 643–655.

O'Donnell, B. V., Tew, D. G., Jones, O. T., and England, P. J. (1993). Studies on the inhibitory mechanism of iodonium compounds with special reference to neutrophil NADPH oxidase. *Biochem. J.* **290**(Pt 1), 41–49.

Pettersen, E. O., Larsen, L. H., Ramsing, N. B., and Ebbesen, P. (2005). Pericellular oxygen depletion during ordinary tissue culturing, measured with oxygen microsensors. *Cell Prolif.* **38,** 257–267.

Ridley, A. J., Paterson, H. F., Johnston, C. L., Diekmann, D., and Hall, A. (1992). The small GTP-binding protein rac regulates growth factor-induced membrane ruffling. *Cell* **70,** 401–410.

Royall, J. A., and Ischiropoulos, H. (1993). Evaluation of 2′,7′-dichlorofluorescin and dihydrorhodamine 123 as fluorescent probes for intracellular H2O2 in cultured endothelial cells. *Arch. Biochem. Biophys.* **302,** 348–355.

Seshiah, P. N., Weber, D. S., Rocic, P., Valppu, L., Taniyama, Y., and Griendling, K. K. (2002). Angiotensin II stimulation of NAD(P)H oxidase activity: Upstream mediators. *Circ. Res.* **91,** 406–413.

Sundaresan, M., Yu, Z. X., Ferrans, V. J., Sulciner, D. J., Gutkind, J. S., Irani, K., Goldschmidt-Clermont, P. J., and Finkel, T. (1996). Regulation of reactive-oxygen-species generation in fibroblasts by Rac1. *Biochem. J.* **318**(Pt 2), 379–382.

Takemoto, M., Sun, J., Hiroki, J., Shimokawa, H., and Liao, J. K. (2002). Rho-kinase mediates hypoxia-induced downregulation of endothelial nitric oxide synthase. *Circulation* **106,** 57–62.

Thannickal, V. J., and Fanburg, B. L. (2000). Reactive oxygen species in cell signaling. *Am. J. Physiol. Lung Cell. Mol. Physiol.* **279,** L1005–L1028.

Turcotte, S., Desrosiers, R. R., and Beliveau, R. (2003). HIF-1alpha mRNA and protein upregulation involves Rho GTPase expression during hypoxia in renal cell carcinoma. *J. Cell Sci.* **116,** 2247–2260.

Vanhaesebroeck, B., and Waterfield, M. D. (1999). Signaling by distinct classes of phosphoinositide 3-kinases. *Exp. Cell Res.* **253,** 239–254.

Wang, Z., Jin, N., Ganguli, S., Swartz, D. R., Li, L., and Rhoades, R. A. (2001). Rho-kinase activation is involved in hypoxia-induced pulmonary vasoconstriction. *Am. J. Respir. Cell Mol. Biol.* **25,** 628–635.

Wojciak-Stothard, B., and Ridley, A. J. (2002). Rho GTPases and the regulation of endothelial permeability. *Vascul. Pharmacol.* **39,** 187–199.

Wojciak-Stothard, B., Tsang, L. Y., and Haworth, S. G. (2005). Rac and Rho play opposing roles in the regulation of hypoxia/reoxygenation-induced permeability changes in pulmonary artery endothelial cells. *Am. J. Physiol. Lung Cell. Mol. Physiol.* **288,** L749–L760.

Wojciak-Stothard, B., Tsang, L. Y., Paleolog, E., Hall, S. M., and Haworth, S. G. (2006). Rac1 and RhoA as regulators of endothelial phenotype and barrier function in hypoxia-induced neonatal pulmonary hypertension. *Am. J. Physiol. Lung Cell. Mol. Physiol.* **290,** L1173–L1182.

CHAPTER TWENTY-ONE

ROLE OF RHO GTPASES IN THE MORPHOGENESIS AND MOTILITY OF DENDRITIC SPINES

Ayumu Tashiro* *and* Rafael Yuste[†]

Contents

Abstract

Dendritic spines are major sites to receive synapses in the mammalian brain. Spines with abnormal morphologies are found in different brain diseases, suggesting that malformation of dendritic spines could be causally linked to those diseases. Rho GTPase-signaling pathways are implicated in the regulation of spine morphology and also in some forms of mental retardation.

* Centre for the Biology of Memory, Norwegian University of Science and Technology, Trondheim, Norway
† Howard Hughes Medical Institute, Department of Biological Sciences, Columbia University, New York, New York

Methods in Enzymology, Volume 439
ISSN 0076-6879, DOI: 10.1016/S0076-6879(07)00421-1

Therefore, understanding the dynamic regulation of spine morphology by Rho GTPases may provide insights into the etiology and therapeutic strategy of brain diseases. This chapter describes methods used to examine the molecular mechanisms regulating the morphological features of dendritic spines, including slice cultures, biolistic transfections, and live imaging techniques, and summarizes our findings made using these methods.

1. INTRODUCTION

Dendritic spines are small protrusions that reach out from dendrites to receive synapses from axon terminals (Fig. 21.1A; Gray, 1959). In many types of neurons in the central nervous system, most excitatory synapses are formed on dendritic spines (Harris and Kater, 1994). Because synapses are primary sites for the regulation of information flow in neural circuits, the formation and the morphology of spines have been thought to have important roles in brain function.

The morphology of spines is highly variable. Although it is not yet clear whether morphological categorization reflects important functional features, spines have been traditionally categorized into three groups: thin (with a neck and a small head), mushroom (with a neck and a large head), and stubby spines (without a neck)(see Fig. 21.1B; Jones and Powell, 1969; Peters and Kaiserman–Abramof, 1969). Also, dendritic filopodia, thin protrusions without clear heads, are thought to be precursors of dendritic spines (Miller and Peters, 1981; Ziv and Smith, 1996). These categories are dynamic: indeed, during the peak period of spine formation, morphological conversion among these categories occurs within hours, while a systematic trend from filopodia to spines was detected (Parnass *et al.*, 2000).

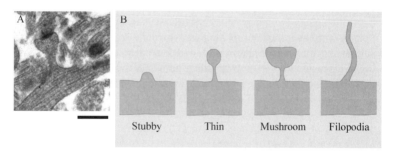

Stubby   Thin   Mushroom   Filopodia

Figure 21.1 Morphology of dendritic spines. (A) An electron micrograph of a dendritic spine in hippocampal slice culture. Scale bar: 500 nm. (B) Schematic drawing of spine morphologies in the categories described by Peters and Kaiserman–Abramof (1970).

The morphology of spines is correlated with its synaptic functions. The size of the spine head is correlated with presynaptic release probability (Harris and Stevens, 1989; Rosenmund and Stevens, 1996; Schikorski and Stevens, 2001) and the number of AMPA receptors in the postsynaptic density (Matsuzaki et al., 2001; Nusser et al., 1998), indicating that larger spines have stronger synapses. Indeed, it has been shown that long-term potentiation (LTP) is associated with spine head growth, confirming this correlation between spine head size and synaptic strength, in single spines before and after LTP induction (Matsuzaki et al., 2004). The spine neck seems to isolate synapses biochemically (Noguchi et al., 2005; Svoboda et al., 1996; Yuste and Denk, 1995; Yuste et al., 2000) and electrically (Araya et al., 2006). The spine neck length is inversely correlated with diffusional coupling between spine heads and dendrites (Majewska et al., 2000a), indicating that spine neck length is an index of biochemical compartmentalization of spine heads, which is important in the regulation of synaptic functions, such as LTP. In addition, it has been shown that the attenuation of synaptic potentials from spines to soma is correlated with spine neck length, suggesting that the spine neck filters membrane potential and can directly regulate synaptic strength (Araya et al., 2006).

Until recently, spines and synapses were considered relatively stable structures. However, Fisher et al. (1998) found that spines are highly motile structure in cell culture. This finding was later validated in more intact preparations, such as brain slices (Dunaevsky et al., 1999) and in vivo (Lendvai et al., 2000). There appears to be two types of spine motility: head morphing and protrusive motility (Tashiro and Yuste, 2004). While protrusive motility may represent searching behavior for presynaptic partners, head morphing could be related to spine maturation or synaptic competition (Bonhoeffer and Yuste, 2002; Konur and Yuste, 2004).

As described earlier, morphology of dendritic spines is thought to be important in normal brain functions. Therefore, the misregulation of spine morphology could be a pathogenesis of brain diseases (Fiala et al., 2002). Indeed, many mental retardation syndromes in humans are associated with abnormal spine morphology. Particularly, X-linked mental retardations are associated with the mutation of PAK3 (Allen et al., 1998), oligophrenin (Billuart et al., 1998), and αPIX (Kutsche et al., 2000), which are components of the Rho GTPase cascade. Therefore, dendritic spines and the Rho GTPase pathway could become possible therapeutic targets for mental retardations (Ramakers, 2002).

Following the pioneering work of Luo and colleagures (1996), we have characterized the roles of Rho GTPases, Rac1 and RhoA in the regulation of spine morphology and motility (Tashiro and Yuste, 2004; Tashiro et al., 2000). This chapter describes the techniques used to examine the roles of Rho GTPases in the regulation of spines and summarizes the results of our studies.

2. METHODS

2.1. Slice cultures

We use cultured slice preparations to maintain healthy slices long enough to allow the expression of genes of interest. Slice cultures also enable us to follow long-term changes in neuronal morphology over several days. Although we initially used older mice (>postnatal day 7), we found that slice cultures from neonatal mice are more reliable. A fast preparation is a critical factor to obtain good slice cultures. Therefore, we slice the brain with a tissue chopper (TC-2 tissue sectioner; Smith & Farquhar) and quickly produce up to 20 slices from individual mice. Serum is a particularly important variable. We find that horse serum from Hyclone is the best to keep slice cultures in good condition. While we prepare slices with 300 μm thickness, slices normally spread and become flattened into a 150- to 250-μm thickness during the first few days in culture. Too thin or whitish slices are generally bad signs for their eventual health. The best way to check the health of slices is to visualize the morphology of individual neurons, for example, by biolistic transfection of green fluorescent protein (GFP), as described later. Fast preparation and gentle handling without mechanical damage are critical factors to good slice cultures. Therefore, we recommend extensive training for at least a couple of weeks for experimenters who have no experience with dissecting small tissues before starting these experiments.

2.2. Reagents

2.2.1. Culture medium (100 ml)

50 ml basal medium Eagle (GIBCO)
25 ml Hank's balanced salt solution (with Ca^{2+}, Mg^{2+} GIBCO)
25 ml heat-inactivated horse serum (Hyclone)
0.65 g dextrose/D-glucose (Sigma)
0.5 ml L-glutamine (200mM GIBCO)
1.0 ml HEPES (1M GIBCO)
1.0 ml 100× penicillin–streptomycin (10000 units/ml, 10000 microgram/ml GIBCO)

2.2.2. Sucrose-artificial cerebrospinal fluid (ACSF)

222 mM sucrose
2.6 mM KCl
27 mM $NaHCO_3$
1.5 mM NaH_2PO_4
2 mM $CaCl_2$

$2 \ mM \ MgSO_4$

All chemicals are purchased from Sigma.

2.3. Protocol for cultured slices

1. Cryoanesthetize a neonatal mouse (P0–3). Cover the mouse with ice and wait for 1 min.
2. Clean the mouse with 70% ethanol.
3. Decapitate the mouse with scissors.
4. Transfer the head to a tissue culture hood.
5. Cut skin and skull with scissors and part sideways with forceps.
6. Remove the brain and place into a 35-mm tissue culture dish filled with cold sucrose-ACSF.
7. Under a dissection microscope , separate the two hemispheres with a surgical blade and orient such that the medial surface faces down.
8. Remove the cerebellum and the mesencephalon carefully.
9. Rotate the hemisphere such that its medial side faces up.
10. Remove the diencephalon with a surgical blade and a flat-ended spatula.
11. Trim remaining piece of tissue, representing the cortex and the hippocampus, into a rectangular block along the anterior edge of hippocampus parallel to the septotemporal axis.
12. Transfer the tissue onto a tissue chopper with two spatulas.
13. Position the rectangular block of tissue such that the chopping orientation is perpendicular to the septotemporal axis of the hippocampus.
14. Slice the tissue 300 μm thick with the tissue chopper.
15. With two flat-ended spatulas, transfer the slices to a fresh culture dish containing cold sucrose-ACSF.
16. Separate the slices from each other with a surgical blade and a flat-ended spatula as soon as possible.
17. Pour 1 ml culture medium on and under culture inserts (Millipore) in a sterile hood. The membrane of the inserts should be completely submerged in culture medium.
18. Transfer individual slices onto the membrane with a flat-ended spatula. Cultivate three to six slices on single inserts.
19. Remove most of the medium (but not all) from the top of the inserts with a pipette. This makes it easier to position the slices.
20. Position the slices, with the spatula, at the center of the insert with at least 2 to 3 mm of space between slices.
21. Remove all remaining medium on the insert.
22. Fill each well of six-well culture plates with 1 ml culture medium.
23. Transfer the inserts into six-well culture plates.
24. Incubate the culture plates in the incubator (5% CO_2, 37°C).
25. Repeat procedures from step 8 onward for the other hemisphere.
26. Exchange 0.6 ml of culture medium with fresh medium every other day.

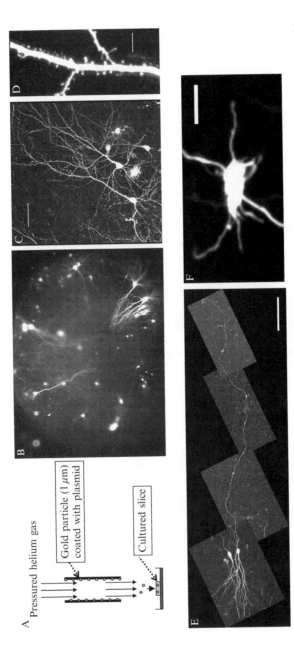

Figure 21.2 Live hippocampal neurons visualized by biolistic transfection of GFP. (A) Diagram of experiment. (B) Transfected hippocampus in a slice culture. A number of pyramidal neurons are visualized with GFP. (C) Two-photon image of transfected hippocampal pyramidal neurons. Scale bar: 50 μm. (D) High magnification image of a dendritic segment in C. Scale bar: 5 μm. (E) Two-photon image of dentate granule cells. Note that their cell bodies, dendrites, and axons are clearly visible. Scale bar: 150 μm. (F) High magnification image of a mossy fiber terminal and its filopodial extensions. Scale bar: 5 μm. (C, D) Reprinted from Dunaevsky *et al.* (1999). Copyright (1999) National Academy of Sciences, USA. (E, F) Modified with permission from Tashiro *et al.* (2003).

2.4. Biolistic transfection

To visualize neuronal microstructures such as dendritic spines (Dunaevsky et al., 1999; Majewska et al., 2000a; Parnass et al., 2000; Tashiro and Yuste, 2004; Tashiro et al., 2000) and axonal filopodia (Tashiro et al., 2003), we transfect GFP using biolistics (so-called gene gun) (Fig. 21.2). We use the Helios gun system from Bio-Rad. The principle of this technique is to physically transfer gold microcarrier particles coated with DNA into the nucleus by pressured helium gas (see Fig. 21.2A). The expression of GFP driven by strong promoters such as the CMV promoter then enables the whole neuronal microstructures to be visualized (see Figs. 21.2B–21.2F).

One of the advantages of biolistics over other transfection methods is the highly efficient and simple cotransfection of multiple genes. Two (or more) separate plasmids encoding different genes can be cotransfected with high cotransfection efficiency (>90% in our hands) simply by coating gold particles with multiple vectors. Using this approach, we have studied the roles of Rho GTPases in regulating the morphology and motility of dendritic spines (Tashiro and Yuste, 2004; Tashiro et al., 2000).

2.5. Protocol for preparing DNA "bullets"

The following protocol is modified from the procedure described in the manual for the Helios gene gun system (Bio-Rad).

1. Purify plasmids with a Maxiprep kit (Qiagen).
2. Weigh 12.5 mg gold microcarrier (1 μm diameter, Bio-Rad) in 1.5-ml tube.
3. Add 100 μl of 0.05 M spermidine (Sigma).
4. Sonicate the tube for 5 s to dissociate aggregated gold particles.
5. Add solution containing a desired amount of plasmid. For the eGFP-C1 plasmid (Clontech), we use 100 μg.
6. Add 100 μl of 1 mM CaCl$_2$ to precipitate plasmid onto gold particles.
7. Incubate for 10 min. Gold particles precipitate to the bottom of the tube during this incubation.
8. Remove supernatant solution without disturbing the particles.
9. Wash particles three times with 100% EtOH. If precipitated gold particles are disturbed, briefly spin the tube with a table-top centrifuge.
10. Suspend the particles in 100% EtOH.
11. Transfer the particles/EtOH into a new 15-ml tube.
12. Repeat steps 9 and 10 to transfer particles remaining in the original tube.
13. Add 100% EtOH to make final volume to 3 ml.
14. Sonicate the particles/EtOH briefly.
15. Submerge one end of Tefzel tube (Bio-Rad) in the particles/EtOH and transfer the particles/EtOH into the Tefzel tube by suction from the syringe connected to the other end of the Tefzel tube.

16. Insert Tefzel tube into the tubing prep station.
17. Incubate for 90 s. Gold particles precipitate to the bottom of the tube during this incubation.
18. Remove EtOH from the tube slowly without disturbing the particles.
19. Slowly rotate the Tefzel tube 180° by hand to spread the particles on the inside wall of the tube.
20. Start rotating the Tefzel tube in the tubing prep station.
21. After 30 s, turn on the nitrogen flow to dry the Tefzel tube. Wait until the inside surface of the tube becomes completely dry.
22. Cut the tube into small pieces with a tubing cutter (Bio-Rad).
23. Store the tubing sets with desiccant at 4° for up to a month and −80° for longer storage.

2.6. Protocol for "shooting bullets" into cultured slices

1. Adjust the helium pressure to 100–150 psi.
2. Fire two or three "preshots" with an empty cartridge holder to clean the helium pathway and to make sure that pressure is stable after each shot. In order to reduce the damage of slices caused by high-pressure flow, cover the tips of barrel liners with nylon mesh (90 μm diameter, Small Parts, Inc.).
3. Load the bullets into the cartridge holder.
4. Take the culture plate from the incubator.
5. Remove the cover of the culture plate.
6. Fire the gun perpendicularly to the plate with a distance of 10 mm between the tip of the barrel liner and the insert.
7. Put the culture plates back into the incubator.

Incubate slices for 2 to 5 days before imaging.

3. IMAGING

Imaging is carried out with a custom-made, two-photon microscope (Majewska et al., 2000b; Nikolenko et al., 2003) using Fluoview (Olympus) software or the Ultraview spinning-disk confocal system (Perkin Elmer). While higher resolution in the z direction is achievable with the two-photon microscope, the spinning-disk confocal microscope has an advantage in fast image acquisition. With the two-photon microscope, images are acquired with a 60× (0.9 NA) water-immersion objective at 6.5× digital zoom or 40× (0.8 NA) at 10× zoom, resulting in a nominal spatial resolution of 20 pixels/μm, using 790–850 nm excitation. With the spinning-disk confocal microscope, images are acquired with a 100× (1.0 NA) water-immersion objective, resulting in a spatial resolution of 15.2 pixels/μm.

3.1. Imaging conditions and procedures

We use a temperature-controlled imaging chamber (Series 20 imaging chamber, Warner Instrument). The chamber is perfused with standard ACSF containing (in mM): 126 NaCl, 3 KCl, 2 CaCl$_2$, 1 MgSO$_4$, 1.1 NaH$_3$PO$_4$, 26 NaHCO$_3$, and 10 dextrose and is saturated with 95% O$_2$ and 5% CO$_2$. The flow of ACSF is driven by raising a container above the chamber and is adjusted by a flow regulator. At the same time, ACSF is removed continuously from the chamber by suction from a vacuum pump. Medium flow in the chamber can cause movement of the slices. For time-lapse imaging of small structures such as dendritic spines, even tiny movement is a serious problem for its analysis. To avoid movement, we minimize medium flow to $\approx$1 ml/min and stabilize the slices with a slice anchor (Warner Instrument). ACSF is heated at 37.5°C before flowing into the chamber by an in-line heater (SH-27B, Warner instrument), and the base of chamber is also heated at 37°C by a platform heater (Series 20 platform, Warner Instrument). These heaters are controlled by a dual channel heater controller (TC-344B, Warner Instrument). If ACSF is saturated with O$_2$ and CO$_2$ at room temperature, these gases come out of the solution and produce small bubbles in the chamber. These bubbles optically deteriorate image acquisition. To prevent this, we keep an ACSF container in a 37°C bath and make ACSF saturated with the gases at 37°C.

To transfer cultured slices to the imaging chamber, we cut membranes of culture inserts surrounding cultured slices using a razor blade and keep slices from any mechanical damage. By holding the surrounding membrane with tweezers, the slices are mounted on a flat spatula. Then the spatula with the slices is dipped into ACSF in the imaging chamber. The slices are positioned to the center of the imaging chamber by tweezers holding the surrounding membrane.

3.2. Fast time-lapse imaging to record rapid spine motility

To examine rapid spine motility, images are taken at 30-s intervals over 15 min, and movies are composed of 30 frames. At each time point, 4 to 9 focal planes, 1 μm apart (two photon), or 10 to 20 focal planes, 0.6 μm apart (spinning-disk confocal), are scanned. Later, these are projected into a single image. Images with these focal distances have enough overlap to achieve a good projection.

3.3. One-day time-lapse analysis

To examine the formation/disappearance and morphological changes of dendritic spines over longer timescales, we image the same dendritic segments over 24 h. The first imaging session is performed as described earlier to acquire a single projected image from a Z stack. Immediately after the end of the session, cultured slices with surrounding membrane are transferred onto new culture inserts. It is important not to damage slices mechanically. We hold surrounding membrane with tweezers and gently transfer the slices onto a new culture insert using a flat spatula as performed when cultured slices are transferred into the imaging chamber (see earlier discussion). The cultured slice has to face up and the surrounding membrane down. Slices are kept in the incubator for 24 h before the second imaging session is performed.

4. ANALYSIS

The morphology of spines is analyzed by measuring the length, width, and density from individual still images. Although they are time-consuming, these measurements have been done by manually selecting individual spines in NIH image or ImageJ software (NIH).

We use the spine motility index to quantify the degree of spine motility (Dunaevsky et al., 1999)(Fig. 21.3). The absolute amount of morphological changes in spines is quantified by the difference between accumulated area swept by a spine over the whole imaging period and smallest spine area among all the time points. This value is normalized according to spine size by dividing the value with average spine area from all time points.

Initially, we quantified all spine motility by the spine motility index. However, we realized later that protrusive motility has a large influence on the index because the difference between accumulated area and smallest area becomes very large if spine length is changed substantially. Therefore, we decided to quantify the two types of motility separately. To measure protrusive motility, the length-change rate of spines is calculated. The length of spines is measured manually for each frame and each spine, and the absolute values of length change between consecutive frames are summed over the entire period of image acquisition, and this total length change is divided by time. To avoid noise arising from movement of the specimens, fluctuations in length of less than

A

$$\text{Spine motility index} = \frac{(\text{the accumulated spine area}) - (\text{the smallest spine area})}{(\text{average spine area})}$$

B

Figure 21.3 Measuring spine motility. (A) Definition of spine motility index. (B) Measurement of the accumulated spine area. First, images of spines from all time points are binarized into black-and-white images. Second, outline images of the spines are created from the binarized images. Third, the outline images are projected into a single image. Then the area occupied by the spine during the time-lapse movie (white area in the right-most image) is measured as accumulated spine area.

0.3 μm are removed from the analysis. Head morphing is measured using the motility index. To prevent protrusive motility from affecting the measurement, we only analyze spines without significant length change (<0.06 μm/min).

5. ROLE OF RHOA AND RHO KINASE

Tashiro and colleagues (2000) cotransfected RhoAV14 (a constitutively active mutant) or C3 transferase (an inhibitor of Rho) with GFP in cortical and hippocampal neurons to analyze the role of RhoA in spine morphogenesis. We found that spine density is lower and spine length is shorter in RhoAV14-expressing neurons than in control neurons (Fig. 21.4). In contrast, the spine density was higher, and spine length longer, in C3 transferase-expressing neurons (see Fig. 21.4). These opposite effects indicate that spine formation/maintenance and spine extension are inhibited by RhoA in normal conditions.

A later study (Tashiro and Yuste, 2004) examined the role of Rho kinase, a downstream effector of RhoA, using its chemical blocker Y-27632. First, we found that spine density was reduced with bath application of Y-27632 (Figs. 21.5A and 21.5B), contrasting with the results in C3 transferase-expressing neurons. These opposite effects may be because a different effector is involved in the effect of RhoA on spine density or because bath application inhibits RhoA pathway, not only in the imaged neurons, but also in surrounding neurons. However, spine length was increased in Y-27632-treated neurons (see Figs. 21.5A and 21.5C), consistent with the results described earlier with RhoAV14 and C3 transferase. Furthermore, in rapid time-lapse analysis, we found that a small subset of spines showed protrusive motility within the first few hours after bath application of Y-27632 (see Figs. 21.5D and 21.5E), indicating that disinhibition of the RhoA/Rho kinase pathway is a key switch for inducing protrusive motility.

These results demonstrate that RhoA is involved in determining spine length in a process of spine morphogenesis. RhoA seems to induce shortening of spines and to stabilize them in some stages of spine maturation, while inactivation of RhoA may disinhibit protrusive motility and increase spine length.

6. ROLE OF RAC1

The role of Rac1 in spine morphogenesis was first suggested by the study using transgenic mice expressing a constitutively active mutant, RacV12, in cerebellar Purkinje cells (Luo et al., 1996). This study found smaller spines with higher density in Purkinje cells from transgenic mice than control mice. Inspired by this finding, we examined the effects of RacV12

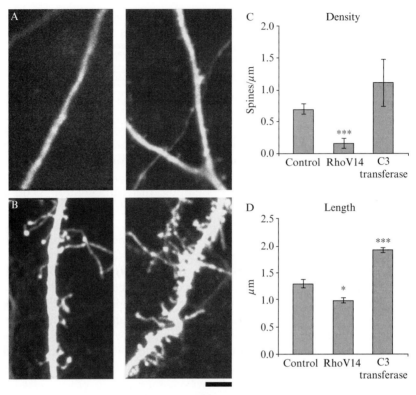

Figure 21.4 Effects of RhoA manipulations on spine density and length. (A) Dendritic segments from RhoAV14 (a constitutively active mutant)-transfected neurons. In many neurons, spines were undetectable (left) or existed with very low density (right). Most of the spines had short necks. (B) Dendritic segments from C3 transferase (Rho blocker)-transfected neurons. Spines have longer necks. Some neurons have abnormally high spine density (right). Scale bars: 5 μm. Mean values of spine density (C) and length (D) in control, RhoV14-transfected, and C3 transferase-transfected neurons. Error bars are SEM. $\star\star\star P < 0.001, \star P < 0.05$, t test against control. Modified with permission from Tashiro *et al.* (2000).

and a dominant-negative mutant of Rac1, RacN17, in slice culture to address the role of Rac1 in spine morphogenesis and motility in more detail. We found that, in Rac1N17-expressing neurons, spine length is longer and spine width is smaller (Figs. 21.6A and 21.6B), indicating that Rac1 is involved in the maturational process of spines, normally associated with shorter spines with wider heads. However, RacV12 induced enlargement of spine heads shortly ($\approx$12 h) after transfection (see Figs. 21.6C and 21.6D), although, 2 days after transfection, "veil"-like structures, all around the dendrites, were found in RacV12-expressing neurons (Tashiro *et al.*, 2000). This finding is in contrast with the previous study, which found smaller spines. However, together with results with RacN17, it is likely that Rac1

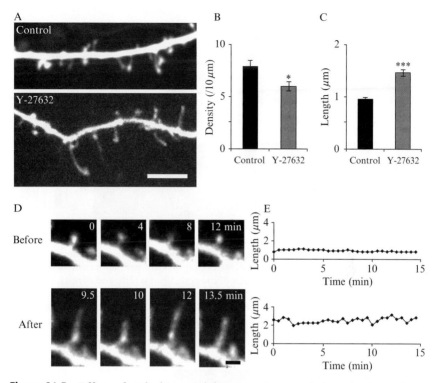

Figure 21.5 Effects of a Rho kinase inhibitor on spine morphology, density, and protrusive motility. (A) Dendritic segments from control and Y-27632-treated neurons. Scale bar: 5 μm. Mean density (B) and length (C) of spines. Error bars are SEM. $\star\star\star P <$ 0.005, $\star P < 0.05$, t test against control. (D) Time-lapse images showing protrusive motility induced by Y-237632. Scale bar: 1 μm. Numbers marked in the images correspond to the time in E. The length in each time point is plotted in E. Reproduced with permission from Tashiro and Yuste (2004).

induces the enlargement of spine heads, which in turn suggests that Rac1 may be a critical regulator of synaptic strength and be involved in LTP.

We also found that spine density was reduced in RacN17-transfected neurons (see Fig. 21. 6B), consistent with an earlier study by Nakayama *et al.* (2000). This decrease of spine density can be due to an abnormal balance between spine formation and elimination or decreased stability of existing spines. Therefore, we performed 1–day time-lapse analysis to follow the existence of the same spines (Fig. 21.7). This analysis showed that the stability of spines is impaired in RacN17-transfected neurons, whereas neither the rate of spine formation nor elimination is affected.

Further, we examined the role of Rac1 in two types of spine motility. While there was no effect of RacN17 on protrusive motility, we found that head morphing is blocked in RacN17-expressing neurons (Fig. 21.8).

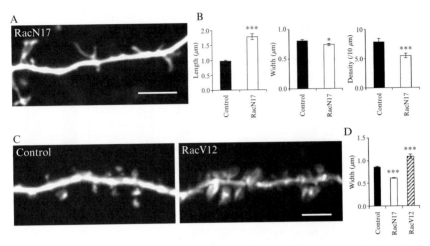

Figure 21.6 Effects of Rac1 manipulations on spine morphology and density. (A) A dendritic segment from RacN17-transfected neurons. Compare with control in Fig. 21. 5A. Scale bar: 5 μm. (B) Mean values of spine length, head width, and density. (C) Dendritic segments from control and RacV12-transfected neurons. Scale bar: 3 μm. (D) Mean spine head width. Error bars are SEM. ★★★$P < 0.005$, ★$P < 0.05$, t test against control. Reproduced with permission from Tashiro and Yuste (2004).

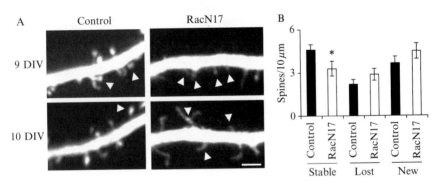

Figure 21.7 Blockade of Rac1 affects spine stability. (A) The same dendritic segments in control and RacN17-transfected neurons, imaged at a 1-day interval. Many spines were stable between the two time points, whereas a subset of spines was lost (arrowheads) or newly formed (arrow). (B) Density of stable, lost, and new spines. ★$P < 0.05$, t test against control. Reproduced with permission from Tashiro and Yuste (2004).

Thus, we found that Rac1 is involved in the maturation, head enlargement, stability, and head morphing of spines. Spine head size is correlated with synaptic strength (Harris and Stevens, 1989; Matsuzaki *et al.*, 2001; Nusser *et al.*, 1998; Rosenmund and Stevens, 1996; Schikorski and Stevens, 2001) and spine stability (Tashiro and Yuste, 2004; Trachtenberg *et al.*, 2002), suggesting that these three parameters are different aspects of spine/synapse maturation

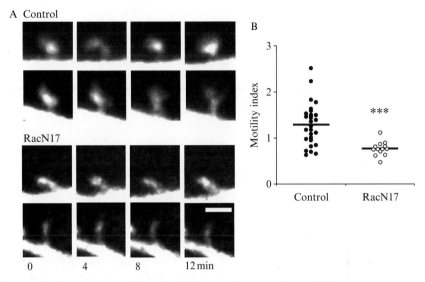

Figure 21.8 Blockade of Rac1 inhibits spine head motility. (A) Time-lapse images of spines in control and RacN17-transfected neurons. To examine the effect of RacN17 on spine head motility, spines that did not show significant length changes were analyzed. Scale bar: 1 μm. (B) Motility index of spines in control and RacN17-transfected neurons. Each circle represents single spines. Black bar represents mean values. ***$P <$ 0.005, t test against control. Reproduced with permission from Tashiro and Yuste (2004).

and that this process is mediated by Rac1. Although the function of head morphing is still not clear, involvement of Rac1 suggests that head morphing may be a process required for spine maturation (Tashiro and Yuste, 2004), for example, to compete with other neighboring spines for presynaptic terminals (Konur and Yuste, 2004).

In summary, our studies indicate that Rac1 and RhoA/Rho kinase pathways are involved in different aspects of spine morphogenesis (Fig. 21.9). Initially, filopodia-like, long, thin protrusions are converted to short spine-like protrusions with heads. RhoA/Rho kinase and Rac1 seem to be involved cooperatively in this process, probably at slightly different stages. In addition, data suggest that Rac1 is involved in the maturation of spines, which is associated with the head enlargement and stability of spines. These roles regulating spine morphology in turn suggest that Rac1 and RhoA/Rho kinase may be critical regulators of the function of spine synapses by regulating synaptic strength and biochemical compartmentalization through controlling the head size and neck length of spines. Additional studies are required to fully understand the process and mechanism regulating spine morphogenesis mediated by Rho GTPases. This line of research could provide important information in understanding the pathophysiology of mental retardation and developing its effective therapeutic strategy.

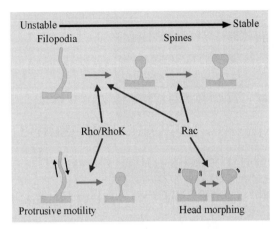

Figure 21.9   Model of the regulation of spine morphology and motility by RhoA/Rho kinase and Rac1. Rac1 and RhoA/Rho kinase are involved in different processes of spine morphogenesis, motility, and stability. Both of them are required to transform long, filopodia-like protrusions into relatively stable, short spines and to maintain morphology of these short spines. Further, Rac1 promotes spine head growth and stabilization. RhoA/Rho kinase inhibits protrusive motility to convert dynamic filopodia into stable spines, whereas Rac1 is essential in head morphing, which may be important in spine head growth and stabilization. Reproduced with permission from Tashiro and Yuste (2004).

ACKNOWLEDGMENTS

We thank members of the laboratory for help and the National Eye Institute and the NYSTAR program for funding.

REFERENCES

Allen, K. M., Gleeson, J. G., Bagrodia, S., Partington, M. W., MacMillan, J. C., Cerione, R. A., Mulley, J. C., and Walsh, C. A. (1998). PAK3 mutation in nonsyndromic X-linked mental retardation. *Nat. Genet.* **20,** 25–30.

Araya, R., Jiang, J., Eisenthal, K. B., and Yuste, R. (2006). The spine neck filters membrane potentials. *Proc. Natl. Acad. Sci. USA* **103,** 17961–17966.

Billuart, P., Bienvenu, T., Ronce, N., des Portes, V., Vinet, M. C., Zemni, R., Roest Crollius, H., Carrie, A., Fauchereau, F., Cherry, M., Briault, S., Hamel, B., *et al.* (1998). Oligophrenin-1 encodes a rhoGAP protein involved in X-linked mental retardation. *Nature* **392,** 923–926.

Bonhoeffer, T., and Yuste, R. (2002). Spine motility: Phenomenology, mechanisms, and function. *Neuron* **35,** 1019–1027.

Dunaevsky, A., Tashiro, A., Majewska, A., Mason, C., and Yuste, R. (1999). Developmental regulation of spine motility in the mammalian central nervous system. *Proc. Natl. Acad. Sci. USA* **96,** 13438–13443.

Fiala, J. C., Spacek, J., and Harris, K. M. (2002). Dendritic spine pathology: Cause or consequence of neurological disorders? *Brain Res. Brain Res Rev.* **39,** 29–54.

Fischer, M., Kaech, S., Knutti, D., and Matus, A. (1998). Rapid actin-based plasticity in dendritic spines. *Neuron* **20**, 847–854.

Gray, E. G. (1959). Electron microscopy of synaptic contacts on dendrite spines of the cerebral cortex. *Nature* **183**, 1592–1593.

Harris, K. M., and Kater, S. B. (1994). Dendritic spines: Cellular specializations imparting both stability and flexibility to synaptic function. *Annu Rev Neurosci.* **17**, 341–371.

Harris, K. M., and Stevens, J. K. (1989). Dendritic spines of CA 1 pyramidal cells in the rat hippocampus: Serial electron microscopy with reference to their biophysical characteristics. *J. Neurosci.* **9**, 2982–2997.

Jones, E. G., and Powell, T. P. (1969). Morphological variations in the dendritic spines of the neocortex. *J. Cell Sci.* **5**, 509–529.

Konur, S., and Yuste, R. (2004). Imaging the motility of dendritic protrusions and axon terminals: Roles in axon sampling and synaptic competition. *Mol. Cell Neurosci.* **27**, 427–440.

Kutsche, K., Yntema, H., Brandt, A., Jantke, I., Nothwang, H. G., Orth, U., Boavida, M. G., David, D., Chelly, J., Fryns, J. P., Moraine, C., Ropers, H. H., et al. (2000). Mutations in ARHGEF6, encoding a guanine nucleotide exchange factor for Rho GTPases, in patients with X-linked mental retardation. *Nat Genet.* **26**, 247–250.

Lendvai, B., Stern, E. A., Chen, B., and Svoboda, K. (2000). Experience-dependent plasticity of dendritic spines in the developing rat barrel cortex *in vivo*. *Nature* **404**, 876–881.

Luo, L., Hensch, T. K., Ackerman, L., Barbel, S., Jan, L. Y., and Jan, Y. N. (1996). Differential effects of the Rac GTPase on Purkinje cell axons and dendritic trunks and spines. *Nature* **379**, 837–840.

Majewska, A., Tashiro, A., and Yuste, R. (2000a). Regulation of spine calcium dynamics by rapid spine motility. *J. Neurosci.* **20**, 8262–8268.

Majewska, A., Yiu, G., and Yuste, R. (2000b). A custom-made two-photon microscope and deconvolution system. *Pflüg. Arch.* **441**, 398–408.

Matsuzaki, M., Ellis-Davies, G. C., Nemoto, T., Miyashita, Y., Iino, M., and Kasai, H. (2001). Dendritic spine geometry is critical for AMPA receptor expression in hippocampal CA1 pyramidal neurons. *Nat. Neurosci.* **4**, 1086–1092.

Matsuzaki, M., Honkura, N., Ellis-Davies, G. C., and Kasai, H. (2004). Structural basis of long-term potentiation in single dendritic spines. *Nature* **429**, 761–766.

Miller, M., and Peters, A. (1981). Maturation of rat visual cortex. II. A combined Golgi-electron microscope study of pyramidal neurons. *J. Comp. Neurol.* **203**, 555–573.

Nakayama, A. Y., Harms, M. B., and Luo, L. (2000). Small GTPases Rac and Rho in the maintenance of dendritic spines and branches in hippocampal pyramidal neurons. *J. Neurosci.* **20**, 5329–5338.

Nikolenko, V., Nemet, B., and Yuste, R. (2003). A two-photon and second-harmonic microscope. *Methods* **30**, 3–15.

Noguchi, J., Matsuzaki, M., Ellis-Davies, G. C., and Kasai, H. (2005). Spine-neck geometry determines NMDA receptor-dependent Ca^{2+} signaling in dendrites. *Neuron* **46**, 609–622.

Nusser, Z., Lujan, R., Laube, G., Roberts, J. D., Molnar, E., and Somogyi, P. (1998). Cell type and pathway dependence of synaptic AMPA receptor number and variability in the hippocampus. *Neuron* **21**, 545–559.

Parnass, Z., Tashiro, A., and Yuste, R. (2000). Analysis of spine morphological plasticity in developing hippocampal pyramidal neurons. *Hippocampus* **10**, 561–568.

Peters, A., and Kaiserman-Abramof, I. R. (1969). The small pyramidal neuron of the rat cerebral cortex: The synapses upon dendritic spines. *Z. Zellforsch. Mikrosk Anat.* **100**, 487–506.

Peters, A., and Kaiserman-Abramof, I. R. (1970). The small pyramidal neuron of the rat cerebral cortex: The perykarion, dendrites and spines. *J. Anat.* **127**, 321–356.

Ramakers, G. J. (2002). Rho proteins, mental retardation and the cellular basis of cognition. *Trends Neurosci.* **25**, 191–199.

Rosenmund, C., and Stevens, C. F. (1996). Definition of the readily releasable pool of vesicles at hippocampal synapses. *Neuron* **16**, 1197–1207.

Schikorski, T., and Stevens, C. F. (2001). Morphological correlates of functionally defined synaptic vesicle populations. *Nat. Neurosci.* **4**, 391–395.

Svoboda, K., Tank, D. W., and Denk, W. (1996). Direct measurement of coupling between dendritic spines and shafts. *Science* **272**, 716–719.

Tashiro, A., Dunaevsky, A., Blazeski, R., Mason, C. A., and Yuste, R. (2003). Bidirectional regulation of hippocampal mossy fiber filopodial motility by kainate receptors: A two-step model of synaptogenesis. *Neuron* **38**, 773–784.

Tashiro, A., Minden, A., and Yuste, R. (2000). Regulation of dendritic spine morphology by the rho family of small GTPases: Antagonistic roles of Rac and Rho. *Cereb Cortex* **10**, 927–938.

Tashiro, A., and Yuste, R. (2004). Regulation of dendritic spine motility and stability by Rac1 and Rho kinase: Evidence for two forms of spine motility. *Mol. Cell Neurosci.* **26**, 429–440.

Trachtenberg, J. T., Chen, B. E., Knott, G. W., Feng, G., Sanes, J. R., Welker, E., and Svoboda, K. (2002). Long-term *in vivo* imaging of experience-dependent synaptic plasticity in adult cortex. *Nature* **420**, 788–794.

Yuste, R., and Denk, W. (1995). Dendritic spines as basic functional units of neuronal integration. *Nature* **375**, 682–684.

Yuste, R., Majewska, A., and Holthoff, K. (2000). From form to function: Calcium compartmentalization in dendritic spines. *Nat Neurosci.* **3**, 653–659.

Ziv, N. E., and Smith, S. J. (1996). Evidence for a role of dendritic filopodia in synaptogenesis and spine formation. *Neuron* **17**, 91–102.

CHAPTER TWENTY-TWO

RHO GTPASES IN ALVEOLAR MACROPHAGE PHAGOCYTOSIS

Henry Koziel

Contents

Abstract

Alveolar macrophages are "professional phagocytes" and critical effector cells that protect the lungs from a broad array of microbes that can cause severe respiratory tract infections such as pneumonia. The molecular mechanisms that mediate microbial phagocytosis in alveolar macrophages are not fully known, and the specific role of small Rho GTPases has not been established. Most studies of Rho GTPase and phagocytosis focus on cell lines and transfected cells, and results may not accurately represent mechanisms operant in the lungs of humans. The use of clinically relevant primary human lung macrophages to examine phagocytosis in the context of host defense function may provide data that translate more readily to human conditions in health and disease. This chapter provides a description of methods and techniques for

Division of Pulmonary, Critical Care and Sleep Medicine, Department of Medicine, Beth Israel Deaconess Medical Center and Harvard Medical School, Boston, Massachusetts

Methods in Enzymology, Volume 439
ISSN 0076-6879, DOI: 10.1016/S0076-6879(07)00422-3

isolating, culturing, and assaying human alveolar macrophages for studies of small Rho GTPases and phagocytosis in the context of host defense. Data support the concept that different macrophage phagocytic receptors may exhibit distinct molecular mechanisms of small Rho GTPase activation that mediate phagocytosis.

1. INTRODUCTION

Alveolar macrophages represent the predominant resident host defense cells in the alveolar airspace and function as critical innate immune effector cells in response to infectious or environmental challenge in the lungs (Arredouani *et al.*, 2004; Zhou and Kobzik, 2007). Experimental data support the concept that alveolar macrophages contribute to an effective host response to a variety of potential lung pathogens (Knapp *et al.*, 2003; Limper *et al.*, 1997). As a "professional" tissue phagocyte, many alveolar macrophage effector cell responses are directly coupled to or associated with phagocytosis, requiring cytoskeleton rearrangement. Macrophage recognition of potential pathogens is mediated through a variety of surface molecular receptors, which in general promote microbe internalization and pathogen elimination through oxidative and enzymatic mechanisms. The best-characterized phagocytic receptors, immune globulin receptor (FcγR) and complement receptor 3 (CR3; or integrin αMβ2), may mediate opsonin-dependent phagocytosis through specific activation patterns of small Rho guanosine triphosphatases (Rho GTPases), designated type I and type II, respectively (Caron and Hall, 1998). Data suggest that opsonin-independent receptor-mediated phagocytosis (such as mannose receptor) in alveolar macrophages also involves small Rho GTPases (Zhang *et al.*, 2005), but the molecular activation pattern may be distinct from FcγR- and CR3-mediated phagocytosis.

2. ALVEOLAR MACROPHAGES

Alveolar macrophages are the most abundant nonparenchymal cells in the lungs (Bezdicek and Crystal, 1997; Brain, 1990; Fels and Cohn, 1986; Lohmann-Matthes *et al.*, 1994; Sibille and Reynolds, 1990; Zhang and Koziel, 2002), accounting for >85% of nonadherent immune cells in the alveolar airspace (Fels and Cohn, 1986) with an estimated 50 to 100 alveolar macrophages per alveolus (Crapo *et al.*, 1982). Alveolar macrophages are derived predominantly from circulating peripheral blood monocytes (Thomas *et al.*, 1976), and recruitment is regulated by integrin-mediated interactions and monocyte-specific chemoattractants such as macrophage inflammatory protein-1 (MIP-1) and MIP-2 (Brieland *et al.*, 1993; Iyonaga *et al.*, 1994). Alveolar macrophages are generally long lived with estimated life spans of months to years (Marques *et al.*, 1997), and local proliferation accounts for ≤2% expansion of the alveolar macrophage population (Bitterman *et al.*, 1984).

Alveolar macrophages obtained by bronchoalveolar lavage (BAL) likely represent a heterogeneous group of innate immune cells (Brain, 1988) that regulate inflammation, influence the nature of the acquired immune response to antigen (Fearon and Locksley, 1996), and perform a number of critical antimicrobial functions. As "professional" tissue phagocytes, alveolar macrophages ingest or internalize molecules and small particles (<1 μm, endocytosis) and fluid (pinocytosis) by a clathrin-associated actin-independent process, and larger particles (>1 μm) by actin-dependent processes that require assembly and cross-linking of actin filaments (Anderson *et al.*, 1990; Greenberg and Silverstein, 1993; Orosi and Nugent, 1993; Wright and Detmers, 1991). Human alveolar macrophages are unique phagocytes as optimal phagocytosis is dependent on an oxygen tension >25 mm Hg (Hunninghake *et al.*, 1980).

Phagocytosis is facilitated greatly by a variety of receptors expressed on the surface of alveolar macrophages. Opsonin-*dependent* phagocytosis is mediated through two well-characterized families of receptors, including (1) the family of Fcγ receptors (FcγR), including FcγRIa (high affinity), FcγRII (low affinity), and FcγRIIIa, that recognize the Fc domain of immunoglobulin G (IgG) coating the surface of particles, and often trigger superoxide production and release of proinflammatory cytokines such as tumor necrosis factor (TNF)-α; and (2) the family of complement receptors (CR1, CR3, and CR4) that recognize complement-opsonized particles (especially C3bi) and often are not associated with significant superoxide production or proinflammatory cytokine release (Sibille and Reynolds, 1990; Unkeless and Wright, 1988). Opsonin-*independent* internalization is mediated through pattern recognition receptors such as mannose receptor (Ezekowitz *et al.*, 1991), β-glucan receptor (Brown and Gordon, 2001), and scavenger receptor (Palecanda *et al.*, 1999). These receptors recognize certain conserved patterns of molecules expressed on microbes, and thus represent an important component of innate immunity.

3. RHO GTPASE ACTIVATION IN MACROPHAGE PHAGOCYTOSIS

Changes in the cell cytoskeleton represent key features of the biological process of phagocytosis, and the Rho GTPases play an important regulatory role in phagocytosis by linking membrane receptors to the actin cytoskeleton (Etienne-Manneville and Hall, 2002). Of the 20 Rho GTPases, the best characterized and most studied include Rac (three isoforms: Rac1, Rac2, and Rac3), Rho (three isoforms: RhoA, RhoB, and RhoC), and Cdc42. Rac induces actin polymerization at the cell periphery to produce lamellipodia, Rho induces assembly of actin and myosin filaments to produce stress fibers, and Cdc42 induces actin filament assembly and filopodia formation at the cell periphery. Rac also regulates the activity of the NADPH enzyme complex in phagocytic cells to induce the respiratory burst response, an important component of the host cell response to microbial challenge that often accompanies phagocytosis.

Specific patterns of Rho GTPase activation may provide the molecular basis for the observed differences in opsonin-*dependent* receptor-mediated phagocytosis described for FcγR and C3R. In cell lines, FcγR-mediated phagocytosis of appropriately opsonized particles may be mediated predominantly by Cdc42 and Rac, whereas C3R-mediated phagocytosis of appropriately opsonized particles may be mediated primarily by Rho (Caron and Hall, 1998). However, the role of the actin-based cytoskeleton and the role of Rho GTPases in mediating opsonin-*independent* phagocytosis by human alveolar macrophages have only been investigated recently.

4. RHO GTPASES AND ALVEOLAR MACROPHAGE PHAGOCYTOSIS

Investigations in our laboratory focused on alveolar macrophage innate receptor-mediated phagocytosis of opportunistic pathogens in the context of host defense. Alveolar macrophages are critical components of the lung innate immune response to infectious challenge and likely represent important effector cells in the host response to opportunistic pathogens such as *Pneumocystis* (Limper *et al.*, 1997; Steele *et al.*, 2003). Human alveolar macrophages express surface innate immune receptors such as mannose receptors (Greenberg and Grinstein, 2002) that can mediate *Pneumocystis* phagocytosis (Ezekowitz *et al.*, 1991). Our laboratory has focused exclusively on clinically relevant human alveolar macrophages and demonstrated that *Pneumocystis* phagocytosis was associated with focal F-actin polymerization and Cdc42, Rac1, and Rho activation in a time-dependent manner (Zhang *et al.*, 2005). Importantly, *Pneumocystis* phagocytosis was primarily dependent on Cdc42 and RhoB activation, as determined by alveolar macrophage transfection with Cdc42 and RhoB dominant-negative alleles, and mediated predominantly through mannose receptors, as determined by siRNA gene silencing of alveolar macrophage mannose receptors.

Pneumocystis phagocytosis was partially dependent on Cdc42 effector molecule p21-activated kinases activation, but dependent on the Rho effector molecule Rho-associated coiled-coil kinase (Zhang *et al.*, 2005).

These data provide a molecular mechanism for alveolar macrophage mannose receptor-mediated phagocytosis of unopsonized *Pneumocystis* organisms. Furthermore, the pattern of Rho GTPase activation and associated host defense cell responses (such as release of reactive oxygen species and proinflammatory cytokines) for phagocytosis mediated through the mannose receptor appear distinct from either FcγR or CR3 (Table 22.1). These observations suggest that patterns of Rho GTPase activation for opsonin-independent phagocytosis may be distinct from opsonin-dependent phagocytosis. However, investigations for β-glucan receptor (dectin-1) described a macrophage phenotype and pattern of Rho GTPase activation distinct from mannose receptor, FcγR, and CR3 (see Table 22.1). Taken together, these investigations suggest that the molecular mechanism mediating phagocytosis may not be limited to type 1 and type 2 pathways, but support an evolving concept that specific phagocytic receptors activate a unique pattern of Rho GTPases and molecular signaling pathways.

Table 22.1 Unique molecular mechanisms for macrophage receptor-mediated phagocytosis

Macrophage receptor	Phagocytosis characteristic morphology	Oxidative burst response	IL-1β, IL-6, TNF-α release	Small RhoGTPase activation pattern	Reference
Opsonin dependent					
Fcγ	"Spiral" or "coil"	Yes	Yes	Cdc42, Rac	Caron and Hall (1998)
Complement	"Zipper"	No	No	Rho	Caron and Hall (1998)
Opsonin independent					
β-glucan	?	Yes	No	Cdc42, Rac	Herre *et al.* (2004)
Mannose	"Spiral" or "coil"	Yes	No	Cdc42, Rho	Zhang *et al.* (2005)

5. METHODS AND MATERIALS

5.1. Study subject volunteers

All recruitment and procedures are performed on consenting adults following protocols approved by the Beth Israel Deaconess Medical Center institutional review board and Committee for Clinical Investigations. Healthy 18- to 55-year-old adults volunteering for research bronchoscopy are without evidence for active pulmonary disease as determined by an unremarkable medical history, normal lung examination, normal chest radiograph, and normal bedside lung function testing (as measured by spirometry). Healthy individuals are without known risk factors for HIV infection and confirmed to be HIV seronegative by ELISA (performed according to the manufacturer's instructions; Abbott Diagnostics, Chicago, IL). Demographic characteristics recorded for all volunteers included age, gender, and smoking status.

5.2. Research bronchoscopy

Lung immune cells are obtained by BAL using standard technique as described previously (Koziel *et al.*, 1998). Briefly, following topical 2% lidocaine anesthesia to the oropharynx, a fiber optic bronchoscope is passed into the airways (using additional 1- to 2-ml aliquots of 1% topical lidocaine instilled via the bronchoscopy port as needed to suppress cough as the bronchoscope is advanced) and wedged in a subsegment of the right middle lobe. BAL is performed by instilling four to eight 50-ml aliquots of warm nonbacteriostatic normal saline (0.9%), followed by gentle suction after each aliquot is infused, and collected into common sterile traps at room temperature. The sterile traps are sealed, placed in plastic biohazard bags, and transported to the laboratory on wet ice. In general, the pooled BAL fluid retrieved at bronchoscopy represents 50 to 75% of the saline instilled into the airways.

5.3. Isolation of BAL cells

Working in a laminar flow hood (i.e., biosafety cabinet) using a standard sterile technique and following universal precautions, BAL cells are initially separated from the pooled BAL fluid by transfer into sterile 50-ml conical centrifuge tubes (for each tube, use maximum pooled BAL fluid volume of 40 ml, and minimum pooled BAL volume of 20 ml), followed by centrifugation at $100\,g$ for 10 minutes at 4 to 8°. The BAL fluid supernatant is aspirated from each tube (and can be discarded or aliquoted and saved for future use), and individual cell pellets are resuspended in 9 ml cold media RPMI-1640 wash (using a 10-ml sterile pipette), all BAL cell pellets combined, and the total BAL cell pellet transferred into a single 50-ml conical centrifuge tube and resuspended in a total volume of 40 ml cold media RPMI-1640 wash. The media RPMI-1640 wash is prepared with media RPMI-1640 supplemented with penicillin 100 U/ml and streptomycin 100 μg/ml (Sigma, St. Louis, MO) without serum. To preserve cell viability, the time from completion of the bronchoscopy to the resuspension of BAL cells in media should be <1 h.

The BAL cell suspension in cold media RPMI-1640 wash is centrifuged at 100 g for 10 min at 4 to 8°, the supernatant is discarded, the BAL cell pellet is resuspended again in a total volume of 40 ml cold media RPMI-1640 wash, followed by repeat centrifugation at 100 g for 10 min at 4 to 8°, and the supernatant is discarded. For counting, total BAL cells are resuspended in a 1- to 5-ml aliquot of cold media RPMI-1640 wash and an appropriate aliquot counted on a hemacytometer, and cell viability is determined by 0.4% Trypan blue dye exclusion. Following counting, the total BAL cell pellet is routinely resuspended in cold media RPMI-1640 wash at a final concentration of 10×10^6 cells/ml in a 50-ml conical centrifuge tube and maintained on ice. Slides for cell morphology and differential determination are prepared by cytocentrifugation (Shandon, Pittsburgh, PA), stained by the modified Giemsa method (Dif-Quik, Sigma), and examined by light microscopy. In general, for an individual, the total number of BAL cells retrieved by bronchoscopy is approximately 3 to 15×10^6 cells for nonsmokers and 20 to 50×10^6 cells for smokers, with cell viability >85%.

5.4. Human alveolar macrophages

Alveolar macrophages are isolated by adherence to plastic bottom tissue culture plates ($1-3 \times 10^6$ cells/well in 6-well plates, $5-7.5 \times 10^5$ cells/well in 24-well plates, $40-50 \times 10^4$ cells/well in 96-well plates) or 13-mm round glass coverslips ($1-2.5 \times 10^5$ cells/slip in 24-well plates) as described previously (Koziel et al., 1998). For the isolation step, cold media RPMI-1640 wash is placed into each well (in general 3 ml for 6-well plates, 1 ml for 24-well plates, 100 μl for 96-well plates) followed by the addition of an appropriate aliquot of the BAL cell suspension. The plated is rocked gently two to three times to distribute the cells and is then incubated undisturbed at 37° in 5% CO_2 for 2 h. After 2 h, the medium is carefully aspirated from each well, and cells are washed gently twice with warmed media RPMI-1640 wash to remove nonadherent cells, host molecules, and debris. After the final wash and aspiration, warmed *complete* culture media RPMI-1640 is added to each well, and cells are incubated at 37° in 5% CO_2 for 2 to 3 days. Endotoxin-free complete culture media is prepared with RPMI-1640 supplemented with 10% heat-inactivated fetal calf serum (JRH Biosciences, Lanexa, KS), penicillin 100 U/ml, streptomycin 100 μg/ml, amphotericin-B and L-glutamine 2 mM (Sigma). In general, isolation of alveolar macrophages from all healthy volunteers yields cells that are $\geq$98% viable, as determined by Trypan blue dye exclusion, demonstrate >95% positive nonspecific esterase staining and are >90% phagocytic of 1-μm latex beads. All procedures are performed using standard tissue culture techniques to ensure sterile conditions.

5.5. Cdc42, Rac, and Rho activation assay

Alveolar macrophage cytoplasmic extracts are prepared, and activation of Cdc42, Rac (Benard *et al.*, 1999), and Rho (Ren *et al.*, 1999) is determined by affinity precipitation using GST fusion-protein-based kits according to the manufacturer's protocols (Rac1/Cdc42 assay reagent using GST-PBD, and Rhotekin Rho-binding domain using GST-RBD; Upstate Biotechnology, Lake Placid, NY). For Cdc42 and Rac1-GTP, proteins are separated by 12% SDS-PAGE, transferred to a nitrocellulose membrane, and blotted for the appropriate specific monoclonal antibodies (Santa Cruz Biotechnology, Santa Cruz, CA). To identify Rho-GTP, immobilized Rhotekin (Rho-binding domain) is used to selectively precipitate Rho-GTP (Upstate Biotechnology), and precipitated GTP-Rho is detected by immunoblot using a polyclonal antibody (Rho-A, Rho-B, and Rho-C; Upstate Biotechnology). For all assays, the signal is detected by the SuperSignal West Pico protein detection system (Pierce, Rockford, IL).

5.6. Dominant negative Cdc42, Rac, and Rho allele amplification, purification, and transfection

Guanine nucleotide binding-deficient (dominant negative) N-terminal 3$\times$ hemagglutinin (HA)-tagged alleles Cdc42 T 17 N, Rac1 T17 N, Rho A T19 N, and Rho B T19 N subcloned in pcDNA 3.1+ vector (Guthrie cDNA Resource Center, Sayre, PA) are used for alveolar macrophage transfections. The empty vector pcDNA 3.1+ (Invitrogen, Carlsbad, CA) is used as a control. Plasmid amplification is performed using MAX Efficiency DH5α MAX Competent *Escherichia coli* (Invitrogen) as described previously (Miralem and Avraham, 2003). DNA purification is performed using the Geno Pure Plasmid Maxi kit (Roche, Indianapolis, IN) following the manufacturer's protocol, and cDNA is confirmed by resolution on an agarose gel and optical density measurement (260 nm) by spectrophotometry. Transfection of alveolar macrophages (1×10^6 cells/well of 6-well plate) of each dominant negative allele is performed using the Effectene transfection reagent (Qiagen, Valencia, CA) for 60 h. Transfection is confirmed by immunoprecipitation of N-terminal 3$\times$HA-tagged dominant negative Rho GTPase proteins with monoclonal anti-3$\times$HA overnight at 4°, followed by incubation with protein A-agarose (Upstate Biotechnology) for 2 h at room temperature, washed three times with PBST, resolved by SDS-PAGE, and probed for 3$\times$HA.

6. CONCLUSIONS

Although current understanding of the molecular mechanisms that mediate and facilitate alveolar macrophage phagocytosis has been advanced greatly, much remains to be determined. For example, the varied molecular patterns of Rho GTPase activation associated with macrophage phagocytosis may in part reflect the involvement of specific regulatory GEFs and GAPs. Although Rac, Cdc42, and Rho represent the most studied of the small Rho GTPases in the context of phagocytosis, the influence of other small Rho GTPases on macrophage phagocytosis remains to be established. The distinct pattern of small Rho GTPase activation may in part reflect distinct pathways related to opsonin-independent ingestion, ingestion through a specific phagocytic receptor(s), interaction of phagocytic and nonphagocytic receptors, differences in the molecular behavior of primary macrophages compared to cell lines or transfected nonphagocytic cells, or differences in macrophage species of origin. Whether the activation state of macrophages influences small Rho GTPase signaling is not established. Finally, whether certain patterns of small Rho GTPase activation are specific for alveolar macrophages and distinct from other tissue macrophages requires further investigation. Many of these points are underscored by a recent report that bone marrow–derived murine macrophages exhibit markedly different pattern of small Rho GTPase activation for C3R- and FcγR-mediated phagocytosis (Hall *et al.*, 2006) compared to cell lines (Caron and Hall, 1998). The scientific investigations of receptor-mediated phagocytosis and the specific role(s) of small Rho GTPases remain a fertile area of scientific investigation to better understand the complex regulatory mechanisms that characterize phagocytosis (Stuart and Ezekowitz, 2005).

ACKNOWLEDGEMENT

This work was supported by NIH Research Grant RO1 HL 063655.

REFERENCES

Anderson, C. L., Shen, L., Eicher, D. M., Wewers, M. D., and Gill, J. K. (1990). Phagocytosis mediated by three distinct Fcg receptor classes on human leukocytes. *J. Exp. Med.* **171**, 1333–1345.

Arredouani, M., Yang, Z., Ning, Y. Y., Qin, G., Soininen, R., Tryggvason, K., and Kobzik, L. (2004). The scavenger receptor MARCO is required for lung defense against Pneumococcal pneumonia and inhaled particles. *J. Exp. Med.* **200**, 267–272.

Benard, V., Bohl, B. P., and Bokoch, G. M. (1999). Characterization of Rac and Cdc42 activation in chemoattractant-stimulated human neutrophils using a novel assay for active GTPases. *J. Biol. Chem.* **274**, 13198–13204.

Bezdicek, P., and Crystal, R. G. (1997). Pulmonary macrophages. *In* "The Lung: Scientific Foundations" (R. G. Crystal, J. B. West, P. J. Barnes, and E. Weibel, eds.) pp. 859–875. Lipincott-Raven, Philadelphia.

Bitterman, P. B., Saltzman, L. E., Adelberg, S., Ferrans, V. J., and Crystal, R. G. (1984). Alveolar macrophage replication: One mechanism for the expansion of the mononuclear phagocyte population in the chronically inflamed lung. *J. Clin. Invest.* **74**, 460–469.

Brain, J. D. (1988). Lung macrophages: How many kinds are there? What do they do? *Am. Rev. Respir. Dis.* **137**, 507–509.

Brain, J. D. (1990). Macrophages in the respiratory tract. *In* "Handbook of Physiology: The Respiratory System," pp. 447–471. Williams & Wilkins, Baltimore.

Brieland, J. K., Jones, M. L., Flory, C. M., Miller, G. R., Warren, J. S., Phan, S. H., and Fantone, J. C. (1993). Expression of monocyte chemoattractant protein-1 (MCP-1) by rat alveolar macrophages during chronic lung injury. *J. Respir. Cell Mol. Biol.* **9**, 300–305.

Brown, G. D., and Gordon, S. (2001). A new receptor for beta-glucans. *Nature* **413**, 36–37.

Caron, E., and Hall, A. (1998). Identification of two distinct mechanisms of phagocytosis controlled by different Rho GTPases. *Science* **282**, 1717–1721.

Crapo, J. D., Barry, B. E., Gehr, P., Bachofen, M., and Weibel, E. R. (1982). Cell number and characteristics of the normal human lung. *Am. Rev. Respir. Dis.* **125**, 332–337.

Etienne-Manneville, S., and Hall, A. (2002). Rho GTPases in cell biology. *Nature* **420**, 629–635.

Ezekowitz, R. A. B., Williams, D. J., Koziel, H., Armstrong, M. Y. K., Warner, A., Richards, F. F., and Rose, R. M. (1991). Uptake of *Pneumocystis carinii* mediated by the macrophage mannose receptor. *Nature* **351**, 155–158.

Fearon, D. T., and Locksley, R. M. (1996). The instructive role of innate immunity in the acquired immune response. *Science* **272**, 50–54.

Fels, A., and Cohn, Z. (1986). The alveolar macrophage. *J. Appl. Physiol.* **60**, 353–369.

Greenberg, S., and Grinstein, S. (2002). Phagocytosis and innate immunity. *Curr. Opin. Immunol.* **14**, 136–145.

Greenberg, S., and Silverstein, S. C. (1993). Phagocytosis. *In* "Fundamental Immunology" (W. E. Paul, ed.), pp. 941–964. Raven Press, New York.

Hall, A. B., Gakitis, M. A. M., Glogauer, M., Wilsbacher, J. L., Gao, S., Swat, W., and Brugge, J. S. (2006). Requirements for Vav guanine nucleotide exchange factors and Rho GTPases in FcgR- and complement-mediated phagocytosis. *Immunity* **24**, 305–316.

Herre, J., Marshall, A. S. J., Caron, E., Edwards, A. D., Williams, D. L., Schweighoffer, E., Tybulewicz, V., Sousa, C. R., Gordon, S., and Brown, G. D. (2004). Dectin-1 utilizes novel mechanisms for yeast phagocytosis in macrophages. *Blood* **104**, 4038–4045.

Hunninghake, G. W., Gradek, J. E., Szapiel, S. V., Strumpf, I. J., Kawanami, O., Ferrans, V. J., Keogh, B. A., and Crystal, R. G. (1980). The human alveolar macrophage. *In* "Methods in Cell Biology" (C. C. Harris, B. F. Trump, and G. D. Stoner, eds.), Vol. 21A, pp. 95–112. Academic Press, New York.

Iyonaga, K., Takeya, M., Saita, N., Sakamoto, O., Yoshimura, T., Ando, M., and Tacahashi, K. (1994). Monocyte chemoattractant protein-1 in idiopathic pulmonary fibrosis and other interstitial lung diseases. *Hum. Pathol.* **25**, 455–463.

Knapp, S., Leemans, J. C., Florquin, S., Branger, J., Maris, N. A., Pater, J., van Rooijen, N., and van der Poll, T. (2003). Alveolar macrophages have a protective antiinflammatory role during murine pneumococcal pneumonia. *Am. J. Respir. Crit. Care Med.* **167**, 171–179.

Koziel, H., Eichbaum, Q., Kruskal, B. A., Pinkston, P., Rogers, R. A., Armstrong, M. Y. K., Richards, F. F., Rose, R. M., and Ezekowitz, R. A. B. (1998). Reduced binding and phagocytosis of *Pneumocystis carinii* by alveolar macrophages from persons infected with HIV-1 correlates with mannose receptor downregulation. *J. Clin. Invest.* **102,** 1332–1344.

Limper, A. H., Hoyte, J. S., and Standing, J. E. (1997). The role of alveolar macrophages in *Pneumocystis carinii* degradation and clearance from the lung. *J. Clin. Invest.* **99,** 2110–2117.

Lohmann-Matthes, M. L., Steinmuller, C., and Franke-Ullmann, G. (1994). Pulmonary macrophages. *Eur. Respir. J.* **7,** 1678–1689.

Marques, L. J., Teschler, H., Guzman, J., and Costabel, U. (1997). Smoker's lung transplanted to a nonsmoker. *Am. J. Respir. Crit. Care Med.* **156,** 1700–1702.

Miralem, T., and Avraham, H. K. (2003). Extracellular matrix enhances heregulindependent BRCA1 phosphorylation and suppresses BRCA1 expression through its C terminus. *Mol. Cell Biol.* **23,** 579–593.

Orosi, P., and Nugent, K. (1993). Studies of phagocytic and killing activities of alveolar macrophages in patients with sarcoidosis. *Lung* **171,** 225–233.

Palecanda, A., Paulauskis, J., Al-Mutairi, E., Imrich, A., Qin, G., Suzuki, H., Kodama, T., Tryggvason, K., Koziel, H., and Kobzik, L. (1999). Role of the scavenger receptor MARCO in alveolar macrophage binding of unopsonized environmental particles. *J. Exp. Med.* **189,** 1497–1506.

Ren, X. D., Kiosses, W. B., and Schwartz, M. A. (1999). Regulation of the small GTP-binding protein Rho by cell adhesion and the cytoskeleton. *EMBO J.* **18,** 578–585.

Sibille, Y., and Reynolds, H. Y. (1990). Macrophages and polymorphonuclear neutrophils in lung defense and injury. *Am. Rev. Respir. Dis.* **141,** 471–500.

Steele, C., Marrero, L., Swain, S., Harmsen, A. G., Zheng, M., Brown, G. D., Gordon, S., Shellito, J. E., and Kolls, J. K. (2003). Alveolar macrophage-mediated killing of *Pneumocystis carinii* f. sp. muris involves molecular recognition by the dectin-1 beta-glucan receptor. *J. Exp. Med.* **198,** 1677–1688.

Stuart, L. M., and Ezekowitz, R. A. B. (2005). Phagocytosis: Elegant complexity. *Immunity* **22,** 539–550.

Thomas, E. D., Rambergh, R. E., Sale, G. E., Sparkes, R. S., and Golde, D. W. (1976). Direct evidence for bone marrow origin of the alveolar macrophage in man. *Science* **192,** 1016–1018.

Unkeless, J. D., and Wright, S. D. (1988). Phagocytic cells: Fcg and complement receptors. *In* "Inflammation: Basic Principles and Clinical Correlates" (J. E. Gallin, I. M. Goldstein, and R. Snyderman, eds.), pp. 343–362. Raven Press, New York.

Wright, S. D., and Detmers, P. A. (1991). Receptor-mediated phagocytosis. *In* "The Lung: Scientific Foundations" (R. G. Crystal, J. B. West, P. J. Barnes, N. S. Cherniack, and E. R. Weibel, eds.), pp. 539–552. Raven Press, New York.

Zhang, J., and Koziel, H. (2002). Alveolar Macrophages. *In* "Cellular Aspects of HIV Infection" (A. Cossarizza and D. Kaplan, eds.), pp. 207–227. Wiley-Liss, New York.

Zhang, J., Zhu, J., Bu, X., Cushion, M., Kinane, T. B., Avraham, H., and Koziel, H. (2005). Cdc42 and RhoB activation are required for mannose receptor-mediated phagocytosis by human alveolar macrophages. *Mol. Biol. Cell* **16,** 824–834.

Zhou, H., and Kobzik, L. (2007). Effect of concentrated ambient particles on macrophage phagocytosis and killing of *Streptococcus pneumoniae*. *Am. J. Respir. Cell Mol. Biol.* **36,** 460–465.

CHAPTER TWENTY-THREE

ANALYSIS OF THE ELP COMPLEX AND ITS ROLE IN REGULATING EXOCYTOSIS

Peter B. Rahl *and* Ruth N. Collins

Contents

Abstract

The regulation of membrane trafficking events in the secretory and endocytic pathways by Rab GTPases requires the cycling and activation of a Rab protein. The cycle of nucleotide binding and hydrolysis of Rab proteins is accompanied by a physical cycle of membrane translocation. An open question in membrane traffic remains how the cycle of Rab GTPase function is coupled to regulatory inputs from other cellular processes. This chapter describes the principles and methodologies used to identify the physiological regulators that influence Rab-mediated membrane traffic.

1. INTRODUCTION

Classical approaches identifying secretory components have led to a fairly comprehensive set of proteins that are supportive of, or required for, membrane traffic, but have yielded little insight into the mechanisms by which traffic is regulated in response to, or coordination with, other intracellular signals (Brennwald *et al.*, 1994; Elbert *et al.*, 2005; Grosshans *et al.*, 2006; Guo *et al.*, 1999; Ortiz *et al.*, 2002). All pathways of membrane traffic are regulated by the Rab family of small GTPases. These proteins are nucleotide-dependent molecular switches whose interactions are necessary for the trafficking of a vesicle from a donor compartment to its acceptor membrane.

Department of Molecular Medicine, Cornell University College of Veterinary Medicine, Ithaca, New York

Methods in Enzymology, Volume 439
ISSN 0076-6879, DOI: 10.1016/S0076-6879(07)00423-5

The Rab GTPase Sec4p serves as a key focal point in the regulation of post-Golgi vesicular transport in *Saccharomyces cerevisiae* (Salminen and Novick, 1987). Sec4p function is conserved throughout evolution with mammalian orthologs including Rab3A, Rab13, and Rab8 (Chavrier *et al.*, 1990; Collins, 2005; Zahraoui *et al.*, 1989) Sec4p function is regulated by multiple factors, but the guanine nucleotide exchange factor Sec2p provides an essential activation step to stimulate GTP/GDP exchange (Walch-Solimena *et al.*,1997). We reasoned that isolating suppressors of *sec2* mutants would provide insights into the factors or pathways that influence the function of Sec4p and regulate exocytosis. In this scenario, a *sec2* mutant would have lowered intracellular levels of activated Sec4p and therefore have reduced cellular growth (Fig. 23.1). Inactivation of a negative regulator of this pathway may increase the intracellular levels of GTP-bound Sec4p and provide increased cell growth. A null allele of *ELP1* was identified as an extragenic suppressor of a temperature-sensitive mutant of Sec2p, *sec2-59* (Rahl *et al.*, 2005). Elp1p is a component of the Elp complex, a six-subunit complex with a catalytic core that has been implicated in influencing multiple cellular processes. The Elp complex has also been reported to regulate transcription and translation, although the mechanism by which it influences each is presently unknown (Huang *et al.*, 2005; Krogan and Greenblatt, 2001; Otero *et al.*, 1999; Rahl *et al.*, 2005; Winkler *et al.*, 2001) The enhanced survival of secretory mutants by the removal of Elp complex encoding genes suggests that the Elp complex is a negative regulator of polarized secretion and growth in *S. cerevisiae*. The action of Elp1p and the Elp complex differs from that of most established players in that they regulate exocytosis negatively. While perhaps not necessary for vesicle budding, motility, or fusion, a negative regulator can influence secretion in response to cellular events and play an important role in adapting the membrane trafficking machinery to the physiological demands of the cell.

2. GENERATION OF POLYCLONAL ANTIBODIES AGAINST ELP1P

A polyclonal antibody was generated in chickens using a recombinantly produced fragment from the Elp1p COOH terminus as the antigen. The Elp1p COOH-terminal 185 residues were cloned into pET15b with *Xho*I/*Xho*I restriction sites, generating a tagged construct with His$_6$ at the NH$_2$-terminus. The construct is transformed into BL21(DE3) *Escherichia coli* for recombinant expression. Expression tests are performed to assess the relative expression levels under different growth conditions. We found that the Elp1p fragment is expressed well when grown overnight in Superbroth

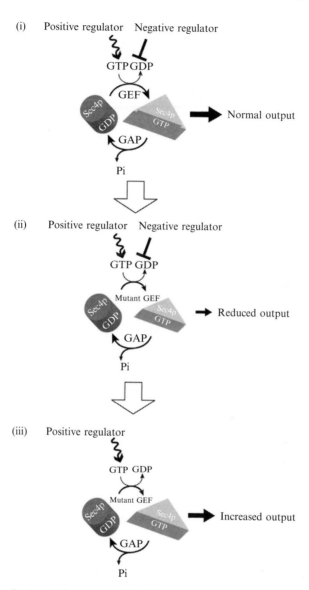

Figure 23.1 Rationale for screen to identify negative regulators of Rab GTPase function. (i) The cycle of Rab GTPase activation is regulated by accessory proteins such as guanine nucleotide exchange factors (GEFs) and GTPase-activating proteins (GAPs) that directly modulate the nucleotide-binding status of a Ras superfamily member. These accessory factors are impacted themselves by signal transduction pathways that regulate membrane trafficking, either positively or negatively. (ii) A mutant GEF is unable to maintain the specific activity of Rab GTP needed for normal physiological responses. (iii) Inactivation, either genetically or pharmacologically, of a negative regulator of the system will increase the output and eliminate the bottleneck resulting from the mutant GEF, resulting in suppression and cellular survival in the case of an essential trafficking pathway.

media (SB, 25 g/liter bacto-tryptone, 15 g/liter yeast extract, 5 g/liter NaCl) at 37° with 750 nM isopropyl-β-D-thiogalactoside present throughout the growth period. Solubility tests are performed on the expressed protein using 250-ml cultures. Cells are lysed in 50 ml lysis buffer (50 mM Tris, pH 8.0, 200 mM NaCl, 0.1% Triton X-100) with protease inhibitors [1 mM phenylmethylsulfonyl fluoride (PMSF), 10 μg/ml pepstatin, 1 mM benzamidine]. The cell suspension is sonicated three times for 1 min, with 1-min intervals on ice, and is centrifuged at 28,300 g for 12 min at 4°. The supernatant is collected, and sample buffer [3% (w/v) sodium dodecyl sulfate (SDS), 12.5% (w/v) sucrose, 37.5 mM Tris, pH 7.0, 1.5% 2-mercaptoethanol] is added to the pellet for solubilization and further analysis. The samples are run with 12% SDS-polyacrylamide gel electrophoresis (SDS-PAGE) to determine which fraction contains the expressed protein fragment.

Because a portion of the recombinant protein is insoluble when the culture is grown at 37°, we took advantage of this characteristic for antigen production.

A 1-liter SB culture is centrifuged, and bacteria are resuspended in 50 ml of lysis buffer (50 mM Tris, pH 8.0, 200 mM NaCl, 0.1% Triton X-100) with protease inhibitors (1 mM PMSF, 10 μg/ml pepstatin, 1 mM benzamidine)(Ortiz et al., 2002). The cell suspension is sonicated three times on ice for 1 min, with 1-min intervals. The sonicate is centrifuged at 28,300 g for 12 min at 4°. The pellet is solubilized in 8 M urea and 2× sample buffer is added [6% (w/v) SDS, 23% (w/v) sucrose, 75 mM Tris, pH 7.0, 3% 2-mercaptoethanol]. The sample is heated and placed on a preparative SDS-PAGE gel. The gel is stained with Coomassie blue, and the Elp1p protein band is excised using a razor blade. This procedure is followed for each injection. Approximately 500 to 750 μg of antigen is used per injection for each chicken. We have found that if enough protein is generated in a single preparation, excising three gel slices and storing at −20° until each injection is suitable.

3. Antigen Injections and IgY Purification

Two white leghorn chickens, maintained in a specific pathogen-free facility, are used with each receiving three total injections of antigen. Typically, the first injection is given when the chicken is between 16 and 18 weeks of age. The first two injections follow the same procedure and the third injection uses a slightly modified procedure. For the first two injections, the gel slice is scooped into a 10-ml syringe and 0.5 ml Freud's adjuvant and 1 ml of 2% Tween-80 [in sterile phosphate-buffered saline (PBS)] is added to the syringe. The sample is mixed between two syringes using a double lewer lock until the gel is highly fragmented and can easily pass through each side. It is important to make sure that the antigen is distributed evenly so that both chickens get an equal dose; 0.5 ml of sample is injected into each chicken breast muscle (1 ml total per chicken). The third injection is done similarly, except only antigen and PBS are injected into each chicken. Each injection is typically done in 2- to 3-week intervals.

When chickens start laying eggs, the antibodies are tested for protein detection. Generally we will test antibodies generated from both chickens as well as testing antibodies from eggs laid at various times. Initially we will test eggs at 2-week intervals to determine if there is a specific window of time that produces the best antibodies. The antibody titer will begin to decrease over time, which generally happens after 2 to 3 months after egg laying has started. The following procedure is used to purify the IgY-containing fraction from the egg yolk.

1. The eggs are cracked and the yolk is isolated and put into a 100–ml Pyrex screw cap bottle (No. 1395).
2. Twenty-five milliliters of sterile PBS, without calcium or magnesium (Cellgro), is added using a sterile technique.
3. Chloroform is added to bring the total volume to 80 ml.
4. The bottle is closed and the yolk is broken and mixed with the PBS/chloroform by vigorous shaking.
5. The bottle is centrifuged at 2500 rpm for 30 min at 4°.
6. The bottle is removed from the centrifuge carefully, ensuring that the supernatant is not disturbed. *Note*: if the supernatant is cloudy, add a little more chloroform and centrifuge again.
7. Collect the supernatant containing the IgY fraction and dispose of the pellet.
8. Store labeled aliquots of the IgY fraction at −80°.

This procedure yields approximately 20 ml of the IgY fraction in PBS. The antibodies are then tested for specificity and detection levels at this stage, generally starting with a working dilution of 1:1000 in Tris-buffered saline Tween-20 (TBST) and adjusting the dilution as needed. The IgY fraction can also be further concentrated and affinity purified if necessary.

4. ANTIBODY TESTING PROCEDURE

The antibody is tested against whole cell lysates created from a wild-type strain (RCY239, *ura3-52 leu2-3,112*) to detect endogenous levels of Elp1p and a deletion strain (RCY242, *ura3-52 leu2-3,112 elp1Δ::URA3*) as a negative control to identify nonspecific antigens detected (Fig. 23.2A). It is also desirable to include a lysate that has overexpressed levels of the antigen. An overnight culture of each strain is grown to midlog phase, harvested, and used to generate cell lysates. The culture is resuspended in TAz buffer (10 mM Tris, pH 7.5, 10 mM azide), and 10 OD_{600} units are transferred to a microfuge tube. Cells are microfuged and the supernatant is aspirated. The cells are resuspended in 150 μl of TAz buffer with protease inhibitors (1 mM PMSF, 10 μg/ml pepstatin, 1 mM benzamidine). An equal volume of glass beads is added and the slurry is vortexed for 2 min at 4° using a Turbo-beater apparatus. After lysis, an additional 160 μl of lysis buffer with protease inhibitors is added in addition to 210 μl of 4× sample buffer [12% (w/v) SDS, 48% (w/v) sucrose, 150 mM Tris, pH 7.0, 6% 2-mercaptoethanol, Serva blue G]. Samples are incubated at 50° for 10 min, and 10 μl of sample is loaded onto a 6% SDS-PAGE gel. The gel is transferred to polyvinylidene fluoride (Pall Gelman Corp., BioTrace 0.45 μm) for 3 h at 200 mA using standard Western blot procedures. Elp1p protein levels are detected using a 1:1000 antibody dilution in TBST

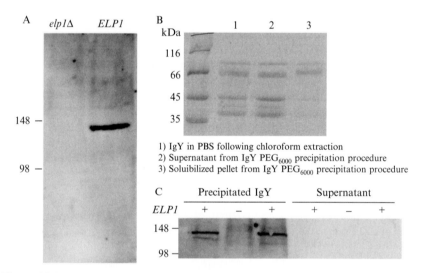

Figure 23.2 Detection of Elp1p using a polyclonal Elp1p antibody. (A) Whole cell lysates of *ELP1* (RCY239) and *elp1Δ* (RCY242) were incubated with sample buffer prior to SDS-PAGE and Western blot analysis. Elp1p protein levels were detected using a 1:1000 antibody dilution in TBST from antibody isolated from chicken egg yolks. (B) Protein levels in chloroform extracted antibody diluted 1:100 (1), PEG$_{6000}$ precipitate supernatant diluted 1:100 (2), and PEG$_{6000}$ precipitate resuspended in 1 ml of sterile PBS, followed by a 1:500 dilution (3). Protein composition was analyzed on a 12% SDS-PAGE gel, followed by Coomassie blue staining. The 66-kDa IgY is concentrated from the PEG$_{6000}$ precipitation step, as described in Bizhanov and Vyshniauskis (2000). (C) Western blot analysis using precipitate (lane 3 from Fig. 23.1B) and supernatant (lane 2 from Fig. 23.1B) fractions to probe for Elp1p from whole cell lysates where a plus sign indicates *ELP1* (RCY239) and a negative sign indicates *elp1Δ* (RCY242). Each fraction is used at 1:1000 dilution in TBST.

overnight at room temperature. The secondary antibody is goat antichicken conjugated to alkaline phosphatase (1:10,000 dilution), blots are developed with chemiluminescence CDP Star using a Fuji Film LAS-3000 imager, and data are collected using Image Reader LAS-3000 software.

An additional purification step that can be used if necessary is antibody precipitation using PEG_{6000}. In this procedure, 1.2 g PEG_{6000} is added to 10 ml of antibody from the aforementioned procedure for a final concentration of 12% PEG_{6000}. The solution is rocked gently at room temperature to solubilize the PEG_{6000}, which generally takes 2 to 3 min. The solution is centrifuged at 15,700 g for 10 min at 4°. The pellet contains the antibody fraction and is resuspended in 1 ml of sterile PBS with gentle pipetting. The 66-kDa chicken IgY is enriched from this purification step [see Fig. 23.2B, comparing lanes 2 and 3 (Bizhanov and Vyshniauskis, 2000)]. This procedure is convenient for concentration, as the chloroform purification step generates roughly 20 ml of antibody per egg and each chicken will lay 30 or more eggs.

5. GENETIC ANALYSIS TO IDENTIFY NEGATIVE REGULATORS OF EXOCYTOSIS

The genetic analysis uses mutant haploid strains from the same genetic background. This will minimize variability in the temperature sensitivity profiles of each mutant allele being tested. We generally create our deletion strain in the background of the *sec* mutants we are testing using standard molecular biology techniques. Figure 23.3 shows two tetrads analyzed for genetic interactions between *elp1Δ* (RCY1366B, *MATa ura3-52 leu2-3,112 elp1ΔKAN^R*) and *sec2–59* (RCY274, *MATα ura3-52 sec2-59*).

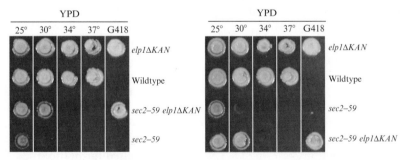

Figure 23.3 Genetic analysis of exocytic regulators. Growth at the indicated temperatures of strains isolated by tetrad dissection of sporulated diploids generated for *elp1ΔKAN* (RCY1366, *MATa ura3-52 leu2-3, 112 elp1ΔKAN^R*) crossed with *sec2-59* (RCY274, *MATα ura3-52 sec2-59*). Two independent tetrads are shown, indicating the suppression of the *sec2-59* temperature sensitivity through deletion of *ELP1*.

The two haploids are crossed on YPD 25° and grown overnight. Diploids are selected by replica plating onto YPD + 2 mM caffeine at 40°. This is a restrictive condition for each haploid strain, where $elp1\Delta$ cells are defective for growth on 2 mM caffeine at 40° and the $sec2-59$ cells are defective for growth at temperatures higher than 30°. The diploid, however, contains one wild-type copy of each protein and therefore survives under this condition. It is convenient to use temperature sensitivity or nutrient deficiencies to select for diploids.

If diploids cannot be isolated using these two selection methods, zygotes can be selected using a micromanipulator mounted on a Zeiss Axiolab microscope. For this procedure, the two haploid strains are mixed on YPD plates and grown at 25° for 4 to 5 h. Zygotes can be observed as dumbbell-shaped cells resulting from the two fused haploid cells (Gammie et al., 1998). Individual zygotes may die due to the stress caused from the micromanipulator needle pressing against the cell and agar while trying to pick up the cell. Therefore, it is recommended to isolate at least four to five zygotes per cross.

The diploid is grown in 5 ml YPD at room temperature. Once the culture is in log phase, approximately 200 μl of culture is added to a separate tube with 5 ml of sterile water. The culture is centrifuged to pellet the cells and the water is aspirated. The cells are then resuspended in 2 ml of SPM media (0.3% KOAc, 0.02% raffinose) to induce sporulation. Depending on the strain, cells will take anywhere from 3 to 7 days for sporulation.

When the diploids have sporulated, approximately 200 μl of the SPM culture is resuspended in 400 μl of zymolase buffer (1.2 M sorbitol, 50 mM KP$_i$, pH 7.2, 10 mM EDTA), 3 μl of 2-mercaptoethanol stock solution, and approximately 15 μg/ml zymolase (Zymolase 20T, US Biological). The cells are incubated for 15 min at 30° to allow for a light zymolase treatment to break down the spore wall. Cells are spread along one end of a YPD plate. Spores are dissected from tetrads using a micromanipulator (mounted on Zeiss Axiolab). The resulting spores are grown at 25° for 3 to 5 days, depending on strain testing.

Once the spores have grown, each is tested for phenotype and genotype. The $sec2-59$ containing spores can be identified using temperature sensitivity. Strains containing $elp1\Delta KAN$ can be identified by positive growth on YPD + G418 at 25°. We test the temperature sensitivity profiles of each strain, generally on YPD media at 25, 30, 32, 34, and 37°. This analysis is used to identify that deletion of $ELP1$ suppresses the temperature sensitivity of $sec2-59$ (see Fig. 23.3).

The resulting $sec2-59$ $elp1\Delta KAN$ strain can be used further to study the nature of the regulation. As shown in Fig. 23.4A, the addition of an Elp1p-GFP to the $sec2-59$ $elp1\Delta KAN$ restores the thermosensitivity of $sec2-59$, indicating that this construct can provide Elp1p function as a negative regulator for $ELP1$. The $sec2-59$ $elp1\Delta KAN$ with Elp1p-GFP can grow

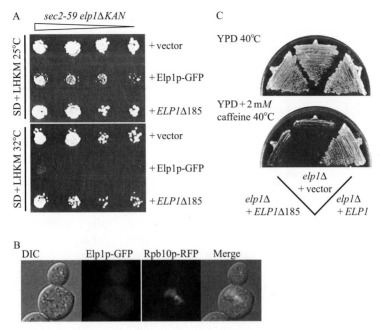

Figure 23.4 Analysis of proteins influencing *ELP1*-dependent regulation of exocytosis. (A) Constructs bearing *ELP1-GFP* and *ELP1*Δ 185 on pRS316 shuttle plasmids were transformed in *sec2-59 elp1ΔKAN* cells and tested for restoration of thermosensitivity. Positive transformants were spotted on SD plates supplemented with the required nutrients and grown for 3 days. Functionality can be assessed through restoration of thermosensitivity of *sec2-59*. (B) Elp1p-GFP localization in *elp1ΔKAN/elp1ΔKAN* cells. Rpb10p-RFP, an RNA polymerase II subunit, is used as a nuclear marker (Woychik and Young, 1990) . The corresponding DIC image is shown. (C) Constructs bearing *ELP1* and *ELP1*Δ185 were transformed into *elp1ΔKAN/elp1ΔKAN* cells, and transformants were tested for growth on YPD 40° or YPD + 2 m*M* caffeine 40°. Functionality can be assessed through growth on YPD + 2 mM caffeine 40°, as vector-only cells are inviable at this condition and cells containing *ELP1* are viable.

at 25° but is temperature sensitive at 32°, compared to vector only, which can grow at both temperatures. Elp1p-GFP is localized to the cytoplasm when expressed as the sole copy (see Fig. 23.4B). A similar analysis can be used to test the functionality of other tagged Elp complex proteins or proteins that may act in the same pathway. This assay utilizes the sensitivity caused by the loss of Elp complex function: sensitivity for growth on YPD + 2 mM caffeine at 40°. With this assay, an *elp1*Δ strain supplemented with wild-type *ELP1* can grow on YPD + 2 mM caffeine at 40° but a vector-only control is inviable. For example, Elp1p fusion proteins, where GFP or 3xHA is encoded at the COOH terminus, are functional; however, encoding GFP at the NH$_2$-terminus results in a nonfunctional protein.

The thermosensitivity analysis can be used to assess the role in secretion of a truncation mutant of *ELP1*. Previous studies in our laboratory found that Elp1p interacts with Sec2p through residues 1165 to 1349 (Rahl *et al.*, 2005). Analysis with the *sec2-59 elp1ΔKAN* strain indicates that a truncated version of Elp1p that lacks its Sec2p-interacting domain (*ELP1Δ185*) cannot restore the temperature sensitivity of *sec2-59*, demonstrating that it requires the COOH-terminal Sec2p-interacting region for its role in exocytosis. Additionally, supplementing an *elp1Δ* strain with the *ELP1Δ185* truncation mutant is inviable on caffeine, displaying the same phenotype as a vector control (see Fig. 23.4C).

To assay for restoration of thermosensitivity, an *elp1ΔKAN sec2-59* strain (RCY3045, *ura3-52 leu2-3,112 sec2-59 elp1ΔKAN^R*) is transformed with the indicated constructs made in a pRS316 shuttle plasmid. It is important to use fresh cells for this assay to minimize any possible revertants that may be generated in cells that have been grown for an extended period of time. Positive transformants are grown for approximately 3 days at 25° and selected for on media lacking uracil. It is critical to grow transformants at 25° because, although the *sec2-59 elp1ΔKAN* can grow at higher temperatures, addition of a negative regulator will restore thermosensitivity and provide an additional selection method and the cells may adapt to higher temperatures, which will alter the results of the assay. A permissive temperature for both *sec2-59 ELP1* and *sec2-59 elp1Δ* cells is 25°.

Growth is tested on minimal media lacking uracil at 25 and 32° using a dilution series in a sterile 96-well plate. Generally, 200-μl of sterile water is used to fill each well in the first column and 150-μl of water for each additional column except for the final column, which is filled with 100-μl of water. Three to four transformation colonies are used to inoculate the well in the first column. A 50-μl serial dilution is done once inoculation is complete. Cells are spotted onto each plate, and cells are tested for growth at the indicated temperatures. It is critical to plate the cells on minimal media lacking the nutrient encoded on the shuttle plasmid to provide a positive selection and ensure that the cells maintain the plasmid.

REFERENCES

Brennwald, P., Kearns, B., Champion, K., Keranen, S., Bankaitis, V., and Novick, P. (1994). Sec9 is a SNAP-25-like component of a yeast SNARE complex that may be the effector of Sec4 function in exocytosis. *Cell* **79**, 245–258.

Bizhanov, G., and Vyshniauskis, G. (2000). A comparison of three methods for extracting IgY from the egg yolk of hens immunized with Sendai virus. *Vet. Res. Commun.* **24**, 103–113.

Chavrier, P., Vingron, M., Sander, C., Simons, K., and Zerial, M. (1990). Molecular cloning of YPT1/SEC4-related cDNAs from an epithelial cell line. *Mol. Cell. Biol.* **10**, 6578–6585.

Collins, R. N. (2005). Application of phylogenetic algorithms to assess Rab functional relationships. *Methods Enzymol.* **403,** 19–28.

Elbert, M., Rossi, G., and Brennwald, P. (2005). The yeast par-1 homologs kin1 and kin2 show genetic and physical interactions with components of the exocytic machinery. *Mol. Biol. Cell* **16,** 532–549.

Gammie, A. E., Brizzio, V., and Rose, M. D. (1998). Distinct morphological phenotypes of cell fusion mutants. *Mol. Biol. Cell* **9,** 1395–1410.

Grosshans, B. L., Andreeva, A., Gangar, A., Niessen, S., Yates, J. R., 3rd, Brennwald, P., and Novick, P. (2006). The yeast lgl family member Sro7p is an effector of the secretory Rab GTPase Sec4p. *J. Cell Biol.* **172,** 55–66.

Guo, W., Roth, D., Walch-Solimena, C., and Novick, P. (1999). The exocyst is an effector for Sec4p, targeting secretory vesicles to sites of exocytosis. *EMBO J.* **18**(4), 1071–1080.

Huang, B., Johansson, M. J., and Bystrom, A. S. (2005). An early step in wobble uridine tRNA modification requires the Elongator complex. *RNA* **11,** 424–436.

Krogan, N. J., and Greenblatt, J. F. (2001). Characterization of a six-subunit holo-elongator complex required for the regulated expression of a group of genes in *Saccharomyces cerevisiae. Mol. Cell. Biol.* **21**(23), 8203–8212.

Ortiz, D., Medkova, M., Walch-Solimena, C., and Novick, P. (2002). Ypt32 recruits the Sec4p guanine nucleotide exchange factor, Sec2p, to secretory vesicles; evidence for a Rab cascade in yeast. *J. Cell Biol.* **157**(6), 1005–1015.

Otero, G., Fellows, J., Li, Y., de Bizemont, T., Dirac, A. M., Gustafsson, C. M., Erdjument-Bromage, H., Tempst, P., and Svejstrup, J. Q. (1999). Elongator, a multisubunit component of a novel RNA polymerase II holoenzyme for transcriptional elongation. *Mol. Cell* **3**(1), 109–118.

Rahl, P. B., Chen, C. Z., and Collins, R. N. (2005). Elp1p, the yeast homolog of the FD disease syndrome protein, negatively regulates exocytosis independently of transcriptional elongation. *Mol. Cell* **17**(6), 841–853.

Salminen, A., and Novick, P. J. (1987). A ras-like protein is required for a post-Golgi event in yeast secretion. *Cell* **49,** 527–538.

Walch-Solimena, C., Collins, R. N., and Novick, P. J. (1997). Sec2p mediates nucleotide exchange on Sec4p and is involved in polarized delivery of post-Golgi vesicles. *J. Cell Biol.* **137**(7), 1495–1509.

Winkler, G. S., Petrakis, T. G., Ethelberg, S., Tokunaga, M., Erdjument-Bromage, H., Tempst, P., and Svejstrup, J. Q. (2001). RNA polymerase II elongator holoenzyme is composed of two discrete subcomplexes. *J. Biol. Chem.* **276**(35), 32743–32749.

Woychik, N. A., and Young, R. A. (1990). RNA polymerase II subunit RPB10 is essential for yeast cell viability. *J. Biol. Chem.* **265,** 17816–17819.

Zahraoui, A., Touchot, N., Chardin, P., and Tavitian, A. (1989). The human Rab genes encode a family of GTP-binding proteins related to yeast YPT1 and SEC4 products involved in secretion. *J. Biol. Chem.* **264,** 12394–12401.

CHAPTER TWENTY-FOUR

RAB-REGULATED MEMBRANE TRAFFIC BETWEEN ADIPOSOMES AND MULTIPLE ENDOMEMBRANE SYSTEMS

Pingsheng Liu, René Bartz, John K. Zehmer, Yunshu Ying, *and* Richard G. W. Anderson

Contents

Abstract

Lipid droplets play a critical role in a variety of metabolic diseases. Numerous proteomic studies have provided detailed information about the protein composition of the droplet, which has revealed that they are functional organelles involved in many cellular processes, including lipid storage and metabolism, membrane traffic, and signal transduction. Thus, the droplet proteome indicates that lipid accumulation is only one of a constellation of organellar functions critical for normal lipid metabolism in the cell. As a result of this new understanding, we suggested the name adiposome for this organelle. The trafficking ability of the adiposome is likely to be very important for lipid uptake, retention, and distribution, as well as membrane biogenesis and lipid signaling. We have taken advantage of the ease of purifying lipid-filled adiposomes to develop a cell-free system for studying adiposome-mediated traffic.

Department of Cell Biology, University of Texas Southwestern Medical Center, Dallas, Texas

Methods in Enzymology, Volume 439
ISSN 0076-6879, DOI: 10.1016/S0076-6879(07)00424-7

Using this approach, we have determined that the interaction between adiposomes and endosomes is dependent on Rab GTPases but is blocked by ATPase. These methods also allowed us to identify multiple proteins that dynamically associate with adiposomes in a nucleotide-dependent manner. An adiposome–endosome interaction *in vitro* occurs in the absence of cytosolic factors, which simplifies the assay dramatically. This assay will enable researchers to dissect the molecular mechanisms of interaction between these two organelles. This chapter provides a detailed account of the methods developed.

1. INTRODUCTION

Imbalances in lipid storage and metabolism in human cells have been linked to the progression of many metabolic diseases, such as morbid obesity, type 2 diabetes, cardiovascular disease, and nonalcoholic fatty liver disease. Generally, these health consequences are driven by the excessive storage of lipids in cells. For example, obesity is the excessive storage of lipids in adipocytes. A common consequence of obesity is the inappropriate sequestration of lipids in liver and skeletal muscle cells, which is thought to drive insulin resistance and the eventual development of overt type 2 diabetes. Lipid accumulation in hepatocytes is also the first stage in the development of fatty liver disease. The accumulation of cholesterol in macrophages transforms the cells into foam cells, which initiates the development of atherosclerosis. However, a lack of neutral lipid storage, especially in adipose tissue, can cause lipodystrophies that are associated with similar metabolic complications. Thus, understanding the regulation of cellular lipid storage is key to preventing and treating these metabolic diseases.

All cells, from bacteria to mammals, are able to store neutral lipids in a cellular structure that has been given many names, including lipid droplets, lipid bodies, oil bodies, and fat bodies. The term lipid droplet, which describes the structure of a lipid storage depot, is the most commonly used appellation (Martin and Parton, 2006). Recently, we and other groups have partially identified the droplet proteome from various cell types and tissues. Unexpectedly the proteome contains proteins involved in lipid metabolism, membrane traffic, and signal transduction. This information has extended our understanding of the droplet and suggests that rather than being a simple storage depot, it is a complex organelle involved in the synthesis, degradation, and transport of cellular lipids that participate in energy balance, membrane biogenesis, and cellular signaling. This suggests that the droplet is only one morphologic form of a complex functional organelle and, therefore, deserves a new term that encompasses all of its properties. We have proposed the name adiposome for this organelle, in keeping with the long-standing tradition in cell biology to name organelles with a unique prefix followed by a common suffix such as endosome, lysosome, and peroxisome (Liu *et al.*, 2004).

Among the newly identified adiposome-associated proteins are ones known to be involved in regulating membrane traffic (e.g., Rab GTPases), which led to the hypothesis that adiposomes are involved in lipid and/or protein traffic (Fujimoto *et al.*, 2004; Liu *et al.*, 2004). The recruitment of Rab proteins to adiposomes when lipolysis is stimulated supports this view (Brasaemle *et al.*, 2004). To test this hypothesis and further investigate the molecular mechanism of adiposome-mediated trafficking, we developed a cell-free system to study the interaction between purified lipid-filled adiposomes and isolated early endosomes. Using this approach, we demonstrated that Rab proteins interact with adiposomes and found that these Rabs exhibit biological properties similar to those when associated with membrane organelles such as endosomes (Liu *et al.*, 2007). As detailed later, this reconstitution system also enables one to study the interaction between adiposomes and early endosomes and to study the role of Rab proteins in this process (Liu *et al.*, 2007).

2. Rab-Mediated Adiposome–Endosome Interaction *in vitro*

Taking advantage of the dramatic density difference between lipid-filled adiposomes and other endomembranes, we developed a cell-free system to study the dynamic behavior of this organelle and its interaction with endosomes. To study the dynamics, we incubated purified adiposomes with isolated cytosol and then reisolated adiposomes by flotation, followed by analysis of its protein profile. Using this assay, we observed the recruitment of small G proteins such as Rabs and Arfs, as well as cytoskeleton proteins, including actin and tubulin (Bartz *et al.*, 2007b; Liu *et al.*, 2007). To study the interaction of adiposomes with endosomes, we incubated the isolated organelles under defined conditions and then reisolated the adiposomes by flotation. The physical interaction between endosomes and adiposomes is detected by immunoblotting the adiposome fraction for endosome marker proteins. The activity of Rab proteins is regulated using GTP and GDP. Rab proteins on adiposomes and endosomes can be removed by treating each with Rab-GDP dissociation inhibitor (RabGDI). Rabs are recruited back to Rab-free adiposomes by incubating in the presence of cytosol and GTP. Using these methods, we discovered that the adiposome interaction with early endosome is dependent on GTPγS. Moreover, removing Rabs from both organelles totally abolished this interaction, demonstrating that Rab proteins are essential for the interaction (Liu *et al.*, 2007). The methods used are detailed next.

2.1. Sample Preparation

2.1.1. Purification of adiposome

The purification method developed can be used to purify lipid-filled adiposomes from virtually any type of animal cell or tissue, including Chinese hamster ovary fibroblasts (CHO K2) (Liu *et al.*, 2004), human fibroblasts (Bartz *et al.*, 2007a), HeLa cells (Bartz *et al.*, 2007b; Liu *et al.*, 2004), 3T3 L1 adipocytes (Bartz *et al.*, 2007a), human B cells (Bartz *et al.*, 2007a), LNCaP cells, and mouse liver. Figure 24.1A shows the signature Coomassie colloidal blue staining pattern of lipid-filled adiposomes isolated from various sources. While the pattern is highly reproducible for each cell type, it varies between different cell types and tissues, which indicate that there are tissue-specific differences in the protein composition of the organelle.

Methods

Cells are cultured to 100% confluence in 150-mm plates containing 25 ml of standard tissue culture media. Cells are washed with 10 ml ice-cold phosphate-buffered saline (PBS) per plate, scraped into 5 ml ice-cold PBS plus 5 μl 100 mM phenylmethylsulfonyl fluoride (PMSF), pooled, and placed in a 50-ml tube. The cell suspension from nine plates is centrifuged at 500 g for 5 min at 4° and the supernatant is discarded. Cells are resuspended by vortexing in 8 ml buffer A (20 mM Tricine, pH 7.8, 250 mM sucrose) plus 8 μl of 100 mM PMSF and then incubated on ice for 20 min. The cells are homogenized with a nitrogen bomb (Parr Instrument Company, Moline, IL). The cells are placed in the bomb at 450 psi for 15 min on ice. The mixture is then released from the bomb chamber drop wise into a 50-ml tube and the lysate is centrifuged at 1000 g for 10 min at 4°. The slow release of the sample from the bomb under pressure is essential for proper homogenization. Seven milliliters of the supernatant fraction (PNS) is added to a SW41 tube, and then 3.5 ml of buffer B (20 mM HEPES, pH 7.4, 100 mM KCl, 2 mM MgCl$_2$) is loaded on top of the PNS fraction using a gradient maker. It is important to be able to see a sharp interface. The gradient is centrifuged at 40,000 rpm for 1 h at 4°. Lipid-filled adiposomes will concentrate in a white band at the top of the gradient. Adiposomes are collected using a 1-ml pipette tip and transferred to a 1.5-ml microfuge tube, making sure to remove the most adiposomes with the least amount of buffer. Adiposomes are separated from buffer by centrifuging at 10,000 g for 5 min at 4°. The solution underlying the adiposome is removed using a gel-loading tip attached to a 1-ml tip. The tip is slid into solution on the tube surface to avoid disturbing the floating adiposomes. Enough buffer is removed until the adiposome fraction reaches the pipette tip opening. Removing the buffer can disturb the adioposme layer. Adiposomes are suspended in buffer form a cloudy solution. It is impossible to remove the buffer from this cloudy solution without losing material. To avoid loss the sample is centrifuged again and the buffer is removed.

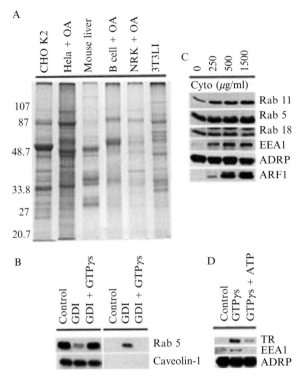

Figure 24.1 (A) Signature protein of adiposomes from different cells. Adiposomes were purified from the indicated cells grown in the presence or absence of oleic acid (OA). Proteins were precipitated with 100% acetone and dissolved in SDS sample buffer, and 10 µg of proteins was separated on 10% SDS-PAGE and stained with Coomassie colloidal blue. Adiposomes were isolated from 3T3 L1 cells after 8 days of differentiation. (B) Release of GDP-bound Rab 5 from adiposomes by RabGDI. Purified adiposomes were incubated in the presence of RabGDI (GDI) plus or minus GTPγs. Adiposomes (left) and reaction buffer (right) were separated and processed for immunoblotting to detect the indicated proteins. (C) Recruitment of Rabs from cytosol to adiposome. Purified adiposomes were processed to remove endogenous Rabs with RabGDI. Rab-depleted adiposomes were then incubated with the indicated concentrations of cytosol for 1 h. At the end of the reaction, adiposomes were reisolated, washed, and processed for immunoblotting to detect the indicated proteins. The adiposome-specific protein ADRP was used as a loading control. (D) Early endosomes bind adiposomes. Purified adiposomes were mixed with purified early endosomes in the presence or absence of GTPγs and ATP. Adiposomes were reisolated and unbound endosomes removed. Adiposomes were then processed for immunoblotting to detect the indicated proteins. ADRP was used as a loading control. TR, transferrin receptor.

The centrifugation step is repeated at least three times to remove all the buffer. To separate out contaminating membranes, the adiposomes are resuspended in 1 ml buffer B by vortexing and centrifuged at 40,000 rpm for 5 min in a Beckman Optima Max ultracentrifuge using a TLA 100.3 rotor at 4°. After centrifugation, some of the adiposomes are retained on the side of the tube while the remainder float at the top of the solution. The floating adiposomes are collected and transferred into a new 1.5-ml Eppendorf tube. The rest of the solution is discarded. The pellet in the bottom of the tube is resuspended using 100 μl buffer B and is discarded. Another 100 μl of buffer B is used to wash the bottom of the tube. Do not touch the adiposomes on the side of the tube during these steps (carefully use a 200-μl tip). Adiposomes are transferred back into the original tube, which is vortexed to wash the adiposomes off the side of tube. When all of the adiposomes are resuspended, transfer them into a new 1.5-ml Eppendorf tube and adjust the volume to 500 μl with buffer B. Centrifuge the sample at 10,000 g for 5 min. Remove the solution using a gel-loading tip attached to a 1-ml tip, making sure not to disturb the floating adiposomes. Repeat the washing step two more times. After the final wash, the adiposomes are ready for biochemical, morphological, and functional analysis. *Note*: Two important clues are needed for good purification. (1) Handle the sample very gently because lipid-filled adiposomes are extremely fragile (especially large ones). Avoid harsh vortexing and use wide-end pipette tips (cut tip) for transferring. (2) Broken adiposomes tend to be retained on plastic surfaces and can be driven into the pellet by centrifugation. Therefore, make sure that all adiposomes are resuspended in buffer B and that any aggregates (broken adiposomes and contaminated membrane structures) are removed using new tubes and tips. Remove the pellet from the bottom of the tube using a gel-loading tip.

2.1.2. Purification of His-tagged RabGDI

The RabGDI is purified from the BL 21(DE3) *Escherichia coli* strain hosting the pRSETa-GDI construct as described previously (Bartz *et al.*, 2003).

Methods

Briefly, 100 ml of cell lysate from a 10-liter culture is mixed and incubated with 10-ml preequilibrated Ni–NTA agarose beads (Qiagen) in 50-ml Falcon tubes. The samples are rotated on a wheel for 1 h at 4°. The beads are centrifuged to a pellet, washed, and then loaded into a column. RabGDI is eluted with 200 mM imidazole, and the sample is dialyzed overnight at 4° against buffer C (20 mM HEPES, pH 7.2, 10 mM 2-mercaptoethanol). Purified RabGDI is analyzed by SDS-PAGE to determine purity and protein concentration. The protein is further analyzed to measure its ability to remove Rab. The rest of the protein is snap frozen as aliquots and stored at −80°.

2.1.3. Isolation of cytosol

PNS (10 ml) from the adiposome purification step is loaded into a SW41 tube and centrifuged at 40,000 rpm for 1 h to remove any membranes. The cytosol is collected between adiposome (top) and membrane pellet (bottom) and is used fresh in each experiment.

2.1.4. Isolation of early endosomes

Early endosomes are purified by the method of Bartz *et al.* (2005) with minor modifications to isolate them from CHO K2 cells.

Methods

CHO K2 cells are cultured to 60% confluence, collected by scraping, and concentrated by pelleting through centrifugation. The cells are then resuspended in buffer B with proteinase inhibitor cocktail and homogenized firmly with a 1-ml syringe (slim) and 22-gauge $\times$ 1.5-in. needle (Aldrich) five times up and down. The homogenate is centrifuged in a 15-ml Falcon tube at 1000 g for 15 min at 4° to remove nuclei and unbroken cells. After centrifugation, the supernatant fraction is collected and adjusted to 40.6% sucrose using 62% sucrose in buffer B. The sample is transferred into a SW28 tube, overlaid with 12 ml of 35% sucrose in buffer B, 8 ml 20% sucrose in buffer B, and 2 ml of buffer B. The gradient is centrifuged at 28,000 rpm (103,745 g) for 3 h. The visible interface (2 ml) between 25 and 30% sucrose is collected in a 15-ml tube, mixed carefully two to three times with a disposable plastic pipette, placed in 200-μl aliquots in an Eppendorf tube (use cut tips), snap frozen in liquid N_2, and stored at $-80°$.

2.2. Verification Proteins Associated with Adiposomes

It is important to verify by morphology that any protein of interest is associated with adiposomes and not a contaminating organelle. This can only be done by immunogold electron microscopy. For example, even though the proteome indicates that lipid-filled adiposomes are enriched in multiple Rabs, they could be in contaminating endoplasmic reticulum (ER) (Liu *et al.*, 2007).

Methods

Purified adiposomes in 200 μl of buffer B are mixed with 200 μl of 6% paraformaldehyde in buffer B containing 0.15% crystallized BSA (buffer D) and incubated for 1 h at room temperature. The fixed sample is washed by flotation (centrifuge at 20,000 g for 3 min) with buffer B three times, 5 min each at room temperature, resuspended in buffer B plus 50 mM NH$_4$Cl (buffer E) for 10 min, and washed again with buffer B two times, 5 min each. The sample is mixed with buffer D and incubated for 30 min at room temperature. Buffer D is removed and the sample is mixed with pAb or mAb, diluted 1:25 with buffer D, and incubated further for 15 h at 4°. The adiposomes are washed by flotation with buffer D and mixed with the appropriate α-IgG conjugated to 10 nm gold diluted 1:30 in buffer D for 2 h at room temperature. Adiposomes are washed again with buffer D three times, 5 min each, at 4°. Finally, samples are washed with 50 mM Na$_2$HPO$_4$, pH 8.0, 0.3 M NaCl, 10 mM 2-mercaptoethanol (buffer F), fixed with 1% GTA, and postfixed in 1% OsO$_4$, all in buffer F. Samples are embedded in Epon.

2.3. Testing for Adiposome Interaction with Proteins

2.3.1. Release of Rab proteins from adiposomes by RabGDI

Rabs are small GTP-binding proteins that regulate membrane traffic by controlling targeting, tethering, docking, and fusion between various membrane systems. Previous studies demonstrate that Rab proteins are activated by GTP. When GTP is hydrolyzed to GDP, Rabs are released from the membrane into the cytosol by RabGDI (Holtta-Vuori *et al.*, 2000). Cytosolic Rabs are recruited back to target membranes through the action of the RabGDI displacement factor, RabGDF (Dirac-Svejstrup *et al.*, 1997). Therefore, a measure of whether an organelle-associated Rab is functional is if it can cycle off and on the organelle in a RabGDI- and RabGDF-dependent fashion. This assay can be carried out on isolated adiposomes.

Methods

Purified adiposomes (100 μl) are incubated in the presence of 2 mM GDP and 10 μM RabGDI in buffer B at 37° for 1 h. The reaction is vortexed briefly every 10 min to keep adiposomes in suspension because they tend to float to the top of solution. Adiposomes are repurified from the reaction buffer by flotation, and the reaction buffer and adiposome fraction are processed. The adiposome fraction is washed three times with 200 μl/time of buffer B using the same flotation method, and the protein is precipitated with 100% acetone. Proteins are precipitated from the reaction buffer with 7.2% trichloroacetic acid. The presence of Rabs or other proteins of interest in either fraction is measured by immunoblotting. The immunoblot is reprobed with an appropriate adiposome resident protein marker that is not released by washing, such as caveolin-1. Ral A, another adiposome-associated small G protein, is used as a negative control because it is not removed by RabGDI. Figure 24.1B (left) shows that RabGDI releases Rab 5 and that release is blocked by GTPγs. Therefore, as expected, RabGDI only removes Rabs from adiposomes that are bound to GDP. RabGDI-released Rab 5 appears in the incubation buffer (see Fig. 24.1, right).

2.3.2. Recruitment of cytosolic Rabs to adiposomes

Rab-depleted adiposomes are used to study the recruitment of cytosolic proteins.

Methods

Rab-deficient adiposomes (50 μl) prepared as described earlier are incubated in the presence of 100 μl cytosol (250–1500 μg/ml) in buffer B plus 1 mM GTPγs for 1 h at 37°. The reaction is vortexed briefly every 10 min to keep adiposomes in suspension. Adiposomes are then reisolated, washed, and processed for immunoblotting to detect both the Rabs of interest and any other proteins. To avoid possible contamination from cytosol, adiposomes are washed five times with buffer B. Liposomes prepared from CHO K2 cell total lipids are used as a nonspecific binding control. To make liposomes, lipids are extracted by the method of Bligh and Dyer (1959) from the total membrane pellet of cultured cells. The solvent is removed from the lipids using a rotary evaporator followed by 20 min under low torr vacuum. Diethyl ether (5 ml) and buffer B (300 μl) are added to the lipid film. The ether is driven off under a stream of nitrogen with bath sonication. The resulting concentrated liposomes are diluted with 1.7 ml of buffer B and drawn through a 22-gauge needle three times.

Figure 24.1C shows an experiment demonstrating that isolated, RabGDI-treated adiposomes can recruit Rabs 5 and 11 plus the Rab 5 effector EEA1 as a function of cytosol concentration in the incubation media. Interestingly, Rab 18 is neither removed by RabGDI nor recruited from cytosol. Arf1 is also recruited to the adiposomes. The peripheral adiposome marker ADRP is not affected.

2.4. Rab-Dependent Interaction between Early Endosomes and Adiposomes

The GTP-dependent interaction of Rabs with adiposomes is similar to how Rabs function in intracellular membrane traffic, which suggests that adiposomes may interact with various endomembranes. Indeed, Rab 18 appears to mediate the close apposition of the ER to adiposomes *in situ* (Martin *et al.*, 2005; Ozeki *et al.*, 2005). Since we found that adiposomes bound cytosolic EEA1 (see Fig. 24.1C), we focused on the endosome interaction with adiposomes. Not only can the adiposome–endosome interaction be reconstituted, but the requirements for GTP, Rabs, and other proteins can be determined (Liu *et al.*, 2007). Interestingly, the presence of cytosol in the assay does not affect the interaction, suggesting that the factors required for the interaction are completely contained in the interacting organelles. Early endosome binding to adiposomes is saturable, which suggests that there are a limited number of binding sites on each adiposome. Finally, in stark contrast to membrane–membrane interactions during membrane traffic, ATP inhibits the binding of endosome to adiposome (see Fig. 24.1D).

Methods

Purified adiposomes (50 μl) are mixed with 50 μl endosomes (100 μg/ml) in buffer B and incubated for 1 h at 37° in the presence or absence of 1 mM nucleotide. The reaction is vortexed briefly every 10 min to keep the adiposomes in suspension. Endosomes bound to adiposomes are separated from free endosomes by flotation using a centrifuge speed of 10,000 g for 4 min at 4°. The centrifuge speed is very critical because too high a speed can strip off specifically bound endosomes, whereas too low a speed will lead to nonspecific binding. Additional nonspecifically bound endosomes are removed by washing adiposomes three times with 200 μl buffer B as mentioned previously. With each spin the membrane pellet gets smaller. After three spins, the pellet is not readily visible unless the bottom of the tube is viewed with a bright light against a dark background. Proteins are precipitated and lipids are removed from the adiposomes using 1 ml/tube of 100% acetone followed by centrifugation at 20,000 g for 10 min. The bound endosomes are detected by immunoblotting using the pAb transferrin receptor and mAb EEA1 IgG. The lipid-filled adiposome marker ADRP is used as a loading control. To determine whether Rabs are required for adiposome–early endosome interaction, it is necessary to remove Rabs from both organelles using RabGDI. We used the method described earlier to remove Rabs from adiposomes. To remove them from endosomes, the isolated endosomes (200 μl) are incubated in the presence of 2 mM GDP and 10 μM RabGDI in buffer B at 37° for 1 h. Then the reaction is mixed with 600 μl of buffer B, loaded on the top of 200 μl of 2 M KCl in an Eppendorf tube, and centrifuged at 20,000 g for 30 min at 4°. Endosomes (400 μl) are collected from the interface between 2 M KCl and buffer B. The ability of Rab-depleted adiposomes and endosomes to interact is determined as described earlier.

3. Conclusion

The development of an *in vitro* method for reconstituting the interaction between lipid-filled adiposomes and either cytosolic or membrane proteins is a rich system for exploring the organellar properties of the adiposome. In conjunction with various genetic, molecular biology, and cell biology techniques, these methods should be very useful not only in understanding the function of adiposome-associated Rabs, but also the function of adiposomes in the synthesis, storage, and distribution of cellular lipids. While the assays are technically straightforward, they demand attention to details and fresh materials. The lipid-filled adiposome is fragile and cannot be frozen for any length of time.

ACKNOWLEDGMENTS

We thank Meifang Zhu for her valuable technical assistance and Brenda Pallares for administrative assistance. This work was supported by grants from the National Institutes of Health, HL 20948, GM 52016, the Cecil H. Green Distinguished Chair in Cellular and Molecular Biology, and the Perot Family Foundation to RGWA and GM070117 to John K. Zehmer.

REFERENCES

Bartz, R., Benzing, C., and Ullrich, O. (2003). Reconstitution of vesicular transport to Rab11-positive recycling endosomes *in vitro*. *Biochem. Biophys. Res. Commun.* **312**, 663–669.

Bartz, R., Benzing, C., and Ullrich, O. (2005). Reconstitution of transport to recycling endosomes *in vitro*. *Methods Enzymol.* **404**, 480–490.

Bartz, R., Li, W. H., Venables, B., Zehmer, J. K., Roth, M. R., Welti, R., Anderson, R. G., Liu, P., and Chapman, K. D. (2007a). Lipidomics reveals that adiposomes store ether lipids and mediate phospholipid traffic. *J. Lipid Res.* **48**, 837–847.

Bartz, R., Zehmer, J. K., Zhu, M., Chen, Y., Serrero, G., Zhao, Y., and Liu, P. (2007b). Dynamic activity of lipid droplets: Protein phosphorylation and GTP-mediated protein translocation. *J. Proteome Res.* **6**, 3256–3265.

Bligh, E. G., and Dyer, W. J. (1959). A rapid method of total lipid extraction and purification. *Can. J. Biochem. Physiol.* **37**, 911–917.

Brasaemle, D. L., Dolios, G., Shapiro, L., and Wang, R. (2004). Proteomic analysis of proteins associated with lipid droplets of basal and lipolytically stimulated 3T3-L1 adipocytes. *J. Biol. Chem.* **279**, 46835–46842.

Dirac-Svejstrup, A. B., Sumizawa, T., and Pfeffer, S. R. (1997). Identification of a GDI displacement factor that releases endosomal Rab GTPases from Rab-GDI. *EMBO J.* **16**, 465–472.

Fujimoto, Y., Itabe, H., Sakai, J., Makita, M., Noda, J., Mori, M., Higashi, Y., Kojima, S., and Takano, T. (2004). Identification of major proteins in the lipid droplet-enriched fraction isolated from the human hepatocyte cell line HuH7. *Biochim. Biophys. Acta* **1644**, 47–59.

Holtta-Vuori, M., Maatta, J., Ullrich, O., Kuismanen, E., and Ikonen, E. (2000). Mobilization of late-endosomal cholesterol is inhibited by Rab guanine nucleotide dissociation inhibitor. *Curr. Biol.* **10**, 95–98.

Liu, P., Bartz, R., Zehmer, J. K., Ying, Y. S., Zhu, M., Serrero, G., and Anderson, R. G. (2007). Rab-regulated interaction of early endosomes with lipid droplets. *Biochim. Biophys. Acta* **1773**, 784–793.

Liu, P., Ying, Y., Zhao, Y., Mundy, D. I., Zhu, M., and Anderson, R. G. (2004). Chinese hamster ovary K2 cell lipid droplets appear to be metabolic organelles involved in membrane traffic. *J. Biol. Chem.* **279**, 3787–3792.

Martin, S., Driessen, K., Nixon, S. J., Zerial, M., and Parton, R. G. (2005). Regulated localization of Rab18 to lipid droplets: Effects of lipolytic stimulation and inhibition of lipid droplet catabolism. *J. Biol. Chem.* **280**, 42325–42335.

Martin, S., and Parton, R. G. (2006). Lipid droplets: A unified view of a dynamic organelle. *Nat. Rev. Mol. Cell. Biol.* **7**, 373–378.

Ozeki, S., Cheng, J., Tauchi-Sato, K., Hatano, N., Taniguchi, H., and Fujimoto, T. (2005). Rab18 localizes to lipid droplets and induces their close apposition to the endoplasmic reticulum-derived membrane. *J. Cell Sci.* **118**, 2601–2611.

CHAPTER TWENTY-FIVE

DETECTION OF COMPOUNDS THAT RESCUE RAB1-SYNUCLEIN TOXICITY

James Fleming,* Tiago F. Outeiro,[†] Mark Slack,[‡] Susan L. Lindquist,[§,¶] *and* Christine E. Bulawa*

Contents

Abstract

Recent studies implicate a disruption in Rab-mediated protein trafficking as a possible contributing factor to neurodegeneration in Parkinson's disease (PD). Misfolding of the neuronal protein α-synuclein (asyn) is implicated in PD. Overexpression of asyn results in cell death in a wide variety of model systems, and in several organisms, including yeast, worms, flies, and rodent primary

* FoldRx Pharmaceuticals, Inc., Cambridge, Massachusetts
[†] Institute of Molecular Medicine, Cellular and Molecular Neuroscience Unit, Lisbon, Portugal
[‡] Evotec AG, Hamburg, Germany
[§] Whitehead Institute for Biomedical Research, Cambridge, Massachusetts
[¶] Howard Hughes Medical Institute, Cambridge, Massachusetts

Methods in Enzymology, Volume 439
ISSN 0076-6879, DOI: 10.1016/S0076-6879(07)00425-9

neurons, this toxicity is suppressed by the overproduction of Rab proteins. These and other findings suggest that asyn interferes with Rab function and provide new avenues for PD drug discovery. This chapter describes two assay formats that have been used successfully to identify small molecules that rescue asyn toxicity in yeast. The 96-well format monitors rescue by optical density and is suitable for screening thousands of compounds. A second format measures viable cells by reduction of the dye alamarBlue, a readout that is compatible with 96-, 384-, and 1536-well plates allowing the screening of large libraries (>100,000 compounds). A secondary assay to eliminate mechanistically undesirable hits is also described.

1. INTRODUCTION

Defects in Rab GTPase function underlie a variety of inherited diseases (Olkkonen and Ikonen, 2006; Seabra *et al.*, 2002). Examples include mutations in Rab proteins (Rab7a/Charcot–Marie–Tooth disease type 2B and Rab27a/Griscelli syndrome) and mutations in proteins that regulate Rab activity (Rab escort protein-1/choroideremia, Rab5 GTPase activating protein/tuberous sclerosis, and Rab GDP dissociation inhibitor α/X-linked mental retardation). A deficiency of Rab function has been implicated in Parkinson's disease (PD), a movement disorder caused by the loss of dopamine-producing neurons in the substantia nigra. In the case of PD, the deficiency of Rab function is caused not by mutations in the Rab regulatory machinery, but rather by overexpression (Cooper *et al.*, 2006) of the aggregation-prone protein α-synuclein (asyn), a critical determinant in familial (Eriksen *et al.*, 2005) and sporadic PD (Goedert, 2001).

Overexpression of wild-type asyn in several model systems (Maries *et al.*, 2003), including yeast (Outeiro and Lindquist, 2003), worms (Lasko *et al.*, 2003), flies (Whitworth *et al.* 2006), rodents (Kirik *et al.*, 2002; Lo Bianco *et al.*, 2002), and primates (Kirik *et al.*, 2003), induces cell death. To investigate the mechanism of asyn toxicity, Cooper *et al.* (2006) identified genetic modifiers of asyn toxicity in yeast. Especially noteworthy was the recovery of a number of genes involved in endoplasmic reticulum (ER) to Golgi protein trafficking, including the ER Rab *YPT1*, suggesting that asyn inhibits ER to Golgi protein trafficking, a hypothesis that was confirmed in yeast by demonstrating the asyn-dependent accumulation of ER forms of secreted proteins (Cooper *et al.*, 2006). To test whether asyn inhibits ER to Golgi trafficking in dopaminergic neurons, the *YPT1* ortholog, Rab1a, was overexpressed in three different models of PD (transgenic worm and fly and rat primary neurons) and found to rescue asyn-mediated toxicity in all. Thus, the mechanism of asyn toxicity discovered in yeast is conserved in dopaminergic neurons.

Current PD therapies achieve their clinical benefit by restoring dopamine levels but ultimately fail, as they do not address the underlying cause of the disease, the degeneration of dopaminergic neurons. To identify candidate neuroprotective agents, we used the yeast PD model to screen for small molecules that rescue asyn toxicity. Because overexpression of *YPT1/ Rab1a* rescues asyn toxicity, our screen can identify small molecules that rescue asyn toxicity by stimulating Rab function and/or increasing Rab levels.

2. STRAINS

Strains suitable for primary screening (rescue of asyn toxicity) and a secondary assay (elimination of mechanistically undesirable hits using induction of *lacZ*) are constructed using standard methods for *Saccharomyces cerevisiae* (Rothstein, 1991) essentially as described (Outeiro and Lindquist, 2003). Briefly, DNA constructs containing asyn or *lacZ* under the control of the galactose promoter (Schneider and Guarente, 1991) are integrated into the genome of a Gal$^+$ host strain. Plasmid-borne asyn expression constructs are unstable under inducing conditions and give inconsistent and/or weak toxicity. Asyn toxicity is strongly dose dependent, and two constructs per haploid genome must be integrated to obtain growth inhibition. The site of integration appears to modestly affect asyn expression. This observation can be used to generate strains with different levels of growth inhibition. We have found that asyn integrated at *URA3* and *TRP1* gives stronger toxicity than integration at *HIS3* and *TRP1*. To test whether positives from the primary screen rescue by reducing transcription or translation, we developed a secondary assay that monitors the induction of *lacZ*. This reporter assay uses a congenic strain containing vector inserts at *URA3* and *TRP1* and *lacZ* under the control of the galactose promoter integrated at *HIS3*.

3. METHODS FOR DETECTING COMPOUNDS THAT RESCUE α-SYNUCLEIN-INDUCED TOXICITY

To identify molecules that rescue the toxicity of asyn we developed two methods of compound screening. The first is a medium throughput 96-well assay that uses optical density (OD) as a measure of yeast growth. The second, a high throughput assay configurable to 96-, 384-, and 1536-well formats, employs alamarBlue reduction to readout cell growth. Both methods make use of the asyn screening strain. As this strain contains integrated copies of asyn at both *URA3* and *TRP1* loci, both protocols use synthetic complete media lacking the nucleoside uracil (-Ura) and the

Table 25.1 Buffered synthetic complete medium

Component	Vendor	Amount per liter	Final concentration
Yeast nitrogen base without amino acids	Difco	6.7 g	0.67% (w/v)
Carbon source: one of glucose, galactose, raffinose (see Table 25.2)	See Table 25.2	See Table 25.2	2 or 4% (w/v)
CSM -Trp -Ura or CSM -His -Trp -Ura	Qbiogene	≈0.8 g[a]	
MOPS (mol. biol. grade)	EMD	20.9 g[b,c]	0.1 M
Milli Q water	—	1 liter	—

[a] According to manufacturer's instructions.
[b] If required.
[c] If making media containing MOPS, pH to 6.0 with 1 M NaOH.

Table 25.2 Carbon sources

Component	Vendor	Amount per liter	Final concentration
Glucose (also known as dextrose)	Fisher	20 g	2% (w/v)
Galactose	Sigma	20 g	2% (w/v)
Raffinose	Difco	40 g	4% (w/v)

amino acid tryptophan (-Trp, see Tables 25.1 and 25.2 and Section 4). Expression of asyn is toxic in yeast (Outeiro and Lindquist, 2003). Therefore, the copies of asyn in the screening strain are under control of the inducible galactose promoter (Schneider and Guarente, 1991). This promoter is controlled by the carbon source supplied in the media. Growth in glucose-containing media represses transcription from the galactose promoter. Growth in raffinose neither induces nor represses transcription of the galactose promoter, whereas growth in galactose-containing media results in high expression. The screening strain is maintained on glucose media. When growing cells for the rescue assay, raffinose media are used so that simple dilution of the cells into galactose media results in the induction of asyn. Four percent instead of the usual 2% raffinose is used in the media as our experiments demonstrate that the asyn screening strain grows optimally under these conditions (data not shown).

When screening compounds in either format, negative and positive controls are tested in each screening plate. This allows for per plate quality control metrics and a way to compare the extent of compound rescue between plates (or experiments) by expressing the extent of rescue as a function of the positive control for that plate. As compounds are dissolved in dimethyl sulfoxide (DMSO), it alone is used as the negative control. A sublethal concentration of the topoisomerase poison daunorubicin (Sigma) is used as a positive control in these assays. In addition to causing double-stranded DNA breaks, daunorubicin inhibits transcription in yeast. Daunorubicin binds preferentially to GC-rich regions of DNA. As the galactose promoter is GC rich in an AT–rich S. cerevisiae genome, sublethal concentrations of daunorubicin preferentially inhibit transcription from this promoter (Marin et al., 2002). This inhibition of galactose-promoted transcription rescues the asyn screening strain by lowering the expression of the toxic asyn protein.

3.1. Synthetic Media

Based on the genotype of the strain to be tested, the appropriate supplementation for synthetic medium is chosen. Strains containing integrated constructs should be grown in medium that maintains selection for the construct (see later). CSM (Qbiogene) is a commercially available amino acid mix for growing S. cerevisiae. It can be obtained lacking one or more amino acids as required.

To make liquid synthetic medium, mix the components listed in Tables 25.1 and 25.2. After the components have dissolved, adjust the pH to 6.0 if necessary (when using alamarBlue as a readout, see later) and sterilize by filtration (Millipore Stericup) into a sterile bottle.

4. Primary Assays

4.1. Medium throughput format (96 well)

The medium throughput protocol allows for screening of compounds in 96-well plates without the need of robotics or detection equipment more sophisticated than a plate reader that can measure absorbance. We perform OD_{600} measurements on a Wallach Victor2 plate reader (Perkin Elmer) with a 600-nm (10-nm band-pass) filter. Flat-bottom tissue culture-treated polystyrene plates are used for assaying yeast growth. Tissue culture-treated flat bottom plates allow for easier mixing of yeast. Compounds are dispensed from V-bottom polypropylene plates. Polypropylene plates are used for compounds, as DMSO and polystyrene are not compatible during long-term storage.

Plates are mixed immediately before OD_{600} reads. This step was incorporated because yeast cells settling in the wells result in increased OD_{600} measurements. This cell settling effect begins shortly after mixing and plateaus at 1 h; final values for settled cells typically exceed initial values by more than 2.5-fold (data not shown). Reading plates in which the yeast cells have been allowed to settle is not advisable, as the movements of the plate in the reader generate foci of yeast at the bottom of wells, which increases the variability and extent of the OD_{600} read (data not shown).

4.2. Medium throughput protocol

1. Inoculate an appropriate volume of SC -Trp -Ura 4% raffinose medium with the asyn screening strain at OD_{600} of 0.05, 0.025, and 0.0125. By inoculating dilutions of the strain, we ensure that at least one culture is in log phase growth the next day.
2. Grow strains overnight 14 to 18 h at 30° with shaking.
3. Measure OD_{600} of the cell cultures and identify one in log phase growth (OD_{600} 0.1–1.0). *Note:* In our spectrophotometer, an OD_{600} of 1.0 equals $\approx 2 \times 10^7$ cells/ml. This ratio varies depending on the spectrophotometer used.
4. Dilute cells into SC -Trp -Ura 2% galactose media to OD_{600} 0.003.
5. Transfer 100 μl of galactose culture to each well of a 96-well flat bottom microtiter assay plate.
6. Transfer 1 μl of compound in DMSO at $100\times$ testing concentration to each well. Add 1 μl DMSO alone to negative controls and 1 μl 2.5 mM daunorubicin in DMSO to positive control wells. The final concentration of DMSO should be $\leq 1\%$. Mix the compound in wells by either pipetting or vortexing the plate. *Note:* Compounds can be added by hand quickly using a multichannel pipettor that can dispense 1 μl accurately.
7. Incubate plates for 48 h at 30°.
8. At the end of the incubation period, mix the plate to suspend yeast and immediately read OD_{600} on a plate reader.

We employed this protocol to screen a 2000 compound library for compounds that rescue asyn toxicity. At a cutoff of 6 SD above the mean value for the DMSO-only samples, we were able to identify 23 compounds that rescue at 12.5 μM (1.2% hit rate). We rescreened hits under identical conditions and found a 70% confirmation rate.

4.3. High throughput format

To screen a large compound library, we devised a high throughput asyn rescue assay that allows for screening in a 96-, 384-, or 1536-well format. Using OD as a readout at this scale is problematic because it is prone to cell settling effects that create highly variable values, especially in small wells that are difficult to mix. As an alternative readout for growth, we have used alamarBlue to assess cell proliferation (Ahmed et al., 1994; To et al., 1995). In the presence of metabolically active cells, alamarBlue (resazurin) is reduced to resorufin, which can be monitored either colormetrically or fluorometrically. As it is more sensitive, we opted to use fluorometric detection of alamarBlue reduction. AlamarBlue is not fluorescent, wheres resorufin excites at 530 to 560 nm and emits at 590 nm. We assayed resorufin production on a Wallach Victor2 plate reader (Perkin Elmer) using a 540-nm filter for excitation and a 590-nm filter to read emission. The fluorescence of resorufin is pH dependent (Bueno et al., 2002). Growing yeast cells acidify the media, and the pH of rescued wells is not optimal for fluorescent detection of resorufin. Therefore, to enhance the signal and reproducibility in the assay we use media buffered with 0.1 M 3-(N-morpholino)propane-sulfonic acid (MOPS) to pH 6.0. *Note*: There is a point with large numbers of yeast cells where resorufin is further reduced to a nonfluorescent product (hydroresorufin). This can be determined by visual inspection of wells or by absorbance reading, as hydroresorufin is clear and colorless.

4.4. High throughput protocol (384 well)

1. Inoculate cultures and grow overnight as described in steps 1 and 2 of Section 5.2 except that medium containing 0.1 M MOPS and pH adjusted to 6.0 is used.
2. Measure OD_{600} of the cell cultures and identify one in log phase growth (OD_{600} 0.1–1.0).
3. Dilute cells into SC -Trp -Ura 2% galactose 0.1 M MOPS, pH 6.0, medium to OD_{600} 0.03.
4. Transfer 25 μl of galactose culture to each well of a 384-well flat bottom microtiter assay plate.
5. Transfer 0.25 μl of compound at 100× testing concentration to each well. Add 0.25 μl DMSO alone to negative controls and 0.25 μl 2.5 mM daunorubicin in DMSO to positive control wells. Mix the compound in wells by either pipetting or vortexing.
6. Incubate plates for 24 h at 32°.
7. At the end of the incubation period, add 2.5 μl of alamarBlue to each well. Mix alamarBlue in wells by either pipetting or vortexing.
8. Incubate plates for 5 h at 32°.
9. After incubation, measure resorufin fluorescence on a Victor2 Plate reader.

The high throughput protocol is employed to screen a 280k compound library at 20 μM. The Z' factor describes the separation of positive and negative controls and is employed as a common screening metric (Zhang *et al.*, 1999). The mean Z' value for the screen (0.8) was greater than the commonly accepted criteria of Z' $\geq$ 0.5. Compounds with alamarBlue signal greater than 3 standard deviations above the mean of the DMSO controls were designationd hits, 0.84% of compound passed this criterion, 77% of the hits test positive in confirmation assays.

5. Secondary Assays

Compounds identified in the high throughput screen are tested in secondary assays to assess the potency and mechanism of action (MOA). Potency is assessed by examining the activity of the compound in a dose–response fashion. Limited MOA testing is performed to identify compounds with an undesirable mechanism of action (see later). *Note*: The majority of compounds identified in the compound screens described previously rescue in both buffered and unbuffered media. However, subsets have been identified that are exclusively active in either condition. In addition, we have discovered that compounds can have differential activity dependent on the final concentration of DMSO in the assay. Therefore, when testing hits from the screening protocols in either secondary assay, it is important to keep these parameters consistent.

5.1. Dose–response testing

Compounds that rescue asyn toxicity are retested in a dose–response format to assess potency. In this protocol, compounds are serially diluted 1:1 (v:v) in DMSO in rows of a 96-well V-bottom polypropylene microtiter plate and then added to yeast in the assay plate. Addition and dilution of compounds in medium give unsatisfactory results. Certain compound classes appear prone to sticking to the sides of tips in the presence of medium and leech off during subsequent dilutions, which leads to falsely high compound potencies. The compound potency is assessed by the extent of rescue compared to the positive control (daunorubicin), by the minimal concentration that results in significant rescue above background, and by the concentration of compound that results in half-maximal rescue (EC_{50}). Example dose–responses are shown in Fig. 25.1.

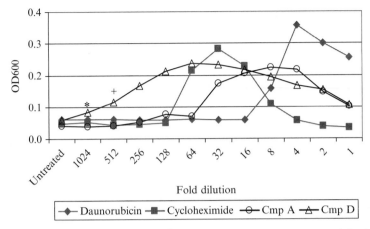

Figure 25.1 Dose-dependent rescue of asyn toxicity. Daunorubicin, cycloheximide, and two hits from the asyn screen [compound (Cmp) A and Cmp D] were examined for rescue using the asyn dose–response protocol. The starting concentrations of compound (fold dilution of 1) were daunorubicin, 50 μM; cycloheximide, 500 ng/ml; compound A, 150 μM; and compound D, 50 μM. Compound concentration increases left to right. The minimal rescue concentration is the lowest concentration that increases the OD_{600} above the DMSO control (indicated by an asterisk for compound D). The EC_{50} is the concentration that gives 50% of the maximal OD_{600} (indicated by a plus sign for compound D).

5.2. Dose–response protocol

1. Inoculate cultures and grow overnight as described in steps 1 and 2 of Section 5.2.
2. Prepare compound plate by adding 25 μl of DMSO to columns 2 to 12 of a 96-well V-bottom microtiter plate.
3. Add 50 μl of compounds at 100× the highest testing concentration to column 1 of the microtiter plate. *Note:* We routinely start at 5 mM compound (50 μM after dilution with cells) and always run daunorubicin as a positive control in the assay.
4. Using a multichannel pipettor, aspirate 25 μl of compound in DMSO from the first column of each row, dispense into the second column, and mix.
5. Repeat step 4 nine additional times on each subsequent set of columns. After mixing in column 11, aspirate 25 μl of the DMSO solution-containing compound and discard. The final plate should contain an approximately 1000-fold dilution of compound in columns 1 to 11 and DMSO alone in column 12.
6. Go to step 3 in Section 5.2 and proceed as described.

5.3. Eliminating compounds that rescue by inhibition of transcription/translation

As exemplified by the use of daunorubicin as a positive control, inhibitors of transcription can rescue in the asyn toxicity assay. Additionally, translational inhibitors such as cycloheximide rescue (see Fig. 25.1) via the same underlying mechanism (lowering asyn expression). To eliminate compounds identified in the screen with either of these MOAs we examined the ability of compounds to inhibit the expression of a reporter gene under control of the galactose promoter. This secondary screen uses a strain that is congenic with the screening strain; it does not express asyn but instead has a galactose promoter *lacZ* fusion (Rupp, 2002) integrated at the *HIS3* locus. Expression of *lacZ* is induced in the presence of the compound, and activity is assessed at 1 and 4 h. β-Galactosidase activity is measured using the gal-Screen kit for yeast (T1030) supplied by Applied Biosystems. Compounds that inhibit transcription and translation result in lowered activity of the reporter (Fig. 25.2). Because the galactose-promoted *lacZ* reporter is integrated at the *HIS3* locus, the strain is grown in synthetic complete media minus uracil, tryptophan, and histidine. We use white microtiter assay plates for this experiment, as they result in a better luminescence signal and minimize signal bleed into adjacent wells. We typically test compounds at concentrations 2- to 10-fold above the EC_{50} of rescue to ensure those with undesirable MOA are identified. Figure 25.2 shows *lacZ* reporter data for two

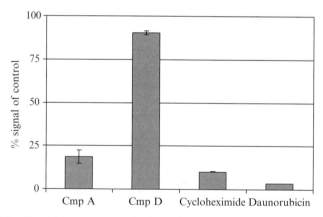

Figure 25.2 Identifying compounds that inhibit transcription or translation. Two novel compounds that rescue asyn toxicity (Cmp A and Cmp D) and two controls (daunorubicin, a transcriptional inhibitor; and cycloheximide, a translational inhibitor) were tested for effects in the *lacZ* reporter activity assay as described in the text. The concentrations tested were Cmp A, 150 μM; Cmp D and daunorubicin, 100 μM; and cycloheximides, 100 ng/ml. Values are the average of four replicates per condition. Data are expressed as the percentage of the DMSO control; this allows for comparison of compound activities across multiple experiments.

compounds identified in the asyn high throughput screen that have distinct effects on β-galactosidase expression.

5.4. lacZ protocol

1. Inoculate an appropriate volume of SC -His -Trp -Ura 4% raffinose medium with the *lacZ* reporter strain.
2. Grow strains overnight at 30° with shaking.
3. Measure OD_{600} of the cell cultures and identify one in log phase growth (OD_{600} 0.1–1.0).
4. Dilute cells into SC -His -Trp -Ura 2% galactose media to OD_{600} 0.03.
5. Transfer 100 μl of galactose culture to each well of a 96-well flat bottom white microtiter assay plate.
6. Transfer 1 μl of compound at 100× testing concentration to each well. Add 1 μl DMSO alone to negative controls and 1 μl 5.0 mM daunorubicin in DMSO to positive control wells. Mix the compound in wells by either pipetting or vortexing. *Note*: Because of the variability intrinsic to reporter assays, we normally test each concentration of compound in triplicate.
7. Incubate plates for 1 to 4 h at 30°.
8. During incubation, prepare the detection reagent by diluting the gal-Screen substrate 1:24 (v:v) into the gal-Screen lysis buffer and allow to warm to room temperature.
9. At the end of the incubation period, add 100 μl of detection reagent to each well of the microtiter plate.
10. Incubate plates for 1 h at room temperature and measure total luminescence for 1.0 s/well on a Wallach Victor2 plate reader (Perkin Elmer).

Nine hundred and eight hits from the 280k screen described earlier were tested for activity in the *lacZ* reporter activity assay. The compounds were tested at 40 μM (twofold the concentration used as described in the primary screen). Nine percent of the compounds tested demonstrated a >50% inhibition of β-galactosiadase reporter activity and were not pursued further.

6. CONCLUSION

FoldRx Pharmaceuticals has performed the aforementioned primary screens on approximately 350,000 compounds in an effort to identify Parkinson's disease therapeutics. These screens have identified multiple compound series that are active not only in the yeast model, but in worms overexpressing asyn, primary neurons expressing the familial PD asyn mutation A53T (Cooper *et al.*, 2006), and in H4 neurogliomal cells overexpressing wild-type asyn. The ER to Golgi trafficking block induced by asyn can be overcome by Rab overexpression. Therefore, among the compounds that result asyn toxicity may be modulators of Rab function. Experiments are underway to address this possibility.

REFERENCES

Ahmed, S. A., Gogal, R. M., Jr., and Walsh, J. E. (1994). A new rapid and simple non-radioactive assay to monitor and determine the proliferation of lymphocytes: An alternative to [^{3}H]thymidine incorporation assay. *J. Immunol. Methods* **170,** 211–224.

Bueno, C., Villegas, M. L., Bertolotti, S. G., Previtali, C. M., Neumann, M. G., and Encinas, M. V. (2002). The excited-state interaction of resazurin and resorufin with amines in aqueous solutions: Photophysics and photochemical reactions. *Photochem. Photobiol.* **76,** 385–390.

Cooper, A. A., Gitler, A. D., Cashikar, A., Haynes, C. M., Hill, K. J., Bhullar, B., Liu, K., Xu, K., Strathearn, K. E., Liu, F., Cao, S., Caldwell, K. A., *et al.* (2006). Alpha-synuclein blocks ER-Golgi traffic and Rab1 rescues neuron loss in Parkinson's models. *Science* **313,** 324–328.

Eriksen, J. L., Przedborski, S., and Petrucelli, L. (2005). Gene dosage and pathogenesis of Parkinson's disease. *Trends Mol. Med.* **11,** 91–96.

Goedert, M. (2001). Alpha-synuclein and neurodegenerative diseases. *Nat. Rev. Neurosci.* **2,** 492–501.

Kirik, D., Annett, L. E., Burger, C., Muzyczka, N., Mandel, R. J., and Bjorklund, A. (2003). Nigrostriatal alpha-synucleinopathy induced by viral vector-mediated overexpression of human alpha-synuclein: A new primate model of Parkinson's disease. *Proc. Natl. Acad. Sci. USA* **100,** 2884–2889.

Kirik, D., Rosenblad, C., Burger, C., Lundberg, C., Johansen, T. E., Muzyczka, N., Mandel, R. J., and Bjorklund, A. (2002). Parkinson-like neurodegeneration induced by targeted overexpression of alpha-synuclein in the nigrostriatal system. *J. Neurosci.* **22,** 2780–2791.

Lakso, M., Vartiainen, S., Moilanen, A. M., Sirvio, J., Thomas, J. H., Nass, R., Blakely, R. D., and Wong, G. (2003). Dopaminergic neuronal loss and motor deficits in *Caenorhabditis elegans* overexpressing human alpha-synuclein. *J. Neurochem.* **86,** 165–172.

Lo Bianco, C., Ridet, J. L., Schneider, B. L., Deglon, N., and Aebischer, P. (2002). alpha-Synucleinopathy and selective dopaminergic neuron loss in a rat lentiviral-based model of Parkinson's disease. *Proc. Natl. Acad. Sci. USA* **99,** 10813–10818.

Maries, E., Dass, B., Collier, T. J., Kordower, J. H., and Steece-Collier, K. (2003). The role of alpha-synuclein in Parkinson's disease: Insights from animal models. *Nat. Rev. Neurosci.* **4,** 727–738.

Marin, S., Mansilla, S., Garcia-Reyero, N., Rojas, M., Portugal, J., and Pina, B. (2002). Promoter-specific inhibition of transcription by daunorubicin in *Saccharomyces cerevisiae*. *Biochem. J.* **368,** 131–136.

Olkkonen, V. M., and Ikonen, E. (2006). When intracellular logistics fails: Genetic defects in membrane trafficking. *J. Cell Sci.* **119,** 5031–5045.

Outeiro, T. F., and Lindquist, S. (2003). Yeast cells provide insight into alpha-synuclein biology and pathobiology. *Science* **302,** 1772–1775.

Rothstein, R. (1991). Targeting, disruption, replacement, and allele rescue: Integrative DNA transformation in yeast. *Methods Enzmol.* **194,** 281–301.

Rupp, S. (2002). *LacZ* assays in yeast. *Methods Enzmol.* **350,** 112–131.

Seabra, M. C., Mules, E. H., and Hume, A. N. (2002). Rab GTPases, intracellular traffic and disease. *Trends Mol. Med.* **8,** 23–30.

Schneider, J. C., and Guarente, L. (1991). Vectors for expression of cloned genes in yeast: Regulation, overproduction, and underproduction. *Methods Enzmol.* **194,** 373–388.

To, W. K., Fothergill, A. W., and Rinaldi, M. G. (1995). Comparative evaluation of macrodilution and alamar colorimetric microdilution broth methods for antifungal susceptibility testing of yeast isolates. *J. Clin. Microbiol.* **33,** 2660–2664.

Whitworth, A. J., Wes, P. D., and Pallanck, L. J. (2006). *Drosophila* models pioneer a new approach to drug discovery for Parkinson's disease. *Drug Discov. Today* **11,** 119–126.

Zhang, J. H., Chung, T. D., and Oldenburg, K. R. (1999). A simple statistical parameter for use in evaluation and validation of high throughput screening assays. *J. Biomol. Screen.* **2,** 67–73.

CHAPTER TWENTY-SIX

ANALYSIS OF RAB GTPASE AND GTPASE-ACTIVATING PROTEIN FUNCTION AT PRIMARY CILIA

Shin-ichiro Yoshimura, Alexander K. Haas, *and* Francis A. Barr

Contents

Abstract

Primary cilia are sensory structures on the cell surface whose formation requires tight integration of microtubules and polarized membrane trafficking events. Rabs are GTP-binding proteins of the Ras superfamily that control directed membrane trafficking by regulating membrane–membrane and membrane–cytoskeleton interactions. This chapter describes cell biological and biochemical methods for the analysis of Rab function and Rab GTPase-activating proteins during primary cilia formation.

1. INTRODUCTION

Primary cilia are sensory structures on the surface of mammalian cells involved in morphogen signaling during development, sensing liquid flow in the kidney, mechanosensation, sight, and smell (Badano *et al.*, 2006; Eggenschwiler and Anderson, 2006; Satir and Christensen, 2007) Mutations that affect primary cilia cause a number of diseases, including neural tube defect, polycystic disease, retinal degeneration, and cancers (Badano *et al.*, 2006;

University of Liverpool, Cancer Research Centre, Liverpool, United Kingdom

Methods in Enzymology, Volume 439
ISSN 0076-6879, DOI: 10.1016/S0076-6879(07)00426-0

Michaud and Yoder, 2006; Zariwala *et al.*, 2007) The primary cilium consists of a ninefold array of doublet microtubules closely wrapped by the plasma membrane. Unlike motile cilia, it lacks a central pair of doublet microtubules and is therefore referred to as a 9+0 axoneme. Cilium formation in mammalian cells and tissues has been described in detail by electron microscopy and can be divided into a number of discrete stages (Sorokin, 1968). First, the basal body moves to the cell surface; following this docking event, microtubules are nucleated from the basal body toward the plasma membrane, and cilium formation is then completed by coordinated microtubule and plasma membrane extension (Sorokin, 1962, 1968). It is therefore an interesting system in which to study the underlying principles of how microtubule and membrane dynamics cooperate to generate specific cellular structures.

This chapter focuses on the role of Rab proteins in the formation of primary cilia. Rab proteins are GTPases of the Ras superfamily that recruit specific effector protein to membrane surfaces when in the active GTP-bound form. These effector complexes are thought to define the identity of the specific membrane compartment due to their involvement in membrane–membrane recognition events and link membranes to the cytoskeleton (Gillingham and Munro, 2003; Munro, 2002). Both these classes of events are often referred to as tethering and are known to be important in vesicle-mediated trafficking. The best understood examples are the role of Rab5 and its effector protein EEA1 in endosome fusion (Christoforidis *et al.*, 1999) and Rab1 with its effector protein p115 in ER to Golgi transport vesicle fusion with the Golgi (Allan *et al.*, 2000) The guanine nucleotide-binding state of Rabs, and hence their specific activation and inactivation, is under the control of additional factors called GDP/GTP exchange factors (GEFs) and GTPase–activating proteins (GAPs). This chapter focuses on the 40 RabGAP proteins encoded in the human genome, each of which is thought to act on a specific Rab.

To identify Rabs implicated in specific cellular processes, this chapter describes the application to primary cilia using three powerful screening methods: (i) RabGAP overexpression screening to specifically inactivate endogenous Rabs, (ii) Rab localization in human telomerase-immortalized retinal-pigmented epithelial (hTERT-RPE1) cells, and (iii) biochemical GAP assay with all Rab proteins. By combining these methods, Rabs that act at or localize to primary cilia could be identified, and their role in primary cilium function further characterized. This resulted in the identification of three Rab proteins, Rab8a, Rab17, and Rab23, that function in primary ciliogenesis (Yoshimura *et al.*, 2007). In the case of Rab8a, its role in ciliogenesis was confirmed independently by the identification of Rabin8, a GEF for Rab8, as a binding partner for the core complex of Bardet–Biedl syndrome proteins known to be mutated in a number of ciliopathies (Nachury *et al.*, 2007). The details of these three screening methods and their application are described next. In addition, variations on these methods are also useful to pinpoint Rab function in other membrane transport pathways (Fuchs *et al.*, 2007; Haas *et al.*, 2007).

2. METHODS

2.1. Cloning of RabGAPs and Rabs

RabGAP proteins are defined by the conserved TBC (Tre2/Bub2/Cdc16) domain containing conserved residues required for catalyzing GTP hydrolysis. By searching the nucleotide, protein, and EST databases held at the National Center for Biotechnology Information with a TBC domain consensus built using yeast proteins, 40 genes encoded TBC domain-containing proteins were identified in humans. Of these, cDNA for 38 RabGAPs were obtained by polymerase chain reaction (PCR) methods using KOD DNA polymerase (Toyobo, Novagen) or pfu DNA polymerase (Stratagene) from human fetus, liver, and testis Marathon-Ready cDNA (Clontech Laboratories, Inc.) according to the manufacturer's standard protocol (Fuchs *et al.*, 2007; Haas *et al.*, 2005, 2007; Yoshimura *et al.*, 2007). TBC1D12 could not be amplified using standard PCR conditions because of its guanosine/cytosine-rich content; however, it could be amplified by PCR with KOD in the presence of 5% (vol/vol) dimethyl sulfoxide (DMSO). The cloning of Rabs is performed with the same protocol as RabGAPs, except for Rab12. The Rab12 gene has also a guanosine/cytosine region, and 7.5% (vol/vol) DMSO with *Taq* DNA polymerase (New England Biolabs.) is used to amplify it successfully, whereas PCR by KOD and pfu polymerase with or without DMSO is not successful.

The recent crystal structure of a Rab, together with a TBC domain, reveals that these two residues are important for the nucleotide hydrolysis reaction (Pan *et al.*, 2006). The glutamine is important for positioning the nucleophilic water molecule required for GTP hydrolysis, whereas the arginine stabilizes the developing charge on the β- and γ-phosphates of the bound GTP. Accordingly, catalytically inactive RabGAP mutants are created by site-directed mutagenesis of either one of these residues to alanine using the Quickchange method and pfu polymerase (Stratagene). Similarly, the conserved glutamine residues of the Rabs are mutated to alanine to generate hydrolysis-defective mutants.

All the primers for cloning and mutagenesis are described in Tables 26.1 and 26.2 (in order to view these tables, please refer to the URL: http://books. elsevier.com/companions/9780123743114); the restriction sites used for cloning are shown as uppercase letters. The wild-type and catalytically inactive mutant RabGAP and Rab cDNA are then subcloned into pCRII TOPO (Invitrogen) prior to transfer into pEGFP-C2 (Clontech) and pcDNA3.1/Myc (Invitrogen) for mammalian cell expression or into pQE32 (Qiagen) and pFAT2 (a His–GST-tagging vector modified from pGAT2 to have a polylinker compatible with pQE32).

2.2. Expression of Rabs and RabGAPs in hTERT-RPE1 cells

2.2.1. Cell culture and primary cilium formation

The hTERT-RPE1 cell line can be used for the analysis of primary cilium formation. The hTERT-RPE1 cell line can be obtained from Clontech Laboratories or the American Type Culture Collection (ATCC). Cells are grown at 37° and 5% CO_2 in a 1:1 mixture of dimethyl ether and Ham's F12 medium containing 10% calf serum, 2.5 mM L–glutamine, and 1.2 g/liter sodium bicarbonate. When these cells are cultured after serum withdrawal, they rapidly form primary cilia with 24 to 48 h. To do this, cells are seeded at 0.5 to 1.0 × 105 cells per well of a six-well plate and grown for 24 h. The growth medium is then removed and the cells are washed in phosphate-buffered saline (PBS) and replaced with growth medium lacking serum. Under these conditions, primary cilium formation is around 60% after 24 h and, after 48 h, has reached a maximum of around 80%. Other cell lines, for example, mouse NIH/3T3 cells and Madine–Darby canine kidney cells, also have primary cilia and can be used for similar experiments with only a slight variation of the methods described later.

2.2.2. RabGAP overexpression screening on primary cilium formation

To test the effects of RabGAPs on cilium formation, 0.5 μg of the pEGFP-C2 vector encoding the 38 RabGAPs is transfected by the Fugene 6 transfection regent (Roche) to 0.5 to 1.0 × 10^5 of hTERT-RPE1 cells on coverslips in separate wells of six-well plates. After 24 h the cells are washed three times with PBS and replaced with new culture medium as described earlier, except without calf serum. The cells are cultured for another 48 h to induce primary cilia formation. At this time point, the cells are treated on ice for 1 h to depolymerize cytoplasmic microtubules. The cells are fixed with ice-cold methanol for 4 min and then washed three times with PBS. The fixed cells are used for indirect immunofluorescence analysis with mouse antiacetylated tubulin antibody (6–11B-1; Sigma-Aldrich) as the primary antibody and Cy3-conjugated donkey antimouse secondary antibody (Jackson ImmunoResearch Laboratories). The cells are observed by fluorescence microscopy [Axioskop 2 with a 63× Plan Apochromat oil-immersion objective of NA 1.4, standard filter sets (Carl Zeiss MicroImaging, Inc.), and a 1300 × 1030 pixel-cooled, charge-coupled device camera (Model CCD-1300-Y; Princeton Instruments)]. RabGAP expression is monitored by EGFP fluorescence, and primary cilium formation is judged by Cy3-acetylated tubulin signals.

For the analysis, 100 RabGAP expressing cells are counted and plotted the number of ciliated cells to the graph. With seven RabGAP proteins, TBC1D3, TBC1D12, RUTBC1, RUTBC2, USP6, AK074305, and KIAA0882 reduce cell viability and increase levels of apoptosis; therefore, we exclude these proteins from the analysis of cilium formation. Using this method, we have shown that TBC1D7, EVI5-like, and XM_037557 cause significant reduction on primary cilia formation, from 80 to 5–40% (Yoshimura et al., 2007).

2.2.3. Rab localization and function in primary cilium in hTERT-RPE1 cells

To identify which Rab proteins localize to primary cilia, hTERT-RPE1 cells are transfected with the pEGFP-tagged Rab constructs described earlier and the same method is used for RabGAPs, except the primary cilium is left induced or induced for 72 h. Of all the Rabs tested, only Rab8a shows cilium localization after serum starvation (Fig. 26.1) (Yoshimura et al., 2007). Notably, Rab8a shows primary cilium localization but not Rab8b (Yoshimura et al., 2007). To examine the subcellular localization of endogenous Rab8a, we use the affinity-purified rabbit anti-Rab8a antibody, a kind gift of Johan Peränen (Institute of Biotechnology, University of Helsinki, Helsinki, Finland). This antibody is also used to check the depletion efficiency of Rab8a protein by siRNA.

For siRNA experiments, 2×10^4 of hTERT-RPE1 cells is plated per well of a six-well plate. The target sequences are as follows: control, CGUACGCGGAAUACUUCGA; Rab8a, CAGGAACGGUUUCGGA CGA, GAAUUAAACUGCAGAUAUG, GAACAAGUGUGAUGUG AAU. All siRNA duplexes are obtained from Dharmacon Inc. Cells are transfected with 3 μl 20 mM of the Rab8a siRNA duplex using 3 μl of the Oligofectamine transfection reagent (Invitrogen) mixed with 100 μl Optimem (Invitrogen). After 15 min of incubation, this mixture is added to the cells, which are then grown for 48 h in normal medium with 10% calf serum. After 48 h, the cells are washed three times with PBS, and fresh culture medium without serum is added and then transfected with the same duplex once again. After culture for a further 48 h, the cells are fixed and then costained with the acetylated tubulin and Rab8a antibodies.

The length of primary cilia in EGFP-Rab8a expression cells or Rab8a-depleted cells indicated by acetylated tubulin and EGFP signals is measured and calculated with the measurement tools in either Metavue (Molecular Devices, MDS Inc.) or Photoshop (Adobe Systems Inc.). The length of cilia is measured in at least 100 independent cells, and the significance is judged by a Students t test (Yoshimura et al., 2007).

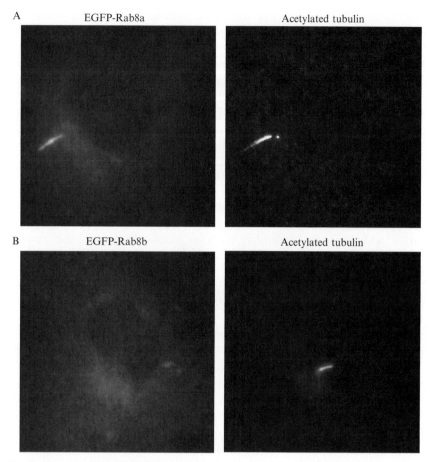

Figure 26.1 Rab8a and Rab8b localization to primary cilium. hTERT-RPE1 cells were transfected with EGFP-Rab8a and Rab8b, cultured for 24 h, and then serum starved for 72 h to induce primary cilia. Cells were fixed and then observed by immuno-fluorescence microscopy. (A) EGFP-Rab8a localized in primary cilia indicated by acetylated tubulin staining, whereas (B) EGFP-Rab8b did not.

2.3. *In vitro* RabGAP assays

2.3.1. Recombinant protein preparation of Rab and RabGAP proteins from bacteria

All cDNA of Rab are subcloned into the pFAT2 vector to generate hexahistidine (6xHis)–GST-tagged fusion proteins when expressed in bacteria. It is important to note that maltose-binding protein (MBP) fusions, while yielding high levels of expression, do not to give rise to Rabs with normal GTP-binding and hydrolysis activities. For this reason, we recommend using either GST or His-tagged Rab proteins. The vector is transformed to *Escherichia coli* BL 21(DE3 lysogen) containing the pRIL plasmid containing rare tRNAs for arginine, isoleucine, and leucine. Bacteria are plated on LB agar with 100 μg/ml of ampicillin to select pFAT2 and 34 μg/ml of chloramphenicol to select pRIL. The plates are kept at 37° for 12 h. Single colonies are then picked from each plate and transferred between 500 ml and 4 liters of LB liquid medium depending on the Rab with 100 μg/ml of ampicillin and 34 μg/ml of chloramphenicol and cultured at 3° to reach an OD_{600} of between 0.5 and 0.6 in a shaking incubator (Infors AG). The cultured is then transferred to an actively cooled 18° incubator and cultured for 1 h, and then protein expression is induced by the addition of 0.25 mM isopropyl-β-D-thiogalactoside (IPTG). The liquid cultures are grown for another 12 to 14 h at 18°. With most Rab proteins, except Rab8a, Rab8b, Rab10, and Rab13, an alternative quick expression protocol can be used. A single colony is picked and transferred into LB liquid medium with 100 μg/ml of ampicillin and 34 μg/ml of chloramphenicol and cultured at 37° to reach an OD_{600} of 0.5 to 0.6. Expression is then induced by adding 0.5 mM of IPTG and culturing for 3 h at 37°. Finally, bacteria are recovered by centrifugation for protein preparation.

For RabGAP proteins, TBC1D7, EVI5-like, and XM_037557 expression and purification from bacteria are tested as either MBP or His-tagged fusion proteins. While the yield of MBP fusion protein is high with each of these three RabGAP proteins, they do not show significant GAP activity with any Rab protein tested. Although MBP fusions of RabGAP-5, RN-tre, and TBC1D20 do show specific activity toward discrete target Rabs (Haas *et al.*, 2005, 2007), we therefore do not recommend use of the MBP tag for RabGAP expression. With the 6xHis-tagged proteins it is difficult to obtain a high yield from bacteria, as highly specifically active proteins are obtained.

TBC1D7, EVI5-like, and XM_037557 in pQE32 fuse the 6xHis tag and cleavage site of TEV protease to the N terminus of RabGAP proteins. These constructs are transformed into the *E. coli* JM109 (pRIL) strain and plated into LB agarose medium with 100 μg/ml of ampicillin and 34 μg/ml of chloramphenicol. The plates are kept overnight (12 h) to form colonies. A single colony is picked and transferred into 4 liters of LB liquid medium with 100 μg/ml of ampicillin and 34 μg/ml of chloramphenicol and then cultured at 37° to reach an OD_{600} of 0.5 to 0.6. Bacteria are subsequently cultured at 18° for 12 to 14 h, without any induction of protein expression by IPTG. The bacterial pellet is recovered by centrifugation, and the pellet is processed to the following protein purification step immediately. Rabs and RabGAPs are purified on 1-ml NTA-agarose columns (Qiagen) as described previously (Fuchs *et al.*, 2005). The peak fractions of eluted protein from NTA-agarose are dialyzed overnight in Tris-buffered saline (TBS: 50 mM Tris-HCl, pH 7.4, 150 mM NaCl) with 2 mM dithiothreitol (DTT). The protein concentration is measured using a standard Bradford protein assay kit (Bio-Rad).

With these methods, between 1 and 10 mg of Rab protein and 0.75 to 3 mg of the RabGAP proteins can be obtained. Rab proteins and RabGAP proteins should be aliquoted and frozen in liquid nitrogen or dry ice and stored at −80°. An example of EVI5 purification is shown (Fig. 26.2A).

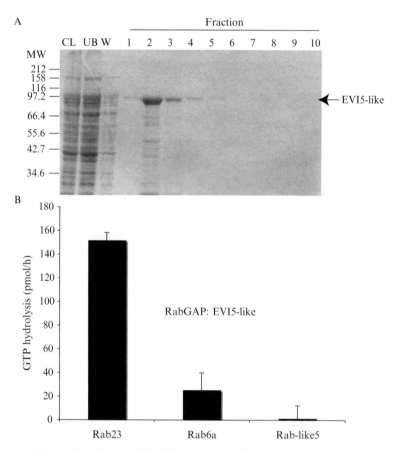

Figure 26.2 Purification profile of EVI5-like and measurement of EVI5-like GAP activity. Bacteria from a 4-liter culture of the JM109 (pRIL) strain transformed with pQE32 EVI5-like were homogenized, and a cleared lysate was prepared by centrifugation (A, CL). The cleared lysate was bound to 1 ml of NTA-agarose beads for 60 min and separated into NTA bead-bound and unbound fractions (A, UB). NTA-agarose beads were then washed with 20 m*M* imidazole (A, W), and the bound protein was eluted by 250 m*M* of imidazole. Ten 1-ml fractions were collected, and 10 μl from each fraction was loaded onto a 10% SDS-PAGE gel stained by Coomassie brilliant blue (A, fractions 1–10.) The proteins in fractions 2 and 3 were pooled, dialyzed, and used for biochemical GAP assays. (B) GAP assays were performed as described. EVI5-like shows specific activity with Rab23, but not with Rab6a and Rab-like 5.

2.3.2. Biochemical RabGAP assay with purified proteins

To measure GAP-stimulated GTP-hydrolysis, 100 pmol of Rab protein is mixed on ice with 10 μl of 10× assay buffer (500 mM HEPES-NaOH, pH 6.8, 10 mM DTT, 2 mg/ml bovine serum albumin), 10 μl of 10 mM EDTA (pH 8.0), 5 μl of 1 mM Mg^{2+}-GTP, and 2 μl of [γ-^{32}P]GTP (GE Healthcare PB10244: 10 mCi/ml; 5000 Ci/mmol or Perkin-Elmer NEG004Z250UC: 10 mCi/ml; 6000 Ci/mmol) and adjusted to 100 μl by dH$_2$O. GTP and [γ-^{32}P]GTP should not be freeze–thawed. The GTP-loading reaction is incubated for 15 min at 30° and is then kept briefly on ice for use in the following GAP reaction. For GAP assays, two 5-μl aliquots from the GTP-loading reaction are taken and immediately added to 795 μl of ice-cold 5% (wt/vol)-activated charcoal slurry in 50 mM NaH$_2$PO$_4$ and left on ice as the $t = 0$ background values. The Rab loading mix is then split into two equal halves. In one half, 0.5 to 10 pmol of GAP is added. This is the GAP reaction. In the other half, the equal volume of TBS is added; this reaction is used to calculate basal or GAP-independent GTP hydrolysis. Reactions are then incubated at 30°, taking 5-μl samples in duplicate at each time point (5, 10, 15, 45, and 60 min). The 5-μl aliquots are added immediately to 795 μl of ice-cold 5% (wt/vol)-activated charcoal slurry in 50 mM NaH$_2$PO$_4$, left for 1 h on ice, and then centrifuged at 16000 g_{av} in a bench-top microcentrifuge (Eppendorf 5417C with 30 space rotor and aerosol-resistant lid) to pellet the charcoal. The supernatant is carefully removed with a pipette, and 400-μl aliquots are scintillation counted in 4 ml of Ultima Gold scintillation liquid (Perkin-Elmer). A 2.5-μl aliquot of the assay mix is also scintillation counted to allow calculation of the specific activity of the reaction.

2.3.3. Calculation of GAP activity

To calculate the number of counts per picomole of GTP in the reaction mixture, the activity measured in 2.5 μl of assay mix is multiplied by 40 to obtain the total activity for the full assay volume of 100 μl. This value is then divided by the total of 5000 pmol GTP in the reaction to obtain the specific activity in cpm/pmol of GTP. The measured values of the 5-μl aliquots taken from each reaction in duplicate at each time point are then averaged and then multiplied by 2, as only 400 μl out of the 800-μl charcoal mix is measured by scintillation counting. Because 5 μl of the 100-μl reaction mix is measured, these values are then multiplied by 20. To calculate the amount of GTP hydrolyzed in picomoles, these values are then divided by the specific activity per picomole of GTP (cpm/pmol). The $t = 0$ values are then subtracted to eliminate the background of the reaction. The amount of GTP hydrolyzed by Rabs alone (-GAP) is then subtracted from the stimulated reactions (+GAP). Via this calculation the stimulation of GTP hydrolysis by all Rabs upon addition of the same GAP can be compared.

An example of EVI5-like activity toward Rab23, Rab6a, and Rabl5 is plotted in Fig. 26.2B. In biochemical RabGAP assays, EVI5-like, TBC1D7, and XM_037557 showed activity with Rab23, Rab17, and Rab8a, respectively, and these Rabs were therefore implicated in cilium formation (Yoshimura *et al.*, 2007).

2.4. Identification and analysis of Rab8a-specific effector proteins

2.4.1. Identification of cenexin/ODF2 variant 3 as Rab8a-specific effector protein

Cenexin was originally identified as a centriolar protein (Lange and Gull, 1995) and as a component of the outer dense fibers (ODF) of rat sperm (Brohmann *et al.*, 1997). These data suggest that cenexin/ODF2 has a function as a centriole/centrosome and microtubule-binding protein. Importantly, it was also reported that cenexin/ODF2 was involved in primary cilium formation in mouse F9 cells (Ishikawa *et al.*, 2005).

To identify effector proteins for Rab8a protein, yeast two-hybrid screening is performed using the Matchmaker system with a human testis cDNA library (Clontech) as described before (Fuchs *et al.*, 2005). With this approach, 13 truncated cenexin clones derived from the third splice variant of this gene are obtained. A full-length clone is reexamined by yeast two-hybrid analysis, and cenexin 3 interacts with Rab8a, but not with Rab8b (Yoshimura *et al.*, 2007).

To assess the subcellular localization of cenexin 3 proteins, cenexin 3 is subcloned into the pcDNA3.1 flag to introduce the flag tag to the N terminus of cenexin 3, and the mixture of 50 ng of this plasmid and 1 μg pBluescript-II is transfected with hTERT-RPE1 cells and analysis the localization by costaining by the rabbit anti-flag antibody (F7425; Sigma-Aldrich) and acetylated tubulin or γ-tubulin antibody (GTU88; Sigma-Aldrich). When cenexin is transfected with normal plasmid amounts (0.5–1 μg), it aggregates and causes abnormal microtubule bundling (Nakagawa *et al.*, 2001). The method described earlier retains high transfection efficiency, but results in much lower expression per cell, and thus avoids these cenexin aggregation artifacts.

2.4.2. Analysis of the Rab8a–cenexin interaction

For binding assays, recombinant purified cenexin 3 protein is required; however, the full length of cenexin 3 could not be obtained in soluble form from bacteria. A C-terminal domain of cenexin 3 (amino acids 397–657) encoded in pQE32 is obtained successfully as a soluble form by the same methods as RabGAP protein preparation. For binding assays, 10 μg of GST-Rab proteins (Rab1b, Rab8a, Rab8b, Rab10, Rab13, and Rab35) is bound to 25 μl of packed glutathione-Sepharose (GE Healthcare) in 1 ml total volume of PBS for 60 min at 4°. The beads are first washed three times in 500 μl with nucleotide exchange buffer [20 mM HEPES-NaOH, pH 7.5, 100 mM NaOAc, 10 mM EDTA, and 0.1% (vol/vol) NP-40], followed by two washes in 500 μl nucleotide loading buffer [NL100: 20 mM HEPES-NaOH, pH 7.5, 100 mM NaOAc, 0.1 mM MgCl$_2$, and 0.1% (vol/vol) NP-40]. The beads are then resuspended in 200 μl NL100, and 20 μl of 100 mM GTP and 10 μg of cenexin 3 (amino acids 397–657) are added. Binding is then allowed to proceed for 60 min at 4°, rotating to mix. The beads are then washed three times with 500 μl NL100, and bound proteins are eluted by the addition of elution buffer [20 mM HEPES-NaOH, pH 7.5, 200 mM NaCl, 20 mM EDTA, and 0.1% (vol/vol) NP-40], rotating at 4° for 15 min. Beads are pelleted by centrifugation at 2000 g for 1 min, and the supernatant is transferred to a fresh tube. To remove contaminating Rabs, 50 μl of packed glutathione-Sepharose is added to the eluate and incubated for 10 min at 4° with mixing. The beads are pelleted by centrifugation at 2000 g for 1 min, and the supernatant is transferred to a fresh tube. This procedure is repeated three times. Eluted proteins are then precipitated using trichloracetic acid and analyzed on 12% minigels stained with Coomassie brilliant blue.

REFERENCES

Allan, B. B., Moyer, B. D., and Balch, W. E. (2000). Rab1 recruitment of p115 into a cis-SNARE complex: Programming budding COPII vesicles for fusion. *Science* **289**(5478), 444–448.

Badano, J. L., Mitsuma, N., Beales, P. L., and Katsanis, N. (2006). The ciliopathies: An emerging class of human genetic disorders. *Annu. Rev. Genomics Hum. Genet.* **7**, 125–148.

Brohmann, H., Pinnecke, S., and Hoyer-Fender, S. (1997). Identification and characterization of new cDNAs encoding outer dense fiber proteins of rat sperm. *J. Biol. Chem.* **272** (15), 10327–10332.

Christoforidis, S., McBride, H. M., Burgoyne, R. D., and Zerial, M. (1999). The Rab5 effector EEA1 is a core component of endosome docking. *Nature* **397**(6720), 621–625.

Eggenschwiler, J. T., and Anderson, K. V. (2006). Cilia and developmental signaling. *Annu. Rev. Cell Dev. Biol.* **23**, 345–373.

Fuchs, E., Haas, A. K., Spooner, R. A., Yoshimura, S., Lord, J. M., and Barr, F. A. (2007). Specific Rab GTPase-activating proteins define the Shiga toxin and epidermal growth factor uptake pathways. *J. Cell Biol.* **177**(6), 1133–1143.

Fuchs, E., Short, B., and Barr, F. A. (2005). Assay and properties of rab6 interaction with dynein-dynactin complexes. *Methods Enzymol.* **403**, 607–618.

Gillingham, A. K., and Munro, S. (2003). Long coiled-coil proteins and membrane traffic. *Biochim. Biophys. Acta* **1641**(2–3), 71–85.

Haas, A. K., Fuchs, E., Kopajtich, R., and Barr, F. A. (2005). A GTPase-activating protein controls Rab5 function in endocytic trafficking. *Nat. Cell Biol.* **7**(9), 887–893.

Haas, A. K., Yoshimura, S., Stephens, D. J., Preisinger, C., Fuchs, E., and Barr, F. A. (2007). Analysis of GTPase-activating proteins: Rab1 and Rab43 are key Rabs required to maintain a functional Golgi complex in human cells. *J. Cell Sci.* **120**(Pt 17), 2997–3010.

Ishikawa, H., Kubo, A., Tsukita, S., and Tsukita, S. (2005). Odf2-deficient mother centrioles lack distal/subdistal appendages and the ability to generate primary cilia. *Nat. Cell Biol.* **7**(5), 517–524.

Lange, B. M., and Gull, K. (1995). A molecular marker for centriole maturation in the mammalian cell cycle. *J. Cell Biol.* **130**(4), 919–927.

Michaud, E. J., and Yoder, B. K. (2006). The primary cilium in cell signaling and cancer. *Cancer Res.* **66**(13), 6463–6467.

Munro, S. (2002). Organelle identity and the targeting of peripheral membrane proteins. *Curr. Opin. Cell Biol.* **14**(4), 506–514.

Nachury, M. V., Loktev, A. V., Zhang, Q., Westlake, C. J., Peränen, J., Merdes, A., Slusarski, D. C., Scheller, R. H., Bazan, J. F., Sheffield, V. C., and Jackson, P. K. (2007). A core complex of BBS proteins cooperates with the GTPase Rab8 to promote ciliary membrane biogenesis. *Cell* **129**(6), 1201–1213.

Nakagawa, Y., Yamane, Y., Okanoue, T., Tsukita, S., and Tsukita, S. (2001). Outer dense fiber 2 is a widespread centrosome scaffold component preferentially associated with mother centrioles: Its identification from isolated centrosomes. *Mol. Biol. Cell* **12**(6), 1687–1697.

Pan, X., Eathiraj, S., Munson, M., and Lambright, D. G. (2006). TBC-domain GAPs for Rab GTPases accelerate GTP hydrolysis by a dual-finger mechanism. *Nature* **442**(7100), 303–306.

Satir, P., and Christensen, S. T. (2007). Overview of structure and function of mammalian cilia. *Annu. Rev. Physiol.* **69**, 377–400.

Sorokin, S. (1962). Centrioles and the formation of rudimentary cilia by fibroblasts and smooth muscle cells. *J. Cell Biol.* **15**, 363–377.

Sorokin, S. P. (1968). Reconstructions of centriole formation and ciliogenesis in mammalian lungs. *J. Cell Sci.* **3**(2), 207–230.

Yoshimura, S., Egerer, J., Fuchs, E., Haas, A. K., and Barr, F. A. (2007). Functional dissection of Rab GTPases involved in primary cilium formation. *J. Cell Biol.* **178**(3), 363–369.

Zariwala, M. A., Knowles, M. R., and Omran, H. (2007). Genetic defects in ciliary structure and function. *Annu. Rev. Physiol.* **69**, 423–450.

CHAPTER TWENTY-SEVEN

RHO GTPASES AND REGULATION OF HEMATOPOIETIC STEM CELL LOCALIZATION

David A. Williams,[‡] Yi Zheng,[*] and Jose A. Cancelas[*,†]

Contents

Abstract

Bone marrow engraftment in the context of hematopoietic stem cell and progenitor (HSC/P) transplantation is based on the ability of intravenously administered cells to lodge in the medullary cavity and be retained in the appropriate marrow space, a process referred to as homing. It is likely that homing is a multistep process, encompassing a sequence of highly regulated events that mimic the migration of leukocytes to inflammatory sites. In leukocyte biology, this process includes an initial phase of tethering and rolling of cells to the endothelium via E- and P-selectins, firm adhesion to the vessel wall via integrins that appear to be activated in an "inside-out" fashion, transendothelial

[*] Division of Experimental Hematology, Cincinnati Children's Research Foundation, Cincinnati Children's Hospital Medical Center, Cincinnati, Ohio
[†] Hoxworth Blood Center, University of Cincinnati College of Medicine, Cincinnati, Ohio
[‡] Division of Hematology/Oncology, Childrens Hospital Harvard Medical School, Boston, MA

Methods in Enzymology, Volume 439
ISSN 0076-6879, DOI: 10.1016/S0076-6879(07)00427-2

migration, and chemotaxis through the extracellular matrix (ECM) to the inflammatory nidus. For HSC/P, the cells appear to migrate to the endosteal space of the bone marrow. A second phase of engraftment involves the subsequent interaction of specific HSC/P surface receptors, such as $\alpha_4\beta_1$ integrin receptors with vascular cell–cell adhesion molecule-1 and fibronectin in the ECM, and interactions with growth factors that are soluble, membrane, or matrix bound. We have utilized knockout and conditional knockout mouse lines generated by gene targeting to study the role of Rac1 and Rac2 in blood cell development and function. We have determined that Rac is activated via stimulation of CXCR4 by SDF-1, by adhesion via β_1 integrins, and via stimulation of c-kit by the stem cell factor—all of which involved in stem cell engraftment. Thus Rac proteins are key molecular switches of HSC/P engraftment and marrow retention. We have defined Rac proteins as key regulators of HSC/P cell function and delineated key unique and overlapping functions of these two highly related GTPases in a variety of primary hematopoietic cell lineages *in vitro* and *in vivo*. Further, we have begun to define the mechanisms by which each GTPase leads to specific functions in these cells. These studies have led to important new understanding of stem cell bone marrow retention and trafficking in the peripheral circulation and to the development of a novel small molecule inhibitor that can modulate stem cell functions, including adhesion, mobilization, and proliferation. This chapter describes the *biochemical footprint* of stem cell engraftment and marrow retention related to Rho GTPases. In addition, it reviews abnormalities of Rho GTPases implicated in human immunohematopoietic diseases and in leukemia/lymphoma.

1. BASIC MECHANISMS OF HEMATOPOIETIC STEM CELL AND PROGENITOR (HSC/P) HOMING AND RETENTION IN BONE MARROW (BM)

Bone marrow engraftment in the context of HSC/P transplantation is based on the ability of intravenously administered cells to lodge in the medullary cavity and be retained in the appropriate marrow space, a process referred to as homing. It is likely that homing is a multistep process, encompassing a sequence of highly regulated events that mimic the migration of leukocytes to inflammatory sites. In leukocyte biology, this process includes an initial phase of tethering and rolling of cells to the endothelium via E- and P-selectins, firm adhesion to the vessel wall via integrins that appear to be activated in an "inside-out" fashion, transendothelial migration, and chemotaxis through the extracellular matrix (ECM) to the inflammatory nidus (Butcher and Picker, 1996; Peled *et al.*, 1999a; Springer, 1994). For HSC/P, the cells appear to migrate to the endosteal space of the bone marrow (Driessen *et al.*, 2003; Gong, 1978; Nilsson *et al.*, 2001; Wilson and Trumpp, 2006). A second phase of engraftment involves the subsequent interaction of specific HSC/P surface receptors, such as $\alpha_4\beta_1$ integrin receptors with vascular cell–cell adhesion molecule-1 (VCAM-1) and fibronectin in the ECM, and interactions with growth factors that are soluble, membrane, or matrix bound (Williams *et al.*, 1991a,b). HSC/P can be temporarily detected in other organs such as liver, lung, and kidneys after intravenous infusion but disappear from these sites within 48 h after transplantation. In contrast, the retention of HSC/P in BM is sustained and appears specific (Papayannopoulou *et al.*, 2001a).

Some of the factors that influence this specific retention of HSC/P in the bone marrow have been defined recently and appear to involve the interplay among chemokines, growth factors, proteolytic enzymes, and adhesion molecules (Papayannopoulou, 2003).

Among the chemokines, stromal derived factor-1α (SDF-1α) and its receptor, the G-protein-coupled seven-span transmembrane receptor, CXCR4, play key roles in HSC trafficking and repopulation (Lapidot and Kollet, 2002). SDF-1α is expressed by both human and murine BM endothelium and stroma (Nagasawa *et al.*, 1998; Peled *et al.*, 1999a) and acts as a powerful chemoattractant of HSC/P (Aiuti *et al.*, 1997; Wright *et al.*, 2002). SDF-1α may also regulate the survival of HSC/P (Broxmeyer *et al.*, 2003; Lataillade *et al.*, 2000). SDF-1α induces the integrin-mediated firm arrest of hematopoietic progenitor cells and facilitates their transendothelial migration (Peled *et al.*, 1999a, 2000) and regulates HSC/P homing (Kollet *et al.*, 2001) and BM engraftment (Peled *et al.*, 1999b). Furthermore, SDF-1α is also required for the retention of murine HSC/P within the BM (Ma *et al.*, 1999; Nagasawa *et al.*, 1996). We have demonstrated that Rac proteins are activated by SDF-1 in HSC/P and that Rac-deficient HSC/P do not respond to SDF-1 (Fig. 27.1).

Among growth factors, a critical component for HSC/P survival and engraftment is the stem cell factor (SCF), which is expressed on BM stromal cells and is the ligand for the receptor tyrosine kinase, c-kit. A transmembrane isoform of SCF, membrane-bound SCF (membrane, mSCF), has been shown to be critical in the lodgment and retention of HSC within the hematopoietic microenvironment, although it does not appear to play a role in the homing of transplanted cells to BM (Driessen *et al.*, 2003). In addition, it appears that c-kit activation is differentially affected by soluble versus membrane-bound SCF and that mSCF appears to enhance maintenance of long-term hematopoiesis *in vitro* (Miyazawa *et al.*, 1995; Toksoz *et al.*, 1992) and induces overexpression of CXCR4 (Kollet *et al.*, 2001; Peled *et al.*, 1999b). Our studies of SCF-stimulated cell proliferation demonstrate that Rac activation is a critical component of c-kit signaling (see Fig. 27.1).

A third factor important for homing and engraftment of HSC/P are the integrin-mediated adhesion molecules. Among them, β_1 integrins are probably the best characterized, and inhibition of integrin function leads to defective medullary engraftment (Papayannopoulou and Craddock, 1997;

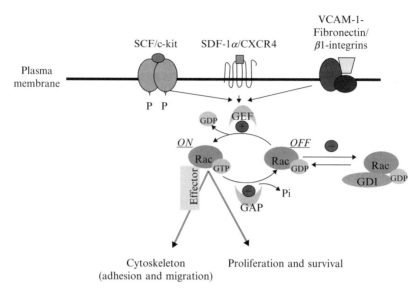

Figure 27.1 Rac GTPases integrate signals from multiple surface receptors involved in HSC engraftment and retention. (See color insert.)

Papayannopoulou *et al.*, 1995, 2001a; Scott *et al.*, 2003; van der Loo *et al.*, 1998; Williams *et al.*, 1991a). As detailed later, Rac-deficient HSC/P show significantly defective adhesion to fibronectin (Cancelas *et al.*, 2005; Gu *et al.*, 2003; Yang *et al.*, 2001) (see Fig. 27.1).

2. BASIC MECHANISMS OF HSC/P MOBILIZATION AND TRAFFICKING

Hematopoietic stem cell and progenitor mobilization is also a dynamic and complex process (Kronenwett *et al.*, 2000; Rafii *et al.*, 2002; Thomas *et al.*, 2002). HSC/P must exit the stem cell niche in the BM (presumably; see comments later), migrate through the marrow sinusoidal endothelium, and gain access to the blood. Circulating HSC/P and BM-adherent HSC/P appear to be interchangeable. Studies utilizing parabiotic mice have demonstrated that HSC/P can leave their niche without induction, traffic through the bloodstream, and finally migrate into BM of the conjoined animal (Abkowitz *et al.*, 2003; Wang *et al.*, 2003; Warren *et al.*, 1960; Wright *et al.*, 2001). This suggests that HSC/P trafficking is a physiological process. If so, circulating HSC/P would be predicted to move into the BM microenvironment through transendothelial migration directed by chemoattractants and ultimately anchor within the extravascular BM space where proliferation and differentiation occur. In this process, adhesion molecules, chemokine receptors, and integrin signaling require signal integration that drives cytoskeleton rearrangements and regulates gene expression, cell survival, and cell cycle activation. An additional proposed HSC location in the marrow is the vascular niche, where HSC would be attached to the fenestrated endothelium of the BM specialized vessels, so-called sinusoids (Kiel et al., 2005). Such an outlook is supported by evidence that Rac-deficient HSC with profoundly defective cell migration due to loss of the combined function of CXCR4, β_1 integrins, and c-kit signaling pathways can be mobilized in large numbers (Cancelas et al., 2005) [for a commentary, see Cancelas et al. (2006)]. The BM sinusoids express molecules important for HSC mobilization, homing, and engraftment, including chemokines such as CXCL12 (ligand for CXCR4) and adhesion molecules such as endothelial-cell (E)-selectin and vascular cell-adhesion molecule 1 (VCAM-1). These findings have given additional microanatomical clarity to the concept of stem cell niches as spatial structures in which HSC reside, self-renew, and differentiate.

At the molecular level, the interaction between SDF-1α and the G-coupled chemokine receptor CXCR4 has been recognized as pivotal in stem cell mobilization. As HSC/P are known to migrate toward a SDF-1α (Sweeney et al., 2002), it has been suggested that treatment with granulocyte colony-stimulating factor-1 (G-CSF), cyclophosphamide, or interleukin (IL)-8 leads to a reduction of SDF-1β in BM, resulting in a positive gradient in blood and induction of HSC/P migration toward PB. Raising the plasma levels of SDF-1α by intravenous injection of SDF-1α-expressing adenovirus (Hattori et al., 2001) or sulfated polysaccharides (Sweeney et al., 2002) or by inhibition of the CXCR4 receptor (Devine et al., 2004; Liles et al., 2003; Tavor et al., 2004) leads to mobilization of HSC/P. G-protein inhibition by pertussis toxin (Papayannopoulou et al., 2003) induces a similar mobilization effect, probably by interfering with the CXCR4 signaling pathway. It has been suggested that bone expression of CXCL12 (another ligand for CXCR4) is regulated by G-CSF-induced β_2-adrenergic signals that modify osteoblast protein expression and shape (Katayama et al., 2006).

Functional blocking of $\alpha_4\beta_1$ integrin (receptor for VCAM-1 and fibronectin) alone or together with $\alpha_1\beta_2$ integrins or the functional blocking of the β_2 integrin leukocyte function-associated antigen-1 by antibodies results in the mobilization of HSC/P (Craddock et al., 1997; Papayannopoulou et al., 2001b). HSC/P accumulate in the PB soon after gene deletion in inducible $\alpha_4\beta_1$ integrin-deficient mice. Although their numbers gradually stabilize at a lower level, progenitor cell influx into the circulation continues at above-normal levels for more than 50 weeks with a concomitant progressive accumulation of spleen HSC/P (Scott et al., 2003).

Playing an important and independent role in HSC/P mobilization is the interaction between SCF and its receptor, c-kit. SCF/c-kit interaction plays a critical role in G-CSF-mediated mobilization (Heissig *et al.*, 2002; Levesque *et al.*, 2003), and SCF in combination with G-CSF has been shown to enhance HSC/P mobilization (McNiece and Briddell, 1995). As mentioned earlier, tm-SCF has been shown to be critical in retaining HSC/P in BM (Driessen *et al.*, 2003).

3. RHO GTPASES

Almost all Rho family GTPases influence actin polymerization within the cell via specific or shared effectors and are thereby implicated in reorganization of the cytoskeleton, migration, and adhesion. However, Rho proteins regulate a multitude of other cellular functions. Among these are apoptosis and survival, cell cycle progression, and genomic stability. The activity of individual Rho proteins can be regulated by multiple guanine exchange factors (GEFs) and GTPase-activating proteins (GAPs), which are cell and agonist specific. Indeed, Rho GTPases such as Rac appear to integrate signaling from multiple receptors in individual cells. For instance, as mentioned earlier and detailed later, we have shown in hematopoietic cells that Rac is activated by stimulation of CXCR4 via SDF-1, adhesion via β_1 integrins, and stimulation of c-kit via SCF—all pathways involved in stem cell engraftment (Cancelas *et al.*, 2005; Gu *et al.*, 2003; Yang *et al.*, 2001). In addition, Rho family members recognize both unique and shared effectors. This, at least in part, explains the diversity of cellular functions influenced by a single Rho GTPase but also presents significant complexities in developing an understanding of the physiological roles of these proteins, particularly if studies utilize cell lines and expression of dominant negative (DN) or constitutive active (CA) mutants, which generally lack specificity among related GTPases. We have exploited mouse knockouts to study the function of Rac GTPases in hematopoiesis in an attempt to circumvent the problems associated with these DN and CA mutants.

Using genetic approaches (primarily gene targeting in mice), we have specifically implicated PAK, POR1, and STAT5 in Rac effector functions in primary hematopoietic cells and, depending on the specific lineage and agonist, found that Rac can activate p42/p44 and p38 ERKs, JNK, and Akt kinases (Cancelas *et al.*, 2005; Carstanjen *et al.*, 2005; Gu *et al.*, 2003; Roberts *et al.*, 1999; Yang *et al.*, 2000, 2001) (and preliminary data). In a similar manner to the Wiskott–Aldrich syndrome protein (WASp), a key downstream target of Rac, WAVE1/2 and insulin receptor substrate (IRS) p53, has been implicated by others (primarily Takenawa and Miki, 2001) in actin polymerization and assembly. IRSp53 is a linker between Rac and WAVE1/2 that adds specificity for the actin-related protein (Arp)2/3 complex activator in actin polymerization. WAVE1 is required for Rac-mediated dorsal membrane ruffling, whereas WAVE2 may be involved in Rac-induced peripheral ruffling during cell migration (Miki et al., 2000; Suetsugu et al., 2003). The induction of actin polymerization by WAVE is dependent on Arp2/3 via the VCA (verprolin homology, cofilin homology, acidic) domain of WAVE1/2 in a manner analogous to WASp. The VCA domain is a G-actin and Arp2/3-binding domain required for Rac/Cdc42-induced de novo actin nucleation and actin polymerization (reviewed in Takenawa and Miki, 2001). While the physiological relevance of these molecular links to Rac remains unknown, de novo actin nucleation appears critical for actin assembly at the leading edge of migrating cells and we hypothesize that this is of particular relevance to the migration of hematopoietic cells. In addition (and likely also relevant to the BM microenvironment), WAVE is essential for cell migration mediated through ECM in mouse embryo fibroblasts (Eden et al., 2002).

Thus, the combination of unique and shared upstream activators and downstream effectors, which may be cell type specific, represents an important mechanism by which the same Rho GTPase regulates a variety of the aforementioned cellular processes. The utilization of a genetic approach and the study of primary cells in murine models have contributed greatly to a new understanding of both unique and overlapping functions of Rho GTPases in hematopoiesis.

A significant proportion of known Rho GTPases is expressed ubiquitously, while two Rho proteins show tissue-specific expression. This is particularly important in hematopoietic cells and has been studied extensively for the Rac subfamily. All members of this subfamily (Rac1, Rac2, and Rac3) show high sequence similarity. Rac1 is expressed ubiquitously, whereas Rac3 is expressed in nearly all cell lines examined thus far and is expressed at high levels in murine heart, placenta, pancreas, and brain. Rac2 is expressed in a hematopoietic-restricted fashion. Thus, hematopoietic cells are unique in that all three Rac proteins are coexpressed and also express RhoH, the only other identified hematopoietic-specific Rho GTPase. RhoH has been shown by us and others to modulate Rac signaling (Gu et al., 2005b, 2006; Li et al., 2002a). Despite this expression pattern and their significant homology, individual Rac proteins are responsible for unique functions in hematopoietic cells, as gene-targeted mice deficient in each protein have measurable and distinct phenotypes. This has been well documented by our group and others using gene-targeted mice to examine the role of Rac1 versus Rac2 in hematopoietic cells (Fig. 27.2) (and see later). Genetic deletion of Rac2 leads to a number of phenotypic changes in multiple hematopoietic lineages, including granulocytes (Abdel-Latif et al., 2004; Carstanjen et al., 2005; Filippi et al., 2004; Glogauer et al., 2003; Kim and Dinauer, 2001; Lacy et al., 2003; Li et al., 2002a; Roberts et al., 1999), B cells (Croker et al., 2002b; Walmsley et al., 2003), T cells (Croker et al., 2002a),

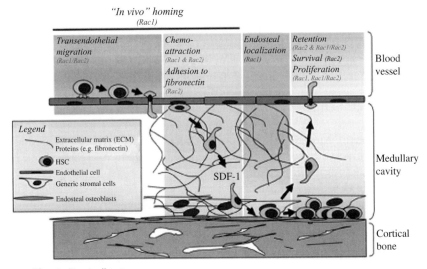

Figure 27.2 Unique and overlapping roles of Rac1 versus Rac2 in hematopoiesis. Putative utilization of Rac1 and Rac2 in processes involved in homing, migration, endosteal localization, retention, and proliferation/survival is shown. (See color insert.)

mast cells (Gu *et al.*, 2002; Tan *et al.*, 2003; Yang *et al.*, 2003), eosinophils (Fulkerson *et al.*, 2005), and platelets (Akbar *et al.*, 2006), despite continued expression (and even a compensatory increase in expression in some cases) of Rac1. Considerable evidence shows that Rac2 and Rac1 regulate both separable and overlapping functions in nearly all lineages on the hematopoietic cells examined (Filippi *et al.*, 2004; Gu *et al.*, 2003; Walmsley *et al.*, 2003). Thus, Rac1 cannot compensate for the loss of Rac2 function in hematopoietic cells and vice versa.

In addition, Rho GTPases from different subgroups appear to demonstrate cross talk to regulate cellular responses. In previous studies, primarily in fibroblasts, introduction of constitutive active or dominant negative mutants of Cdc42, Rac, and RhoA was shown to affect activation or inhibition of each other. Growth factor receptor-induced activation of Cdc42 has been shown to activate Rac, which in turn stimulates Rho activity, resulting in cytoskeletal remodeling. RhoE has been shown to downregulate the activity of RhoA by activating p190RhoGAP. Similarly, as mentioned earlier, RhoH has been shown to repress Rac activity in lymphoid cell lines (Li *et al.*, 2002a) and cytokine-stimulated hematopoietic progenitor cells, resulting in reduced proliferation, increased apoptosis, and defective actin polymerization (Gu *et al.*, 2005a). In addition, we have demonstrated that expression of a patient-derived dominant negative Rac

GTPase, D57N, in hematopoietic cells not only inhibits Rac1 and Rac2 activities, but may also inhibit Cdc42 (Y. Gu and D. Williams, unpublished results). Additional genetic studies have demonstrated cross-talk between Rac and both Cdc42 (Yang *et al.*, 2001) and RhoA (Filippi *et al.*, manuscript in preparation).

4. ROLE OF RAC1 AND RAC2 GTPASES IN HEMATOPOIESIS

Using gene-targeted mice, it has become evident that the Rho family of GTPases plays an important role in hematopoietic stem cell function. Rac activity has been demonstrated to be important for such diverse functions as retention in the bone marrow (Cancelas *et al.*, 2005; Yang *et al.*, 2001), long-term engraftment of HSC (Jansen *et al.*, 2005), and HSC mobilization (Gu *et al.*, 2003). Furthermore, in more committed hematopoietic cells, Rac activity is associated with B-lymphocyte development and signaling (Croker *et al.*, 2002b; Walmsley *et al.*, 2003), granulocyte chemotaxis and superoxide production (Abdel-Latif *et al.*, 2004; Carstanjen *et al.*, 2005; Filippi *et al.*, 2004; Glogauer *et al.*, 2003; Kim and Dinauer, 2001; Lacy *et al.*, 2003; Li *et al.*, 2002a; Roberts *et al.*, 1999), migration and degranulation of mast cells (Gu *et al.*, 2002; Yang *et al.*, 2000), differentiation of mature osteoblasts (Lax *et al.*, 2004), and maturation of TRAP-positive, pro-osteoclasts into multinucleated osteoclasts (Korhonen *et al.*, manuscript in preparation).

Rac1-deficient HSC/P stimulated with SCF demonstrate defective proliferative signaling from the c-kit receptor tyrosine kinase *in vitro* (Gu *et al.*, 2003). In contrast, loss of Rac2 activity leads to a pro-apoptotic phenotype in both mast cells and HSC/P in the presence of SCF (Gu *et al.*, 2002, 2003; Yang *et al.*, 2000). Rac integrates signals from β_1 and β_2 integrins and c-kit in HSC/P and mast cells (Gu *et al.*, 2002, 2003; Tan *et al.*, 2003). Signaling of c-kit to Rac is mediated through the GEF Vav (see later and preliminary data), although the specific Vav responsible for signaling to Rac in nonlymphoid hematopoietic cells remains largely unknown. Thus, overall, studies utilizing mouse mutants implicate Rac proteins downstream of CXCR4, c-kit, and β_1 and β_2 integrins, and Rac-deficient hematopoietic cells show loss of adhesion, migration, degranulation, changes in cell shape consistent with deregulated actin assembly, and defects in cell proliferation and survival linked to alterations in kinase pathways that are both lineage and agonist specific (for a complete review, see Cancelas *et al.*, 2006). Rac GTPases are thus important molecular switches controlling stem cell localization and retention in the marrow microenvironment, engraftment, and reconstitution in transplanted mice. These proteins represent a novel molecular target to modulate hematopoietic cell functions (Nasser *et al.*, 2006), and we have developed a first-generation, small molecule inhibitor, NSC23766, which induces mobilization of HSC/P (Cancelas *et al.*, 2006).

5. RAC3 GTPASE, A NEWLY DEFINED MEMBER OF THE RAC FAMILY CLONED FROM A BCR-ABL TRANSFORMED CELL LINE

Rac3 is a third member of the Rac subfamily, which was originally identified from a chronic myelogenous leukemia cell line, and has been implicated in human breast cancer (Baugher *et al.*, 2005; Mira *et al.*, 2000), ovarian cancer (Morris *et al.*, 2000), cellular transformation (Keller *et al.*, 2005), and tumor invasion (Chan *et al.*, 2005). Rac3 has been shown to interact with the integrin-binding protein CIB and promotes integrin-mediated adhesion and spreading in immortalized cell lines. In addition, Rac3 has been shown to be expressed differentially during myeloid differentiation (U. Knaus, personal communication). Rac3 null mutant mice have been reported (Cho *et al.*, 2005; Corbetta *et al.*, 2005) and are viable, fertile, and without obvious physical anomalies. One group has reported a mild neurological phenotype (Corbetta *et al.*, 2005). In addition, in a p190 Bcr-abl transgenic mouse model of acute lymphoblastic leukemia, Rac3, but not Rac1or Rac2, is activated and Rac3 deficiency attenuates the development of leukemia in female mice (Cho *et al.*, 2005). Data suggest that in the absence of Rac1 and Rac2, Rac3 can mediate an attenuated myelodysplastic phenotype in mice transplanted with p210 Bcr-abl (see preliminary data). However, no systematic analysis of hematopoiesis has been reported in Rac3-/- mice. Using reverse transcriptase polymerase chain reaction (RT-PCR) and Northern blot analysis, Burkhalter and co-workers (2002) have reported that the expression of Rac3 expression is downregulated dramatically during terminal myeloid differentiation. The functional significance of this observation has not been reported. We have demonstrated normal neutrophil differentiation in Rac1-/-; Rac2-/- cells (Filippi *et al.*, 2004; Gu *et al.*, 2003), but observed abnormal myeloid development *in vitro* after transduction of HSC/P with the dominant negative D57NRac2, which most likely inhibits Rac3 in addition to Rac1 and Rac2 (Tao *et al.*, 2002). These data suggest that Rac3 may be important in myelopoiesis, but suffer from the weaknesses of the use of DN mutants, the effects of which are not specific and are determined in part by expression levels. Interestingly, HSC/P expressing D57NRac2 fail to reconstitute hematopoiesis when transplanted into lethally irradiated recipients (Gu *et al.*, 2002).

6. Cdc42 in Hematopoiesis

Cdc42 has been linked for some time with gradient sensing and filopodia (Ridley and Hall, 1992). Until very recently, most work in hematopoietic cells utilized macrophage cell lines or examined the role of Cdc42 in lymphocytes, and interest in lymphocytes is derived in part because of the association of the Cdc42 target WASp in the human immunodeficiency disease of the same name (Symons *et al.*, 1996). In macrophages and myeloid cell lines, DN Cdc42 expression or Cdc42 inhibition in cell lines is associated with a lack of polarization in response to the growth factor/chemotactic factor CSF-1, leading to reduced directed but not random migration (Allen *et al.*, 1998; Srinivasan *et al.*, 2003). In monocytes, either constitutive active or DN Cdc42 expression leads to reduced migration across the endothelium (Weber *et al.*, 1998). In primary T cells, DN Cdc42 reduces chemotaxis in response to SDF-1, a potent chemokine for lymphocytes (del Pozo *et al.*, 1999).

More recent studies have utilized gene-targeted mice. Loss of the Cdc42 GEF PIXα leads to defective G-coupled receptor signaling and PAK activation and reduced migration. Gene-targeted mice deficient in the Cdc42 GAP protein exhibit increased Cdc42 activity in the bone marrow with increased apoptosis in HSC populations. Hematopoietic cells exhibit disorganized actin structure and defective engraftment in stem cell transplant protocols (Wang *et al.*, 2006). Neutrophils from gene-targeted Cdc42-deficient mice show increased random motility but reduced directed migration associated with reduced podosome-like structures at the leading edge of the cells (Szczur *et al.*, 2006). Cdc42-/- neutrophils show increased lateral and tail membrane protrusions. Directed migration appears inhibited by defective p38MAPK activity apparently required for antagonizing these lateral filopodia-like structures. HSC from Cdc42-/- mice show defective migration and adhesion, which is associated with abnormal F-actin assembly, homing, and engraftment/retention in the bone marrow. Cdc42-/- mice show increased numbers of circulating HSC and reduced development of erythrocytes with anemia (Yang *et al.*, 2007a). In contrast to Cdc42 GAP-/- mice, these animals do not show increased apoptosis, but do show abnormalities in cell cycle progression associated with dysregulated p21 and cMyc expression. More recent studies show that Cdc42 regulates the balance between myelopoiesis and erythropoiesis (Yang *et al.*, 2007b). Cdc42-deficient mice developed a fatal myeloproliferative disorder characterized by neutrophilia, myeloid cell proliferation, and infiltration into multiple organs. Early erythroid development was inhibited. Bone marrow of Cdc42-/- mice showed decreased erythroid burst-forming units and erythroid colony-forming units. These changes were associated with upregulation of the myeloid transcription factor PU.1, C/EBP1α, and Gfi-1 and downregulation of GATA-2.

7. RHOA IN HEMATOPOIESIS

The effect of RhoA on hematopoiesis has been less well studied compared with Rac and Cdc42. As noted previously, activation of RhoA leads to stress fiber formation and cell shape changes, although most of these studies have been performed on fibroblasts. In fibroblasts, activation of RhoA has been reported to decrease the expression of Cdk inhibitors and to shorten G1 (Olson *et al.*, 1998). Using the same cell types, inactivation of RhoA has been shown to induce the expression of cyclin D–Cdk4 complexes in early G1 phase and promote a rapid G1/S phase transition (Roovers *et al.*, 2003; Welsh *et al.*, 2001). In mammary gland epithelial cells, transforming growth factor-β-induced activation of RhoA stimulates the nuclear translocation of p160 ROCK, a known target of RhoA, which results in cell cycle arrest by decreasing the activity of Cdc25A phosphatase and decreasing Rb phosphorylation (Bhowmick *et al.*, 2003). Therefore, the effect of RhoA GTPase activity on cell cycle and proliferation appears both cell type and agonist specific.

We have examined the role of RhoA GTPase in hematopoietic stem and progenitor cell functions by expressing DN mutant RhoAN19 in HSC/P via retrovirus-mediated gene transfer (Ghiaur *et al.*, 2006). In contrast with the published role of RhoA in fate determination and differentiation in mesenchymal stem cells (Sordella *et al.*, 2003), inhibition of RhoA activity was associated with a significant enhancement of HSC engraftment and reconstitution *in vivo*. Increased engraftment of HSC expressing RhoAN19 was associated with increased cyclin D1 expression and enhanced proliferation and cell cycle progression of hematopoietic progenitor cells *in vitro*, despite this enhanced engraftment *in vivo*. Consistent with studies reported in fibroblast cells (Hall, 1998), RhoA was essential for normal adhesion and migration of hematopoietic progenitor cells *in vitro*. Decreased activity of RhoA GTPase resulted in defective $\alpha_4\beta_1$ and $\alpha_5\beta_1$ integrin-mediated adhesion and impaired SDF-1α-directed migration of hematopoietic progenitor cells *in vitro*. These results are surprising given the role of adhesion and migration in HSC engraftment. Taken together, these data suggest that RhoA GTPase plays a crucial role in HSC engraftment, although the mechanism of enhanced engraftment seen with expression of the DN RhoA protein is unclear. In the context of previous reports describing Rac GTPase function in HSC (Cancelas *et al.*, 2005; Gu *et al.*, 2003), these studies suggest that inhibition in Rac activity may enhance mobilization, whereas inhibition of RhoA may augment HSC engraftment.

Additional studies using gene-targeted mice are needed to better clarify the role of RhoA, RhoB, and RhoC in hematopoiesis.

8. RhoGTPase in Human Diseases

8.1. Rac deficiency syndrome

The initial report of Rac2-deficient mice described a phagocytic immuno-deficiency syndrome emphasizing neutrophil dysfunctions related to actin cytoskeletal abnormalities. Subsequently, Ambruso *et al.* (2000) and our own group (Williams *et al.*, 2000) reported the identification of a child with serious, life-threatening infections associated with a dominant negative mutation of Rac2 (D57N). The patient exhibited leukocytosis but reduced inflammatory infiltrate in areas of infection. Neutrophils from this patient responded normally with respect to the respiratory burst to phorbol 12-myristate 13-acetate. Normal expression of CD11b, CD11c, and CD18 suggested that the patient did not suffer from leukocyte adhesion deficiency (LAD) or classical chronic granulomatous disease (CGD). However, the patient's neutrophils exhibited decreased chemotaxis in response to N-formyl-methionyl-leucyl-phenylalanine (fMLP) and interleukin-8, reduced rolling on GlyCAM-1 (a ligand for L-selectin), reduced superoxide generation in response to fMLP, mildly reduced phagocytosis, and adhesion to fibrinogen. Thus, the patient appeared clinically to have a phenotype overlapping between LAD and CGD.

At the molecular level, both genomic and cDNA sequencing confirmed the presence of the Asp→Asn mutation at position 57. Genomic sequencing confirmed a mono-allelic change in the gene, and ≈50% of the cloned cDNAs exhibited this mutation. Expression of the mutant protein via retrovirus-mediated gene transfer in normal neutrophils reproduced the cellular phenotype. The mutant protein displayed 10% GTP-binding activity, resulting in a markedly enhanced rate of GTP dissociation and did not respond to GEFs (Gu *et al.*, 2001). When expressed in murine-derived HSC, D57N Rac2 reduced endogenous activities of both Rac1 and Rac2 and led to decreased cell expansion *in vitro* associated with increased apoptosis. Transplantation of transduced bone marrow into lethally irradiated recipients showed a markedly reduced reconstitution of hematopoiesis in mutant-expressing cells over time, consistent with the role of Rac GTPases in marrow engraftment and retention described earlier. Interestingly, prior to successful curative allogeneic transplantation of this patient, his peripheral blood counts were diminishing and he was mildly pancytopenic at the time of marrow ablation in preparation for the transplant.

Taken together, these data suggest that the mutation behaves as a dominant negative mutation, likely by sequestering multiple GEFs in the cell.

This is a highly conserved amino acid in all GTPases and in the Ras superfamily and coordinates the binding of the γ-phosphate to the GTPase. Addition of recombinant Rac2 to cell-free extracts from the patient's neutrophils restored superoxide production, demonstrating the specificity of the molecular mutation in Rac2. This single case is the first reported mutation in humans of a GTPase and provides a fascinating correlation between the basic biology as elucidated in gene targeting models and human disease phenotype. Undoubtedly additional similar cases will become apparent, as many children with recurrent infections and neutrophil dysfunction remain poorly characterized at the molecular level.

8.2. Wiskott–Aldrich disease

The Cdc42 effector protein WASp is defective in the X-linked immunodeficiency disorder Wiskott–Aldrich syndrome (WAS) (Ochs and Thrasher, 2006; Thrasher and Burns, 1999). WASp activation depends on the specific interaction with guanosine triphosphate (GTP)-loaded Cdc42, which is mediated through a Cdc42- and Rac-interactive binding (CRIB) domain (Abdul-Manan *et al.*, 1999; Miki *et al.*, 1998; Rohatgi *et al.*, 1999). WASp is expressed in a hematopoietic-specific fashion. A spectrum of clinical disease is seen that correlates with mutations in specific domains of the WASp protein. Classic WAS patients express no WASp and have severely defective immune function that is characterized by aberrant polarization and directed migration of hematopoietic cells. WASp and the isoform N-WASp activate Arp 2/3, which regulates polymerization of actin from the barbed and branching filaments. WASp-deficient mice have been generated and also show significant hematopoietic defects.

Macrophages and dendritic cells from patients with WAS and from WASp-deficient mice have been shown to be defective in their migratory behavior. Chemotaxis of mutant macrophages in response to macrophage colony-stimulating factor (M–CSF), fMLP, monocyte chemoattractant protein 1 (MCP-1), and macrophage inflammatory protein 1a (MIP1 a) has been shown to be abrogated (Badolato *et al.*, 1998; Zicha *et al.*, 1998). WASp-deficient dendritric cells (DCs) exhibit similar abnormalities of cytoskeletal organization, chemotaxis, and migration (Binks *et al.*, 1998). WASp-deficient murine DCs exhibit multiple defects of trafficking *in vivo* after stimulation, including the emigration of Langerhans cells from the skin to secondary lymphoid tissues and the correct localization of DCs within T-cell areas, which correlated with a deficient migratory response of dendritic cells to the chemokines CCL19 and CCL21 (de Noronha *et al.*, 2005; Snapper *et al.*, 2005). It is therefore possible that DC trafficking abnormalities contribute in a significant way to the immune dysregulation observed in WAS and are responsible for the inflammation initiation and eczema development in this disease.

Defects of migration, anchorage, and localization have been defined more recently for other cell lineages, including T and B lymphocytes, neutrophils, and HSC/P. T lymphocytes from patients with WAS respond less well than normal cells *in vitro* to CXCL12 and CCL19 and demonstrate abrogated homing to secondary lymphoid tissue after adoptive transfer *in vivo* (Haddad *et al.*, 2001; Snapper *et al.*, 2005). There is a defect in the localization and function of the immunologic synapse, as WASp is recruited to lipid rafts immediately after the T-cell receptor and CD28 triggering event and is required for the movements of lipid rafts. T cells from WAS patients, lacking WASp, proliferate poorly after TCR/CD28 activation and have impaired capacities to cluster the lipid raft marker GM1 and to upregulate GM1 cell surface expression (Dupre *et al.*, 2002). Interestingly, cells that are deficient for both WASp and Wiskott–Aldrich syndrome protein-interacting protein (WIP) exhibit much more profound deficiencies than either alone, suggesting the existence of some redundancy (Gallego *et al.*, 2006). Similarly, WASp-deficient B lymphocytes have been shown to have marked morphologic abnormalities, defective migration, and adhesion *in vitro* and impaired homing *in vivo* (Westerberg *et al.*, 2005). This defect is likely to contribute to the observed deficiencies of humoral responses to both T-dependent and T-independent antigens and to the marked deficiency of marginal zone B cells in both murine and human spleens (Facchetti *et al.*, 1998; Westerberg *et al.*, 2005). A further example of defective trafficking *in vivo* originates from the observation that carrier female subjects for classic WAS almost universally exhibit nonrandom X-inactivation patterns in CD34+ bone marrow progenitors (Wengler *et al.*, 1995). This implies that WASp is functional within the HSC/P compartment and is consistent with evidence for WASp expression in this cell type in human adult and embryonic hematopoietic stem cells (Marshall *et al.*, 2000; Parolini *et al.*, 1997). Serial stem cell transplantation and competitive repopulation studies in mice have confirmed a selective homing and engraftment advantage for normal HSC, and hematopoiesis established by means of engraftment of chimeric fetal liver populations results in dominance of normal HSC/P over WASp-deficient hematopoiesis (Lacout *et al.*, 2003), suggesting that throughout development, there be preferential establishment of hematopoiesis by normal rather than mutant HSCs due to an intrinsic homing advantage.

Much can be learned from the study of human patients with naturally occurring mutations. Most molecular defects in the WASp gene result in diminished activity, either because aborted protein production or because of intrinsic instability the mutant mRNA or protein. However, some mutants have been shown to display impaired interaction with key regulators. For instance, patients with X-linked thrombocytopenia express lower levels of WASp and have residual immune function (Lemahieu *et al.*, 1999). Some WASp gene defects result in expression of mutant protein with amino acid substitutions within the Ena/VAS homology 1 domain, predictive of a disturbed interaction with WIP (Volkman *et al.*, 2002). Some clinically relevant mutations have been shown to abolish *in vivo* proper N-WASp localization and actin polymerization (Moreau *et al.*, 2000). Some X-linked neutropenia patients have missense mutations in the Cdc42-binding site of the WASp protein, and a subset of these patients has activating mutations that lead to constitutive activation of the protein. These activating mutations act by preventing autoinhibition of the Cdc42-binding domain of the molecule inducing unregulated actin polymerization and abnormal cytoskeletal structure and dynamics (Ancliff *et al.*, 2006). Interestingly, the phenotype of clinical disease arising from these mutations affecting the Cdc42-binding site is quite unlike that of classical WAS. These mutations lead to myelodysplastic changes in the bone marrow, reduced lymphocyte numbers and function, and increased apoptosis in the myeloid lineage associated with neutropenia and markedly abnormal cytoskeletal structure and dynamics. The mechanism of this defect is unclear but it can be related to abnormalities of cytokinesis affecting the chromosomal separation during mitosis.

8.3. Rac hyperactivation in leukemia

Rac GTPases have been previously implicated in p210-BCR-ABL-mediated transformation (Burridge and Wennerberg, 2004; Harnois *et al.*, 2003; Renshaw *et al.*, 1996; Schwartz, 2004; Sini *et al.*, 2004; Skorski *et al.*, 1998), although the specific role(s) of individual Rac subfamily members in the development of disease *in vivo* has not been defined. Evidence also suggests that Rac3 plays a role in p190-BCR-ABL-mediated ALL, whereas Rac1 and Rac2 do not appear to be hyperactivated in these lymphoma lysates (Cho *et al.*, 2005). This is of particular relevance, as p190-BCR-ABL differs from p210 in potentially important ways as it relates to RhoGTPases. For instance, while p210-BCR-ABL binds to and activates the Rho GTPases, apparently through the Dbl homology domain, p190-BCR-ABL, which lacks this domain, cannot bind to Rho GTPases but can still activate Rac1 and Cdc42 (Harnois *et al.*, 2003) through activation of the GEF Vav1 by BCR-ABL (Bassermann *et al.*, 2002). Rac GTPases have been shown to regulate signaling pathways that are downstream of p210-BCR-ABL (Burridge and Wennerberg, 2004; Schwartz, 2004). Together, these data suggest that Rac GTPases may integrate multiple signaling components of p210-BCR-ABL-activated pathways.

We analyzed whether Rac isoforms were hyperactivated in human chronic phase CML HSC/P. Activation of Rac was determined by p21-activated kinase (PAK) binding domain pull-down assays in isolated CD34$^+$ cells from CML patients. We observed that Rac1, Rac2, and, to a lesser degree, Rac3 were hyperactivated in CD34$^+$ cells purified from peripheral blood of two CML patients at diagnosis (Thomas *et al.*, 2007). We subsequently utilized a retroviral murine model in Rac gene-targeted BM cells to investigate the importance of Rac GTPase activation in the development and progression of p210-BCR-ABL-mediated MPD. We showed that the combined deficiency of Rac1 and Rac2 significantly attenuates p210-BCR-ABL-induced proliferation *in vitro* and MPD *in vivo*. Attenuation of the disease phenotype is associated with severely diminished p210-BCR-ABL-induced downstream signaling in primary hematopoietic cells. These data are consistent with previous reports of Rac3 activation in p190-BCR-ABL expressing malignant precursor B-lineage lymphoid cells (Cho *et al.*, 2005). We then utilized NSC23766, a small molecule antagonist of Rac activation (Gao *et al.*, 2004), to biochemically and functionally validate Rac as a molecular target in both a relevant animal model and in primary human CML cells *in vitro* and in a xenograft model *in vivo*, including in imatinib-resistant p210-BCR-ABL disease. These data demonstrate that Rac is an important signaling molecule in BCR-ABL-induced transformation and an additional therapeutic target in p210-BCR-ABL-mediated myeloproliferative disease. Additional studies in other chronic and acute myelogenous leukemia may define the role of Rac and other GTPases in both chronic and acute leukemias.

8.4. RhoH and lymphomas

The RhoH/TTF (Translocation Three Four) gene was first identified as a fusion protein containing the LAZ3/BCL6 oncogene as a result of the t(3;4)(q27;p11) translocation in a non-Hodgkin's lymphoma (NHL) cell line (Dallery *et al.*, 1995; Dallery-Prudhomme *et al.*, 1997). A chromosomal alteration involving the RhoH/TTF gene in the t(4;14)(p13;q32) translocation has also been found in another patient with multiple myeloma (Preudhomme *et al.*, 2000). In some cases, RT-PCR analyses of NHL patients have shown deregulated expression of both RhoH and BCL6 genes by promoter exchange between these two genes (Preudhomme *et al.*, 2000). The RhoH gene, along with three other oncogenes (PIM1, MYC, and PAX5), has been found to have a more than 45% mutation rate in human diffuse large B-cell lymphomas (DLBCLs) (Pasqualucci *et al.*, 2001). Mapping analyses demonstrated mutations scattered throughout the 1.6 kb of intron 1 in the RhoH gene in 13 of 28 DLBCLs, suggesting potential effects on the regulation of RhoH gene expression with pathophysiological relevance. Similar aberrant hypermutation in the RhoH gene also occurs in AIDS-related non-Hodgkin lymphomas (Gaidano *et al.*, 2003) and primary central nervous system lymphomas (Montesinos-Rongen *et al.*, 2004). However, it remains unclear whether these mutations translate into abnormal levels of RhoH expression in lymphomas and what physiological contribution hypermutation in the RhoH gene plays in lymphomagenesis. p53, a tumor suppressor gene, is a key regulator of apoptosis and cell cycle arrest upon DNA damage in many cells. p53 is the most frequently altered tumor suppressor in human solid tumors and is also altered in hematologic malignancies. Interestingly, p53 inactivation is frequent in transformed follicular lymphomas (80%) (Lo Coco *et al.*, 1993) and Burkitt's lymphoma (28%) (Kaneko *et al.*, 1996; Preudhomme *et al.*, 1995), suggesting that the frequency of p53 mutations in NHL may be higher than in other hematopoietic malignancies. Activating mutants of Rac1 cooperate with p53 deficiency to promote primary mouse embryonic fibroblast transformation and/or invasion, suggesting a possible functional cooperation between loss of the p53 gene and Rho GTPase-mediated signaling pathways in tumorigenesis.

The human RhoH/TTF gene encodes a 191 amino acid protein belonging to the Rho GTPase family. The C-terminal tail of RhoH, CKIF, represents a typical CAAX motif present in the entire Ras superfamily of small GTP-binding proteins. Proteins containing this motif will be geranylated if the C-terminal amino acid (X) is leucine (L) or phenylalanine (F). This post-translational modification plays a critical role in the localization of Ras and Rho proteins to the plasma membrane (Kinsella *et al.*, 1991). Biochemical studies showed that RhoH is GTPase deficient and remains constitutively in the active, GTP-bound state (Li *et al.*, 2002b). Interestingly, RhoH and RhoE are naturally GTPase deficient due to the amino acid substitutions at key residues that are highly conserved among all Rho GTPases (Li *et al.*, 2002b). This suggests that in contrast to many other family members, regulation of RhoH and RhoE may depend on the level of the protein expressed in the cells rather than guanine nucleotide cycling. Possible mechanisms for regulating RhoH and RhoE activity may include transcriptional, translational, and post-translational processes, which have not been well studied.

Like Rac2, RhoH is expressed only in the hematopoietic lineages, reportedly predominantly in T- and B-cell lines (Dallery-Prudhomme *et al.*, 1997; Li *et al.*, 2002b). Studies in Jurkat cells showed that RhoH expression is transcriptionally regulated upon stimulation with cytokines. Under physiological conditions, RhoH transcripts are also found differentially expressed in murine Th1 and Th2 T-cell subpopulations (Li *et al.*, 2002b), suggesting that RhoH may play a role in differentiation or function in Th1 and Th2 cells. However, these studies have been limited to lymphocytes and are mainly based on cell lines.

Alteration of RhoH expression experimentally affects proliferation and engraftment of hematopoietic progenitor cells (Gu *et al.*, 2005a) and integrin-mediated adhesion in Jurkat cells (Cherry *et al.*, 2004). We and others have determined major physiological functions of RhoH using gene-targeted mice deficient in the RhoH protein. $RhoH^{-/-}$ mice demonstrate impaired TCR-mediated thymocyte positive selection and maturation, resulting in T-cell deficiency (Dorn *et al.*, 2007; Gu *et al.*, 2006). Loss of RhoH leads to defective CD3ζ phosphorylation, impaired translocation of ZAP-70 to the immunological synapse, and reduced activation of ZAP-70-mediated pathways in thymic and peripheral T cells. Furthermore, proteomic analysis demonstrated RhoH to be a component of TCR signaling via TCR-activated ZAP-70 SH2-mediated interaction with immunoreceptor tyrosine-based activation motifs (ITAMs) in RhoH. *In vivo* reconstitution studies showed that RhoH function in thymopoiesis is dependent on phosphorylation of the ITAMs. These findings suggest that RhoH is a critical regulator of thymocyte development and TCR signaling by mediating recruitment and activation of ZAP-70. While a direct relationship has yet to be ascertained, taken together these experiments suggest that alterations in the expression and/or function of RhoH may play a role in lymphoma formation. Clearly RhoH is an important signaling molecule in T-cell development, although its exact role in T-cell receptor signaling remains to be elucidated.

9. SUMMARY AND PERSPECTIVES

The development of gene-targeted mice deficient in Rac GTPases and the use of knockout mice to study the role of these important molecular switches have contributed to the understanding of the role of Rho GTPase in normal blood cell development and function and, indeed, have led to the delineation of complex functioning of Rho GTPases in primary cells in physiological settings. This is particularly true with regard to the unique functions of different Rac molecules in HSC/P, which has not previously been studied or appreciated because of the lack of specificity of experimental methods relying on activated or dominant negative mutants and cell lines. To summarize, Rac2 deficiency leads to a variety of cellular phenotypes in hematopoietic cells, including abnormalities in cell adhesion, migration, degranulation, and phagocytosis as a consequence of abnormal F-actin assembly. Surprisingly, Rac2 appears to regulate survival in several cell types via activation of Akt pathways. In contrast, Rac1 regulates both overlapping and unique F-actin functions. and these differences appear, at least in neutrophils, to be because of differences in intracellular localization controlled by sequences in the carboxy-terminal tail and a specific region of the protein not previously implicated in Rac function (Filippi *et al.*, 2004). Rac1 regulates HSC/P cell cycle progression. Significantly, Rac1 is critical to stem cell engraftment and Rac2 is a major determinant of HSC/P retention in the marrow cavity. In addition, Cdc42 and RhoA may play an opposite role in HSC/P homing and engraftment. Thus, it is possible that each Rho family member will be utilized differently in these processes.

Overall, these studies implicate Rac, Cdc42, and RhoA as major regulators of HSC/P engraftment and marrow retention and begin to define the "intracellular signaling profile of the stem cell niche." Indeed, these studies suggest that engraftment and mobilization are separable biochemically and imply that these processes are not "mirror images" functionally.

These studies have also raised several unanswered questions. They include what the specific effector pathways (such as STAT, PAK, or WAVE) are downstream of individual Rho GTPases critical for stem cell adhesion, engraftment, and retention, as well as for stem cell transformation. Whether the altered cell adhesion and migration properties of HSC/Ps are directly responsible for the engraftment effect and for the survival/proliferation will also need to be dissected. Finally, the specificity of upstream guanine exchange factors coupling to the upstream stimuli of SDF1a, SCF, or integrins in the activation of Rho GTPases will need to be determined in hematopoietic cells.

REFERENCES

Abdel-Latif, D., Steward, M., Macdonald, D. L., Francis, G. A., Dinauer, M. C., and Lacy, P. (2004). Rac2 is critical for neutrophil primary granule exocytosis. *Blood* **104**, 832–839.

Abdul-Manan, N., Aghazadeh, B., Liu, G. A., Majumdar, A., Ouerfelli, O., Siminovitch, K. A., and Rosen, M. K. (1999). Structure of Cdc42 in complex with the GTPase-binding domain of the 'Wiskott-Aldrich syndrome' protein. *Nature* **399**, 379–383.

Abkowitz, J. L., Robinson, A. E., Kale, S., Long, M. W., and Chen, J. (2003). Mobilization of hematopoietic stem cells during homeostasis and after cytokine exposure. *Blood* **102**, 1249–1253.

Aiuti, A., Webb, I. J., Bleul, C., Springer, T., and Gutierrez-Ramos, J. C. (1997). The chemokine SDF-1 is a chemoattractant for human CD34+ hematopoietic progenitor cells and provides a new mechanism to explain the mobilization of CD34+ progenitors to peripheral blood. *J. Exp. Med* **185**, 111–120.

Akbar, H., Cancelas, J., Johnson, J., Lee, A., Williams, D. A., and Zheng, Y. (2007). Rac1 GTPase as a novel anti-platelet target. *J. of Thrombosis and Haemostasis* **5**, 1747–1755.

Allen, W. E., Zicha, D., Ridley, A. J., and Jones, G. E. (1998). A role for Cdc42 in macrophage chemotaxis. *J. Cell Biol.* **141**, 1147–1157.

Ambruso, D. R., Knall, C., Abell, A. N., Panepinto, J., Kurkchubasche, A., Thurman, G., Gonzalez-Aller, C., Hiester, A., deBoer, M., Harbeck, R. J., Oyer, R., Johnson, G. L., *et al.* (2000). Human neutrophil immunodeficiency syndrome is associated with an inhibitory Rac2 mutation. *Proc. Natl. Acad. Sci. USA* **97**, 4654–4659.

Ancliff, P. J., Blundell, M. P., Cory, G. O., Calle, Y., Worth, A., Kempski, H., Burns, S., Jones, G. E., Sinclair, J., Kinnon, C., Hann, I. M., Gale, R. E., *et al.* (2006). Two novel activating mutations in the Wiskott-Aldrich syndrome protein result in congenital neutropenia. *Blood* **108**, 2182–2189.

Badolato, R., Sozzani, S., Malacarne, F., Bresciani, S., Fiorini, M., Borsatti, A., Albertini, A., Mantovani, A., Ugazio, A. G., and Notarangelo, L. D. (1998). Monocytes from Wiskott-Aldrich patients display reduced chemotaxis and lack of cell polarization in response to monocyte chemoattractant protein-1 and formyl-methionyl-leucyl-phenyl-alanine. *J. Immunol.* **161**, 1026–1033.

Bassermann, F., Jahn, T., Miething, C., Seipel, P., Bai, R. Y., Coutinho, S., Tybulewicz, V. L., Peschel, C., and Duyster, J. (2002). Association of Bcr-Abl with the proto-oncogene Vav is implicated in activation of the Rac-1 pathway. *J. Biol. Chem.* **277,** 12437–12445.

Baugher, P. J., Krishnamoorthy, L., Price, J. E., and Dharmawardhane, S. F. (2005). Rac1 and Rac3 isoform activation is involved in the invasive and metastatic phenotype of human breast cancer cells. *Breast Cancer Res.* **7,** R965–R974.

Bhowmick, N. A., Ghiassi, M., Aakre, M., Brown, K., Singh, V., and Moses, H. L. (2003). TGF-beta-induced RhoA and p160ROCK activation is involved in the inhibition of Cdc25A with resultant cell-cycle arrest. *Proc. Natl. Acad. Sci. USA* **100,** 15548–15553.

Binks, M., Jones, G. E., Brickell, P. M., Kinnon, C., Katz, D. R., and Thrasher, A. J. (1998). Intrinsic dendritic cell abnormalities in Wiskott-Aldrich syndrome. *Eur. J. Immunol.* **28,** 3259–3267.

Broxmeyer, H. E., Kohli, L., Kim, C. H., Lee, Y., Mantel, C., Cooper, S., Hangoc, G., Shaheen, M., Li, X., and Clapp, D. W. (2003). Stromal cell-derived factor-1/CXCL12 directly enhances survival/antiapoptosis of myeloid progenitor cells through CXCR4 and G(alpha)i proteins and enhances engraftment of competitive, repopulating stem cells. *J. Leukoc. Biol.* **73,** 630–638.

Burkhalter, S., Janssen, E., Tschan, M. P., Torbett, B. E., and Knaus, U. (2002). Differential expression and subcellular localization of the GTPase Rac3. *In* "Proceedings of the Small GTPase Meeting Snowmass" (T. S. R. Institute, ed.). La Jolla, CA.

Burridge, K., and Wennerberg, K. (2004). Rho and Rac take center stage. *Cell* **116,** 167–179.

Butcher, E. C., and Picker, L. J. (1996). Lymphocyte homing and homeostasis. *Science* **272,** 60–66.

Cancelas, J. A., Jansen, M., and Williams, D. A. (2006). The role of chemokine activation of Rac GTPases in hematopoietic stem cell marrow homing, retention and peripheral mobilization. *Exp. Hematol.* **34,** 976–985.

Cancelas, J. A., Lee, A. W., Prabhakar, R., Stringer, K. F., Zheng, Y., and Williams, D. A. (2005). Rac GTPases differentially integrate signals regulating hematopoietic stem cell localization. *Nat. Med.* **11,** 886–891.

Carstanjen, D., Yamauchi, A., Koornneef, A., Zang, H., Filippi, M. D., Harris, C., Towe, J., Atkinson, S., Zheng, Y., Dinauer, M. C., and Williams, D. A. (2005). Rac2 regulates neutrophil chemotaxis, superoxide production, and myeloid colony formation through multiple distinct effector pathways. *J. Immunol.* **174,** 4613–4620.

Chan, A. Y., Coniglio, S. J., Chuang, Y. Y., Michaelson, D., Knaus, U. G., Philips, M. R., and Symons, M. (2005). Roles of the Rac1 and Rac3 GTPases in human tumor cell invasion. *Oncogene* **24,** 7821–7829.

Cherry, L. K., Li, X., Schwab, P., Lim, B., and Klickstein, L. B. (2004). RhoH is required to maintain the integrin LFA-1 in a nonadhesive state on lymphocytes. *Nat. Immunol.* **5,** 961–967.

Cho, Y. J., Zhang, B., Kaartinen, V., Haataja, L., de Curtis, I., Groffen, J., and Heisterkamp, N. (2005). Generation of rac3 null mutant mice: Role of Rac3 in Bcr/Abl-caused lymphoblastic leukemia. *Mol. Cell. Biol.* **25,** 5777–5785.

Corbetta, S., Gualdoni, S., Albertinazzi, C., Paris, S., Croci, L., Consalez, G. G., and de Curtis, I. (2005). Generation and characterization of Rac3 knockout mice. *Mol. Cell. Biol.* **25,** 5763–5776.

Craddock, C. F., Nakamoto, B., Andrews, R. G., Priestley, G. V., and Papayannopoulou, T. (1997). Antibodies to VLA4 integrin mobilize long-term repopulating cells and augment cytokine-induced mobilization in primates and mice. *Blood* **90,** 4779–4788.

Croker, B. A., Handman, E., Hayball, J. D., Baldwin, T. M., Voigt, V., Cluse, L. A., Yang, F. C., Williams, D. A., and Roberts, A. W. (2002a). Rac2-deficient mice display perturbed T-cell distribution and chemotaxis, but only minor abnormalities in T(H)1 responses. *Immunol. Cell Biol.* **80,** 231–240.

Croker, B. A., Tarlinton, D. M., Cluse, L. A., Tuxen, A. J., Light, A., Yang, F. C., Williams, D. A., and Roberts, A. W. (2002b). The Rac2 guanosine triphosphatase regulates B lymphocyte antigen receptor responses and chemotaxis and is required for establishment of B-1a and marginal zone B lymphocytes. *J. Immunol.* **168,** 3376–3386.

Dallery, E., Galiegue-Zouitina, S., Collyn-d'Hooghe, M., Quief, S., Denis, C., Hildebrand, M. P., Lantoine, D., Deweindt, C., Tilly, H., Bastard, C., *et al.* (1995). TTF, a gene encoding a novel small G protein, fuses to the lymphoma-associated LAZ3 gene by t(3;4) chromosomal translocation. *Oncogene* **10,** 2171–2178.

Dallery-Prudhomme, E., Roumier, C., Denis, C., Preudhomme, C., Kerckaert, J. P., and Galiegue-Zouitina, S. (1997). Genomic structure and assignment of the RhoH/TTF small GTPase gene (ARHH) to 4p13 by in situ hybridization. *Genomics* **43,** 89–94.

del Pozo, M. A., Vicente-Manzanares, M., Tejedor, R., Serrador, J. M., and Sanchez-Madrid, F. (1999). Rho GTPases control migration and polarization of adhesion molecules and cytoskeletal ERM components in T lymphocytes. *Eur. J. Immunol.* **29,** 3609–3620.

de Noronha, S., Hardy, S., Sinclair, J., Blundell, M. P., Strid, J., Schulz, O., Zwirner, J., Jones, G. E., Katz, D. R., Kinnon, C., and Thrasher, A. J. (2005). Impaired dendritic-cell homing *in vivo* in the absence of Wiskott-Aldrich syndrome protein. *Blood* **105,** 1590–1597.

Devine, S. M., Flomenberg, N., Vesole, D. H., Liesveld, J., Weisdorf, D., Badel, K., Calandra, G., and DiPersio, J. F. (2004). Rapid mobilization of CD34+ cells following administration of the CXCR4 antagonist AMD3100 to patients with multiple myeloma and non-Hodgkin's lymphoma. *J. Clin. Oncol.* **22,** 1095–1102.

Dorn, T., Kuhn, U., Bungartz, G., Stiller, S., Bauer, M., Ellwart, J., Peters, T., Scharffetter-Kochanek, K., Semmrich, M., Laschinger, M., Holzmann, B., Klinkert, W. E., *et al.* (2007). RhoH is important for positive thymocyte selection and T-cell receptor signaling. *Blood* **109,** 2346–2355.

Driessen, R. L., Johnston, H. M., and Nilsson, S. K. (2003). Membrane-bound stem cell factor is a key regulator in the initial lodgment of stem cells within the endosteal marrow region. *Exp. Hematol.* **31,** 1284–1291.

Dupre, L., Aiuti, A., Trifari, S., Martino, S., Saracco, P., Bordignon, C., and Roncarolo, M. G. (2002). Wiskott-Aldrich syndrome protein regulates lipid raft dynamics during immunological synapse formation. *Immunity* **17,** 157–166.

Eden, S., Rohatgi, R., Podtelejnikov, A. V., Mann, M., and Kirschner, M. W. (2002). Mechanism of regulation of WAVE1-induced actin nucleation by Rac1 and Nck. *Nature* **418,** 790–793.

Facchetti, F., Blanzuoli, L., Vermi, W., Notarangelo, L. D., Giliani, S., Fiorini, M., Fasth, A., Stewart, D. M., and Nelson, D. L. (1998). Defective actin polymerization in EBV-transformed B-cell lines from patients with the Wiskott-Aldrich syndrome. *J. Pathol.* **185,** 99–107.

Filippi, M. D., Harris, C. E., Meller, J., Gu, Y., Zheng, Y., and Williams, D. A. (2004). Localization of Rac2 via the C terminus and aspartic acid 150 specifies superoxide generation, actin polarity and chemotaxis in neutrophils. *Nat. Immunol.* **5,** 744–751.

Fulkerson, P. C., Zhu, H., Williams, D. A., Zimmermann, N., and Rothenberg, M. E. (2005). CXCL9 inhibits eosinophil responses by a CCR3- and Rac2-dependent mechanism. *Blood* **106,** 436–443.

Gaidano, G., Pasqualucci, L., Capello, D., Berra, E., Deambrogi, C., Rossi, D., Maria Larocca, L., Gloghini, A., Carbone, A., and Dalla-Favera, R. (2003). Aberrant somatic hypermutation in multiple subtypes of AIDS-associated non-Hodgkin lymphoma. *Blood* **102,** 1833–1841.

Gallego, M. D., de la Fuente, M. A., Anton, I. M., Snapper, S., Fuhlbrigge, R., and Geha, R. S. (2006). WIP and WASp play complementary roles in T cell homing and chemotaxis to SDF-1alpha. *Int. Immunol.* **18**, 221–232.

Gao, Y., Dickerson, J. B., Guo, F., Zheng, J., and Zheng, Y. (2004). Rational design and characterization of a Rac GTPase-specific small molecule inhibitor. *Proc. Natl. Acad. Sci. USA* **101**, 7618–7623.

Ghiaur, G., Lee, A., Bailey, J., Cancelas, J. A., Zheng, Y., and Williams, D. A. (2006). Inhibition of RhoA GTPase activity enhances hematopoietic stem and progenitor cell proliferation and engraftment. *Blood* **108**, 2087–2094.

Glogauer, M., Marchal, C. C., Zhu, F., Worku, A., Clausen, B. E., Foerster, I., Marks, P., Downey, G. P., Dinauer, M., and Kwiatkowski, D. J. (2003). Rac1 deletion in mouse neutrophils has selective effects on neutrophil functions. *J. Immunol.* **170**, 5652–5657.

Gong, J. K. (1978). Endosteal marrow: A rich source of hematopoietic stem cells. *Science* **199**, 1443–1445.

Gu, Y., Byrne, M. C., Paranavitana, N. C., Aronow, B., Siefring, J. E., D'Souza, M., Horton, H. F., Quilliam, L. A., and Williams, D. A. (2002). Rac2, a hematopoiesis-specific Rho GTPase, specifically regulates mast cell protease gene expression in bone marrow-derived mast cells. *Mol. Cell. Biol.* **22**, 7645–7657.

Gu, Y., Chae, H., Siefring, J., Jast, I. J., Hildeman, D., and Williams, D. A. (2006). RhoH, a GTPase recruits and activates Zap70 required for T cell receptor signaling and thymocyte development. *Nat. Immunol.* **7**, 1182–1190.

Gu, Y., Filippi, M. D., Cancelas, J. A., Siefring, J. E., Williams, E. P., Jasti, A. C., Harris, C. E., Lee, A. W., Prabhakar, R., Atkinson, S. J., Kwiatkowski, D. J., and Williams, D. A. (2003). Hematopoietic cell regulation by Rac1 and Rac2 guanosine triphosphatases. *Science* **302**, 445–449.

Gu, Y., Jasti, A. C., Jansen, M., and Siefring, J. E. (2005a). RhoH, a hematopoietic-specific Rho GTPase, regulates proliferation, survival, migration, and engraftment of hematopoietic progenitor cells. *Blood* **105**, 1467–1475.

Gu, Y., Jia, B., Yang, F. C., D'Souza, M., Harris, C. E., Derrow, C. W., Zheng, Y., and Williams, D. A. (2001). Biochemical and biological characterization of a human Rac2 GTPase mutant associated with phagocytic immunodeficiency. *J. Biol. Chem.* **276**, 15929–15938.

Gu, Y., Zheng, Y., and Williams, D. A. (2005b). RhoH GTPase: A key regulator of hematopoietic cell proliferation and apoptosis? *Cell Cycle* **4**, 201–202.

Haddad, E., Zugaza, J. L., Louache, F., Debili, N., Crouin, C., Schwarz, K., Fischer, A., Vainchenker, W., and Bertoglio, J. (2001). The interaction between Cdc42 and WASp is required for SDF-1-induced T-lymphocyte chemotaxis. *Blood* **97**, 33–38.

Hall, A. (1998). G Proteins and small GTPases: Distant relatives keep in touch. *Science* **280**, 2074–2075.

Harnois, T., Constantin, B., Rioux, A., Grenioux, E., Kitzis, A., and Bourmeyster, N. (2003). Differential interaction and activation of Rho family GTPases by p210bcr-abl and p190bcr-abl. *Oncogene* **22**, 6445–6454.

Hattori, Y., Kato, H., Nitta, M., and Takamoto, S. (2001). Decrease of L-selectin expression on human CD34+ cells on freeze-thawing and rapid recovery with short-term incubation. *Exp. Hematol.* **29**, 114–122.

Heissig, B., Hattori, K., Dias, S., Friedrich, M., Ferris, B., Hackett, N. R., Crystal, R. G., Besmer, P., Lyden, D., Moore, M. A., Werb, Z., and Rafii, S. (2002). Recruitment of stem and progenitor cells from the bone marrow niche requires MMP-9 mediated release of kit-ligand. *Cell* **109**, 625–637.

Jansen, M., Yang, F. C., Cancelas, J. A., Bailey, J. R., and Williams, D. A. (2005). Rac2-deficient hematopoietic stem cells show defective interaction with the hematopoietic microenvironment and long-term engraftment failure. *Stem Cells* **23**, 335–346.

Kaneko, H., Sugita, K., Kiyokawa, N., Iizuka, K., Takada, K., Saito, M., Yoshimoto, K., Itakura, M., Kokai, Y., and Fujimoto, J. (1996). Lack of CD54 expression and mutation of p53 gene relate to the prognosis of childhood Burkitt's lymphoma. *Leuk. Lymphoma* **21,** 449–455.

Katayama, Y., Battista, M., Kao, W. M., Hidalgo, A., Peired, A. J., Thomas, S. A., and Frenette, P. S. (2006). Signals from the sympathetic nervous system regulate hematopoietic stem cell egress from bone marrow. *Cell* **124,** 407–421.

Keller, P. J., Gable, C. M., Wing, M. R., and Cox, A. D. (2005). Rac3-mediated transformation requires multiple effector pathways. *Cancer Res.* **65,** 9883–9890.

Kiel, M. J., Yilmaz, O. H., Iwashita, T., Terhorst, C., and Morrison, S. J. (2005). SLAM family receptors distinguish hematopoietic stem and progenitor cells and reveal endothelial niches for stem cells. *Cell* **121,** 1109–1121.

Kim, C., and Dinauer, M. C. (2001). Rac2 is an essential regulator of neutrophil nicotinamide adenine dinucleotide phosphate oxidase activation in response to specific signaling pathways. *J. Immunol.* **166,** 1223–1232.

Kinsella, B. T., Erdman, R. A., and Maltese, W. A. (1991). Carboxyl-terminal isoprenylation of ras-related GTP-binding proteins encoded by rac1, rac2, and ralA. *J. Biol. Chem.* **266,** 9786–9794.

Kollet, O., Spiegel, A., Peled, A., Petit, I., Byk, T., Hershkoviz, R., Guetta, E., Barkai, G., Nagler, A., and Lapidot, T. (2001). Rapid and efficient homing of human CD34(+) CD38(−/low)CXCR4(+) stem and progenitor cells to the bone marrow and spleen of NOD/SCID and NOD/SCID/B2m(null) mice. *Blood* **97,** 3283–3291.

Kronenwett, R., Martin, S., and Haas, R. (2000). The role of cytokines and adhesion molecules for mobilization of peripheral blood stem cells. *Stem Cells* **18,** 320–330.

Lacout, C., Haddad, E., Sabri, S., Svinarchouk, F., Garcon, L., Capron, C., Foudi, A., Mzali, R., Snapper, S. B., Louache, F., Vainchenker, W., and Dumenil, D. (2003). A defect in hematopoietic stem cell migration explains the nonrandom X-chromosome inactivation in carriers of Wiskott-Aldrich syndrome. *Blood* **102,** 1282–1289.

Lacy, P., Abdel-Latif, D., Steward, M., Musat-Marcu, S., Man, S. F., and Moqbel, R. (2003). Divergence of mechanisms regulating respiratory burst in blood and sputum eosinophils and neutrophils from atopic subjects. *J. Immunol.* **170,** 2670–2679.

Lapidot, T., and Kollet, O. (2002). The essential roles of the chemokine SDF-1 and its receptor CXCR4 in human stem cell homing and repopulation of transplanted immune-deficient NOD/SCID and NOD/SCID/B2m(null) mice. *Leukemia* **16,** 1992–2003.

Lataillade, J. J., Clay, D., Dupuy, C., Rigal, S., Jasmin, C., Bourin, P., and Le Bousse-Kerdiles, M. C. (2000). Chemokine SDF-1 enhances circulating CD34(+) cell proliferation in synergy with cytokines: Possible role in progenitor survival. *Blood* **95,** 756–768.

Lax, A. J., Pullinger, G. D., Baldwin, M. R., Harmey, D., Grigoriadis, A. E., and Lakey, J. H. (2004). The *Pasteurella multocida* toxin interacts with signalling pathways to perturb cell growth and differentiation. *Int. J. Med. Microbiol.* **293,** 505–512.

Lemahieu, V., Gastier, J. M., and Francke, U. (1999). Novel mutations in the Wiskott-Aldrich syndrome protein gene and their effects on transcriptional, translational, and clinical phenotypes. *Hum. Mutat.* **14,** 54–66.

Levesque, J. P., Hendy, J., Winkler, I. G., Takamatsu, Y., and Simmons, P. J. (2003). Granulocyte colony-stimulating factor induces the release in the bone marrow of proteases that cleave c-KIT receptor (CD117) from the surface of hematopoietic progenitor cells. *Exp. Hematol.* **31,** 109–117.

Li, S., Yamauchi, A., Marchal, C. C., Molitoris, J. K., Quilliam, L. A., and Dinauer, M. C. (2002a). Chemoattractant-stimulated Rac activation in wild-type and Rac2-deficient murine neutrophils: Preferential activation of Rac2 and Rac2 gene dosage effect on neutrophil functions. *J. Immunol.* **169,** 5043–5051.

Li, X., Bu, X., Lu, B., Avraham, H., Flavell, R. A., and Lim, B. (2002b). The hematopoiesis-specific GTP-binding protein RhoH is GTPase deficient and modulates activities of other Rho GTPases by an inhibitory function. *Mol. Cell. Biol.* **22**, 1158–1171.

Liles, W. C., Broxmeyer, H. E., Rodger, E., Wood, B., Hubel, K., Cooper, S., Hangoc, G., Bridger, G. J., Henson, G. W., Calandra, G., and Dale, D. C. (2003). Mobilization of hematopoietic progenitor cells in healthy volunteers by AMD3100, a CXCR4 antagonist. *Blood* **102**, 2728–2730.

Lo Coco, F., Gaidano, G., Louie, D. C., Offit, K., Chaganti, R. S., and Dalla-Favera, R. (1993). p53 mutations are associated with histologic transformation of follicular lymphoma. *Blood* **82**, 2289–2295.

Ma, Q., Jones, D., and Springer, T. A. (1999). The chemokine receptor CXCR4 is required for the retention of B lineage and granulocytic precursors within the bone marrow microenvironment. *Immunity* **10**, 463–471.

Marshall, C. J., Kinnon, C., and Thrasher, A. J. (2000). Polarized expression of bone morphogenetic protein-4 in the human aorta-gonad-mesonephros region. *Blood* **96**, 1591–1593.

McNiece, I. K., and Briddell, R. A. (1995). Stem cell factor. *J. Leukoc. Biol.* **58**, 14–22.

Miki, H., Suetsugu, S., and Takenawa, T. (1998). WAVE, a novel WASp-family protein involved in actin reorganization induced by Rac. *EMBO J.* **17**, 6932–6941.

Miki, H., Yamaguchi, H., Suetsugu, S., and Takenawa, T. (2000). IRSp53 is an essential intermediate between Rac and WAVE in the regulation of membrane ruffling. *Nature* **408**, 732–735.

Mira, J. P., Benard, V., Groffen, J., Sanders, L. C., and Knaus, U. G. (2000). Endogenous, hyperactive Rac3 controls proliferation of breast cancer cells by a p21-activated kinase-dependent pathway. *Proc. Natl. Acad. Sci. USA* **97**, 185–189.

Miyazawa, K., Williams, D. A., Gotoh, A., Nishimaki, J., Broxmeyer, H. E., and Toyama, K. (1995). Membrane-bound Steel factor induces more persistent tyrosine kinase activation and longer life span of *c-kit* gene encoded protein than its soluble form. *Blood* **85**, 641–649.

Montesinos-Rongen, M., Van Roost, D., Schaller, C., Wiestler, O. D., and Deckert, M. (2004). Primary diffuse large B-cell lymphomas of the central nervous system are targeted by aberrant somatic hypermutation. *Blood* **103**, 1869–1875.

Moreau, V., Frischknecht, F., Reckmann, I., Vincentelli, R., Rabut, G., Stewart, D., and Way, M. (2000). A complex of N-WASp and WIP integrates signalling cascades that lead to actin polymerization. *Nat. Cell Biol.* **2**, 441–448.

Morris, C. M., Haataja, L., McDonald, M., Gough, S., Markie, D., Groffen, J., and Heisterkamp, N. (2000). The small GTPase RAC3 gene is located within chromosome band 17q25.3 outside and telomeric of a region commonly deleted in breast and ovarian tumours. *Cytogenet. Cell Genet.* **89**, 18–23.

Nagasawa, T., Hirota, S., Tachibana, K., Takakura, N., Nishikawa, S., Kitamura, Y., Yoshida, N., Kikutani, H., and Kishimoto, T. (1996). Defects of B-cell lymphopoiesis and bone-marrow myelopoiesis in mice lacking the CXC chemokine PBSF/SDF-1. *Nature* **382**, 635–638.

Nagasawa, T., Tachibana, K., and Kishimoto, T. (1998). A novel CXC chemokine PBSF/SDF-1 and its receptor CXCR4: Their functions in development, hematopoiesis and HIV infection. *Semin. Immunol.* **10**, 179–185.

Nasser, N., Cancelas, J., Zheng, J., Williams, D. A., and Zheng, Y. (2006). Structure-function based design on small molecule inhibitors targeting Rho family GTPases. *Curr. Top. Med. Chem.* **6**, 1109–1116.

Nilsson, S. K., Johnston, H. M., and Coverdale, J. A. (2001). Spatial localization of transplanted hemopoietic stem cells: Inferences for the localization of stem cell niches. *Blood* **97**, 2293–2299.

Ochs, H. D., and Thrasher, A. J. (2006). The Wiskott-Aldrich syndrome. *J. Allergy Clin. Immunol* **117**, 725–738; quiz 739.

Olson, M. F., Paterson, H. F., and Marshall, C. J. (1998). Signals from Ras and Rho GTPases interact to regulate expression of p21Waf1/Cip1. *Nature* **394**, 295–299.

Papayannopoulou, T. (2003). Bone marrow homing: The players, the playfield, and their evolving roles. *Curr. Opin. Hematol.* **10**, 214–219.

Papayannopoulou, T., and Craddock, C. (1997). Homing and trafficking of hemopoietic progenitor cells. *Acta Haematol.* **97**, 97–104.

Papayannopoulou, T., Craddock, C., Nakamoto, B., Priestley, G. V., and Wolf, N. S. (1995). The VLA4/VCAM-1 adhesion pathway defines contrasting mechanisms of lodgement of transplanted murine hemopoietic progenitors between bone marrow and spleen. *Proc. Natl. Acad. Sci. USA* **92**, 9647–9651.

Papayannopoulou, T., Priestley, G. V., Bonig, H., and Nakamoto, B. (2003). The role of G-protein signaling in hemopoietic stem/progenitor cell mobilization. *Blood* **101**, 4739–4747.

Papayannopoulou, T., Priestley, G. V., Nakamoto, B., Zafiropoulos, V., and Scott, L. M. (2001a). Molecular pathways in bone marrow homing: Dominant role of alpha(4)beta(1) over beta(2)-integrins and selectins. *Blood* **98**, 2403–2411.

Papayannopoulou, T., Priestley, G. V., Nakamoto, B., Zafiropoulos, V., Scott, L. M., and Harlan, J. M. (2001b). Synergistic mobilization of hemopoietic progenitor cells using concurrent beta1 and beta2 integrin blockade or beta2-deficient mice. *Blood* **97**, 1282–1288.

Parolini, O., Berardelli, S., Riedl, E., Bello-Fernandez, C., Strobl, H., Majdic, O., and Knapp, W. (1997). Expression of Wiskott-Aldrich syndrome protein (WASp) gene during hematopoietic differentiation. *Blood* **90**, 70–75.

Pasqualucci, L., Neumeister, P., Goossens, T., Nanjangud, G., Chaganti, R. S., Kuppers, R., and Dalla-Favera, R. (2001). Hypermutation of multiple proto-oncogenes in B-cell diffuse large-cell lymphomas. *Nature* **412**, 341–346.

Peled, A., Grabovsky, V., Habler, L., Sandbank, J., Arenzana-Seisdedos, F., Petit, I., Ben-Hur, H., Lapidot, T., and Alon, R. (1999a). The chemokine SDF-1 stimulates integrin-mediated arrest of CD34(+) cells on vascular endothelium under shear flow. *J. Clin. Invest.* **104**, 1199–1211.

Peled, A., Kollet, O., Ponomaryov, T., Petit, I., Franitza, S., Grabovsky, V., Slav, M. M., Nagler, A., Lider, O., Alon, R., Zipori, D., and Lapidot, T. (2000). The chemokine SDF-1 activates the integrins LFA-1, VLA-4, and VLA-5 on immature human CD34(+) cells: Role in transendothelial/stromal migration and engraftment of NOD/SCID mice. *Blood* **95**, 3289–3296.

Peled, A., Petit, I., Kollet, O., Magid, M., Ponomaryov, T., Byk, T., Nagler, A., Ben-Hur, H., Many, A., Shultz, L., Lider, O., Alon, R., *et al.* (1999b). Dependence of human stem cell engraftment and repopulation of NOD/SCID mice on CXCR4. *Science* **283**, 845–848.

Preudhomme, C., Dervite, I., Wattel, E., Vanrumbeke, M., Flactif, M., Lai, J. L., Hecquet, B., Coppin, M. C., Nelken, B., Gosselin, B., *et al.* (1995). Clinical significance of p53 mutations in newly diagnosed Burkitt's lymphoma and acute lymphoblastic leukemia: A report of 48 cases. *J. Clin. Oncol.* **13**, 812–820.

Preudhomme, C., Roumier, C., Hildebrand, M. P., Dallery-Prudhomme, E., Lantoine, D., Lai, J. L., Daudignon, A., Adenis, C., Bauters, F., Fenaux, P., Kerckaert, J. P., and Galiegue-Zouitina, S. (2000). Nonrandom 4p13 rearrangements of the RhoH/TTF gene, encoding a GTP- binding protein, in non-Hodgkin's lymphoma and multiple myeloma. *Oncogene* **19**, 2023–2032.

Rafii, S., Heissig, B., and Hattori, K. (2002). Efficient mobilization and recruitment of marrow-derived endothelial and hematopoietic stem cells by adenoviral vectors expressing angiogenic factors. *Gene Ther.* **9,** 631–641.

Renshaw, M. W., Lea-Chou, E., and Wang, J. Y. (1996). Rac is required for v-Abl tyrosine kinase to activate mitogenesis. *Curr. Biol.* **6,** 76–83.

Ridley, A. J., and Hall, A. (1992). The small GTP-binding protein rho regulates the assembly of focal adhesions and actin stress fibers in response to growth factors. *Cell* **70,** 389–399.

Roberts, A. W., Kim, C., Zhen, L., Lowe, J. B., Kapur, R., Petryniak, B., Spaetti, A., Pollock, J. D., Borneo, J. B., Bradford, G. B., Atkinson, S. J., Dinauer, M. C., *et al.* (1999). Deficiency of the hematopoietic cell-specific Rho family GTPase Rac2 is characterized by abnormalities in neutrophil function and host defense. *Immunity* **10,** 183–196.

Rohatgi, R., Ma, L., Miki, H., Lopez, M., Kirchhausen, T., Takenawa, T., and Kirschner, M. W. (1999). The interaction between N-WASp and the Arp2/3 complex links Cdc42-dependent signals to actin assembly. *Cell* **97,** 221–231.

Roovers, K., Klein, E. A., Castagnino, P., and Assoian, R. K. (2003). Nuclear translocation of LIM kinase mediates Rho-Rho kinase regulation of cyclin D1 expression. *Dev. Cell* **5,** 273–284.

Schwartz, M. (2004). Rho signalling at a glance. *J. Cell Sci.* **117,** 5457–5458.

Scott, L. M., Priestley, G. V., and Papayannopoulou, T. (2003). Deletion of alpha4 integrins from adult hematopoietic cells reveals roles in homeostasis, regeneration, and homing. *Mol. Cell. Biol.* **23,** 9349–9360.

Sini, P., Cannas, A., Koleske, A. J., Di Fiore, P. P., and Scita, G. (2004). Abl-dependent tyrosine phosphorylation of Sos-1 mediates growth-factor-induced Rac activation. *Nat. Cell Biol.* **6,** 268–274.

Skorski, T., Wlodarski, P., Daheron, L., Salomoni, P., Nieborowska-Skorska, M., Majewski, M., Wasik, M., and Calabretta, B. (1998). BCR/ABL-mediated leukemo-genesis requires the activity of the small GTP- binding protein Rac. *Proc. Natl. Acad. Sci. USA* **95,** 11858–11862.

Snapper, S. B., Meelu, P., Nguyen, D., Stockton, B. M., Bozza, P., Alt, F. W., Rosen, F. S., von Andrian, U. H., and Klein, C. (2005). WASp deficiency leads to global defects of directed leukocyte migration *in vitro* and *in vivo. J. Leukoc. Biol.* **77,** 993–998.

Sordella, R., Jiang, W., Chen, G. C., Curto, M., and Settleman, J. (2003). Modulation of Rho GTPase signaling regulates a switch between adipogenesis and myogenesis. *Cell* **113,** 147–158.

Springer, T. A. (1994). Traffic signals for lymphocyte and recirculation and leukocyte emigration: The multistep paradigm. *Cell* **76,** 301–314.

Srinivasan, S., Wang, F., Glavas, S., Ott, A., Hofmann, F., Aktories, K., Kalman, D., and Bourne, H. R. (2003). Rac and Cdc42 play distinct roles in regulating PI(3,4,5)P3 and polarity during neutrophil chemotaxis. *J. Cell Biol.* **160,** 375–385.

Suetsugu, S., Yamazaki, D., Kurisu, S., and Takenawa, T. (2003). Differential roles of WAVE1 and WAVE2 in dorsal and peripheral ruffle formation for fibroblast cell migration. *Dev. Cell* **5,** 595–609.

Sweeney, E. A., Lortat-Jacob, H., Priestley, G. V., Nakamoto, B., and Papayannopoulou, T. (2002). Sulfated polysaccharides increase plasma levels of SDF-1 in monkeys and mice: involvement in mobilization of stem/progenitor cells. *Blood* **99,** 44–51.

Symons, M., Derry, J. M., Karlak, B., Jiang, S., Lemahieu, V., McCormick, F., Francke, U., and Abo, A. (1996). Wiskott-Aldrich syndrome protein, a novel effector for the GTPase CDC42Hs, is implicated in actin polymerization. *Cell* **84,** 723–734.

Szczur, K., Xu, H., Atkinson, S., Zheng, Y., and Filippi, M. D. (2006). Rho GTPase CDC42 regulates directionality and random movement via distinct MAPK pathways in neutrophils. *Blood* **108,** 4205–4213.

Takenawa, T., and Miki, H. (2001). WASp and WAVE family proteins: Key molecules for rapid rearrangement of cortical actin filaments and cell movement. *J. Cell Sci.* **114,** 1801–1809.

Tan, B. L., Yazicioglu, M. N., Ingram, D., McCarthy, J., Borneo, J., Williams, D. A., and Kapur, R. (2003). Genetic evidence for convergence of c-Kit and α4 integrin-mediated signals on class IA PI-3kinase and Rac pathway in regulating integrin directed migration in mast cells. *Blood* **101,** 4725–4732.

Tao, W., Filippi, M. D., Bailey, J. R., Atkinson, S. J., Connors, B., Evan, A., and Williams, D. A. (2002). The TRQQKRP motif located near the C-terminus of Rac2 is essential for Rac2 biologic functions and intracellular localization. *Blood* **100,** 1679–1688.

Tavor, S., Petit, I., Porozov, S., Avigdor, A., Dar, A., Leider-Trejo, L., Shemtov, N., Deutsch, V., Naparstek, E., Nagler, A., and Lapidot, T. (2004). CXCR4 regulates migration and development of human acute myelogenous leukemia stem cells in transplanted NOD/SCID mice. *Cancer Res.* **64,** 2817–2824.

Thomas, E. K., Cancelas, J., Chae, H. D., Cox, A. D., Keller, P. J., Perrotti, D., Neviani, P., Druker, B., Setchell, K. D. R., Zheng, Y., Harris, C., and Williams, D. A. (2007). Rac guanosine triphosphatases represent integrating molecular therapeutic targets for BCR-ABL-induced myeloproliferative disease. *Cancer Cell* **12,** 467–478.

Thomas, J., Liu, F., and Link, D. C. (2002). Mechanisms of mobilization of hematopoietic progenitors with granulocyte colony-stimulating factor. *Curr. Opin. Hematol.* **9,** 183–189.

Thrasher, A. J., and Burns, S. (1999). Wiskott-Aldrich syndrome: A disorder of haematopoietic cytoskeletal regulation. *Microsc. Res. Tech.* **47,** 107–113.

Toksoz, D., Zsebo, K. M., Smith, K. A., Hu, S., Brankow, D., Suggs, S. V., Martin, F. H., and Williams, D. A. (1992). Support of human hematopoiesis in long-term bone marrow cultures by murine stromal cells selectively expressing the membrane-bound and secreted forms of the human homolog of the steel gene products, stem cell factor. *Proc. Natl. Acad. Sci. USA* **89,** 7350–7354.

van der Loo, J. C., Xiao, X., McMillin, D., Hashino, K., Kato, I., and Williams, D. A. (1998). VLA-5 is expressed by mouse and human long-term repopulating hematopoietic cells and mediates adhesion to extracellular matrix protein fibronectin. *J. Clin. Invest.* **102,** 1051–1061.

Volkman, B. F., Prehoda, K. E., Scott, J. A., Peterson, F. C., and Lim, W. A. (2002). Structure of the N-WASp EVH1 domain-WIP complex: Insight into the molecular basis of Wiskott-Aldrich syndrome. *Cell* **111,** 565–576.

Walmsley, M. J., Ooi, S. K., Reynolds, L. F., Smith, S. H., Ruf, S., Mathiot, A., Vanes, L., Williams, D. A., Cancro, M. P., and Tybulewicz, V. L. (2003). Critical roles for Rac1 and Rac2 GTPases in B cell development and signaling. *Science* **302,** 459–462.

Wang, J., Kimura, T., Asada, R., Harada, S., Yokota, S., Kawamoto, Y., Fujimura, Y., Tsuji, T., Ikehara, S., and Sonoda, Y. (2003). SCID-repopulating cell activity of human cord blood-derived CD34- cells assured by intra-bone marrow injection. *Blood* **101,** 2924–2931.

Wang, L., Yang, L., Filippi, M. D., Williams, D. A., and Zheng, Y. (2006). Genetic deletion of Cdc42GAP reveals a role of Cdc42 in erythropoiesis and hematopoietic stem/progenitor cell survival, adhesion, and engraftment. *Blood* **107,** 98–105.

Warren, S., Chute, R. N., and Farrington, E. M. (1960). Protection of the hematopoietic system by parabiosis. *Lab. Invest.* **9,** 191–198.

Weber, K. S., Klickstein, L. B., Weber, P. C., and Weber, C. (1998). Chemokine-induced monocyte transmigration requires cdc42-mediated cytoskeletal changes. *Eur. J. Immunol.* **28,** 2245–2251.

Welsh, C. F., Roovers, K., Villanueva, J., Liu, Y., Schwartz, M. A., and Assoian, R. K. (2001). Timing of cyclin D1 expression within G1 phase is controlled by Rho. *Nat. Cell Biol.* **3**, 950–957.

Wengler, G., Gorlin, J. B., Williamson, J. M., Rosen, F. S., and Bing, D. H. (1995). Nonrandom inactivation of the X chromosome in early lineage hematopoietic cells in carriers of Wiskott-Aldrich syndrome. *Blood* **85**, 2471–2477.

Westerberg, L., Larsson, M., Hardy, S. J., Fernandez, C., Thrasher, A. J., and Severinson, E. (2005). Wiskott-Aldrich syndrome protein deficiency leads to reduced B-cell adhesion, migration, and homing, and a delayed humoral immune response. *Blood* **105**, 1144–1152.

Williams, D. A., Rios, M., Stephens, C., and Patel, V. P. (1991a). Fibronectin and VLA-4 in haematopoietic stem cell-microenvironment interactions. *Nature* **352**, 438–441.

Williams, D. A., Tao, W., Yang, F. C., Kim, C., Gu, Y., Mansfield, P., Levine, J. E., Petryniak, B., Derrow, C. W., Harris, C., Jia, B., Zheng, Y., *et al.* (2000). Dominant negative mutation of the hematopoietic-specific RhoGTPase, Rac2, is associated with a human phagocyte immunodeficiency. *Blood* **96**, 1646–1654.

Williams, D. E., Fletcher, F. A., Lyman, S. D., and de Vries, P. (1991b). Cytokine regulation of hematopoietic stem cells. *Semin. Immunol.* **3**, 391–396.

Wilson, A., and Trumpp, A. (2006). Bone-marrow haemoatopoietic stem-cell niches. *Nat. Rev. Immunol.* **6**, 93–106.

Wright, D. E., Bowman, E. P., Wagers, A. J., Butcher, E. C., and Weissman, I. L. (2002). Hematopoietic stem cells are uniquely selective in their migratory response to chemokines. *J. Exp. Med.* **195**, 1145–1154.

Wright, D. E., Wagers, A. J., Gulati, A. P., Johnson, F. L., and Weissman, I. L. (2001). Physiological migration of hematopoietic stem and progenitor cells. *Science* **294**, 1933–1936.

Yang, F. C., Atkinson, S. J., Gu, Y., Borneo, J. B., Roberts, A. W., Zheng, Y., Pennington, J., and Williams, D. A. (2001). Rac and Cdc42 GTPases control hematopoietic stem cell shape, adhesion, migration, and mobilization. *Proc. Natl. Acad. Sci. USA* **98**, 5614–5618.

Yang, F. C., Ingram, D. A., Chen, S., Hingtgen, C. M., Ratner, N., Monk, K. R., Clegg, T., White, H., Mead, L., Wenning, M. J., Williams, D. A., Kapur, R., *et al.* (2003). Neurofibromin-deficient Schwann cells secrete a potent migratory stimulus for Nf1+/- mast cells. *J. Clin. Invest.* **112**, 1851–1861.

Yang, F. C., Kapur, R., King, A. J., Tao, W., Kim, C., Borneo, J., Breese, R., Marshall, M., Dinauer, M. C., and Williams, D. A. (2000). Rac 2 stimulates Akt activation affecting BAD/Bcl-X$_L$ expression while mediating survival and actin-based cell functions in primary mast cells. *Immunity* **12**, 557–568.

Yang, G., Zhang, F., Hancock, C. N., and Wessler, S. R. (2007a). Transposition of the rice miniature inverted repeat transposable element mPing in. *Arabidopsis thaliana. Proc. Natl. Acad. Sci. USA* **104**, 10962–10967.

Yang, L., Wang, L., Kalfa, T., Cancelas, J. A., Shang, X., Pushkaran, S., Mo, J., Williams, D. A., and Zheng, Y. (2007b). Cdc42 critically regulates the balance between myelopoiesis and erythropoiesis. *Blood.*

Zicha, D., Allen, W. E., Brickell, P. M., Kinnon, C., Dunn, G. A., Jones, G. E., and Thrasher, A. J. (1998). Chemotaxis of macrophages is abolished in the Wiskott-Aldrich syndrome. *Br. J. Haematol.* **101**, 659–665.

CHAPTER TWENTY-EIGHT

In Vitro and In Vivo Assays to Analyze the Contribution of Rho Kinase in Angiogenesis

Kenjiro Sawada, Ken-ichirou Morishige, Seiji Mabuchi, Seiji Ogata, Chiaki Kawase, Masahiro Sakata, *and* Tadashi Kimura

Contents

Abstract

Therapeutic targeting of angiogenesis has become a novel approach to cancer therapy. The recent discovery of specific molecular targets that modulate the endothelial cell response and the development of suitable methods for assessing the contribution of these targets have given further impetus for the development and therapeutic application of an angiogenesis-targeted therapy. The small GTPase Rho and the major downstream effector, Rho kinase, is well established to regulate cell migration by the formation of stress fibers and the

Department of Obstetrics and Gynecology, Osaka University Graduate School of Medicine, Osaka, Japan

Methods in Enzymology, Volume 439
ISSN 0076-6879, DOI: 10.1016/S0076-6879(07)00428-4

turnover of focal adhesions and plays a pivotal role in endothelial cell organization in angiogenesis. Several approaches have been developed to analyze the contribution of Rho kinase so far, and this chapter describes the *in vitro* and *in vivo* protocols used routinely in our laboratory to analyze the contribution in angiogenesis and shows the possibility of Rho kinase inhibitors in the clinical setting.

1. INTRODUCTION

Angiogenesis, the formation of new blood vessels from existing ones, is involved in tissue repair and a variety of diseases, including tumor development (Folkman, 1995; van Nieuw Amerongen *et al.*, 2003). One of the most important proangiogenic factors is vascular endothelial growth factor (VEGF) (Hoeben *et al.*, 2004). VEGF induces endothelial permeability, migration, proliferation, and angiogenesis by multiple mechanisms, including alteration of the F-actin cytoskeleton and induction of gene expression (Hoeben *et al.*, 2004).

Among those, reorganization of the F-actin cytoskeleton and cell–matrix adhesion play crucial roles in endothelial cell (EC) migration in angiogenesis and the repair of injuries along the endothelium. For cells to migrate, they must form new lamellipodia, adhere to the substratum at the front of the cell, detach from the substratum at the tail of the cell, and retract their tail (Kiosses *et al.*, 1999). Formation of adhesive structures and cellular contraction are essential in this process. ECs contain cytoskeletal "cables" of F-actin and nonmuscle myosin filaments that can contract and exert tension (Katoh *et al.*, 1998; Ridley, 1999). A prominent group of these F-actin cables are called stress fibers, which are linked to the cell membrane at focal adhesions. VEGF is known to induce the formation of stress fibers and focal adhesions *in vitro* (Huot *et al.*, 1998). Many studies have shown that stress fibers develop during EC adaptation to unfavorable or pathological situations (White *et al.*, 1983).

It has become well established that the formation of stress fibers and focal adhesions is regulated by the small GTPase Rho (Narumiya, 1996). Numerous effectors for Rho have been identified so far, and the best-characterized downstream effector of Rho is Rho kinase (ROCK), which mediates various cellular functions. ROCK was identified in the mid-1990s as one of the downstream effectors of Rho, consisting of two isoforms of ROCK: ROCK1 and ROCK2 (Fukata *et al.*, 2001; Shimokawa and Takeshita, 2005). ROCK is well demonstrated to regulate the formation of stress fibers and focal adhesion complexes and to increase myosin light-chain (MLC) phosphorylation, which results in an increase in actomyosin-based contractility (Kimura *et al.*, 1996; Sawada *et al.*, 2002a). Accumulating evidence indicates that ROCK inhibitors could not only cover the wide range of pharmacological effects of conventional cardiovascular drugs, but also have the potential to treat osteoporosis, renal disease, erectile dysfunction, and cancers (Shimokawa and Takeshita, 2005). As a pharmacological inhibitor of ROCK, fasudil and Y-27632 have been developed, which inhibit ROCK activity in a competitive manner with ATP (Asano *et al.*, 1987; Uehata *et al.*, 1997). It has been demonstrated that hydroxyfasudil, a major active metabolite of fasudil after oral administration, has a more specific inhibitory effect on ROCK (Shimokawa *et al.*, 1999). The K_i values (μmol) of hydroxyfasudil and Y-27632 are 0.17 and 0.14 for ROCK, 18 and 26 for protein kinase C, and 140 and >250 for MLC kinase, respectively (Shimokawa *et al.*, 1999; Uehata *et al.*, 1997). The development of these drugs has made it possible to analyze the contribution of ROCK in various cellular events.

This chapter describes protocols used routinely in our laboratory to investigate the contribution of ROCK in angiogenesis induced by VEGF and presents data that show pivotal roles of ROCK in angiogenesis.

2. *In Vitro* Assays Used to Analyze the Contribution of Rho Kinase in Angiogenesis

2.1. Protocol for primary cultures of human umbilical vein endothelial cells (HUVECs)

Human umbilical vein endothelial cells represent a model for a large community of researchers all around the world, although properties of HUVECs certainly cannot represent all the metabolic capacities and the responses in physiopathology and toxicity related to the different types of ECs found in an organism (Baudin *et al.*, 2007). Nevertheless, HUVECs are the most simple and available human EC type and are accurate for the preparation of large quantities of cells. This section describes the protocol for easy isolation and culture of HUVECs based on the method of Jaffe *et al.* (1973).

2.1.1. Culture medium

HuMedia-EG2 (Kurabo Industries, Japan) supplemented with 20 fetal bovine serum (BSA), 10 ng/ml human epidermal growth factor, 5 ng/ml human fibroblast growth factor B, 1 μg/ml hydrocortisone, 50 μg/ml gentamicin, 50 ng/ml amphotericin B, and 10 μg/ml heparin is used to maintain HUVECs. M199 (Sigma) supplemented with 1% BSA and penicillin/streptomycin is used during experiments.

2.1.2. Cord manipulation and culture

1. Tidily cut the two ends of the fresh umbilical cords with a scalpel.
2. Introduce a cannula at each extremity of the vein (the widest vessel) and tightly maintain it with string.
3. Wash the cord with phosphate-buffered saline (PBS) without calcium or magnesium using a 50-ml syringe.
4. Inject the trypsin (0.25%)/PBS at one end of the vein using a 30-ml syringe.
5. When leaking out the other end, clamp it tightly with a surgical clip.
6. Incubate the cord for 20 min at room temperature.
7. Gently squeeze the cord to facilitate cell detachment.
8. Fill up a 50-ml sterile Falcon with 10 ml of "full" culture medium.
9. Collect the cells in this Falcon by washing the vein with 40 ml of PBS.
10. Centrifuge at 1000 rpm for 10 min.
11. Carefully discard the supernatant and suspend cells in HuMedia-EB2 medium.

Maintain cultures in a humidified 95% air and 5% CO_2 atmosphere at 37°. Obtain and use subcultures by trypsinization for experiments at passages 3 to 5. Typically, HUVECs present a "cobblestone appearance" in phase-contrast microscopy when they are confluent. Before experiments, make cells quiescent under serum-free or 0.5 to 1% fetal bovine serum conditions in M199 medium for 24 h.

2.2. *In vitro* scratch assay

The *in vitro* scratch assay is an easy, low-cost, and well-developed method used to measure cell migration *in vitro*. The basic steps involve creating a "scratch" in a cell monolayer, capturing the images at the beginning and at regular intervals during cell migration to close the scratch, and comparing the images to quantify the migration rate of the cells (Liang *et al.*, 2007). One of the major advantages of this simple method is that it mimics to some extent the migration of cells *in vivo*. For example, removal of part of the endothelium in the blood vessels will induce migration of ECs into the denuded area to close the wound (Haudenschild and Schwartz, 1979). This section describes the protocol used in our laboratory.

2.2.1. Materials

Recombinant human VEGF (Wako, Japan)
Paraformaldehyde (Sigma)

2.2.2. Procedures

1. Culture HUVECs to subconfluency in 24-well culture plates in M199 medium.
2. Make one straight wound per well with a plastic tip (P-200).

3. Culture the cells further in serum-free M199 medium in the presence or absence of VEGF.
4. The optimal concentration of VEGF is 3 to 10 ng/ml (Hashimoto *et al.*, 2007; Yin *et al.*, 2007).
5. After 16 h of incubation, fix the cells in 3.7% paraformaldehyde in PBS and stain with Giemsa solution.
6. Assess cell migration by examining photographs of cells that have migrated inside the wound area.

An example of the *in vitro* scratch assay is shown in Fig. 28.1. In this assay system, VEGF stimulated the cell migration of HUVECs (see Fig. 28.1, middle). This study used alendronate (ALN), a nitrogen-containing bisphosphonate, which is a potent inhibitor of bone resorption used for the

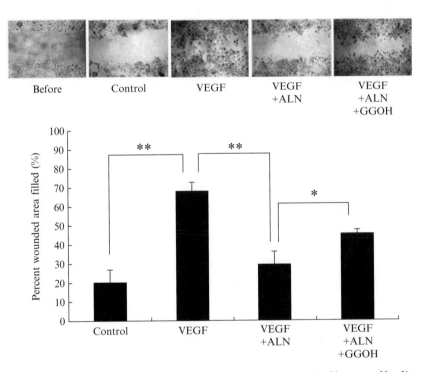

Figure 28.1 Effect of alendronate on HUVEC migration examined by wound healing assay. EC monolayers were wounded at time 0, and cultures were incubated with M199/0.1% BSA in the absence or presence of VEGF, ALN, and GGOH. Micrographs of HUVEC cultures recorded at the time of wounding (time 0) and 16 h after wounding are shown. Percentage wound closure was measured and compared with that at time 0. Experiments were repeated three times and values are means ($\pm$SD) of triplicates. $\star P <$ 0.05, $\star\star P <$ 0.01. The concentrations of agents used are as follows: 10 ng/ml VEGF, 30 μM ALN, and 30 μM GGOH. Reproduced from Hashimoto *et al.* (2007).

treatment and prevention of osteoporosis (Halasy-Nagy *et al.*, 2001; Sawada *et al.*, 2002b). Alendronate inhibits the biosynthesis pathway, as well as isoprenylation (farnesylation and geranylgeranylation), by inhibiting isopentanyl diphosphate synthase, the downstream enzyme farnesyl pyrophosphate synthase, or both (Fisher *et al.*, 1999). Protein targets of isoprenylation include small G proteins such as Rho, Ras, Rac, and Rab, which require post-translational modification to undergo a series of changes that lead to their attachment to the plasma membranes and their full function. Alendronate inhibited this VEGF-induced HUVEC mobility in a dose-dependent manner (1–30 μM). However, incubation with ALN plus geranylgeraniol (GGOH), which is metabolized to geranylgeranyl pyrophosphate and is used to form geranylgeranylated (active) Rho, restored alendronate-reduced HUVEC motility (see Fig. 28.1, right).

2.3. Tube formation assay

Morphologic differentiation of ECs to form tubes is essential for the process of angiogenesis. VEGF and VEGFR2 constitute a paracrine signaling system crucial for the differentiation of ECs and the development of the vascular system (Risau and Flamme, 1995). The process of angiogenesis is completed by the formation of microvascular tubes. *In vitro* assays of the tube formation of ECs have been developed and used to assess this critical step of angiogenesis (Hashimoto *et al.*, 2007).

2.3.1. Materials

Growth factor-reduced Matrigel (BD Biosciences)
Y-27632 (Sigma)
Fasudil (Sigma, HA-1077 dihydrochloride)

2.3.2. Procedures

1. Coat the surface of 96-well plates with 30 μl of growth factor-reduced Matrigel matrix.
2. Allow the surface to polymerize at 37° for 1 h.
3. Pretreat various reagents that affect ROCK activation (Y-27632, fasudil) in serum-starved HUVECs for 1 h.
4. Release the cells from culture dishes using trypsin, wash, and resuspend in serum-free M199 media (1% BSA) with the same concentration of regents plus VEGF (10 ng/ml).
5. Add the cells (2 × 10^4 per well) onto the Matrigel-coated wells.
6. After 8 to 16 h of incubation, photograph the wells randomly at 50× magnification.

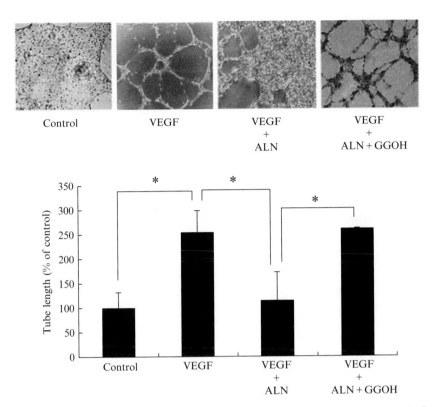

Figure 28.2 Effect of alendronate on tube formation by HUVECs cultured on Matrigel. HUVECs were cultured in M199/0.1% BSA (control) with VEGF and various agents. Representative endothelial capillary-like tubes are shown. Tube length was quantified using NIH image software. Experiments were repeated three times and values are means ($\pm$SD) of triplicates. $*P < 0.05$. The concentrations of agents used are as follows: 10 ng/ml VEGF, 30 μM ALN, and 30 μM GGOH. Reproduced from Hashimoto et al. (2007).

Figure 28.2 shows the effect of ALN on tube formation on Matrigel, where ECs take only several hours to associate with each other and to form microtubes under VEGF stimulation. Alendronate prevents the VEGF-induced tube formation on Matrigel. However, incubation with alendronate plus GGOH abrogated the alendronate-induced inhibition of tube formation. These observations suggest that ALN inhibits the tube formation of microvascular ECs by blocking geranylgeranylation, which is essential for Rho activation.

2.4. Assessment of focal adhesion and stress fiber

Cell migration is regulated by a combination of different processes: forces generated by the contraction of actomyosin, the formation of stress fibers, and the turnover of focal adhesions (Horwitz and Parsons, 1999).

To analyze these cellular events, visualization of the formation of stress fibers and focal adhesions with the immunocytochemical study is well established.

2.4.1. Materials

Eight-well chamber slides (BD Biosciences)
Type 1 collagen (BD Biosciences)
Antipaxillin mouse monoclonal antibody (Biosource)
Antiphospho-paxillin rabbit polyclonal (Tyr118) antibody (Biosource)
Alexa Fluor 546-labeled phalloidin (Invitrogen)
Alexa Fluor 488-labed goat antimouse or rabbit IgG (Invitrogen)

2.4.2. Procedures

1. Dilute collagen type 1 with 0.1 N HCl in PBS at a concentration of 50 μg/ml.
2. Coat eight-well chamber slides with type 1 collagen at 37° for 1 h.
3. After washing with PBS, plate HUVECs (5×10^4/well) onto the chamber slides.
4. Culture the cells in serum-free M199 media (plus 0.1% BSA) with various agents for 24 h.
5. Stimulate the cells with VEGF for 30 min.
6. Fix the cells with 3.7% paraformaldehyde and permeabilize with 0.1% Triton X-100.
7. After blocking with 5% BSA and 5% goat serum, incubate the cells with antipaxillin (1:500) or antiphospho-paxillin (1:500) at 4° overnight.
8. After washing, incubate the cells with the corresponding secondary antibodies labeled with Alexa Fluor-488 (1:1000) for 1 h at room temperature.
9. Double stain the specimens with Alexa Fluor 546-labeled phalloidin (1:200) for 30 min at room temperature.
10. Record and analyze the images with confocal microscopy.

In HUVECs, VEGF induces an accumulation of stress fibers associated with new actin polymerization and rapid formation of focal adhesions at the ventral surface of the cells thorough the Rho signaling pathway (Le Boeuf *et al.*, 2004; Rousseau *et al.*, 2000). ROCK is well known to play a pivotal role in this process. The specific ROCK inhibitors are used to analyze the involvement of this pathway. As shown in Fig. 28.3B, VEGF treatment caused a drastic increase in the actin bundles and changed the localization of paxillin to the focal adhesions. Treatment with fasudil significantly inhibited the VEGF-induced formation of stress fibers and focal adhesion assembly (see Fig. 28.3C), as well as the treatment with Y-27632 (see Fig. 28.3E). Also, VEGF induced a remarkable increase in phosphorylated paxillin in the focal adhesions compared with the level in the quiescent cells

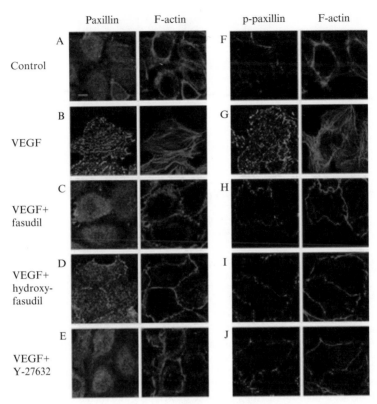

Paxillin F-actin p-paxillin F-actin

Figure 28.3 (A) Fasudil and hydroxyfasudil inhibited VEGF–induced stress fiber formation and focal adhesion assembly associated with the tyrosine phosphorylation of paxillin. After pretreatment with fasudil (10 μM), hydroxyfasudil (10 μM), or Y-27632 (10 μM) for 30 min, HUVECs were incubated in the absence (A, F) or presence (B-E, G-J) of 3 ng/ml VEGF with the addition of fasudil (C, H), hydroxyfasudil (D, I), or Y-27632 (E, J) for 30 min and double stained with Alexa 546-labeled phalloidin and antipaxillin antibody (A-E) or antiphosphospecific paxillin (Try 118) antibody (F-J) followed by Alexa 488-labeled goat antimouse or rabbit IgG. Fluorescence microscopy images focused near the bottom of the cells are shown. Scale bar: 10 μm. Reproduced from Yin, et al. (2007).

(see Fig. 28.3F). When the ROCK inhibitors were present during VEGF stimulation, the staining of phosphorylated paxillin was diminished drastically (see Figs. 28.3H–28.3J). These data show that ROCK regulates VEGF–induced stress fiber formation and focal adhesion complex assembly.

Furthermore, to confirm that these immnunocytochemical data correlated with the biochemical analysis of focal adhesion proteins in HUVECs, the effects of ROCK inhibitors on the tyrosine phosphorylation of focal adhesion proteins FAK and paxillin were analyzed by Western blotting. As shown in Fig. 28.4, the tyrosine phosphorylation of FAK and paxillin increased remarkably after VEGF stimulation and was inhibited significantly by ROCK inhibitors.

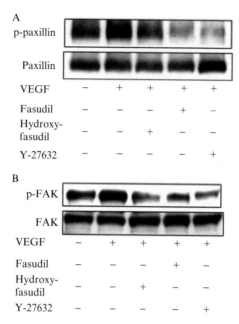

Figure 28.4 Western blot analysis of the effects of fasudil and hydroxyfasudil on VEGF-stimulated tyrosine phosphorylation of paxillin (A) and FAK (B). After pretreatment with fasudil (10 μM), hydroxyfasudil (10 μM), or Y-27632 (10 μM) for 30 min, HUVECs were stimulated with 3 ng/ml VEGF for 30 min. The tyrosine phosphorylation of paxillin and FAK was detected by immunoblotting with antiphospho-paxillin or antiphospho-FAK antibody (top), followed by stripping and then reprobed antipaxillin monoclonal antibody or anti-FAK antibody (bottom). Reproduced from Yin *et al.* (2007).

2.5. Cell transfection

To examine the direct effect of ROCK, the transfection method of CA-ROCK is used.

2.5.1. Materials

pEF-BOS-myc-Rho kinase-CAT (the catalytic domain of Rho kinase, acting as the constitutively active form of Rho kinase, CA-ROCK) was kindly provided by Kozo Kaibuchi (Nagoya University, Nagoya, Japan) (Amano *et al.*, 1998)

Anti-myc monoclonal antibody (9E10) (Santa Cruz Biotechnology)

jetPEI-HUVEC (Polyplus-transfection), which has been optimized for the transfection of primary human endothelial cells such as HUVECs.

2.5.2. Procedures

1. Seed HUVECs (2 × 10⁴/well) in HuMedia-EG2 medium on eight-well chamber slides coated with type I collagen.
2. Dilute 1 μg of DNA into 25 μl of 150 mM NaCl, vortex gently, and spin down briefly.

3. Dilute 2 μl of jetPEI-HUVEC into 25 μl of 150 mM NaCl. Vortex gently and spin down briefly.
4. Add the 25 μl containing the jetPEI-HUVEC to the 25-μl DNA solution at once and mix gently.
5. Incubate for 30 min at room temperature.
6. Add the 50 μl jetPEI-HUVEC/DNA mix onto the wells that contain M199-2% fetal bovine serum and homogenize by swirling the plate gently.
7. Incubate at 37° in a humidified atmosphere for 4 h.
8. Remove the transfection medium and replace with HuMedia-EG2 medium.
9. After 24 h, fix and double stain HUVECs with anti-myc monoclonal mouse antibody (1:100) and antiphopho-paxillin polyclonal rabbit antibody as described earlier, followed by staining with Alexa-Fluor 546-labeled goat antimouse IgG and Alexa Fluor 488-labeled goat antirabbit IgG.
10. Record and analyze the images with a confocal microscopy.

An example of the transfected cells is shown in Fig. 28.5. In CA-ROCK plasmid-transfected cells, focal adhesion assembly with a noticeable increase of paxillin phosphorylation was observed, similar to the treatment of VEGF in nontransfected cells. Data that focal adhesion assembly could not be blocked by fasudil in CA-ROCK plasmid-transfected cells suggest that

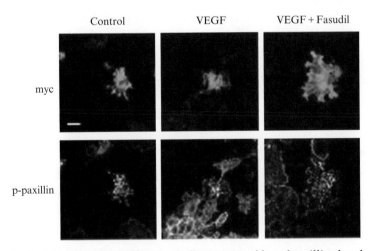

Figure 28.5 The effect of fasudil on focal adhesion assembly and paxillin phosphorylation after HUVECs were transfected with CA-ROCK plasmid (pEF-BOS-myc-Rho kinase-CAT). After transfection with the plasmid, cell were pretreated with or without 10 μM fasudil for 30 min and then stimulated with or without VEGF (3 ng/ml) for 30 min. Cells were double stained with anti-myc monoclonal antibody and antiphospho-paxillin polyclonal antibody followed by Alexa 546-labeled goat antimouse IgG and Alexa 488-labeled goat antirabbit IgG. Scale bar: 30 μm. Reproduced from Yin *et al.* (2007).

the effect of fasudil is ROCK dependent and that VEGF-induced focal adhesion assembly is regulated by ROCK.

2.6. Rho pull-down assay

The Rho family of small GTPases acts as molecular switches that transmit cellular signals through an array of effector proteins. This family mediates a wide range of cellular responses, including cytoskeletal reorganization (Ridley and Hall, 1992; Ridley et al., 1992), regulation of transcription (Waterman-Storer et al., 1999), DNA synthesis, membrane trafficking, and apoptosis (Hill et al., 1995; Hotchin et al., 2000). Traditionally, the direct assay using a pull-down method, wherein a Rho-GTP-binding domain (RBD) of a Rho effector is coupled to agarose beads to affinity purify the active Rho in a biological sample, is performed.

2.6.1. Materials

See Hashimoto et al. (2007)

Rho activation assay kit (Upstate Biotechnology)

Mg^{2+} lysis/wash buffer, $5\times$ ($5\times$ MLB: 125 mM HEPES, pH 7.5, 750 mM NaCl, 5% Igepal CA-630, 50 mM MgCl$_2$, 5 mM EDTA, and 10% glycerol)

Anti-Rho (-A, -B, -C), clone 55

Rho assay reagent (Rhotekin RBD, agarose): one vial containing 650 μg of recombinant protein in 1 ml glutathione-agarose slurry of 50 mM Tris, pH 7.5, 0.5% Triton X-100, 150 mM NaCl, 5 mM MgCl$_2$, 1 mM dithiothreitol, 0.1 mM phenylmethylsulfonyl fluoride, and 1 μg/ml each of aprotinin and leupeptin before the addition of glycerol to 10%

2.6.2. Procedures

1. Culture HUVECs (3×10^5/well) in six-well plates under serum-free M199 (1% BSA) medium with various agents.
2. After the stimulation with VEGF for 10 min, lyze the cells in 250 μl of Mg^{2+} lysis/wash buffer.
3. Add 20 to 30 μg of the Rho assay reagent slurry (Rhotekin RBD agarose beads) and incubate the reaction mixtures for 45 min at 4° with gentle agitation.
4. Pellet the agarose beads by brief centrifugation (10 s, 14,000 g, 4°).
5. Remove, discard the supernatant, and wash the beads three times with MLB. Add 0.5 ml MLB per wash, mix gently, pellet beads, and remove supernatant. Take care to minimize loss of beads when supernatant is removed.
6. Resuspend the agarose beads in 25 μl of $2\times$ reducing sample buffer and boil for 5 min.

7. Perform SDS-PAGE and transfer to a nitrocellulose membrane.
8. Detect the active form of Rho (bound Rho proteins) by Western blot using the anti-Rho antibody.

An example of a pull-down assay with the fusion protein GST-RBD, which recognizes only Rho-GTP, the active form of Rho, is shown in Fig. 28.6. An increase in Rho-GTP was observed in HUVECs treated for 10 min with VEGF. Alendronate inhibits Rho activation induced by VEGF and GGOH abrogates its inhibition.

3. IN VIVO ASSAYS USED TO ANALYZE THE CONTRIBUTION OF RHO KINASE IN ANGIOGENESIS

3.1. CD31 staining

CD31 immunostaining is one of the most authentic methods to visualize the intratumor vasculature. For frozen sections of mouse xenografts, the following protocol is used (Hashimoto *et al.*, 2007).

3.1.1. Materials

Goat serum (Wako, Japan)
Rat antimouse CD31 antibody (Pharminogen)
Alexa Fluor 488-labeled goat antirat IgG (Invitrogen)

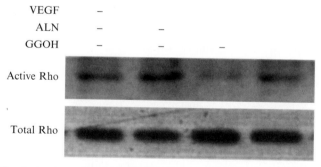

Figure 28.6 Activation of Rho by VEGF is suppressed by alendronate and is restored by the addition of GGOH in HUVECs. HUVECs were cultured under serum-free conditions with or without the various agents indicated for 24 h. After incubation, cells were stimulated with 10 ng/ml VEGF. The cell lysates were incubated with Rhotekin RBD-agarose beads. Bound Rho proteins were detected by Western blotting using the polyclonal antibody against RhoA (top). Western blotting of the total amount of Rho in cell lysates (bottom) was performed for comparison with the Rho activity (level of GTP-bound Rho) in the same lysates. The concentrations of agents used are as follows: 30 μM ALN and 30 μM GGOH. Reproduced from Hashimoto *et al.* (2007).

3.1.2. Procedures

1. Cut frozen tissues into a 10-μm section and fix with acetone at $4°$ for 10 min.
2. Wash with PBS twice, and block with 5% goat serum/PBS at room temperature for 45 min.
3. Add the primary antibody (rat antimouse CD31, Pharminogen) at 1:50 in 2% goat serum in PBS at $37°$ for 30 min.
4. After washing, add the secondary antibody (Alexa Fluor 488-labeled goat antirat IgG) at 1:1000 for 45 min.
5. After washing, mount with 4′,6-diamidino-2-phenylindole and observe with a photomicroscope. The serial section adjacent to each immunofluorescence-stained section should be subjected to hematoxylin-eosin staining to confirm tumors are loaded.

3.2. Directed *in vivo* angiogenesis assay (DIVAA)

One of the major problems in angiogenesis research remains the lack of suitable methods for quantifying the angiogenic response *in vivo*. Guedez *et al.* (2003) developed the DIVAA and demonstrated that it is reproducible and quantitative. We herein describe the DIVAA method based on their method.

3.2.1. Materials

Athymic nude mouse (BALB/cA nu/nu; CLEA Japan, Inc.)
Directed *in vivo* angiogenesis assay (Trevigen, Inc.)
10× wash buffer, dilute 25 ml of 10× wash buffer in 225 ml of sterile, deionized water
25× FITC-lectin diluent, dilute 400 μl of 25× FITC-lectin diluent in 9.6 ml of deionized water
200× FITC-lectin, dilute 50 μl of 200× FITC-lectin in 10 ml of 1× FITC-lectin diluent
CellSperse

3.2.2. Procedures

1. Thaw growth factor-reduced basement membrane extract (BME) at $4°$ on ice.
2. Prechill all pipette tips, angioreactors, and AngioRack.
3. Place angioreactors in the AngioRack. Add 18 μl of BME with or without VEGF to each angioreactor using a prechilled, sterile gel-loading tip.
4. Once the angioreactors are filled, immediately invert angrioreactors, transfer to a sterile microtube, and place at $37°$ for 1 h to polymerize.

5. An approximately 1-cm incision should be made on the dorsal–lateral surface of a nude mouse; implant the angioreactor into the dorsal flank of the mouse with an open end opposite incision.

6. Maintain mice for 2 weeks.

7. After the maintenance period, euthanize mice humanely and remove angioreactors.

8. Carefully remove the bottom cap of the angioreactors with a sterile razor blade and, using a sterile 200-μl pipette tip, push the BME/vessel complex out of the angioreactor into a microtube.

9. Rinse the inside of each angioreactor with 300 μl of CellSperse and transfer into a microtube.

10. Incubate at 37° to digest BME and create a single cell suspension for 1 to 3 h.

11. Centrifuge digested BME at 250 g for 5 min at room temperature to collect cell pellets and insoluble fractions, and discard supernatant. Resuspend pellet in 500 μl of Dulbecco's modified Eagle's medium and incubate at 37° for 1 h.

12. Centrifuge cells at 250 g for 10 min at room temperature to collect cell pellets and resuspend pellet in 500 μl of DIVAA wash buffer.

13. Resuspend pellet in 200 μl of DIVAA FITC–lectin and incubate at 4° overnight.

14. Centrifuge at 250 g, and remove supernatant. Wash pellet three times in DIVAA wash buffer.

15. Suspend pellet in 100 μl of DIVAA wash buffer for fluorometric determination.

16. Measure fluorescence in 96-well plates (excitation 485 nm, emission 510 nm).

An example of DIVAA is shown in Fig. 28.7. Here we confirmed that 500 ng/ml VEGF induced *in vivo* angiogenesis. An obvious red part in the angioreactors was observed in the VEGF treatment group (see Fig. 28.7A), in which new vessel formation was confirmed, and the fluorescence was also increased significantly compared with the control. In the presence of 100 μM fasudil, new vessel formation was prevented and the fluorescence was reduced to the basic level (see Fig. 28.7B). These data show that fasudil, the ROCK inhibitor, inhibits VEGF-induced angiogenesis *in vivo*.

4. CONCLUDING REMARKS

Although the importance of VEGF for angiogenesis is well established, the mechanisms that regulate the morphogenic processes by which proliferating ECs are organized into blood vessels are poorly understood. It is likely that the mechanisms that regulate EC morphogenesis act through

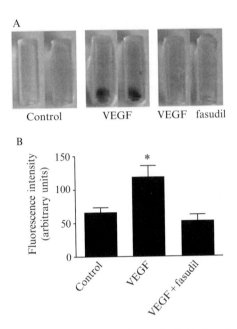

Figure 28.7 The inhibitory effects of fasudil on VEGF-induced angiogenesis *in vivo*.
(A) The directed *in vivo* angiogenesis assay was performed to determine the effect of
fasudil on VEGF-induced angiogenesis *in vivo*. After implantation with angioreactors
containing BME with or without 500 ng/ml VEGF (or 500 ng/ml VEGF + 100 μM
fasudil) for 2 weeks, angioreactors were removed from the mice; cells were collected
from BME and labeled with FITC-lectin. Then fluorescence was measured by spectro-
fluorometry. Photographs of paired angioreactors were taken using a digital camera.
Angioreactors were oriented with the open end at the bottom and the sealed end at the
top. (B) Results were expressed as the fluorescence intensity (arbitrary units), mean ±
SEM ($n = 3$), ⋆$P < 0.05$ compared with control. Reproduced from Yin *et al.* (2007).

cytoskeletal elements, and therefore cytoskeletal regulatory molecules are
involved. Recent development of the specific ROCK inhibitors and
improvement of the suitable experimental model systems make it possible
to analyze the contribution of ROCK in these complicated cellular events
as described in this chapter.

Ying *et al.* (2006) reported that fasudil inhibits human tumor migration
and anchorage-independent growth and inhibits tumor progression in
animal models. Considering the findings of the direct antiangiogenic effect
of ROCK inhibitors described here, we expect that they are attractive
anticancer drug candidates to be explored in the near future.

ACKNOWLEDGMENTS

This work was supported in part by a grant-in-aid for scientific research from the Ministry of
Education, Science, Sports, and Culture of Japan. The authors are grateful to Mrs. Remina
Emoto for her secretarial assistance.

REFERENCES

Amano, M., Chihara, K., Nakamura, N., Fukata, Y., Yano, T., Shibata, M., Ikebe, M., and Kaibuchi, K. (1998). Myosin II activation promotes neurite retraction during the action of Rho and Rho-kinase. *Genes Cells* **3**, 177–188.

Asano, T., Ikegaki, I., Satoh, S., Suzuki, Y., Shibuya, M., Takayasu, M., and Hidaka, H. (1987). Mechanism of action of a novel antivasospasm drug, HA1077. *J Pharmacol Exp Ther.* **241**, 1033–1040.

Baudin, B., Bruneel, A., Bosselut, N., and Vaubourdolle, M. (2007). A protocol for isolation and culture of human umbilical vein endothelial cells. *Nat Protoc.* **2**, 481–485.

Fisher, J. E., Rogers, M. J., Halasy, J. M., Luckman, S. P., Hughes, D. E., Masarachia, P. J., Wesolowski, G., Russell, R. G., Rodan, G. A., and Reszka, A. A. (1999). Alendronate mechanism of action: Geranylgeraniol, an intermediate in the mevalonate pathway, prevents inhibition of osteoclast formation, bone resorption, and kinase activation *in vitro. Proc Natl Acad Sci USA* **96**, 133–138.

Folkman, J. (1995). Angiogenesis in cancer, vascular, rheumatoid and other disease. *Nat Med.* **1**, 27–31.

Fukata, Y., Amano, M., and Kaibuchi, K. (2001). Rho-Rho-kinase pathway in smooth muscle contraction and cytoskeletal reorganization of non-muscle cells. *Trends Pharmacol Sci.* **22**, 32–39.

Guedez, L., Rivera, A. M., Salloum, R., Miller, M. L., Diegmueller, J. J., Bungay, P. M., and Stetler-Stevenson, W. G. (2003). Quantitative assessment of angiogenic responses by the directed in vivo angiogenesis assay. *Am J Pathol.* **162**, 1431–1439.

Halasy-Nagy, J. M., Rodan, G. A., and Reszka, A. A. (2001). Inhibition of bone resorption by alendronate and risedronate does not require osteoclast apoptosis. *Bone* **29**, 553–559.

Hashimoto, K., Morishige, K., Sawada, K., Tahara, M., Shimizu, S., Ogata, S., Sakata, M., Tasaka, K., and Kimura, T. (2007). Alendronate suppresses tumor angiogenesis by inhibiting Rho activation of endothelial cells. *Biochem Biophys Res Commun.* **354**, 478–484.

Haudenschild, C. C., and Schwartz, S. M. (1979). Endothelial regeneration. II. Restitution of endothelial continuity. *Lab Invest.* **41**, 407–418.

Hill, C. S., Wynne, J., and Treisman, R. (1995). The Rho family GTPases RhoA, Rac1, and CDC42Hs regulate transcriptional activation by SRF. *Cell* **81**, 1159–1170.

Hoeben, A., Landuyt, B., Highley, M. S., Wildiers, H., Van Oosterom, A. T., and De Bruijn, E. A. (2004). Vascular endothelial growth factor and angiogenesis. *Pharmacol Rev.* **56**, 549–580.

Horwitz, A. R., and Parsons, J. T. (1999). Cell migration: Movin' on. *Science* **286**, 1102–1103.

Hotchin, N. A., Cover, T. L., and Akhtar, N. (2000). Cell vacuolation induced by the VacA cytotoxin of *Helicobacter pylori* is regulated by the Rac1 GTPase. *J Biol Chem.* **275**, 14009–14012.

Huot, J., Houle, F., Rousseau, S., Deschesnes, R. G., Shah, G. M., and Landry, J. (1998). SAPK2/p38-dependent F-actin reorganization regulates early membrane blebbing during stress-induced apoptosis. *J Cell Biol.* **143**, 1361–1373.

Jaffe, E. A., Nachman, R. L., Becker, C. G., and Minick, C. R. (1973). Culture of human endothelial cells derived from umbilical veins: Identification by morphologic and immunologic criteria. *J Clin Invest.* **52**, 2745–2756.

Katoh, K., Kano, Y., Masuda, M., Onishi, H., and Fujiwara, K. (1998). Isolation and contraction of the stress fiber. *Mol Biol Cell.* **9**, 1919–1938.

Kimura, K., Ito, M., Amano, M., Chihara, K., Fukata, Y., Nakafuku, M., Yamamori, B., Feng, J., Nakano, T., Okawa, K., Iwamatsu, A., and Kaibuchi, K. (1996). Regulation of myosin phosphatase by Rho and Rho-associated kinase (Rho-kinase). *Science* **273**, 245–248.

Kiosses, W. B., Daniels, R. H., Otey, C., Bokoch, G. M., and Schwartz, M. A. (1999). A role for p21-activated kinase in endothelial cell migration. *J Cell Biol.* **147**, 831–844.

Le Boeuf, F., Houle, F., and Huot, J. (2004). Regulation of vascular endothelial growth factor receptor 2-mediated phosphorylation of focal adhesion kinase by heat shock protein 90 and Src kinase activities. *J Biol Chem.* **279**, 39175–39185.

Liang, C. C., Park, A. Y., and Guan, J. L. (2007). In vitro scratch assay: A convenient and inexpensive method for analysis of cell migration in vitro. *Nat Protoc.* **2**, 329–333.

Narumiya, S. (1996). The small GTPase Rho: Cellular functions and signal transduction. *J Biochem. (Tokyo)* **120**, 215–228.

Ridley, A. J. (1999). Stress fibres take shape. *Nat Cell Biol.* **1**, E64–E66.

Ridley, A. J., and Hall, A. (1992). The small GTP-binding protein rho regulates the assembly of focal adhesions and actin stress fibers in response to growth factors. *Cell* **70**, 389–399.

Ridley, A. J., Paterson, H. F., Johnston, C. L., Diekmann, D., and Hall, A. (1992). The small GTP-binding protein rac regulates growth factor-induced membrane ruffling. *Cell* **70**, 401–410.

Risau, W., and Flamme, I. (1995). Vasculogenesis. *Annu Rev Cell Dev Biol.* **11**, 73–91.

Rousseau, S., Houle, F., Kotanides, H., Witte, L., Waltenberger, J., Landry, J., and Huot, J. (2000). Vascular endothelial growth factor (VEGF)-driven actin-based motility is mediated by VEGFR2 and requires concerted activation of stress-activated protein kinase 2 (SAPK2/p38) and geldanamycin-sensitive phosphorylation of focal adhesion kinase. *J Biol Chem.* **275**, 10661–10672.

Sawada, K., Morishige, K., Tahara, M., Ikebuchi, Y., Kawagishi, R., Tasaka, K., and Murata, Y. (2002a). Lysophosphatidic acid induces focal adhesion assembly through Rho/Rho-associated kinase pathway in human ovarian cancer cells. *Gynecol Oncol.* **87**, 252–259.

Sawada, K., Morishige, K., Tahara, M., Kawagishi, R., Ikebuchi, Y., Tasaka, K., and Murata, Y. (2002b). Alendronate inhibits lysophosphatidic acid-induced migration of human ovarian cancer cells by attenuating the activation of rho. *Cancer Res.* **62**, 6015–6020.

Shimokawa, H., Seto, M., Katsumata, N., Amano, M., Kozai, T., Yamawaki, T., Kuwata, K., Kandabashi, T., Egashira, K., Ikegaki, I., Asano, T., Kaibuchi, K., *et al.* (1999). Rho-kinase-mediated pathway induces enhanced myosin light chain phosphorylations in a swine model of coronary artery spasm. *Cardiovasc Res.* **43**, 1029–1039.

Shimokawa, H., and Takeshita, A. (2005). Rho-kinase is an important therapeutic target in cardiovascular medicine. *Arterioscler Thromb Vasc Biol.* **25**, 1767–1775.

Uehata, M., Ishizaki, T., Satoh, H., Ono, T., Kawahara, T., Morishita, T., Tamakawa, H., Yamagami, K., Inui, J., Maekawa, M., and Narumiya, S. (1997). Calcium sensitization of smooth muscle mediated by a Rho-associated protein kinase in hypertension. *Nature* **389**, 990–994.

van Nieuw Amerongen, G. P., Koolwijk, P., Versteilen, A., and van Hinsbergh, V. W. (2003). Involvement of RhoA/Rho kinase signaling in VEGF-induced endothelial cell migration and angiogenesis *in vitro*. *Arterioscler Thromb Vasc Biol.* **23**, 211–217.

Waterman-Storer, C. M., Worthylake, R. A., Liu, B. P., Burridge, K., and Salmon, E. D. (1999). Microtubule growth activates Rac1 to promote lamellipodial protrusion in fibroblasts. *Nat Cell Biol.* **1**, 45–50.

White, G. E., Gimbrone, M. A., Jr., and Fujiwara, K. (1983). Factors influencing the expression of stress fibers in vascular endothelial cells in situ. *J Cell Biol.* **97**, 416–424.

Yin, L., Morishige, K., Takahashi, T., Hashimoto, K., Ogata, S., Tsutsumi, S., Takata, K., Ohta, T., Kawagoe, J., Takahashi, K., and Kurachi, H. (2007). Fasudil inhibits vascular endothelial growth factor-induced angiogenesis *in vitro* and *in vivo*. *Mol Cancer Ther.* **6**, 1517–1525.

Ying, H., Biroc, S. L., Li, W. W., Alicke, B., Xuan, J. A., Pagila, R., Ohashi, Y., Okada, T., Kamata, Y., and Dinter, H. (2006). The Rho kinase inhibitor fasudil inhibits tumor progression in human and rat tumor models. *Mol Cancer Ther.* **5**, 2158–2164.

CHAPTER TWENTY-NINE

ANALYSIS OF CELL MIGRATION AND ITS REGULATION BY RHO GTPASES AND P53 IN A THREE-DIMENSIONAL ENVIRONMENT

Stéphanie Vinot, Christelle Anguille, Mrion de Toledo, Gilles Gadea, *and* Pierre Roux

Contents

Abstract

Cell migration plays a key role both in physiological conditions, such as tissue repair or embryonic development, and in pathological processes, including tumor metastasis. Understanding the mechanisms that allow cancer cells to invade tissues during metastasis requires studying their ability to migrate. While spectacular, the movements observed in cells growing on two-dimensional supports are likely only to represent a deformation of the physiological migratory behavior. In contrast, the analysis of cell migration on a support, which resembles the three-dimensional (3D) extracellular matrix, provides a more pertinent model of physiological relevance. This chapter provides protocols to assay the ability of cells to migrate or to invade a 3D matrix and to analyze their phenotypes. The invasion assay allows the quantification of tumor cell invasiveness, and the 3D migration assay permits the visual observation of the movements and

Universités Montpellier 2 et 1, CRBM, CNRS, UMR 5237, Montpellier, France

Methods in Enzymology, Volume 439
ISSN 0076-6879, DOI: 10.1016/S0076-6879(07)00429-6

morphology of migrating cells. This chapter also describes a method to examine the localization of different markers during 3D migration. Because Rho GTPases are clearly involved in migration and invasion, a protocol is supplied to evaluate their activation during cell migration. These techniques are especially suitable to elucidate the type of motility in a 3D matrix, particularly to discriminate between two different modes of migration adopted by cancer cells: blebbing versus elongation. Indeed, the way a cell moves may have important consequences for its invasiveness, as, for example, cancer cells adopt a rounded blebbing movement when deficient in p53.

1. INTRODUCTION

Cell migration is a multistep process that is crucial at many stages of embryonic morphogenesis (Locascio and Nieto, 2001). Cell migration is also essential not only for physiological processes, including tissue repair and immune response, but also for pathological processes, especially during the final steps of carcinogenesis (Luster *et al.*, 2005; Ridley *et al.*, 2003; Van Haastert and Devreotes, 2004; Vicente-Manzanares *et al.*, 2005). Our understanding of the mechanistic features of cell migration relies on planar and flat, two-dimensional (2D) culture studies. However, these artificial substrates can potentially distort the experimental outcomes by forcing cells to adjust their appearance to the constraints of these artificial, flat, and rigid surfaces. Hence, cell motility on 2D is not necessarily applicable to the study of the movement of cells in tissues and organs, as cells must adapt their migratory behavior in order to compensate for the lack of a three-dimensional (3D) microenvironment (Kim, 2005). Indeed, spatial and biochemical information issued from a 3D extracellular matrix (ECM) influence cell behavior, particularly cell migration.

Only the three-dimensional scaffold of 3D *in vitro* models can provide a genuine environment for moving cells that will take into account the mechanical forces generated by the environment as well as the biochemical complexity of the extracellular matrix molecular composition. Crucially, in 3D cultures, cell shape and cell environment closely match those of an *in vivo* situation. Furthermore, 3D tissue culture models provide a well-defined and controlled environment in contrast to the complex host environment of an *in vivo* model. Thus, 3D models faithfully mimic *in vivo* cell movement, without the complexity of a real organ or tissue, while representing a more physiological approach than the traditional cell culture models, therefore providing a valuable compromise between the two (Bissell *et al.*, 2002; Condeelis and Segall, 2003; Cukierman *et al.*, 2001; Ingber, 2002; Yamada and Cukierman, 2007).

The value of 3D ECM models is well documented for epithelial cells, where the 3D environment promotes normal epithelial polarity and differentiation and can recapitulate the different stages of cell movements. Moreover, 3D cultures have been instrumental in providing important insights on the mechanisms of cancer cell migration during carcinogenesis (Debnath and Brugge, 2005; Kim, 2005). Indeed, 3D matrices are particularly useful in distinguishing between the different modes of motility used by cancer cells of epithelial origin. The elongated mode of migration requires the activity of metalloproteinases (MT-MMP) and the use of membrane protrusions that tract the cell and drive its motility. Inhibition of extracellular proteases will shift these cancer cells toward a rounded mode of motility that is akin to the "amoeboid" movement of cells (Friedl and Brocker, 2000; Friedl and Wolf, 2003; Sahai and Marshall, 2003; Wyckoff et al., 2006). Another important advantage of 3D tissue culture models is the possibility of generating molecular gradients that mimic those naturally occurring in tissues and, therefore, to induce chemotactic cell migration (Griffith and Swartz, 2006).

Rho GTPase family members play an essential role in cell migration by reorganizing their cytoskeleton (Raftopoulou and Hall, 2004; Ridley, 2001; Yamazaki et al., 2005). Correct coordination of Rho GTPase activities is required to generate cell movement. To achieve this adequate dosage, Rho GTPase activities are strictly regulated. In addition to these regulators, the tumor suppressor p53 has been demonstrated to be essential for cell migration (Gadea et al., 2002, 2004; Guo et al., 2003; Guo and Zheng, 2004a,b; Roger et al., 2006; Xia and Land, 2007). Loss of p53 function causes cells cultured in 3D matrices to shift from an elongated to a rounded, blebbing morphology (Gadea et al., 2007).

This chapter focuses on the different procedures commonly used to study migration in 3D matrices to identify the mode of cell migration.

2. Experimental Procedure

To study the movement and polarity of modified cells in a 3D microenvironment, cells need to be transfected in a classical 2D culture. Then, after having checked that the transgene(s) of choice is (are) expressed, cells are transferred into a 3D matrix for analysis.

2.1. Transient transfection of green fluorescent protein (GFP)-tagged constructs

We favor the use of jetPei (Qbiogen) to transfect cells because it offers some strong advantages over lipid-based reagents. jetPei is a powerful transfection reagent that ensures effective and reproducible transfections with low toxicity and without cell membrane permeabilization. Antibiotics can be present in the cell medium during transfection, as jetPei does not permeabilize the cell membrane. High transfection efficiencies can be obtained with jetPei in serum-containing media in a manner almost fully independent of serum concentration. jetPei is a linear polyethylenimine derivative in a stable aqueous solution. The polymere condenses with nucleic acids to form compact, positively charged particles that adhere to ubiquitous cell surface proteoglycans, which are then internalized by endocytosis. jetPei protects its cargo from degradation by strongly buffering the endosomal pH.

In preparation for transfection, colon carcinoma cells (Hct116) are plated in a six-well plate with 2 ml plating medium/well [McCoy's 5A medium from Sigma supplemented with 10% fetal calf serum (FCS)]. Cells are ready for transfection when they have attached to the tissue culture dish and have reached 70% confluency. Two 1.5-ml tubes are needed: in the first tube, 1.5 μg DNA is added to 200 μl of filtered 150 mM NaCl, and in the second tube, 8 μl of jetPei is added to 200 μl of filtered 150 mM NaCl. After mixing, the jetPei solution is added to the DNA. After mixing gently, the DNA–jetPei solution is incubated at room temperature for 15 to 20 min to allow the formation of transfection complexes (i.e., transfection solution). Just before adding this solution to the cells, 2 ml of medium is replaced by 800 μl of fresh medium. The transfection solution is subsequently added to each well and mixed with the 800-μl medium by swirling the tissue culture dish gently.

After 12 to 24 h, the medium containing the transfection solution should be replaced with fresh medium. At this point, the efficiency of transfection can be evaluated by a rapid observation of the GFP fluorescence under a microscope. Most of our experiments have been carried out 48 h after transfection. This implies a passage and dilution at 1:2 of the cells 24 h after transfection to avoid the use of confluent cell cultures. Because we work with GFP-tagged constructs, control cells are transfected with an empty pEGFP vector (pEGFP-C1, Clontech).

2.2. Migration and invasion assays

Two different assays are commonly used to study cell migration: a migration assay, in which cells are prompted to migrate in a 2D environment, and an invasion assay, which reproduces physiological conditions as cells are stimulated to migrate in a 3D matrix. The migration assay is easier to perform. Comparison of both assays may reveal interesting differences that reflect the ability of cells to migrate differently in a 2D or 3D environment.

Both assays are carried out in Transwell cell culture chambers that contain fluorescence-blocking polycarbonate porous membrane inserts (Fluoroblock; BD Biosciences; pore size: 8 μm). The fluorescence-blocking membrane is very important to ensure that the fluorescent cells detected during the observation by microscopy have crossed the pores. Cells that did not migrate through the membrane cannot be detected.

For the invasion assay, a commercially prepared reconstituted basement membrane from Engelbreth–Holm–Swarm mouse sarcomas, that is, the Matrigel matrix (BD Biosciences), is diluted to a final concentration of 2 mg/ml in serum-free medium; this manipulation must be done at 4 °C to avoid polymerization of the Matrigel. The fluorescence-blocking porous inserts are placed in 24-well cell culture insert companion plates (BD Biosciences), and 100 μl of a 2 mg/ml Matrigel solution is added delicately to the upper chamber of the Transwell plate. Plates must be placed in a 37 °C incubator for at least 2 h to ensure polymerization of the Matrigel.

Meanwhile, transfected cells are trypsinized and counted. Cells (10×10^4) are plated on top of the thick layer of Matrigel in the upper chamber of the Transwell for invasion assays or directly onto the porous membrane insert for migration assays. The lower chamber is filled with 700 μl of medium containing 10% FCS, thus establishing a soluble gradient of chemoattractant that permits cell migration or invasion.

Cells not used for migration or invasion assays are centrifuged at 1200 rpm for 5 min, rinsed in 2 ml of phosphate-buffered saline (PBS), and then centrifuged again. These cells are then fixed by adding 1 ml of 70% EtOH and are stored in EtOH at -20 °C to be analyzed by fluorescence-activated cell sorting to evaluate the number of GFP-expressing cells accurately.

Because we do not know how the transfection procedure can influence cell migration and invasion, it is imperative to do a kinetic analysis of both invasion and migration. Therefore, we fix cells at specific times during the experiment: for migration assays at 1 and at 2 h and for invasion assays at 2, 4, 6, 8, 12, and 24 h. During the assay, cells are kept in a tissue culture incubator (37 °C and 5% CO_2), and at the established times, cells are fixed by replacing the culture medium in the bottom and top of the chambers with 1 ml of 3.7% formaldehyde dissolved in PBS for 15 min at room temperature. Fixed cells are then rinsed with 1 ml of PBS three times.

Cells that have migrated through the membrane insert or have invaded the Matrigel can be detected on the lower side of the filter by observing the GFP fluorescence using an inverted microscope equipped with a 10× objective. A specific journal in the Metamorph software allows scanning the whole surface of the well as a mosaic of 81 images (9 × 9) (Fig. 29.1).

We use the Metamorph software to count cells that have migrated or invaded the membrane. After opening the file containing the scan of the well, we delimit the region containing the cells by excluding the edges of the well, which are very refractive. The software counts the fluorescent cells. This automatic method has to be tightly controlled: we always verify that each cell has been counted only once and that no cell has been forgotten. After counting, the results are normalized: we divide

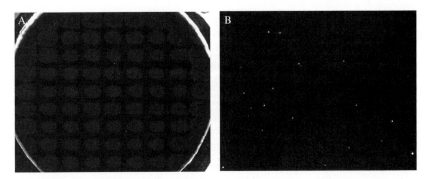

Figure 29.1   (A) pEGFP-transfected cells that have invaded through the Matrigel after 18 h. Reconstituted photograph of the mosaic of the 81 images representing the entire surface of the well. GFP cells are visualized by white points. (B) Higher magnification of a part of the well shown in A.

the number of GFP-positive cells for each experimental condition by the number of GFP-positive cells in the control. Using this method, all experiments can be compared and statistical analysis can be performed.

2.3. Three-dimensional migration

Although the invasion assay is quantitative, it does not allow recording of the movements and the form of the cells during migration. Therefore, we have developed another method to investigate the form of invasive cells that is based on their ability to migrate in a 3D matrix and requires recording by video microscopy. In order to study cell movements and mode of migration, cells have to be motile and, therefore, need to be exposed to a chemoattractant. During the time-lapse recording of the progression of cells, the focus has to be maintained to make sure that the restituted images will be neat. The video microscopy analysis is, thus, strongly facilitated if cells migrate horizontally and for this reason we apply a horizontal gradient of chemoattractant. For this purpose, we have devised home-made bicompartmental chambers, which include an 8-mm metal ring placed in the center of a 35-mm culture petri dish. After transfection and/or treatment with GTPase-inhibiting drugs (see later), cells are trypsinized and mixed with serum-free Matrigel with or without drugs. This suspension is poured into the well formed by the metal ring. After a 30-min incubation at 37 °C to allow solidification of the Matrigel, the metal ring is removed. The resulting disc of Matrigel is surrounded by 10% FCS containing medium, and one has to take care that no medium will flow on top of the Matrigel disc. In this way, cells embedded in the Matrigel are exposed to a lateral (and horizontal) serum gradient. The dish is placed over the 63× objective of an inverted microscope that has been supplied with a sample heater (37 °C) and a CO_2

incubation chamber. Cells are time lapse recorded to monitor their ability to migrate along the serum gradient. Using this procedure, the form of the cells can be observed during the dynamic process of migration: if cells increase the number of membrane protrusions typical of extension movements, they are considered adopting an elongated type of movement. Conversely, if cells exhibit membrane blebbing and maintain a rounded morphology, they are thought to assume an amoeboid-like mode of migration (Fig. 29.2). Information on cell velocity can also be determined by measuring the distance covered by a cell in a certain amount of time and by comparing data obtained in the different conditions used. Figure 29.2 shows an example of still differential interference contrast (DIC) images issued from a time-lapse video microscopy analysis. Whereas wild-type, unstressed mouse embryonic fibroblasts (MEF) exhibit a spindle-shaped morphology during motion, p53-deficient MEF show a translatory movement accompanied by strong membrane deformability, which is typical of amoeboid-like migration.

As blebbing movements depend on the activity of the small GTPase RhoA and its effector Rho kinase (ROCK), it is possible to first approximate the mode of motility employed by invasive cells by preventing the activation either of (1) RhoA by using TAT-C3, the *Clostridium botulinum* exoenzyme C3 fused to the TAT protein-transduction domain of HIV to allow the delivery of C3 to the cells (Gadea *et al.*, 2005); or of (2) ROCK by means of two distinct, structurally unrelated pharmacological inhibitors, Y27632 (Calbiochem) at $10 \mu M$ or H1152 (Calbiochem) at $5 \mu M$, or both. To evaluate the function of these proteins, monolayer cells must be cultured in the presence of the drugs for 12 h before trypsinization (drugs are diluted directly in the culture medium). Moreover, the drugs must also

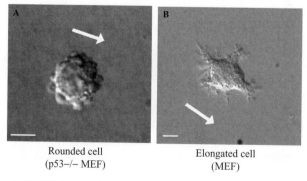

Rounded cell
(p53−/− MEF)

Elongated cell
(MEF)

Figure 29.2   Still DIC images from a video microscopy analysis of fibroblasts migrating in a horizontal gradient of serum in a 3D Matrigel matrix. (A) Rounded migrating cells. p53-deficient fibroblasts exhibit a typical rounded morphology decorated with numerous membrane blebs. (B) Elongated migrating cells. Unstressed normal fibroblasts extend protrusions leading to a spindle-shaped morphology. Arrows show the direction of cell migration. Bars: 10 μm.

be added to the Matrigel matrix during its dilution. An indication of the mode of migration used and consequently of the GTPase involved can also be obtained using specific protease inhibitors (Sigma) at 25 μg/ml (Sahai and Marshall, 2003; Wolf *et al.*, 2003). If invasiveness is independent of protease activity, cells will invade the matrix using blebbing movements, which depend on the activity of RhoA and its kinase effector, ROCK.

2.4. GTPase activity

To confirm the mode of migration of transfected cells (i.e., elongated or blebbing movements), an analysis of the activity of Rho GTPases can be performed. According to the results of the observation on cell shape during 3D migration, we can hypothesize which GTPase may be involved in this process. RhoA and ROCK are generally involved in blebbing movements; Cdc42 and Rac play a role in elongated movements by their action on filopodia and lamellipodia formation. For RhoA activity assay, cells are lysed in 500 μl of lysis buffer [50 mM Tris, pH 7.2, 1% Triton X-100, 0.5% sodium deoxycholate, 500 mM NaCl, 10 mM MgCl$_2$, 1 mM phenyl-methylsulfonyl fluoride (PMSF), and a cocktail of protease inhibitors]. The lysis buffer can be directly applied to the cell monolayer after a PBS wash. If necessary, cells may be centrifuged after trypsinization, rinsed with PBS, and stocked at $-80°$ before the GTPase activity assay. In this case, the lysis buffer must be applied directly on the frozen pellet. Cells in the monolayer are scraped, and the supernatant is transferred to a 1.5-ml tube. After lysis, samples are homogenized by passing them at least five times through a 0.45-mm needle fitted to a syringe and centrifuged at 14,000 rpm for 1 to 2 min. Twenty microliters of each sample is frozen and stored at $-20°$C to be used as a control of the total RhoA protein levels (total lysate). Four hundred microliters of the cleared lysates is incubated with 25 μg of commercial GST-RBD beads (a fusion protein containing the RhoA-binding domain of Rhotekin, cytoskeleton) for 30 min at 4°C. Precipitated complexes are washed twice in 750 μl of Tris buffer containing 1% Triton X-100, 150 mM NaCl, 10 mM MgCl$_2$, and 0.1 mM PMSF, eluted in 20 μl of SDS sample buffer, immuno-blotted, and analyzed with specific anti-RhoA antibodies (Santa Cruz Biotechnology, Inc. diluted at 1:500). An aliquot of the total lysates is run alongside to quantify the total RhoA present in the cell lysates. Scanned autoradiographs are then quantified using the Aida/2D densitometry software and normalized as a function of the expression of the total RhoA.

For Cdc42 and Rac1 activity assays, the protocol is approximately the same, except for the proteins (GST-Crib of PAK for Cdc42 and GST-WASP for Rac1) and the buffers. The lysis buffer consists of 25 mM HEPES, pH 7.5, 1% Nonidet P-40 (NP-40), 10 mM MgCl$_2$, 100 mM NaCl, 5% glycerol, 5 mM NaF, 1 mM NaOVa, 1 mM PMSF, and a cocktail of protease inhibitors, and the washing buffer is made up of 25 mM HEPES, 30 mM MgCl$_2$, 40 mM NaCl, 0.5% NP-40, and 1 mM dithiothreitol.

2.5. Immunofluorescence in three-dimensional matrices

Migration in a 3D matrix allows not only observing the way cells move and how they change their shape, but also studying the localization of different cell markers by immunofluorescence. Cells are fixed during their migration in Matrigel; in this way, cell polarity is maintained because cells migrate through a gradient of serum. Consequently, data obtained can be considered representative of the physiological conditions. For immunofluorescence of cells in the 3D Matrigel, cells are transfected in a monolayer culture. Forty-eight hours after transfection, cells are trypsinized and embedded into the Matrigel in Transwell cell culture chambers as described previously for the invasion assays. Cells are allowed to invade the matrix for 6 h; 50-μm cryosections are then cut at $-20\,^{\circ}$C before fixation in 4% paraformaldehyde in PBS. Then, sections are permeabilized in 0.5% Triton X-100 for 2 min and incubated in PBS containing 1.5% BSA at room temperature for 10 min. Immunolabelling with various antibodies is performed using classical immunofluorescence protocols. Cells are analyzed using a microscope (LSM510 Meta; Carl Zeiss MicroImaging, Inc.) with a 63$\times$ NA 1.32 Apochromat water-immersion objective (Carl Zeiss MicroImaging, Inc.) and a photomultiplicator. Stacks of images are captured with LSM510 expert mode, restored, and then deconvolved with Huygens software (Scientific Volume Imaging) using a maximum likelihood estimation algorithm. Restored stacks of images are processed with Imaris software (Bitplane) for visualization. As exemplified in Fig. 29.3, actin and β_1 integrin are useful for discriminating between the two modes of migration. Actin staining allows observation of the overall cell morphology during migration. Whereas blebbing cells elicit a typically rounded morphology (see Fig. 29.3A), actin marks their extended protrusions in elongated cells (see Fig. 29.3B). Subcellular distribution of β_1 integrin also differs according to the mode of migration in a 3D matrix. Rounded migrating cells redistribute β_1 integrin within the forward sector facing the moving front of the cell (see Fig. 29.3A). In contrast, β_1 integrin is localized throughout the elongated cells and shows no specific intracellular staining concentration (see Fig. 29.3B).

3. Concluding Remarks

This chapter provided useful methods to study the dynamic of cell migration in 3D matrices: invasion assay through Matrigel, 3D migration with a horizontal chemoattractant, and immunofluorescence in 3D matrices.

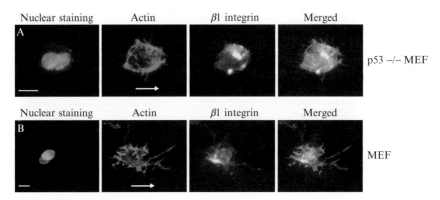

Figure 29.3   Immunofluorescence of cells cultured in a 3D Matrigel matrice network. (A) Example of rounded migrating cells. Deficiency in p53 leads to rounded migration as exemplified by p53-/- MEF. The merged image depicts the covisualization of nuclear staining (green), actin (red), and integrin $\beta1$ (yellow). (B) Unstressed cell harboring wild-type p53 exhibits an elongated-type migration. Visualizations of nucleus, actin, and integrin β_1 are identical to A. Arrows indicate the direction of cell migration. Bars: 10 μm. (See color insert.)

These techniques closely mimic the *in vivo* situation allowing a better understanding of the mode of migration used by cancer cells for invasion during the metastatic process. These methods can be used with different cell types such as fibroblasts, tumor cells originating from solid cancers, and cells of the immune system. Studying migration in 3D matrices and defining the type of the locomotory activity open up new avenues for comparing the efficacy of different chemical agents on the treatment of invasive cancer cells.

ACKNOWLEDGMENTS

We are grateful to MRI for constructive microscopy and E. Andermarcher for critical comments on the manuscript. This work was supported by the Fondation de France, Association pour la Recherche contre le Cancer (Contract 4028), Ligue Regionale contre le Cancer, INSERM, and CNRS. S. Vinot is a recipient of a fellowship from the ARC.

REFERENCES

Bissell, M. J., Radisky, D. C., Rizki, A., Weaver, V. M., and Petersen, O. W. (2002). The organizing principle: Microenvironmental influences in the normal and malignant breast. *Differentiation* **70,** 537–546.

Condeelis, J., and Segall, J. E. (2003). Intravital imaging of cell movement in tumours. *Nat. Rev. Cancer* **3,** 921–930.

Cukierman, E., Pankov, R., Stevens, D. R., and Yamada, K. M. (2001). Taking cell-matrix adhesions to the third dimension. *Science* **294,** 1708–1712.

Debnath, J., and Brugge, J. S. (2005). Modelling glandular epithelial cancers in three-dimensional cultures. *Nat. Rev. Cancer* **5,** 675–688.

Friedl, P., and Brocker, E. B. (2000). The biology of cell locomotion within three-dimensional extracellular matrix. *Cell. Mol. Life Sci.* **57,** 41–64.

Friedl, P., and Wolf, K. (2003). Tumour-cell invasion and migration: Diversity and escape mechanisms. *Nat. Rev. Cancer* **3,** 362–374.

Gadea, G., Boublik, Y., Delga, S., and Roux, P. (2005). Efficient production of *Clostridium botulinum* exotoxin C3 in bacteria: A screening method to optimize production yields. *Protein Expr. Purif.* **40,** 164–168.

Gadea, G., de Toledo, M., Anguille, C., and Roux, P. (2007). Loss of p53 promotes RhoA-ROCK-dependent cell migration and invasion in 3D matrices. *J. Cell Biol.* **178,** 23–30.

Gadea, G., Lapasset, L., Gauthier-Rouviere, C., and Roux, P. (2002). Regulation of Cdc42-mediated morphological effects: A novel function for p53. *EMBO J.* **21,** 2373–2382.

Gadea, G., Roger, L., Anguille, C., de Toledo, M., Gire, V., and Roux, P. (2004). TNFα induces sequential activation of Cdc42- and p38/p53-dependent pathways that antagonistically regulate filopodia formation. *J. Cell Sci.* **117,** 6355–6364.

Griffith, L. G., and Swartz, M. A. (2006). Capturing complex 3D tissue physiology in vitro. *Nat. Rev. Mol. Cell Biol.* **7,** 211–224.

Guo, F., Gao, Y., Wang, L., and Zheng, Y. (2003). p19Arf-p53 tumor suppressor pathway regulates cell motility by suppression of phosphoinositide 3-kinase and Rac1 GTPase activities. *J. Biol. Chem.* **278,** 14414–14419.

Guo, F., and Zheng, Y. (2004a). Involvement of Rho family GTPases in p19Arf- and p53-mediated proliferation of primary mouse embryonic fibroblasts. *Mol. Cell. Biol.* **24,** 1426–1438.

Guo, F., and Zheng, Y. (2004b). Rho family GTPases cooperate with p53 deletion to promote primary mouse embryonic fibroblast cell invasion. *Oncogene* **23,** 5577–5585.

Ingber, D. E. (2002). Cancer as a disease of epithelial-mesenchymal interactions and extracellular matrix regulation. *Differentiation* **70,** 547–560.

Kim, J. B. (2005). Three-dimensional tissue culture models in cancer biology. *Semin. Cancer Biol.* **15,** 365–377.

Locascio, A., and Nieto, M. A. (2001). Cell movements during vertebrate development: Integrated tissue behaviour versus individual cell migration. *Curr. Opin. Genet. Dev.* **11,** 464–469.

Luster, A. D., Alon, R., and von Andrian, U. H. (2005). Immune cell migration in inflammation: Present and future therapeutic targets. *Nat. Immunol.* **6,** 1182–1190.

Raftopoulou, M., and Hall, A. (2004). Cell migration: Rho GTPases lead the way. *Dev. Biol.* **265,** 23–32.

Ridley, A. J. (2001). Rho GTPases and cell migration. *J. Cell Sci.* **114,** 2713–2722.

Ridley, A. J., Schwartz, M. A., Burridge, K., Firtel, R. A., Ginsberg, M. H., Borisy, G., Parsons, J. T., and Horwitz, A. R. (2003). Cell migration: Integrating signals from front to back. *Science* **302,** 1704–1709.

Roger, L., Gadea, G., and Roux, P. (2006). Control of cell migration: A tumour suppressor function for p53? *Biol. Cell* **98,** 141–152.

Sahai, E., and Marshall, C. J. (2003). Differing modes of tumour cell invasion have distinct requirements for Rho/ROCK signalling and extracellular proteolysis. *Nat. Cell Biol.* **5,** 711–719.

Van Haastert, P. J., and Devreotes, P. N. (2004). Chemotaxis: Signalling the way forward. *Nat. Rev. Mol. Cell. Biol.* **5,** 626–634.

Vicente-Manzanares, M., Webb, D. J., and Horwitz, A. R. (2005). Cell migration at a glance. *J. Cell Sci.* **118,** 4917–4919.

Wolf, K., Mazo, I., Leung, H., Engelke, K., von Andrian, U. H., Deryugina, E. I., Strongin, A. Y., Brocker, E. B., and Friedl, P. (2003). Compensation mechanism in tumor cell migration: Mesenchymal-amoeboid transition after blocking of pericellular proteolysis. *J. Cell Biol.* **160,** 267–277.

Wyckoff, J. B., Pinner, S. E., Gschmeissner, S., Condeelis, J. S., and Sahai, E. (2006). ROCK- and myosin-dependent matrix deformation enables protease-independent tumor-cell invasion *in vivo*. *Curr. Biol.* **16,** 1515–1523.

Xia, M., and Land, H. (2007). Tumor suppressor p53 restricts Ras stimulation of RhoA and cancer cell motility. *Nat. Struct. Mol. Biol.* **14,** 215–223.

Yamada, K. M., and Cukierman, E. (2007). Modeling tissue morphogenesis and cancer in 3D. *Cell* **130,** 601–610.

Yamazaki, D., Kurisu, S., and Takenawa, T. (2005). Regulation of cancer cell motility through actin reorganization. *Cancer Sci.* **96,** 379–386.

CHAPTER THIRTY

USE OF *CAENORHABDITIS ELEGANS* TO EVALUATE INHIBITORS OF RAS FUNCTION *IN VIVO*

David J. Reiner,*,†,¶ Vanessa González-Pérez,‡,¶
Channing J. Der,*,†,‡ *and* Adrienne D. Cox*,†,‡,§

Contents

* Lineberger Comprehensive Cancer Center, University of North Carolina at Chapel Hill, Chapel Hill, North Carolina
† Department of Pharmacology, University of North Carolina at Chapel Hill, Chapel Hill, North Carolina
‡ Curriculum in Genetics and Molecular Biology, University of North Carolina at Chapel Hill, Chapel Hill, North Carolina
§ Department of Radiation Oncology, University of North Carolina at Chapel Hill, Chapel Hill, North Carolina
¶ Both Authors Contributed Equally

Methods in Enzymology, Volume 439
ISSN 0076-6879, DOI: 10.1016/S0076-6879(07)00430-2

Abstract

The human RAS genes constitute the most frequently mutated oncogenes in human cancers, and the critical role of aberrant Ras protein function in oncogenesis is well established. Consequently, considerable effort has been devoted to the development of anti-Ras inhibitors for cancer treatment. An important facet of molecularly targeted cancer drug discovery is the validation of a target-based mechanism of action, as well as the identification of potential off-target effects. This chapter describes the use of the nematode worm *Caenorhabditis elegans* for simple, inexpensive pharmacogenetic analysis of candidate molecularly targeted inhibitors of mutationally activated Ras, with a focus on the Ras>Raf>MEK>ERK mitogen-activated protein kinase pathway. This protein kinase cascade is well conserved from worms to humans and is well established as a critical player in the signaling events leading to vulval formation in *C. elegans*. Excess activity results in the development of a multivulva (Muv) phenotype, whose inhibition by test compounds can be characterized genetically as to the specific step of the pathway that is blocked. In addition, off-target activities can also be identified and characterized further using different strains of mutant worms. This chapter presents proof-of-principle analyses using the well-characterized MEK inhibitor U0126 to block the Muv phenotype caused by the constitutively activated Ras homolog *C. elegans* LET-60. It also provides a detailed description of protocols and reagents that will enable researchers to analyze on- and off-target effects of other candidate anti-Ras inhibitors using this system.

1. INTRODUCTION

The nematode worm *Caenorhabditis elegans* is an organism commonly used by researchers wishing to model human biology. The simple lifestyle and body plan of *C. elegans* allow many facets of its development and behavior to be perturbed genetically (Brenner, 1974). Coupled with the extensive evolutionary and functional conservation of signaling proteins, the tractable genetics of *C. elegans* allow the creation of many valuable genetic tools for manipulating signaling pathways. These tools can be isolated in screens or created transgenically, and the resulting phenotypes can be assayed relatively easily. Perhaps the best studied of such pathways in *C. elegans* is the epidermal growth factor receptor (EGFR) receptor tyrosine kinase pathway that activates Ras, which subsequently activates the Raf>MEK>ERK mitogen–activated protein kinase (MAPK) cascade (Fig. 30.1). This EGFR–stimulated Ras pathway is very well conserved in evolution, with all components found in *C. elegans* and *Drosophila*, as well as in all vertebrate species. In *C. elegans*, this pathway controls the development of the vulva, an epithelial aperture through which fertilized eggs are laid.

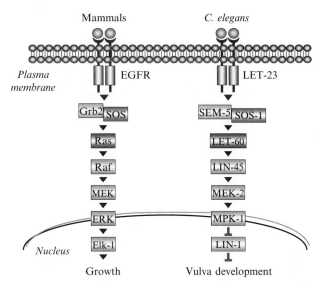

Figure 30.1 A conserved Ras>Raf>ERK MAPK pathway specifies growth in mammalian cells and vulval cell fate in *C. elegans*. In the worm, an EGFR–ligand–like signal, LIN-3 (released from the anchor cell in the gonad; see Fig. 30.2), binds to the LET-23 receptor tyrosine kinase, which becomes activated and binds the adaptor/exchange factor complex SEM-5/SOS-1 (Grb-2/SOS) to activate the Ras small GTPase LET-60. Activated Ras then triggers the equivalent of the Raf>MEK>ERK MAPK cascade of kinases. ERK/MPK-1 enters the nucleus to phosphorylate and activate or inactivate, respectively, Ets family transcription factors such as LIN-1, which negatively regulates the induction of vulval precursor cells. (See color insert.)

Mutations perturbing vulval development are easy to identify (Fig. 30.2). Defective vulval induction results in a vulvaless (Vul) phenotype, and embryos hatch inside the animal. Excessive vulval induction results in the development of nonfunctional ectopic pseudovulvae, resulting in the multivulva (Muv) phenotype. The molecular pathways that regulate these developmental events are highly conserved among all metazoans (Moghal and Sternberg, 2003). By screening for these phenotypes and modifiers of these phenotypes, mutations in over 50 genes that govern vulval development have been identified. These studies have demonstrated that the worm LET-23>LET-60>LIN-45>MEK-2>MPK-1 MAPK cascade, which is highly conserved with the homologous EGFR>Ras>Raf>MEK>ERK MAPK pathway in mammals, is a critical signaling pathway that governs cell fate decisions needed for proper vulval formation.

Studies in model systems such as *C. elegans* and *Drosophila* have been instrumental in identifying previously unknown components of these pathways, or determining the functional role of known pathway components. For example, the human Raf scaffolding protein Ksr was originally identified in mutant screens in *C. elegans* and *Drosophila* (Kornfeld *et al.*, 1995;

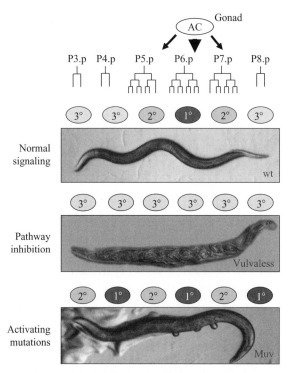

Figure 30.2 Normal or aberrant vulval formation in *C. elegans* is determined by the exposure of vulval precursor cells to graded levels of activation of the Ras>Raf> MEK>ERK MAPK signaling pathway. Under "normal signaling" conditions (top worm photograph), the anchor cell (AC) of the gonad releases signals (*e.g.*, LIN-3 ligand) to activate the Ras pathway in adjacent VPCs (P6.p, P7.p, etc.). In wild-type animals, VPCs adjacent to the AC receive the strongest signal. These cells assume 1° and 2° fates and together develop into a functional vulva. The more distal VPCs are normally exposed to lower levels of signaling inputs from the Ras pathway. These cells remain uninduced and therefore assume a 3° fate and fuse to the hypodermal syncytium rather than contributing to vulval formation. Under conditions of "pathway inhibition" (middle worm photograph), whether because of loss-of-function mutations or drug treatments, blockade of critical elements of the pathway leads to the absence of inductive signals. All VPCs therefore remain undifferentiated and assume a 3° cell fate that results in a failure to develop a functional vulva (vulvaless phenotype) and hence the inability to lay eggs. As a consequence, the mature eggs hatch inside the parent, causing the "bag-of-worms" phenotype shown. Conversely, constitutive signaling ("activating mutations," bottom worm photograph) leads to hyperinduction of all VPCs, which adopt only 1° and 2° fates, resulting in the formation of both a single functional vulva and additional protruding nonfunctional vulva-like structures called pseudovulvae. Worms with such pseudovulvae are described as exhibiting a multivulva (Muv) phenotype. (See color insert.)

Sundaram and Han, 1995; Therrien *et al.*, 1995), and the functional relationship of the SEM-5/Grb2 adaptor protein to other pathway components was originally determined in *C. elegans* (Clark *et al.*, 1992). Thus, regulation of vulval cell fate in *C. elegans* is a useful differential biological readout of EGFR>Ras>Raf>MEK>ERK pathway activity, such as the R7 photoreceptor in the *Drosophila* eye or cell proliferation in human epithelia.

Because the signaling module itself, from the EGFR ligands (*e.g.*, EGF) to the Ets-like transcription factors, is highly conserved and similarly regulated in both worms and humans, it is likely to have the same pharmacological targets and be modified by the same pharmacological treatments in both worms and humans. For example, many human small GTPases, such as the LET-60 orthologs K-Ras, H-Ras, and N-Ras, require post-translational modification by farnesylation at their C-termini to target the proteins to membranes and to promote correct biological activity. First-generation FTase inhibitors (FTIs) manumycin and gliotoxin were capable of blocking excessive LET-60 signaling in worm vulval induction (Hara and Han, 1995), although subsequent studies found these and other FTIs to be ineffective against the Ras isoforms most commonly mutated in human cancers (Rowinsky, 2006).

In addition, the degree of functional conservation of *C. elegans* pathways is such that off-target drug effects can also be conserved across species and identified in *C. elegans*. When a newer generation FTI, BMS-214662, was unexpectedly found to have off-target proapoptotic activity (Rose *et al.*, 2001), many attempts to identify the mechanism for this property failed in mammalian cells. However, the proapoptotic function was conserved in *C. elegans*. Genetic screens then identified genes whose loss of function caused the same phenotype as the p53-independent, caspase-dependent germline apoptosis induced by a panel of proapoptotic FTIs (Lackner *et al.*, 2005). These genes control endosome–lysosome and autophagosome–lysosome docking and fusion, and the conserved Rab prenylation enzyme, Rab geranylgeranyltransferase (RabGGTase/GGTase II), was shown to be the key FTI target responsible for the induction of apoptosis. Target identification was then confirmed in mammalian cells. This study illustrates the utility of using *C. elegans* to identify novel pharmacological targets of known drugs. Several conserved pharmacological targets in the nervous system have been identified in *C. elegans*. For example, in *C. elegans* nicotinic acetylcholine agonists activate acetylcholine receptors (Lewis *et al.*, 1980) and GABA-A receptor agonists activate GABA-A receptors (McIntire *et al.*, 1993). Furthermore, tricyclic (Horvitz *et al.*, 1982) and selective serotonin reuptake inhibitor (SSRI) (Weinshenker *et al.*, 1995) antidepressants promote serotonin signaling, and putative molecular targets of off-target effects of fluoxetine were identified using *C. elegans* genetics (Choy and Thomas, 1999). Together, these observations suggest that *C. elegans* is an excellent system for studying both intended and off-target effects of pharmacological agents.

Because of the potential for detailed genetic manipulation of conserved pathways, the use of *C. elegans* allows the researcher to determine at which level in the pathway a drug acts *in vivo*. For example, the mutant allele *let-60(n1046gf)* introduces a gain-of-function (gf) G13E mutation into the worm Ras protein LET-60, thereby inactivating its intrinsic and GAP-stimulated GTP hydrolysis activity and trapping the protein in its active GTP-bound form. Hyperinduction of vulval tissues causes a Muv phenotype that is easily scored by counting the ectopic ventral tissue protrusions on the normally smooth surface of the animal (see Fig. 30.2). The strong Muv phenotype caused by *let-60(n1046gf)* can be suppressed genetically or pharmacologically by compromising pathway function either at or downstream of *let-60*, whereas perturbation of upstream pathway function has little effect (Beitel *et al.*, 1990). Similar types of manipulation can be performed at most levels of the pathway, either by use of isolated strains with *in situ* mutations or by generation of transgenic strains in which the pathway is manipulated in *trans*. Table 30.1 lists activated pathway reagents that cause a Muv phenotype.

Lacking this extensive genetic tool kit for analyzing EGFR>Ras>Raf>MEK>ERK MAPK signaling, other model systems have their own advantages and disadvantages. Human cell culture can be readily manipulated genetically and has the advantage of excellent biochemical readouts, but it is an *ex vivo* system, and many cancer cell lines have extensive genetic abnormalities that are distinct from those found in the original tumors. Mice have the advantage of being an *in vivo* and mammalian system, but they are expensive, lack the reagents capable of altering each pathway component, and extensive manipulation of this pathway would result in lethality; further, despite their widespread use in preclinical antitumor efficacy studies, they are surprisingly nonpredictive of antitumor and off-target normal cell toxicity in the cancer patient (Sharpless and Depinho, 2006). Zebrafish are relatively inexpensive and have the benefit of complex organ systems, but they are also subject to some of the caveats of the mouse system. For our purposes, although no system is perfect, *C. elegans* is particularly useful because these animals are inexpensive, the genetic tools are currently available and simple to work with, and one can rapidly assay different inhibitors.

2. Experimental Protocol

The experimental protocol described in this section has been developed to study both on-target and off-target effects of pharmacological inhibitors of Ras function in the EGFR>Ras>Raf>MEK>ERK MAPK pathway, in which inhibition of the Muv phenotype serves as a readout for inhibitor activity. The following section illustrates the use of this protocol by describing a proof-of-principle experiment to inhibit the pathway

Table 30.1 Mutant strains of *C. elegans* available for characterization of Ras/ERK MAPK signaling pathway inhibitors

Worm gene	Ref.	Mammalian homolog	Type of mutation[a]	Allele	Strain[b]	Basal Muv[c] (%)	Notes
lin-3	Hill and Sternberg (1992)	EGF	Transgenic overexpression	*syIs1*	PS1123	90	Integrated transgene containing wild-type *lin-3* genomic DNA. Confers a dominant multivulva (Muv) phenotype.
lin-15	Ferguson and Horvitz (1985); Huang *et al.* (1994)	None known	lf	*n765*	MT8189	77	Temperature-sensitive mutation: animals are Muv at 25°C, wild type at 15°C. LIN-3/EGF ectopically expressed in mutant.
let-23	Katz *et al.* (1996)	EGFR	gf (C359Y)	*sa62*	PS1524	89	Activating mutation in the extracellular domain of the LET-23 receptor; Muv phenotype is ligand independent.
let-23	Moghal and Sternberg (2003)	EGFR	gf (C359Y, G270E)	*sa62, sy621*	PS4064*	100	Double extracellular domain mutations confer a stronger

(*continued*)

Table 30.1 (*continued*)

Worm gene	Ref.	Mammalian homolog	Type of mutation[a]	Allele	Strain[b]	Basal Muv[c] (%)	Notes
sos-1	Modzelewska et al. (2007)	Sos	gf (G322R)	sy262, let-23 (sy1)	ND91**	68	ligand-independent Muv phenotype. Muv phenotype due to sy262 is visible only in the sensitized let-23(sy1) background.
let-60	Ferguson and Horvitz (1985)	Ras	gf (G13E)	n1046	MT2124	57–90[d]	Mutation predicted to disrupt intrinsic GTPase activity, resulting in constitutive activation (similar to G12V but predicted to be weaker).
lin-45 (AA)	Chong et al. (2001); Rocheleau et al. (2005)	Raf	TG overexp. gf (S312A, S435A)	kuIs57	MH2209	91	Integrated transgene driving full-length lin-45(AA) with mutational loss of the Akt negative regulatory phosphorylation serines.
lin-45 (TM)	Sieburth et al. (1998); Dickson et al. (1992)	Raf	Conditional TG overexp. gf	kuIs17	UP1154	13	Integrated transgene in which hsp16-41 heat-shock promoter drives expression of

Gene	Reference	Pathway	Type	Allele	Strain/No.		Description
lin-45 (ED)	Chong et al. (2001); Rocheleau et al. (2005)	Raf	ConditionalTG overexp. gf (T626E, S629D)	csEx72	UP1226	37	*Drosophila* Raf kinase domain (411 C-terminal amino acids) fused to the Torso transmembrane domain. Activation is independent of Ras. Nonintegrated transgene (unstable) with *hsp16-41* heat-shock promoter driving *lin-45(ED)* containing T626E, S629D phosphomimetic activating mutations.
mek-2 + mpk-1	Lackner and Kim (1998)	MEK/ERK	Conditional overexp. gf	gaIs36 or gaIs37	SD418*** or SD470***	Not available	Integrated transgene with both *Drosophila* MEK [*Dsor1* gf mutation] and C. *elegans* ERK [*mpk-1* gf mutation (D324N)]. Both constructs are driven by the *hsp16-41*

(continued)

Table 30.1 (*continued*)

Worm gene	Ref.	Mammalian homolog	Type of mutation[a]	Allele	Strain[b]	Basal Muv[c] (%)	Notes
							heat-shock promoter. Muv phenotype is observed only when animals are grown at 25°C, but not at 15–20°C.
lin-1	Beitel *et al.* (1995)	Ets-related transcription factor	lf	*sy254*	MT7567	90–100	Loss of function of *lin-1* confers Muv phenotype.
lin-31	Hill and Sternberg (1992); Miller *et al.* (1993); Tam *et al.* (1998)	Winged helix transcription factors	lf	*n301*	MT301	≈70	Identified in a general screen for Muv or Vul enhancement. Loss of function of *lin-31* confers Muv phenotype. Healthier than *lin-1* (*sy254*).

[a] overexp, overexpression; gf, gain of function; lf, loss of function.

[b] Most strain names were identified using the public domain C. elegans database Wormbase (http://www.wormbase.org/). This database also provides basic information on each strain, as well as links to literature in which the strain of interest has been cited. Not all the mutant strains mentioned here were publicly available at the time of preparation of this document. We obtained marked strain numbers from the original laboratory sources as follows: ★ Paul W. Sternberg, and ★★ Nadeem Moghal, and ★★★ Stuart K. Kim.

[c] The penetrance of the multivulva (Muv) phenotype among mutant strains of the Ras/ERK MAPK pathway varies. This variation may be due to the type of genetic lesion and/or to the role of the mutated gene in Ras/ERK MAPK signaling that regulates worm vulval development.

[d] For this strain, we routinely observe 85–90% Muv worms when grown from fresh cultures.

downstream of Ras. We used the well-characterized MEK inhibitor U0126 (Favata *et al.*, 1998) to block the Muv phenotype induced by the constitutively activated worm Ras homolog LET-60. This protocol can be adapted to analyze other pharmacologic inhibitors and other signaling pathways.

2.1. Overview of experimental procedure

As shown in the schematic overview (Fig. 30.3), each experiment involves pouring the fresh agar plates on which the experiment will be performed, adding inhibitor or vehicle to the agar, growing a lawn of bacteria on the agar to provide food for the worms, growing the worms in the presence or absence of the inhibitor, and finally scoring the phenotype of the worms under each condition. Each experiment therefore generally takes 5 to 8 days to complete, depending on the developmental stage to be scored and on any growth delay induced by the inhibitors tested. For consistency and accuracy, it is critical to follow the recommended time course every time an experiment is run.

2.2. Preparation of materials and reagents

Instructions were adapted from the following sources:

- Alkaline buffer and bleaching for synchronized L1 larvae (Bianchi and Driscoll, 2006)
- NGM agar, M9 buffer, OP50 strain of *Escherichia coli* (Brenner, 1974)

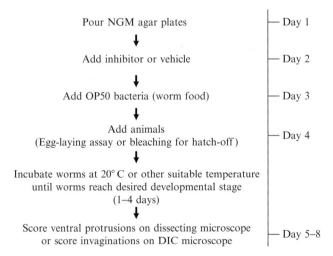

Figure 30.3 Schematic overview and time line for a typical experiment testing inhibitors of the Muv phenotype in robust or sickly animals. Details are provided in the text.

- Mounting worms on agar pads and differential interference contrast (DIC) microscopy (Shaham, 2006)
- General *C. elegans* culturing techniques (Stiernagle, 2006)

2.2.1. 2% neutral growth medium (NGM) agar (1 liter)

- 20 g Bacto agar
- 3 g NaCl
- 2.5 g Bacto peptone
- 1 ml of 5 mg/ml (dissolved in 100% ethanol) cholesterol
- 975 ml dH$_2$O

    Autoclave for 60 min. When cooled to approximately 50°C, add sterile

- 1 ml of 1 M MgSO$_4$
- 1 ml of 1 M CaCl$_2$
- 25 ml of 1 M KPO$_4$, pH 6.0

To make 1 M KPO$_4$, pH 6.0, first prepare and autoclave solutions of 1 M KH$_2$PO$_4$ (monobasic) and 1 M K$_2$HPO$_4$ (dibasic). Adjust the KH$_2$PO$_4$ pH to 6.0 by gradually adding 1 M K$_2$HPO$_4$.

This agar is used to pour plates. Make only enough 2% NGM agar for use in pouring plates that day. If excess agar remains after pouring plates for the inhibitor experiment, it can be used to pour stock plates to maintain worm strains. Store the other solutions (salts, buffers) at room temperature in 250-ml glass bottles, each containing 100 ml.

2.2.2. 3% NGM agar (1 liter)

This agar is used for mounting worms for microscopy. To prepare 3% NGM agar, follow the same recipe as for 2% agar, but increase the Bacto agar to 30 g per liter. Aliquot 2 to 3 ml into sterile disposable glass tubes. Seal the opening of each tube with Parafilm and store at room temperature until needed. Each aliquot is sufficient to make enough agar pads to mount 10 to 15 slides (see later), so it is probably unnecessary to make more than 250 ml of 3% NGM agar at any given time.

2.2.3. 1× M9 buffer (minimal salts) (1 liter)

- 5.8 g Na$_2$HPO$_4$
- 3.0 g KH$_2$PO$_4$
- 0.5 g NaCl
- 1.0 g NH$_4$Cl
- Add dH$_2$O to adjust the final volume to 1 liter and autoclave (no longer than 30 min) before use.

Sterile 1× M9 buffer is commonly used for the handling of worms in many different protocols. Here it is used to dilute the test drugs and to wash the worms after the bleaching protocol.

2.2.4. Inhibitor/drug dilutions

Inhibitors are generally made up and stored as stock solutions of 10, 20, or 50 mM in dimethyl sulfoxide (DMSO) or ethanol solvent, and each of these solvents can be used as vehicle-only controls as appropriate. All working dilutions of stock solutions of inhibitors (whether containing control or test inhibitor, or vehicle only) should be prepared at the time of use in sterile 1× M9 buffer, which is compatible with worm development and can be distributed evenly in the agar plates. Working dilutions of each inhibitor should be planned such that the desired final concentration can be achieved by distributing 150 μl of inhibitor over 5 ml of agar. The rationale for choosing a range of working concentrations is described later in the proof-of-principle experiment. It is important to note that the sensitivity of different worm strains to a given inhibitor may vary considerably for reasons that may or may not be related to the gene of interest for the pathway being tested. Therefore, each strain must be subjected to control dose-finding experiments and not assumed to be equally sensitive to the same inhibitor dose as another strain. Furthermore, a consistent method for obtaining progeny (synchronous egg-laying versus bleaching parental worms) should be used for each strain, as the method may alter the results of some experiments. The 150-μl volume of inhibitor in 1× M9 buffer was chosen as optimal for uniform absorption into 5 ml of NGM agar solidified in one well of a standard six-well tissue culture plate, in a 24 h period.

2.2.5. Alkaline buffer/bleaching solution (5 ml)

- 1.25 ml 1 M NaOH
- 1 ml commercial hypochlorite bleach
- 2.75 ml ddH$_2$O

2.2.6. Worm strains

Table 30.1 lists the properties of various worm strains suitable for evaluating Ras function, particularly with respect to the EGFR>Ras>Raf>MEK>ERK MAPK pathway. We note that because independent strains containing *in situ* mutations or transgenes are not necessarily otherwise genetically uniform, appropriate controls should be performed for each strain. For each strain, it is also particularly important to note the basal distribution of worms expected to display the Muv phenotype.

Standard worm husbandry steps are taken to maintain stock worm strains. For each generation, two or three adult worms are transferred to standard 60-mm dishes containing 12 ml 2% NGM agar and spotted with approximately 250 μl OP50 bacteria (see later). It is also important to pursue best practices to avoid genetic drift of strains. Strains with a growth disadvantage are particularly prone to acquiring modifiers. *C. elegans* strains survive 2 to 3 months on their NGM plates if Parafilmed after starvation. All strains should be kept as Parafilmed stocks, and active cultures can be renewed as necessary from the Parafilmed plates. *C. elegans* strains can also be frozen, and active cultures should be renewed periodically from frozen reserves (Stiernagle, 2006).

3. Experimental Procedure

Investigators new to *C. elegans* are highly recommended to also consult the references cited in Bianchi and Driscoll (2006), Shaham (2006), Stiernagle (2006) for excellent visual aids and basic worm handling tips.

3.1. Neutral growth medium agar plate preparation

Neutral growth medium agar plates must be prepared fresh for each experiment, as they tend to dehydrate over time, which affects the final drug concentration. Therefore, pouring fresh plates for each experiment ensures consistent volume and drug concentration and decreases interexperiment variability. To further improve consistency and minimize dehydration, we recommend the use of six-well tissue culture plates, in which each well is the equivalent of a single 35-mm tissue culture dish, and filling the wells about half full with agar. On day 1, prepare and pour NGM agar (5 ml per well) into only the four outer wells of the six-well tissue culture plates, leaving the two center wells empty. This approach further avoids variability due to uneven dehydration across the plates and ensures that all wells will continue to contain an equal volume of agar throughout a given experiment. Freshly poured plates should be stored in a tightly covered container at room temperature; plastic food storage containers do nicely. To improve the even dissemination of inhibitors into the agar, wait approximately 24 h before the next step.

3.2. Inhibitor addition to agar plates

Add inhibitor 24 h after pouring the agar (i.e., on day 2). Each plate is generally devoted to a single concentration of a given inhibitor, with one well containing the DMSO or ethanol vehicle control and the three remaining wells containing inhibitor, all at the same concentration in order to generate triplicate data points. Each inhibitor concentration should be tested in its own plate. It is strongly recommended to use the same layout for all plates in the experiment, both for consistency and to decrease the chances of error. Add 150 μl of the working dilution of inhibitor or vehicle, freshly prepared in sterile 1× M9 buffer as indicated earlier, drop wise to the agar surface of the appropriate well. Immediately swirl and then rock the plate in perpendicular directions to ensure uniform drug distribution. Apply the drug at a sufficient concentration that will result in the desired final concentration, once the drug has been absorbed into the entire 5 ml volume of agar. Return the plates to their container and allow the inhibitors to absorb into the agar for 24 h.

3.3. Seeding NGM plates with bacterial food for worms

It is best to use OP50, a laboratory strain of *E. coli*, to feed the worms for these experiments (Brenner, 1974; Stiernagle, 2006). Grow OP50 to stationary phase in LB medium without antibiotics. It is not necessary to use fresh OP50 cultures, but the same culture should be used for all experimental plates. On day 3, spot *E. coli* OP50 bacteria onto the plate to provide food for the worms by adding 80 μl of bacteria to the center of each well. Swirl to distribute. This lawn of OP50 will be sufficient to feed the animals for the duration of the experiment. Store plates as before in tightly covered containers at room temperature.

3.4. Adding animals to the bacterial lawn

Add animals to the plates 24 h after spotting the bacteria. On day 4, begin either the egg-laying assay or the bleaching for hatch-off protocol as described later. Healthy strains grown at 20 °C will form L4 larvae between 48 and 60 h post-egg-laying or bleaching, and adults thereafter (Brenner, 1974). The timing for sickly, slower growing strains will need to be adjusted accordingly.

3.5. Egg-laying assay to obtain semisynchronized populations

In one generation, worms from a single stock 60-mm dish will grow to produce sufficient progeny to test one experimental condition, that is, four wells of a six-well plate. To start the egg-laying assay, transfer 12 to 15 adult hermaphrodite worms to each well. More parents can be added for less fecund strains. Place the plates in a sealed plastic container at 20 °C (or a viable temperature appropriate for the strain). Allow them to lay eggs for 3 h. Then, remove all the parents and return plates to the incubator until the progeny hatch and reach the desired developmental stage. This can be as long as 4 days [*e.g.*, for *lin-1(null)* worms that will be scored as L4 larvae] or as short as 60 h [*e.g.*, *let-60*(gf) worms that will be scored as adults]. It is not necessary to begin with exactly the same number of adult animals in each well because the eventual worm progeny will be scored according to the percentages of all the resulting worms that display a given phenotype, not according to absolute numbers.

3.6. Bleaching worms for hatch-off to obtain a synchronized population of L1 larvae

This protocol is used to obtain eggs from an asynchronous culture of adult hermaphrodite worms that will hatch to produce a culture of synchronized L1 larvae. It is generally used when a mutant worm strain lays eggs poorly, such as strains harboring null mutations in the Ets family transcription factor *lin-1* or any strain with a vulvaless phenotype.

- Grow worms as usual on a standard 60-mm stock 2% NGM plate seeded with OP50 bacteria, until the plate is overgrown with adult worms. Unlike the egg-laying assay, to obtain enough worms from the bleaching protocol, it is necessary to start with two overgrown 60-mm plates of stock worms per four wells of each six-well plate.
- Add 2 ml of sterile 1× M9 solution to the plate. Use a sterile Pasteur pipette to transfer the worm suspension to a 15-ml conical tube.
- Spin 30 s in a clinical centrifuge (≈1000 rpm) and remove the supernatant.
- Add 5 ml of freshly made alkaline buffer (bleach solution). This kills all adult animals, but leaves eggs (protected by shells) alive, thus producing a culture of animals that are all within a ≈12-h developmental time window. It is important to make the bleach solution fresh for each experiment in order to assure good lysis of the adult worms.
- Incubate at room temperature for 3 min, with occasional gentle agitation.
- Quickly repeat the centrifugation step and remove the supernatant.
- Add 5 ml of 1× M9 buffer (this step *must* be completed within 5 min of the addition of alkaline buffer) to resuspend the eggs. Mix, spin, and remove supernatant.
- Repeat this wash step twice by adding 5 ml of 1× M9 buffer each time.
- At the final wash step, leave ≈50 μl of 1× M9 buffer in the tube and resuspend the eggs in it.
- Use a Pasteur pipette to distribute a few drops of the suspension containing the eggs (and adult worm carcasses) to each well containing inhibitor or vehicle. Similar to the egg-laying assay given earlier, it is not necessary to distribute exactly the same number of eggs into each well, because data are collected as percentages of all worms in the well displaying a given phenotype, rather than absolute numbers of worms in the well.
- Place the plates, again in a sealed plastic container, at 20 °C (or a viable temperature appropriate for the strain) and continue as with the egg-laying assay described earlier until the worms reach the desired developmental stage for scoring.

3.7. Mounting worms for DIC microscopy

This mounting procedure creates a small "pad" of anesthetic-containing agar on a microscope slide, onto which the animals can be placed before a coverslip is applied. It is necessary to do this to immobilize the worms in order to score their phenotypes by using DIC/Nomarski optics. When ready to score worms:

- Melt one aliquot at a time of 3% NGM agar by placing the glass tube first into a 100 °C block; then partially cool by transferring it to 65 °C. Once an aliquot has been melted, any excess should be discarded.
- Add 1 M sodium azide, which will serve as an immobilizing agent for the worms, to a final concentration of 10 mM. Vortex briefly to mix.
- On both sides of a clean slide place two "spacer" slides, each of which has a single piece of laboratory tape running the length of the slide. The thickness of the labeling tape will determine the thickness of the agar pad. To the center slide add two drops of the melted agar with a Pasteur pipette, avoiding bubbles. Immediately drop another clean slide crosswise on it. Allow the agar to solidify and then remove the top slide, thereby uncovering the resulting agar pad.
- Add 5 μl 1× M9 buffer to the top of the agar pad.
- With a pick, collect the worms to be scored and gently swish them off into the buffer solution on the pad.
- Immediately and gently place an appropriate coverslip on top of the agar pad. For the Nikon Eclipse DIC microscope used for these experiments, a No. 1.5 (18 mm^2) coverslip is recommended.
- If a long session (>1 h) is planned, seal the edges of the coverslip with petroleum jelly or VALAP to avoid desiccation.

4. SCORING WILD-TYPE VERSUS MUV PHENOTYPES

Accurate quantification of the Muv phenotype of animals treated with vehicle or inhibitor requires understanding how to identify the ventral protrusions or invaginations that represent pseudovulvae and consistently applying these criteria at the appropriate developmental stage. It is also important to recognize that throughput is inversely related to resolution. Thus, rapid, high throughput scoring is possible at low resolution using a dissecting microscope, whereas slow, low throughput scoring is possible at higher resolution using DIC/Nomarski optics, but it is not possible to achieve both high throughput and high resolution. Therefore, a reasonable sized assay will include four or five conditions (i.e., four or five plates) for worms that will be scored under a dissecting microscope or one or two conditions for worms that will be scored under a DIC microscope.

Criteria for distinguishing Muv animals from wild-type animals always include the appearance of one or more ventral protrusions in Muv animals, that is, pseudovulvae, or their precursors, that are in addition to the normal wild-type vulva (Figs. 30.2 and 30.4). However, the ease and manner of detecting these pseudovulvae may differ, depending on the strain of worms being tested. Worm strains displaying robust Muv phenotypes at the adult stage of development, for example, *ras/let-60(n1046gf)* mutants, can be scored rapidly and accurately under the dissecting microscope, allowing one person to score large samples (hundreds of worms) in a single day. Because the vulva and pseudovulvae are easily observed through a dissecting microscope only at the adult stage, it is critical that all the scoring done simply by counting these protrusions be performed only on adult worms, and thus the experiment must be timed accordingly. At 20 °C, the time from egg-laying to adulthood is approximately 60 h (Brenner, 1974), although certain sickly strains can take longer, so their developmental timing must be measured empirically.

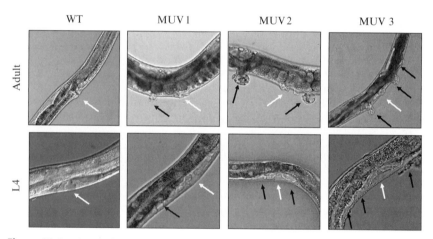

Figure 30.4 Quantification of vulval induction through the Ras>Raf>ERK MAPK pathway at adult and L4 larval stages. In the adult worm (top row), the Muv phenotype of the activated Ras homolog *let-60(n1046gf)* is very robust and is easily identifiable by the appearance of one or more of the excess ventral protrusions called pseudovulvae (black arrows; Muv1, Muv2, Muv3, etc.) that develop in addition to the normal wild-type vulva (white arrows). The Muv phenotype of these worms can be scored easily and rapidly by simply counting the pseudovulvae present at the end of a standard egg-laying assay. In some sickly mutant strains with a Muv phenotype, such as *lin-1(null)* worms, pseudovulvae are hardly distinguishable in adults. These worms must be subjected to the bleaching protocol and the Muv phenotype quantified instead at the L4 larval stage (bottom row), utilizing DIC microscopy. In L4 worms the future wild-type vulva appears as a "Christmas tree" structure (white arrow), whereas future pseudovulvae (black arrows) form characteristic rounded and frequently asymmetric invaginations. All images were obtained using DIC microscopy at 400× magnification.

Data should be collected not only on the numbers of worms that are wild type vs Muv, but also on the numbers of protrusions that appear in the Muv worms: Muv+1, Muv+2, Muv+3, etc. (Fig. 30.4). This is because some inhibitor treatments may reduce the percentage of worms that are Muv, whereas others simply reduce the numbers of protrusions in Muv worms, and still others do both. This situation is analogous to anchorage-independent growth assays using mammalian cells, in which colony number, colony size, or both can be affected. For an example, see analyses of Raf and MEK inhibitors in human tumor cell lines (Hao *et al.*, 2007). The biological mechanisms underlying these different effects remain unclear, however.

While some mutant strains displaying a Muv phenotype, such as *let-60*(gf), are scored easily under a dissecting microscope, other strains of worms that display a Muv phenotype cannot be scored by this means. An example of such a strain is the *lin-1*(*null*) mutant (see Table 30.1), mentioned earlier in the context of poor egg-laying that necessitates the use of the bleaching protocol to obtain synchronous populations of L1 larvae. The *lin-1*(*null*) mutant worms are relatively unhealthy in other ways as well. These worms can undergo significant developmental delays (more than 3 days) and become unsynchronized, which causes the scoring process to be both less efficient and less accurate. In addition, vulval protrusions are poorly distinguished in adult *lin-1*(*null*) mutant worms. Therefore, instead of scoring treated *lin-1*(*null*) worms at the adult stage, they should be scored at the earlier L4 larval stage. Although highly accurate, this procedure is considerably more complex and time-consuming than counting very obvious protrusions under a dissecting microscope.

The numbers of ventral protrusions present at the adult stage (see Fig. 30.4, top row) can be measured very accurately at the L4 stage by counting under high resolution DIC microscopy the structures formed at that point by the vulval precursor cells (VPCs) (see Fig. 30.4, bottom row) (Sulston and Horvitz, 1977). When worms are scored under DIC, the main (wild type) prevulval structure can be differentiated easily from the additional ventral protrusions or invaginations seen in adult or L4 worms, respectively, that display a Muv phenotype. VPCs are induced by the Ras>MAPK pathway, and the VPCs that will form the wild-type vulva first form a "Christmas tree"-like structure. In contrast, those cells inappropriately adopting vulval fates, because of activating mutations or excess activity in this pathway, form a rounded, generally asymmetrical invagination that will eventually become a ventral protrusion or pseudovulva (see Fig. 30.4). The high resolution of DIC microscopy thus makes possible easy and accurate identification and distinction of the invaginations that will become either the normal vulva or the pseudovulvae. However, the processes of worm synchronization and mounting required prior to quantification at the DIC microscope make this scoring method rate limiting because of the low number of worms that can be analyzed at one sitting (<100 worms per experiment). In unhealthy worm strains, loss of synchronous growth is common. Therefore, not all worms present will be at the appropriate developmental stage suitable for scoring. To achieve the best accuracy, it is always important to count as many worms as are present at the appropriate (in this case, L4) developmental stage.

5. PROOF-OF-PRINCIPLE EXPERIMENT

We show here a proof-of-principle experiment designed to demonstrate the utility of this protocol for evaluating pharmacological inhibitors of activated Ras pathway function. The widely used and well-characterized MEK inhibitor U0126 is a potent and selective small molecule inhibitor of both of the dual specificity kinases MEK1 and MEK2; it blocks MEK-induced phosphorylation and activation of ERK MAPKs both in mammalian cells and in *in vitro* enzymatic assays (Favata *et al.*, 1998). U0126 should thus inhibit the Muv phenotype of worm strains in which the activating mutation is at or upstream from MEK (in *C. elegans*, MEK-2; see Fig. 30.1 and Table 30.1), but should not affect the Muv phenotype of worms in which the activating mutation is downstream of MEK. We therefore tested the ability of U0126 to inhibit the Muv phenotype of worm strains with mutations affecting either Ras or a downstream (transcription factor) step in the Ras>Raf>MEK>ERK MAPK pathway.

As indicated in the protocol described previously, we diluted both the DMSO vehicle and the U0126 inhibitor to equivalent concentrations in sterile M9 buffer, applied the working dilutions to separate wells containing NGM agar, and added a lawn of OP50 bacteria. We then performed an egg-laying assay in the vehicle- or inhibitor-impregnated agar using *let-60* (*n1046gf*) worms expressing an endogenously activated LET-60 mutant protein (G13E). This mutation causes activation of the LIN-45-MEK-2-MPK-1 MAPK pathway and hence we expected it to be sensitive to Muv inhibition by U0126. To make sure that any Muv inhibition seen was mechanism-based and not just a consequence of general VPC toxicity, we also grew L1 larvae following a bleach hatch of *lin-1(null)* worms lacking an Ets-like inhibitory transcription factor that regulates the induction of VPCs [recall that the Muv phenotype of *lin-1(null)* worms must be evaluated following bleach hatch-off due to poor egg-laying properties]. Because the Muv phenotype of these worms is driven instead by alterations at the transcription factor level, which is clearly downstream of MEK-2 (see Fig. 30.1 and Table 30.1), we expected it to be insensitive to MEK inhibition.

The IC$_{50}$ of U0126 for MEK1 *in vitro* was reported to be ≈0.07 μM and ≈1 μM in COS-7 cells (Favata *et al.*, 1998). Although it is not clear whether there is a definable relationship between the IC$_{50}$ of U0126 or other inhibitors as identified *in vitro* or in mammalian cell systems and an effective dose in *C. elegans*, this information on U0126 did provide us with a starting point to select doses for assay *in vivo*. Knowing that the presence of their cuticle barrier means that worms do not efficiently take up most small molecules, we chose to test U0126 at 1, 3, 10, and 30 μM. Treatment with U0126 reverted the Muv phenotype of *let-60*(gf) worms in a dose-dependent manner (Fig. 30.5). The IC$_{50}$ of U0126 in *this strain* was ≈7 μM, and ≈30 μM was required to completely block the Ras>MAPK signaling pathway. In contrast, even 30 μM had no effect whatsoever on the Muv phenotype of *lin-1* null animals (see Fig. 30.5, top). These results demonstrate that the Muv inhibition by U0126 was selective, and are consistent

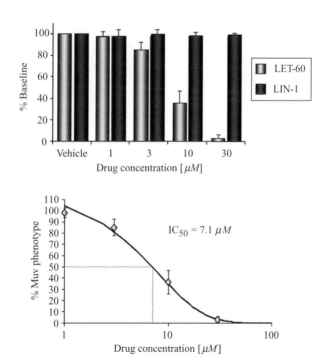

Figure 30.5 The MEK inhibitor U0126 dose dependently reduces the Muv phenotype induced by constitutively activated Ras in *let-60(n1046gf)* worms (top, green), but does not affect the Muv phenotype induced by loss of the downstream Ets family transcription factor in *lin-1(null)* animals (top, red). "Percent baseline" refers to the percentage of animals displaying a Muv phenotype in the presence of inhibitor compared to vehicle-only control. The dose–response curve (bottom) demonstrates that the IC$_{50}$ for Muv inhibition in *let-60(gf)* worms is approximately 7.1 μM, and essentially maximal inhibition can be achieved at the 30 μM dose to which *lin-1(null)* worms are insensitive. (See color insert.)

with the mechanism of U0126 acting as a MEK inhibitor. In additional experiments (not shown) using worm strains expressing other activating mutations, U0126 also acted as expected by completely blocking the Muv phenotype induced by the activated Raf ortholog LIN-45(AA), but not by the ERK MAPK ortholog MPK-1. Although U1026 is a highly specific non–ATP-competitive inhibitor of MEK activation of ERK (Ahn *et al.*, 2001; Davies *et al.*, 2000), it has poor pharmacologic properties, and newer generation MEK inhibitors are now under clinical evaluation (Roberts and Der, 2007). If their anti–MEK activities are also mechanism-based, we would expect similar results, albeit perhaps at a shifted dose range. However, off-target effects may not be shared with U0126 and may be identifiable by these assays, as seen with the unexpectedly proapoptotic FTIs (Lackner *et al.*, 2005).

The same type of experiments described in this chapter can be applied to previously characterized or novel inhibitors of Ras function, or to inhibitors of other pathways, by utilizing appropriate strains of worms. Consult Table 30.1 for additional information about available worm strains with activating mutations in signaling components that function both upstream and downstream of Ras. These strains provide more focused systems to study inhibitors that target the Ras signaling network at distinct nodal points.

 ## 6. Conclusions and Future Directions

Despite intensive efforts by researchers and the pharmaceutical industry to develop inhibitors of Ras for molecularly targeted therapeutics, to date no anti–Ras therapeutics have made the successful passage through the long and winding road of drug discovery. The inability to identify pharmacologic approaches that selectively target mutationally activated Ras directly has contributed to this failure, with the most advanced efforts targeting proteins involved in the post-translational lipid modification of Ras (FTase) or signaling components downstream of Ras (Raf and MEK). Because FTase activity is critical for the function of >50 other human proteins (Reid *et al.*, 2004) and because the Raf>MEK>ERK pathway is not the simple linear cascade once imagined (McKay and Morrison, 2007), inhibitors of these proteins may have considerable off-target activities and cellular consequences. Our validation and application of inhibitor analyses of Ras signaling in *C. elegans* provide another model system for ongoing and future development of anti–Ras inhibitors. Overall, this genetic system, together with more physiologically relevant preclinical models of cancer, including new human cell culture and genetically engineered mouse models (Hahn and Weinberg, 2002; Sharpless and Depinho, 2006), will produce improved preclinical analyses to facilitate greater clinical success for target-based drug discovery.

ACKNOWLEDGMENTS

Our research studies were supported by NIH grants to A.D.C. (CA42978, CA67771, and CA109550) and to C.J.D. (CA42978, CA67771 and CA106991). V.G.P. was supported by an NIH IMSD training grant fellowship.

REFERENCES

Ahn, N. G., Nahreini, T. S., Tolwinski, N. S., and Resing, K. A. (2001). Pharmacologic inhibitors of MKK1 and MKK2. *Methods Enzymol.* **332,** 417–431.

Beitel, G. J., Clark, S. G., and Horvitz, H. R. (1990). *Caenorhabditis elegans* ras gene let-60 acts as a switch in the pathway of vulval induction. *Nature* **348,** 503–509.

Beitel, G. J., Tuck, S., Greenwald, I., and Horvitz, H. R. (1995). The *Caenorhabditis elegans* gene lin-1 encodes an ETS-domain protein and defines a branch of the vulval induction pathway. *Genes Dev.* **9,** 3149–3162.

Bianchi, L., and Driscoll, M. (2006). Culture of embryonic *C. elegans* cells for electrophysiological and pharmacological analyses. The *C. elegans* Research Community, Wormbook, doi/10.1895/wormbook.1.122.1, http://www.wormbook.org

Brenner, S. (1974). The genetics of *Caenorhabditis elegans*. *Genetics* **77,** 71–94.

Chong, H., Lee, J., and Guan, K. L. (2001). Positive and negative regulation of Raf kinase activity and function by phosphorylation. *EMBO J.* **20,** 3716–3727.

Choy, R. K., and Thomas, J. H. (1999). Fluoxetine-resistant mutants in *C. elegans* define a novel family of transmembrane proteins. *Mol. Cell* **4,** 143–152.

Clark, S. G., Stern, M. J., and Horvitz, H. R. (1992). *C. elegans* cell-signalling gene sem-5 encodes a protein with SH2 and SH3 domains. *Nature* **356,** 340–344.

Davies, S. P., Reddy, H., Caivano, M., and Cohen, P. (2000). Specificity and mechanism of action of some commonly used protein kinase inhibitors. *Biochem. J.* **351,** 95–105.

Dickson, B., Sprenger, F., Morrison, D., and Hafen, E. (1992). Raf functions downstream of Ras1 in the Sevenless signal transduction pathway. *Nature* **360,** 600–603.

Favata, M. F., Horiuchi, K. Y., Manos, E. J., Daulerio, A. J., Stradley, D. A., Feeser, W. S., Van Dyk, D. E., Pitts, W. J., Earl, R. A., Hobbs, F., Copeland, R. A., Magolda, R. L., *et al.* (1998). Identification of a novel inhibitor of mitogen-activated protein kinase. *J. Biol. Chem.* **273,** 18623–18632.

Ferguson, E. L., and Horvitz, H. R. (1985). Identification and characterization of 22 genes that affect the vulval cell lineages of the nematode *Caenorhabditis elegans*. *Genetics* **110,** 17–72.

Hahn, W. C., and Weinberg, R. A. (2002). Rules for making human tumor cells. *N. Engl. J. Med.* **347,** 1593–1603.

Hao, H., Muniz-Medina, V. M., Mehta, H., Thomas, N. E., Khazak, V., Der, C. J., and Shields, J. M. (2007). Context-dependent roles of mutant B-Raf signaling in melanoma and colorectal carcinoma cell growth. *Mol. Cancer Ther.* **6,** 2220–2229.

Hara, M., and Han, M. (1995). Ras farnesyltransferase inhibitors suppress the phenotype resulting from an activated ras mutation in *Caenorhabditis elegans*. *Proc. Natl. Acad. Sci. USA* **92,** 3333–3337.

Hill, R. J., and Sternberg, P. W. (1992). The gene lin-3 encodes an inductive signal for vulval development in *C. elegans*. *Nature* **358,** 470–476.

Horvitz, H. R., Chalfie, M., Trent, C., Sulston, J. E., and Evans, P. D. (1982). Serotonin and octopamine in the nematode *Caenorhabditis elegans*. *Science* **216,** 1012–1014.

Huang, L. S., Tzou, P., and Sternberg, P. W. (1994). The lin-15 locus encodes two negative regulators of *Caenorhabditis elegans* vulval development. *Mol. Biol. Cell* **5,** 395–411.

Katz, W. S., Lesa, G. M., Yannoukakos, D., Clandinin, T. R., Schlessinger, J., and Sternberg, P. W. (1996). A point mutation in the extracellular domain activates LET-23, the *Caenorhabditis elegans* epidermal growth factor receptor homolog. *Mol. Cell. Biol.* **16**, 529–537.

Kornfeld, K., Hom, D. B., and Horvitz, H. R. (1995). The ksr-1 gene encodes a novel protein kinase involved in Ras-mediated signaling in *C. elegans*. *Cell* **83**, 903–913.

Lackner, M. R., and Kim, S. K. (1998). Genetic analysis of the *Caenorhabditis elegans* MAP kinase gene mpk-1. *Genetics* **150**, 103–117.

Lackner, M. R., Kindt, R. M., Carroll, P. M., Brown, K., Cancilla, M. R., Chen, C., de Silva, H., Franke, Y., Guan, B., Heuer, T., Hung, T., Keegan, K., *et al.* (2005). Chemical genetics identifies Rab geranylgeranyl transferase as an apoptotic target of farnesyl transferase inhibitors. *Cancer Cell* **7**, 325–336.

Lewis, J. A., Wu, C. H., Levine, J. H., and Berg, H. (1980). Levamisole-resistant mutants of the nematode *Caenorhabditis elegans* appear to lack pharmacological acetylcholine receptors. *Neuroscience* **5**, 967–989.

McIntire, S. L., Jorgensen, E., Kaplan, J., and Horvitz, H. R. (1993). The GABAergic nervous system of *Caenorhabditis elegans*. *Nature* **364**, 337–341.

McKay, M. M., and Morrison, D. K. (2007). Integrating signals from RTKs to ERK/MAPK. *Oncogene* **26**, 3113–3121.

Miller, L. M., Gallegos, M. E., Morisseau, B. A., and Kim, S. K. (1993). lin-31, a *Caenorhabditis elegans* HNF-3/fork head transcription factor homolog, specifies three alternative cell fates in vulval development. *Genes Dev.* **7**, 933–947.

Modzelewska, K., Elgort, M. G., Huang, J., Jongeward, G., Lauritzen, A., Yoon, C. H., Sternberg, P. W., and Moghal, N. (2007). An activating mutation in sos-1 identifies its Dbl domain as a critical inhibitor of the epidermal growth factor receptor pathway during *Caenorhabditis elegans* vulval development. *Mol. Cell Biol.* **27**, 3695–3707.

Moghal, N., and Sternberg, P. W. (2003). The epidermal growth factor system in *Caenorhabditis elegans*. *Exp. Cell Res.* **284**, 150–159.

Reid, T. S., Terry, K. L., Casey, P. J., and Beese, L. S. (2004). Crystallographic analysis of CaaX prenyltransferases complexed with substrates defines rules of protein substrate selectivity. *J. Mol. Biol.* **343**, 417–433.

Roberts, P. J., and Der, C. J. (2007). Targeting the Raf-MEK-ERK mitogen-activated protein kinase cascade for the treatment of cancer. *Oncogene* **26**, 3291–3310.

Rocheleau, C. E., Ronnlund, A., Tuck, S., and Sundaram, M. V. (2005). *Caenorhabditis elegans* CNK-1 promotes Raf activation but is not essential for Ras/Raf signaling. *Proc. Natl. Acad. Sci. USA* **102**, 11757–11762.

Rose, W. C., Lee, F. Y., Fairchild, C. R., Lynch, M., Monticello, T., Kramer, R. A., and Manne, V. (2001). Preclinical antitumor activity of BMS-214662, a highly apoptotic and novel farnesyltransferase inhibitor. *Cancer Res.* **61**, 7507–7517.

Rowinsky, E. K. (2006). Lately, it occurs to me what a long, strange trip it's been for the farnesyltransferase inhibitors. *J. Clin. Oncol.* **24**, 2981–2984.

Shaham, S. (2006). Methods in Cell Biology. The *C. elegans* Research Community, Wormbook doi/10.1895/wormbook.1.49.1, http://www.wormbook.org

Sharpless, N. E., and Depinho, R. A. (2006). The mighty mouse: Genetically engineered mouse models in cancer drug development. *Nat. Rev. Drug Discov.* **5**, 741–754.

Sieburth, D. S., Sun, Q., and Han, M. (1998). SUR-8, a conserved Ras-binding protein with leucine-rich repeats, positively regulates Ras-mediated signaling in *C. elegans*. *Cell* **94**, 119–130.

Stiernagle, T. (2006). Maintenance of *C. elegans*. *Wormbook*, ed. The *C. elegans* Research Community Wormbook, doi/10.1895/wormbook.1.101.1, http://www.wormbook.org

Sulstam, J. E., and Horvitz, H. R. (1977). Post-embryonic cell lineages of the nematode, *Caenorhabditis elegans*. *Dev. Biol.* **56**, 110–156.

Sundaram, M., and Han, M. (1995). The *C. elegans* ksr-1 gene encodes a novel Raf-related kinase involved in Ras-mediated signal transduction. *Cell* **83,** 889–901.

Tan, P. B., Lackner, M. R., and Kim, S. K. (1998). MAP kinase signaling specificity mediated by the LIN-1 Ets/LIN-31 WH transcription factor complex during *C. elegans* vulval induction. *Cell* **93,** 569–580.

Therrien, M., Chang, H. C., Solomon, N. M., Karim, F. D., Wassarman, D. A., and Rubin, G. M. (1995). KSR, a novel protein kinase required for RAS signal transduction. *Cell* **83,** 879–888.

Weinshenker, D., Garriga, G., and Thomas, J. H. (1995). Genetic and pharmacological analysis of neurotransmitters controlling egg-laying in *C. elegans*. *J. Neurosci.* **15,** 6975–6985.

CHAPTER THIRTY-ONE

RAS-DRIVEN TRANSFORMATION OF HUMAN NESTIN-POSITIVE PANCREATIC EPITHELIAL CELLS

Paul M. Campbell,* Kwang M. Lee,[†] Michel M. Ouellette,[†] Hong Jin Kim,* Angela L. Groehler,* Vladimir Khazak,[‡] and Channing J. Der*

Contents

* Lineberger Comprehensive Cancer Center, University of North Carolina at Chapel Hill, Chapel Hill, North Carolina
[†] Eppley Institute for Research in Cancer and Allied Diseases, University of Nebraska Medical Center, Omaha, Nebraska
[‡] NexusPharma, Inc., Langhorne, Pennsylvania

Methods in Enzymology, Volume 439
ISSN 0076-6879, DOI: 10.1016/S0076-6879(07)00431-4

Abstract

Mutational activation of the K-Ras oncogene is well established as a key genetic step in the development and growth of pancreatic adenocarcinomas. However, the means by which aberrant Ras signaling promotes uncontrolled pancreatic tumor cell growth remains to be fully elucidated. The recent use of primary human cells to study Ras-mediated oncogenesis provides important model cell systems to dissect this signaling biology. This chapter describes the establishment and characterization of telomerase-immortalized human pancreatic duct-derived cells to study mechanisms of Ras growth transformation. An important strength of this model system is the ability of mutationally activated K-Ras to cause potent growth transformation *in vitro* and *in vivo*. We have utilized this cell system to evaluate the antitumor activity of small molecule inhibitors of the Raf-MEK-ERK mitogen-activated protein kinase cascade. This model will be useful for genetic and pharmacologic dissection of the contribution of downstream effector signaling in Ras-dependent growth transformation.

1. INTRODUCTION

Pancreatic ductal adenocarcinoma (PDA) represents one of the most lethal forms of cancer, with a 5-year mortality risk of greater than 95% (Jemal *et al.*, 2005; Warshaw and Fernandez-del Castillo, 1992). While the prognosis is better for patients with surgically resectable disease, this cohort unfortunately represents only about 10% of all PDA patients (Ahrendt and Pitt, 2002). The vast majority of PDA cases present in later stages with metastatic or recurrent disease (Allison *et al.*, 1998). As such, it is imperative to understand the nature of PDA so that better therapeutic interventions can be uncovered.

A common progression demonstrated for PDA links changes in the genetic makeup of tumor cells with alterations in cancer stage (Hruban *et al.*, 2000a,b). A mutation of the K-Ras gene is seen early in the initiation of pancreatic ductal cell dysplasia at the pancreatic intraepithelial neoplasia (PanIN) 1 stage. Single base pair substitutions at the 12, 13, or 61 codons of K-Ras result in constitutive activation of the protein and continuous signaling (Malumbres and Barbacid, 2003). These mutations are seen in more than 90% of PDA cases (Malumbres and Barbacid, 2003), but because the signaling cascades downstream of Ras activation are so varied (Repasky *et al.*, 2004), it remains unclear which effector pathways are required for both initiation of PanINs and subsequent development of PDA (Yeh and Der, 2007). As a result, experimental models designed to elucidate the pertinent role of K-Ras are necessary to help guide future molecularly targeted anti-Ras therapy (Cox and Der, 2002).

One approach to understanding cancer cell behavior is to look at the genetic events in a stepwise manner in differentiated but untransformed epithelial cells to determine the specific cellular consequences of defined genetic lesions that initiate and maintain the cancerous phenotype. Much progress in this cell-based tactic has been facilitated by the work of Weinberg and colleagues, who helped design the immortalization of epithelial cells via expression of the catalytic subunit of human telomerase (hTERT) (Elenbaas *et al.*, 2001; Hahn and Weinberg, 2002; Hahn *et al.*, 1999). After primary cells have been made immortal, subpopulations of these cells can be further manipulated by exogenous expression of putative oncogenes or other genes, or selected elimination of potential tumor suppressors to reveal which changes are required for transformation. This chapter discusses one such system that has been devised as a human cell model for studying PDA progression and growth. Methodologies that detail the isolation of ductal cells from the pancreas, their immortalization, and the steps required to make these cells tumorigenic, as well as elements of Ras signaling that contribute to this transformation, are presented.

2. Isolation and Immortalization of Primary Pancreatic Ductal Cells

The partial digestion of a healthy pancreas (Lee *et al.*, 2003), excised following the accidental death of a 52-year-old man, was accomplished with the use of liberase (Linetsky *et al.*, 1997). This enzymatic treatment reduces the organ to large structural components, which are separated by centrifugation on a Ficoll gradient. The band containing large ducts is harvested and further sorted under microscopy to remove pancreatic islets, stroma, acini, and blood vessels. The ductal tissue is rocked in growth medium D [125 ml M3F base medium (InCell), 375 ml Dulbecco's modified Eagle's medium (DMEM) with L-glutamine and low glucose, 25 ml fetal calf serum (FCS), 50 μl of epidermal growth factor (EGF; 100 μg/ml in dH_2O), and 500 μl gentamicin (50 mg/ml, Invitrogen)] at 37° for 2 weeks, with the larger ducts removed by handpicking every other day. The remaining ducts are harvested and seeded onto tissue culture dishes in growth medium at a density of 500 fragments/25 cm^2. Epithelial cell sheets are propagated from these fragments, and cultures are cleared of fibroblasts by trypsinizing the sheets, replating, and treating the new cultures with a rabbit antibody raised against human fibroblasts and a preparation of rabbit complement (Jesnowski *et al.*, 1999). Cells are grown in growth medium at 37° and 5% CO_2. To maintain log-phase growth, cells are trypsinized once a week and replated at a density of 100,000 cells/25 cm^2.

Initially, the primary pancreatic duct-derived cells are transduced with an expression vector for hTERT (Ouellette *et al.*, 1999). First, a portion of the hTERT cDNA (Nakamura *et al.*, 1997) lacking 5′ and 3′ UTRs is cloned into the *Eco*RI site of the retroviral vector pBabe-puro (Morgenstern and Land, 1990). φNX-A packaging cells are transiently transfected with calcium phosphate (MBS kit; Stratagene) to generate infectious virus supernatants. The resulting amphotropic virus particle supernatants are supplemented with 4 mg/ml polybrene (Sigma) to facilitate infection efficiency, and primary pancreatic duct-derived cells are infected at population doubling day (PD) 19. Mass populations of cells with stable viral integration are then selected for by maintaining the infected cultures in growth medium supplemented with 750 ng/ml puromycin. Multiple drug-resistant colonies are then pooled together to establish cells stably expressing ectopic hTERT.

Telomerase gene transcription is assayed by reverse transcription polymerase chain reaction (RT-PCR) using 5′ ACTCGACACCGTGTCACCTA 3′ and 5′ GTGACAGGGCTGCTGGTGTC 3′ as primers for hTERT. Total RNA is obtained from cells using the guanidium/acid phenol procedure. In a final volume of 20 μl, each reverse transcription reaction contains 0.5 μg of heat-denatured RNA, 50 ng of random hexamers, 10 mM dithiothreitol, 500 μM of each dNTP, 50 mM Tris-HCl, pH 8.3, 75 mM KCl, 3 mM MgCl$_2$, and either 1 μl of SuperScript III reverse transcriptase (200 U/μl; Gibco-BRL) or 1 μl of water (mock reactions). After 1 h at 37°, RT reactions are diluted to 100 μl and denatured at 95° for 10 min. In a final volume of 50 μl, each PCR contains 5 μl of the diluted RT reaction, 1 μM of each PCR primer, 200 μM of each dNTP, 1.5 mM MgCl$_2$, 20 mM Tris-HCl, pH 8.4, and 50 mM KCl. PCR are initiated at 95° by the addition of *Taq* DNA polymerase (2.5 U; Gibco-BRL) and allowed to cycle as follows: 45 s at 94°, 45 s at the annealing temperature, and 1 min at 72°. Ten microliters of the PCR products is resolved by agarose gel electrophoresis and visualized with ethidium bromide staining (0.5 μg/ml). Additionally, telomerase activity is verified as described previously (Ouellette *et al.*, 1999) using the TRAP-eze telomerase detection kit (Intergen).

Control primary pancreatic duct cells normally undergo senescence at around PD 25, but those populations expressing exogenous hTERT are immortalized and show unlimited proliferative capacity (Lee *et al.*, 2003). While cancer-related phenotypic changes were not seen in these immortal cells, a loss of the epithelial cell markers carbonic anhydrase II and cytokeratin 19 (Vila *et al.*, 1994) (using primers 5′-AAGGAACCCATCAGCG TCAG-3′/5′-AAAGCACCAACCAGCCACAG-3′ and 5′-GCCACT ACTACACGACCATCC-3′/5′-GAATCCACCTCCACACTGACC-3′, respectively) was evident. Concurrently, immortal clones analyzed at all time points expressed nestin, a putative marker for pluripotent stem cells (Hunziker and Stein, 2000; Zulewski *et al.*, 2001). These cells are now denoted as human pancreatic nestin-expressing (hTERT-HPNE) cells.

3. ADDITIONAL GENETIC STEPS REQUIRED TO TRANSFORM PANCREATIC DUCT-DERIVED CELLS

Because the hTERT-HPNE cells showed no phenotypic changes characteristic of cancer cells [soft agar colony formation, p53 and p21^{WAF1} mutation or p16^{INK4a} loss (Lee *et al.*, 2003)], we sought to drive these cells toward a cancerous transformation by the introduction of additional genes known to be functionally perturbed in PDA. The E6 and E7 proteins of the HPV16 virus were used to emulate the loss of p53 and inactivation of the p16^{INK4a}/Rb pathway, respectively. Oncogenic versions of H-Ras, K-Ras, and N-Ras were compared for their capacity to emulate the activation of K-Ras seen in early PanINs. Finally, the SV40 small t (st) antigen, which had been reported to be required for the malignant transformation of primary human cells, was also used. To prevent Ras-induced senescence, as well as toxicities associated with the expression of the SV40 st antigen, E6 and E7 were transduced prior to the introduction of these oncogenes. In all, eight derivatives of hTERT-HPNE cells were made expressing the following combinations of oncogenes: E6/E7, E6/E7/st, E6/E7/K-Ras12D, E6/E7/H-Ras12V, E6/E7/N-Ras12D, E6/E7/K-Ras12D/st, E6/E7/H-Ras12V/st, or E6/E7/N-Ras12D/st (Fig. 31.1). A brief description of the construction of the retrovirus expression plasmid DNAs used in our studies is presented next.

Plasmid pBabeZeo-st: A *Bam*HI–*Bgl*II fragment from pCMV5-small t (Sontag *et al.*, 1993) is cloned into the *Bam*HI site of pBabe Zeo. Plasmid pBabe Zeo is a derivative of pBabe Puro (Morgenstern and Land, 1990) in which the Puro^r cassette is inactivated by deletion of an *Eco*RI–*Cla*I fragment and replacement with an *Eco*RI–*Acc*I segment from pKS-Zeo. Plasmid pKS-Zeo is made by insertion in the *Bam*HI site of pBluescript KS II of a *Bgl*II–*Bam*HI fragment from plasmid pZeoSV (Invitrogen).

Plasmid pLXSH-N-RasG12D: A *Bam*HI fragment from pBabePuro-N-Ras(G12D) (Fiordalisi *et al.*, 2001) is inserted in the *Bam*HI site of pLXSH. Plasmid pLXSH is a derivative of pLXSN (Miller and Rosman, 1989) in which the neo^r cassette is inactivated by deletion of a *Hind*III–*Nae*I fragment and replacement with a *Hind*III–*Bgl*II segment from pSV2-Hygro (Maione *et al.*, 1992).

Plasmid pLXSH-K-RasG12D: A cDNA fragment encoding K-RasG12D is RT-PCR amplified from PANC-1 cells with primers 5′-CTTGCTA GGATCCTGCTGAAAATGACTGAATATA-3′ and 5′-GCTAGGA TCCGTATGCCTTAAGAAAAAAGTACAA-3′. The resultant PCR product is digested with *Bam*HI and inserted into the *Bam*HI site of plasmid pLXSH.

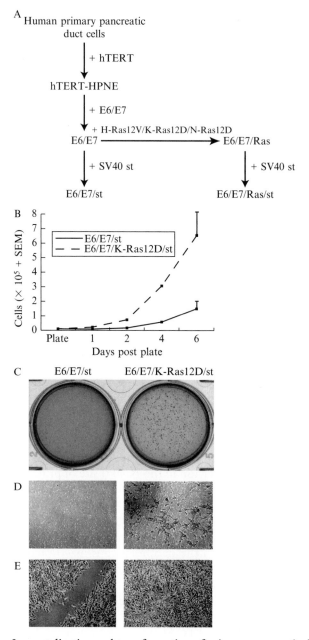

Figure 31.1 Immortalization and transformation of primary pancreatic duct-drived cells. (A) Ectopic expression of hTERT was used to immortalize a primary pancreatic duct cell culture, generating hTERT-HPNE cells. These cells were then sequentially infected to express first the human papilloma virus proteins E6 and E7 (E6/E7) and then constitutively active forms of H-Ras12V, K-Ras12D, or N-Ras12D. E6/E7 and E6/E7/Ras

3.1. Preparing viral supernatants carrying HPV16 E6/E7

PA317 LXSN 16E6E7 cells expressing HPV16 E6/E7 [ATCC (Halbert *et al.*, 1991)] are thawed in DMEM supplemented with 10% FCS and 50 μg/ml gentamicin. Fresh medium is replaced the following day. Upon confluence, cells are split with trypsin-EDTA to split ratios of 1:6 to 1:12. Cells are plated in a 150-mm dish and grown until 80% confluent. These are washed once with Medium D, and then a minimal volume of Medium D (12 ml/150-mm plate) is used to collect viruses overnight. The supernatant is cleared through a 0.45-μm polysulfone filter to remove any suspended cells, and polybrene is added to a final concentration of 4 μg/ml. Supernatants are used immediately or frozen for future experiments.

3.2. Preparing viral supernatants carrying oncogenic Ras and the SV40 small t antigen

ϕNX-A cells are grown in DMEM supplemented with 10% FCS and 50 μg/ml gentamicin and are split with trypsin-EDTA 1:4 to 1:6, avoiding >90% confluence. Cells are seeded into two 100-mm plates at 5×10^6 cells/ plate. On the following day, ϕNX-A cells ($\approx$80% confluent) are transfected with retroviral vectors (pBabeZeo-st, pBabeHygro ras, pLXSH-N-RasG12D, or pLXSH-K-RasG12D). ϕNX-A cells are transfected most easily using the CaPO$_4$ method, but other methods can be used as well. To calcium transfect the cells, we use Stratagene's MBS mammalian transfection kit following the manufacturer's instructions. In a 5-ml Falcon polystyrene tube, a calcium–DNA precipitate is prepared by mixing plasmid DNA (10 μg in 450 μl of sterile water) with 50 μl of solution I (2.5 M CaCl$_2$) and then adding 500 μl of solution II (N,N-bis(2-hydroxyethyl)-2-aminoethanesulfonic acid in buffered saline). The calcium–DNA mix is incubated at room temperature for 10 to 20 min. ϕNX-A cells are washed once with warm phosphate-buffered saline (PBS) and fed with 10 ml/dish of DMEM supplemented with 6% modified bovine serum (as provided by the kit). The precipitated DNA is gently resuspended and added to the ϕNX-A cells drop wise in a circular motion to distribute the DNA evenly. Dishes of ϕNX-A cells are swirled once and incubated at 37° and 5% CO$_2$ for 3 h. Cells are washed once with warm PBS and fed DMEM supplemented with 10% FCS and 50 μg/ml gentamicin.

Viruses are collected in three consecutive harvests over the next 72 h. To collect the viruses, 4 to 5 ml of fresh Medium D is used for each harvest,

cells were then infected with retrovirus encoding SV40 st to establish E6/E7/st and E6/E7/ Ras/st cells, respectively. Examples of K-Ras12D-transformed cells show increased proliferation (B), contact-independent growth (C), invasion (D), and migration (E).

with supernatants exposed to virus-producing cells for 8 to 16 h. Supernatants are cleared through 0.45-μm filters and either kept on ice and used within the hour or frozen at $-80°$, where the virus titer is stable for approximately 6 months.

3.3. Tranducing hTERT-HPNE cells

hTERT-HPNE cells are thawed in Medium D and expanded by trypsinizing and replating at split ratios of 1:3 to 1:4. Viral E6/E7 supernatants (2 ml/100-mm plate) are added to 80% confluent cultures, and cells are incubated overnight at 37°. The viral supernatant is replaced with fresh growth medium, and cells are split at a ratio of 1:3 to 1:4 over 2 to 3 days. Selection for viral gene integration is accomplished by treating infected cultures with 400 μg/ml G418 for 2 to 3 weeks, changing the medium every 3 to 4 days. Cohorts of these resultant cell populations are further transduced in the same manner for expression of oncogenic Ras isoforms (using 200 μg/ml hygromycin B for selection) or SV40 st antigen (using 50 μg/ml zeocin).

To verify expression of these cDNA constructs, total RNA is extracted from cells with TRIzol (1 ml/100-mm plate, Invitrogen), and RT-PCR is performed as described earlier using the following forward/reverse 5′-3′ primers: K-Ras— CTTGCTGAATTCCTGCTGAAAATGACTG AATATA/GCTACTCGAGGTATGCCTTAAGAAAAAAGTACAA; β-\actin—CGGGACCTGACTGACTACCT/CAGCACTGTGTTGGCG TACA; E6—GAACAGCAATACAACAAACCG/GCAACAAGAC ATACATCGACC; E7— AGGAGGAGGATGAAATAGATGG/TGG TTTCTGAGAACAGATGGG; and SV40 st— GAAGCAGTAGCAA TCAACCC/GCTTCTTCCTTAAATCCTGGTG. Annealing temperatures are set to either 55° (E6, st, K-Ras) or 57° (E7). Amplifications are done for 25 to 35 cycles, depending on the primer pair. The sizes of the expected PCR products are 351 bp (β-actin), 156 bp (E6), 197 bp (E7), 171 bp (SV40 st), and 250 bp (K-Ras). Results show equal signal for β-actin and confirm the presence or absence of each oncogene in the different hTERT-HPNE derivatives. PCR products of K-Ras cDNA are digested with *Bcl*I and resolved on ethidium bromide-agarose gels to differentiate between the expression of wild type and exogenous mutated K-Ras. The G12D mutation creates an additional *Bcl*I site, which gives digestion patterns that differ for wild type (154 + 96 bp) and oncogenic (104 + 50 + 96 bp) K-Ras. Digestion is done overnight at 37° in PCR buffer. K-Ras protein expression and activation are assayed by pull-down assays using the pGEX-Raf-RBD glutathione-*S*-transferase-bound fusion protein containing the Ras-binding domain of Raf-1 as described previously (Peterson *et al.*, 1996); expression of β-actin serves as a loading control (antibody AC-15; Sigma).

The HPV proteins E6 and E7 inactivate the cell cycle genes p53 and pRb, elements that are often disrupted in cancer (Munger *et al.*, 1989; Werness *et al.*, 1990), but the exogenous expression of these genes in hTERT-HPNE cells did not cause morphological or phenotypical changes. Because the mutation of Ras is an early event in more than 90% of PDA cases, we introduced oncogenic forms of H-, K-, and N-Ras into E6/E7 cells. Even the expression of these oncogenes failed to lead to soft agar colony growth. However, expression of SV40 st in E6/E7/Ras cells leads to significant transformation, as illustrated by contact-independent growth in soft agar. Because K-Ras is mutated in almost all PDA patients' tumors, we decided to focus on this cell line for further studies. The amount of total K-Ras expressed was similar in the matched E6/E7/st and E6/E7/Ras/st pair of cell lines; however, cells with mutant K-Ras12D expression exhibited significantly more GTP-bound K-Ras. This observation contrasts with other human model cell systems described, where the exogenous Ras gene resulted in significantly greater protein expression than is typically seen in human tumor cells. Thus, the matched pair of E6/E7/st and E6/E7/Ras/st cells provided us with a well-defined system to assess the signaling repercussions of mutant K-Ras expressed at physiologic levels.

4. ANALYSIS OF K-RAS EFFECTOR PATHWAYS IN PANCREATIC CELL TRANSFORMATION

4.1. Growth transformation assays

Initially, we looked at the growth of E6/E7/st and E6/E7/Ras/st under normal culture conditions [four parts high glucose DMEM to one part M3F (InCell) supplemented with 5% FCS, 100 U/ml penicillin, and 100 μg/ml streptomycin]. Ten thousand cells are seeded onto plastic in triplicate, and cells are trypsinized and counted after 24, 48, 96, and 144 h. For analysis of anchorage-independent growth, six-well plates are initially coated with 1 ml of autoclaved 1.8 g/ml agar and allowed to cool and gel. This thin dense bottom layer of agar prevents cell adherence to the plastic. Log-phase growing cells are trypsinized, and triplicates of 3×10^3 cells per well are suspended in 3 ml of enriched medium (supplemented with an additional 10% FCS), mixed quickly with 1 ml of warm (65°) sterile 1.5% agar, and plated onto the agar-coated six-well plates (Campbell and Szyf, 2003). This layer is allowed to set at room temperature for 30 min before 1 ml of standard growth medium is added to the top of the gelled matrix and colonies are grown for 21 days.

To assess the contributions of different effectors downstream of oncogenic K-Ras12D signaling, we utilized established pharmacologic inhibitors. MCP110 disrupts the physical interaction between Ras and Raf, with

MCP122 as a negative control inactive analog (Kato-Stankiewicz *et al.*, 2002), and U0126 is a highly specific inhibitor of MEK1/2. LY294004 blocks the activity of PI3K to prevent activation of Akt and other signaling proteins. Stock solutions of inhibitors are dissolved in dimethyl sulfoxide (DMSO) and added to both the agar containing the cells and the feeding medium at the following final concentrations: MCP110/122 (Nexus-Pharma) at 10 μM, LY294002 (Promega) at 10 μM, or U0126 (Promega) at 30 μM. Plates are refed with fresh growth medium including inhibitors or DMSO once per week. After 21 days in culture, colonies are stained with 0.5 ml/well of 1 mg/ml 3-(4,5-dimethylthiazol-2-yl)-2,5-diphenyltetrazolium bromide, counted in five random three-dimensional microscopic fields per well, and photographed. Cells with K-Ras12D expression show extensive colony formation within a week of suspension and large colonies visible to the naked eye within 14 days.

For tumorigenicity analysis, E6/E7/st or E6/E7/Ras/st cells are injected subcutaneously into the flanks of Hsd:Athymic Nude-*Foxn1^nu* nude mice (Harlan) at seeding densities of 0.2, 0.5, 1, or 2×10^6 cells per site. Tumor measurements are taken by calipers three times per week over 8 weeks or until the tumor burden reaches 1 cm^3. Tumor volumes are calculated by estimation of an ellipsoid using $(4/3)\pi(xyx/6)$, where x is length, y is width, and z is depth of the tumor. For histology analysis, tumors are excised immediately following euthanization and fixed in 10% buffered formalin. Tumor tissue is embedded in paraffin, and sections are cut and stained for cytoarchitecture with hematoxylin and eosin. Sections are additionally stained for cytokeratins with AE1/AE3 antibodies (Listrom and Dalton, 1987) and vimentin (Fig. 31.2). In agreement with Western blot analysis, xenograft tumors arising from E6/E7/Ras/st cells show low levels of cytokeratin expression and strong vimentin expression, and histology reveals that the tumors resemble more of a sarcomatoid pancreatic cancer than classic PDA.

4.2. Signaling protein expression and activation

Cells growing in log phase are incubated for 24 h in starvation medium (normal medium with serum replaced by 1 g/ml bovine serum albumin, 10 mM HEPES) before adding inhibitors for an additional 24 h. Western blot analyses are done using primary antibodies against ERK1/2 (9102, Cell Signaling), phospho-ERK1/2 (9106, Cell Signaling), phospho-MEK1/2 (9121, Cell Signaling), MKP-2 (sc-1200, Santa Cruz), Akt (9272, Cell Signaling), and phospho-Akt (9271, Cell Signaling). Pull-down assays used to detect formation of active GTP-bound RalA are performed as described previously (Wolthuis *et al.*, 1998), with glutathione-*S*-transferase fusion protein containing the Ral-GTP-binding domain (pGEX RalBD). Beads sequestering activated Ral proteins are separated by SDS-PAGE and

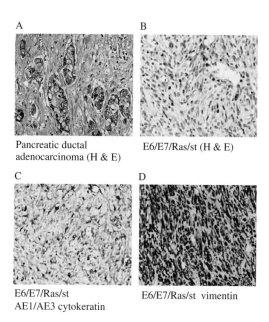

A — Pancreatic ductal adenocarcinoma (H & E)

B — E6/E7/Ras/st (H & E)

C — E6/E7/Ras/st AE1/AE3 cytokeratin

D — E6/E7/Ras/st vimentin

Figure 31.2 Histology of E6/E7Ras/st xenografts. Xenograft tumors derived from E6/E7/Ras/st cells were excised and fixed in formalin. Following postfixation embedding in paraffin, sections were stained for hematoxylin and eosin (H&E, B), AE1/AE3-sensitive cytokeratins (C), or vimentin (D). E6/E7/st cells did not form tumors in mice. A section of human PDA is shown for cytoarchitecture comparison (A).

probed with anti–RalA antibodies (BD Biosciences). The total cell lysate is also blotted to determine total Ral protein levels.

4.3. Migration assays

Cells grown to 90% confluence are starved as described earlier for 24 h. At $t = 0$, cells are treated with inhibitors of the Raf-MEK-ERK or PI3K-Akt pathways as described previously, and a "wound" is created by scoring a line across the monolayer of cells with a pipette tip (Valster *et al.*, 2005). At 12 h, cells are fixed and stained with Diff-Quik (Dade-Behring), a modified fixation and Wright Giemsa stain protocol, and digital micrographs are taken with a 40× objective. Wounds fixed and stained at $t = 0$ are used as the reference against which migration of treated cells is compared. Cells that have migrated into the space of by the wound are counted manually. Triplicate wells for each treatment group are stained and scored.

For transwell invasion assays, cells are starved and treated with inhibitors as described earlier for 24 h. Because trypsin can activate extracellular enzymes such as metallomatrix proteases and promote invasion (Soreide *et al.*, 2006), we avoid its use so as to not obscure the inherent invasive ability of the cells. Cells are suspended from adherent cultures with 1 ml TrypLE Express trypsin-free dissociation solution (Invitrogen) and counted. Twenty thousand cells are loaded into 500 μl of starvation medium and inhibitors in the top well of Biocoat chambers. These wells contain a layer of growth factor-reduced Matrigel extracellular basement membrane proteins over a polyethylene terephthalate membrane with 8-μm pores (BD Biosciences). The bottom chamber contains the same medium and inhibitor concentration. Cells are allowed to invade through the Matrigel for 24 h before the Matrigel is removed with moistened cotton swabs, and cells that have migrated to the underside of the membrane are fixed and stained with Diff-Quik. Cells adhering to the bottom surface of the membrane are counted under microscopy.

5. DISCUSSION

This chapter focused on the stepwise transformation of the nestin-expressing cells from the human pancreas and the steps required to make them tumorigenic, as well as elements of Ras signaling that contribute to their transformation. In recent reports, this cell type was identified as both an adult stem cell and a putative precursor of PanIN lesions (Carriere *et al.*, 2007). Most significant were results of knock-in experiments, which took advantage of endogenous promoters to drive the expression of oncogenic K-Ras in selected mouse tissues. Despite its ductal characteristics, PDA failed to be recapitulated in mice engineered to express oncogenic K-Ras in pancreatic ductal epithelial cells (Brembeck *et al.*, 2003). Forced expression under the control of the endogenous nestin promoter, however, results in a phenotype that reconstitutes the course of the disease from PanIN lesions to invasive PDA (Carriere *et al.*, 2007). Similar results had been reported previously in a mouse model that instead used promoters of the p48 and PDX-1 genes (Aguirre *et al.*, 2003; Hingorani *et al.*, 2005). Nestin, PDX-1, and p48 are expressed in developmental precursors of the exocrine pancreas (Dutta *et al.*, 2001; Herrera *et al.*, 2002; Hunziker and Stein, 2000; Selander and Edlund, 2002). In the adult pancreas, reexpression of these markers accompanied regenerative processes that follow tissue injury (Delacour *et al.*, 2004; Fernandes *et al.*, 1997). Previously, we used the catalytic subunit of telomerase (hTERT) to immortalize nestin-expressing cells isolated from the pancreatic ducts of a 52-year-old organ donor (Lee *et al.*, 2003). We subsequently described in these cells the expression of other markers of putative stem cells, as well as a capacity to give rise to pancreatic ductal cells (Lee *et al.*, 2005). Using oncogenic insults designed to emulate those detected in PDA, Campbell *et al.* (2007) described the stepwise transformation of these cells to a tumorigenic phenotype. This chapter detailed the procedures used for manipulation of these cells and characterization of their transformed phenotype, including growth properties, motility, invasiveness, and expression of tumor markers. These cells provide an additional tool to be used to discover the relative roles of Ras and other oncogenes in the initiation and progression of PDA and to elucidate the signaling mechanisms caused by oncogenic Ras necessary and sufficient for oncogenic growth transformation.

ACKNOWLEDGMENTS

Our research studies were supported by National Institutes of Health grants to C.J.D. (CA42978, CA71341, and CA109550), the Lustgarten Foundation for Pancreatic Cancer Research (LF 01–040) and Early Detection Research Network (U01 CA111294) to M.M.O., and the National Cancer Institute Gastrointestinal SPORE (P50-CA106991–02) to P.M.C.

REFERENCES

Aguirre, A. J., Bardeesy, N., Sinha, M., Lopez, L., Tuveson, D. A., Horner, J., Redston, M. S., and DePinho, R. A. (2003). Activated Kras and Ink4a/Arf deficiency cooperate to produce metastatic pancreatic ductal adenocarcinoma. *Genes Dev.* **17,** 3112–3126.

Ahrendt, S. A., and Pitt, H. A. (2002). Surgical management of pancreatic cancer. *Oncology* **16,** 725–734; discussion 734, 736–738, 740, 743.

Allison, D. C., Piantadosi, S., Hruban, R. H., Dooley, W. C., Fishman, E. K., Yeo, C. J., Lillemoe, K. D., Pitt, H. A., Lin, P., and Cameron, J. L. (1998). DNA content and other factors associated with ten-year survival after resection of pancreatic carcinoma. *J. Surg. Oncol.* **67,** 151–159.

Brembeck, F. H., Schreiber, F. S., Deramaudt, T. B., Craig, L., Rhoades, B., Swain, G., Grippo, P., Stoffers, D. A., Silberg, D. G., and Rustgi, A. K. (2003). The mutant K-ras oncogene causes pancreatic periductal lymphocytic infiltration and gastric mucous neck cell hyperplasia in transgenic mice. *Cancer Res.* **63,** 2005–2009.

Campbell, P. M., Groehler, A. L., Lee, K. M., Ouellette, M. M., Khazak, V., and Der, C. J. (2007). K-Ras promotes growth transformation and invasion of immortalized human pancreatic cells by Raf and phosphatidylinositol 3-kinase signaling. *Cancer Res.* **67,** 2098–2106.

Campbell, P. M., and Szyf, M. (2003). Human DNA methyltransferase gene DNMT1 is regulated by the APC pathway. *Carcinogenesis* **24,** 17–24.

Carriere, C., Seeley, E. S., Goetze, T., Longnecker, D. S., and Korc, M. (2007). The Nestin progenitor lineage is the compartment of origin for pancreatic intraepithelial neoplasia. *Proc. Natl. Acad. Sci. USA* **104,** 4437–4442.

Cox, A. D., and Der, C. J. (2002). Ras family signaling: Therapeutic targeting. *Cancer Biol. Ther.* **1,** 599–606.

Delacour, A., Nepote, V., Trumpp, A., and Herrera, P. L. (2004). Nestin expression in pancreatic exocrine cell lineages. *Mech. Dev.* **121,** 3–14.

Dutta, S., Gannon, M., Peers, B., Wright, C., Bonner-Weir, S., and Montminy, M. (2001). PDX:PBX complexes are required for normal proliferation of pancreatic cells during development. *Proc. Natl. Acad. Sci. USA* **98,** 1065–1070.

Elenbaas, B., Spirio, L., Koerner, F., Fleming, M. D., Zimonjic, D. B., Donaher, J. L., Popescu, N. C., Hahn, W. C., and Weinberg, R. A. (2001). Human breast cancer cells generated by oncogenic transformation of primary mammary epithelial cells. *Genes Dev.* **15,** 50–65.

Fernandes, A., King, L. C., Guz, Y., Stein, R., Wright, C. V., and Teitelman, G. (1997). Differentiation of new insulin-producing cells is induced by injury in adult pancreatic islets. *Endocrinology* **138,** 1750–1762.

Fiordalisi, J. J., Johnson, R. L., 2nd, Ulkü, A. S., Der, C. J., and Cox, A. D. (2001). Mammalian expression vectors for Ras family proteins: Generation and use of expression constructs to analyze Ras family function. *Methods Enzymol.* **332,** 3–36.

Hahn, W. C., Counter, C. M., Lundberg, A. S., Beijersbergen, R. L., Brooks, M. W., and Weinberg, R. A. (1999). Creation of human tumour cells with defined genetic elements. *Nature* **400,** 464–468.

Hahn, W. C., and Weinberg, R. A. (2002). Rules for making human tumor cells. *N. Engl. J. Med.* **347,** 1593–1603.

Halbert, C. L., Demers, G. W., and Galloway, D. A. (1991). The E7 gene of human papillomavirus type 16 is sufficient for immortalization of human epithelial cells. *J. Virol.* **65,** 473–478.

Herrera, P. L., Nepote, V., and Delacour, A. (2002). Pancreatic cell lineage analyses in mice. *Endocrine* **19,** 267–278.

Hingorani, S. R., Wang, L., Multani, A. S., Combs, C., Deramaudt, T. B., Hruban, R. H., Rustgi, A.K, Chang, S., and Tuveson, D. A. (2005). Trp53R172H and KrasG12D cooperate to promote chromosomal instability and widely metastatic pancreatic ductal adenocarcinoma in mice. *Cancer Cell.* **7,** 469–483.

Hruban, R. H., Goggins, M., Parsons, J., and Kern, S. E. (2000a). Progression model for pancreatic cancer. *Clin. Cancer Res.* **6,** 2969–2972.

Hruban, R. H., Wilentz, R. E., and Kern, S. E. (2000b). Genetic progression in the pancreatic ducts. *Am. J. Pathol.* **156,** 1821–1825.

Hunziker, E., and Stein, M. (2000). Nestin-expressing cells in the pancreatic islets of Langerhans. *Biochem. Biophys. Res. Commun.* **271,** 116–119.

Jemal, A., Murray, T., Ward, E., Samuels, A., Tiwari, R. C., Ghafoor, A., Feuer, E. J., and Thun, M. J. (2005). Cancer statistics, 2005. *CA Cancer J. Clin.* **55,** 10–30.

Jesnowski, R., Müller, P., Schareck, W., Liebe, S., and Löhr, M. (1999). Immortalized pancreatic duct cells *in vitro* and *in vivo*. *Ann. N.Y. Acad. Sci.* **880,** 50–65.

Kato-Stankiewicz, J., Hakimi, I., Zhi, G., Zhang, J., Serebriiskii, I., Guo, L., Edamatsu, H., Koide, H., Menon, S., Eckl, R., Sakamuri, S., Lu, Y., *et al.* (2002). Inhibitors of Ras/Raf-1 interaction identified by two-hybrid screening revert Ras-dependent transformation phenotypes in human cancer cells. *Proc. Natl. Acad. Sci. USA* **99,** 14398–14403.

Lee, K. M., Nguyen, C., Ulrich, A. B., Pour, P. M., and Ouellette, M. M. (2003). Immortalization with telomerase of the Nestin-positive cells of the human pancreas. *Biochem. Biophys. Res. Commun.* **301,** 1038–1044.

Lee, K. M., Yasuda, H., Hollingsworth, M. A., and Ouellette, M. M. (2005). Notch 2-positive progenitors with the intrinsic ability to give rise to pancreatic ductal cells. *Lab. Invest.* **85,** 1003–1012.

Linetsky, E., Bottino, R., Lehmann, R., Alejandro, R., Inverardi, L., and Ricordi, C. (1997). Improved human islet isolation using a new enzyme blend, liberase. *Diabetes* **46,** 1120–1123.

Listrom, M. B., and Dalton, L. W. (1987). Comparison of keratin monoclonal antibodies MAK-6, AE1:AE3, and CAM-5.2. *Am. J. Clin. Pathol.* **88,** 297–301.

Maione, R., Fimia, G. M., and Amati, P. (1992). Inhibition of in vitro myogenic differentiation by a polyomavirus early function. *Oncogene* **7,** 85–93.

Malumbres, M., and Barbacid, M. (2003). RAS oncogenes: The first 30 years. *Nat. Rev. Cancer* **3**, 459–465.

Miller, A. D., and Rosman, G. J. (1989). Improved retroviral vectors for gene transfer and expression. *Biotechniques* **7**, 980–982, 984–986, 989–990.

Morgenstern, J. P., and Land, H., (1990). Advanced mammalian gene transfer: High titre retroviral vectors with multiple drug selection markers and a complementary helper-free packaging cell line. *Nucleic Acids Res.* **18**, 3587–3596.

Münger, K., Werness, B. A., Dyson, N., Phelps, W. C., Harlow, E., and Howley, P. M. (1989). Complex formation of human papillomavirus E7 proteins with the retinoblastoma tumor suppressor gene product. *EMBO J.* **8**, 4099–4105.

Nakamura, T. M., Morin, G. B., Chapman, K. B., Weinrich, S. L., Andrews, W. H., Lingner, J., Harley, C. B., and Cech, T. R. (1997). Telomerase catalytic subunit homologs from fission yeast and human. *Science* **277**, 955–959.

Ouellette, M. M., Aisner, D. L., Savre-Train, I., Wright, W. E., and Shay, J. W. (1999). Telomerase activity does not always imply telomere maintenance. *Biochem. Biophys. Res. Commun.* **254**, 795–803.

Peterson, S. N., Trabalzini, L., Brtva, T. R., Fischer, T., Altschuler, D. L., Martelli, P., Lapetina, E. G., Der, C. J., and White, G. C., 2nd (1996). Identification of a novel RalGDS-related protein as a candidate effector for Ras and Rap1. *J. Biol. Chem.* **271**, 29903–29908.

Repasky, G. A., Chenette, E. J., and Der, C. J. (2004). Renewing the conspiracy theory debate: Does Raf function alone to mediate Ras oncogenesis? *Trends Cell Biol.* **14**, 639–647.

Selander, L., and Edlund, H. (2002). Nestin is expressed in mesenchymal and not epithelial cells of the developing mouse pancreas. *Mech. Dev.* **113**, 189–192.

Sontag, E., Fedorov, S., Kamibayashi, C., Robbins, D., Cobb, M., and Mumby, M. (1993). The interaction of SV40 small tumor antigen with protein phosphatase 2A stimulates the map kinase pathway and induces cell proliferation. *Cell* **75**, 887–897.

Soreide, K., Janssen, E. A., Körner, H., and Baak, J. P. (2006). Trypsin in colorectal cancer: Molecular biological mechanisms of proliferation, invasion, and metastasis. *J. Pathol.* **209**, 147–156.

Valster, A., Tran, N. L., Nakada, M., Berens, M. E., Chan, A. Y., and Symons, M. (2005). Cell migration and invasion assays. *Methods* **37**, 208–215.

Vilá, M. R., Balagué, C., and Real, F. X. (1994). Cytokeratins and mucins as molecular markers of cell differentiation and neoplastic transformation in the exocrine pancreas. *Zentralbl. Pathol.* **140**, 225–235.

Warshaw, A. L., and Fernandez-del Castillo, C. (1992). Pancreatic carcinoma. *N. Engl. J. Med.* **326**, 455–465.

Werness, B. A., Levine, A. J., and Howley, P. M. (1990). Association of human papillomavirus types 16 and 18 E6 proteins with p53. *Science* **248**, 76–79.

Wolthuis, R. M., Zwartkruis, F., Moen, T. C., and Bos, J. L. (1998). Ras-dependent activation of the small GTPase Ral. *Curr. Biol.* **8**, 471–474.

Yeh, J. J., and Der, C. J. (2007). Targeting signal transduction in pancreatic cancer treatment. *Expert Opin. Ther. Targets* **11**, 673–694.

Zulewski, H., Abraham, E. J., Gerlach, M. J., Daniel, P. B., Moritz, W., Müller, B., Vallejo, M., Thomas, M. K., and Habener, J. F. (2001). Multipotential nestin-positive stem cells isolated from adult pancreatic islets differentiate *ex vivo* into pancreatic endocrine, exocrine, and hepatic phenotypes. *Diabetes* **50**, 521–533.

CHAPTER THIRTY-TWO

THE RAS INHIBITOR FARNESYLTHIOSALICYLIC ACID (SALIRASIB) DISRUPTS THE SPATIOTEMPORAL LOCALIZATION OF ACTIVE RAS: A POTENTIAL TREATMENT FOR CANCER

Barak Rotblat,* Marcello Ehrlich,[†] Roni Haklai,* *and* Yoel Kloog*

Contents

* Department of Neurobiochemistry, George S. Wise Faculty of Life Sciences, Tel Aviv University, Tel Aviv, Israel
† Department of Cell Biology and Immunology, George S. Wise Faculty of Life Sciences, Tel Aviv University, Tel Aviv, Israel

Methods in Enzymology, Volume 439
ISSN 0076-6879, DOI: 10.1016/S0076-6879(07)00432-6

Abstract

Chronic activation of Ras proteins by mutational activation or by growth factor stimulation is a common occurrence in many human cancers and was shown to induce and be required for tumor growth. Even if additional genetic defects are present, "correction" of the Ras defect has been shown to reverse Ras-dependent tumorigenesis. One way to block Ras protein activity is by interfering with their spatiotemporal localization in cellular membranes or in membrane microdomains, a prerequisite for Ras signaling and biological activity. Detailed reports describe the use of this method in studies employing farnesylthiosalicylic acid (FTS, Salirasib), a Ras farnesylcysteine mimetic, which selectively disrupts the association of chronically active Ras proteins with the plasma membrane. FTS competes with Ras for binding to Ras-escort proteins, which possess putative farnesyl-binding domains and interact only with the activated form of Ras proteins, thereby promoting Ras nanoclusterization in the plasma membrane and robust signals. This chapter presents three-dimensional time-lapse images that track the FTS-induced inhibition of membrane-activated Ras in live cells on a real-time scale. It also describes a mechanistic model that explains FTS selectivity toward activated Ras. Selective blocking of activated Ras proteins results in the inhibition of Ras transformation *in vitro* and in animal models, with no accompanying toxicity. Phase I clinical trials have demonstrated a safe profile for oral FTS, with minimal side effects and promising activity in hematological malignancies. Salirasib is currently undergoing trials in patients with pancreatic cancer and with nonsmall cell lung cancer, with or without identified K-Ras mutations. The findings might indicate whether with the disruption of the spatiotemporal localization of oncogenic Ras proteins and the targeting of prenyl-binding domains by anticancer drugs is worth developing as a means of cancer treatment.

1. INTRODUCTION

Mutations in *ras* genes occur in 30% of all human cancers, with the highest incidence of mutational activation of Ras being detected in pancreatic (90%) and colon (50%) cancers (Baines *et al.*, 2005; Chin *et al.*, 1999; Cox and Der, 2002; Downward, 2003b; Hahn *et al.*, 1999; Hanahan and Weinberg, 2000). *K-ras* and *N-ras* are the most frequently mutated genes (in most cases displaying a point mutation at codon 12), yet all three of the highly homologous Ras isoforms (H-Ras, K-Ras, and N-Ras) promote malignant transformation (Baines *et al.*, 2005; Chin *et al.*, 1999; Cox and Der, 2002; Downward, 2003b; Hahn *et al.*, 1999; Hanahan and Weinberg, 2000). Oncogenic Ras isoforms (e.g., G12V mutants) are constitutively active (bound to GTP) and are critically required for both initiation and maintenance of the transformed phenotype of cancer cells that harbor the mutated Ras (Baines *et al.*, 2005; Chin *et al.*, 1999; Cox and Der, 2002;

Hahn *et al.*, 1999; Hanahan and Weinberg, 2000). Thus, for example, constitutively active Ras causes growth transformation of primary human cells, but other genetic alterations are required to facilitate the Ras transformation (Hahn *et al.*, 1999). Ras activation induces and is required for tumor growth, and even if many genes are defective, "correction" of the Ras defect alone is sufficient to reverse the process (Chin *et al.*, 1999). Blockage of capan-1 (a human pancreatic cell line that harbors K-RasG12V tumor growth in mice) by siRNA for K-RasG12V provides additional strong support for the notion that blockage of Ras will be of clinical benefit (Baines *et al.*, 2005). The aforementioned experimental findings highlight the importance of Ras proteins as a target for cancer drugs. One way to block the functions of Ras proteins is by interfering with their trafficking to and from cellular membranes, as well as with their proper localizations in different cellular localities and microdomains (Kloog and Cox, 2000; Philips and Cox, 2007)—all of which are required for Ras signaling and biological activities (Cox and Der, 2002; Dong *et al.*, 2003; Hancock and Parton, 2005; Kloog and Cox, 2000; Philips and Cox, 2007; Silvius *et al.*, 2006). This chapter describes the methods employed to study the effects of farnesylthiosalicylic acid (FTS; Salirasib), a Ras farnesylcysteine mimetic shown to selectively disrupt the association of chronically active Ras proteins with the plasma membrane, on Ras membrane anchorage and function (Elad *et al.*, 1999; Goldberg and Kloog, 2006; Haklai *et al.*, 1998; Yaari *et al.*, 2005).

2. METHODS

2.1. Preparation of FTS solutions for cell culture experiments

Farnesylthiosalicylic acid is a relatively stable compound (molecular weight 357) that appears as a whitish-yellowish powder. It can be stored dry at room temperature, but we recommend keeping it in a foil-covered closed tube at $-70°$ (Marciano *et al.*, 1995). We have found that under such conditions FTS remains stable for at least 3 years. The simplest way to verify that FTS has not been altered during storage is by subjecting it to thin-layer chromatography (results are obtained within 10–15 min) as described (Marciano *et al.*, 1995). Once the storage tube is opened it should be flushed gently with a stream of nitrogen.

Farnesylthiosalicylic acid is hydrophobic and does not dissolve readily in water or other aqueous solutions such as tissue culture media. Special (although very simple) procedures are therefore needed for the preparation of FTS solutions. To avoid FTS precipitation during cell culture or animal experiments, the use of such solutions is an absolute requirement. This section outlines the procedures used in many of the experiments described in this chapter. Because experiments *in vitro* need relatively small amounts of FTS, for any given set of experiments it is convenient and economical to prepare a stock solution of FTS in a solvent in which it is readily soluble and stable. Our standard procedure is to prepare a solution of 0.1 M FTS in chloroform. FTS dissolves readily in chloroform, and the solution (kept on ice to minimize evaporation) can be divided accurately into the required portions. We usually divide the solution into 10-, 20-, or 40-μl portions in 1.5-ml microtubes kept on ice, then closed tightly, covered with foil, and kept at $-70°$. These solutions remain stable for several weeks at least. The advantage of this procedure is that each tube contains precisely the required amount of FTS and the chloroform can be removed easily (by a gentle stream of nitrogen, or in the air when the tube is opened for a few minutes at room temperature).

One or more of the stored FTS/chloroform solutions are used in each experiment. As an example, from a tube containing 10 μl of solution the chloroform is removed as described earlier, dimethyl sulfoxide (DMSO; 40 μl) is added, the tube is vortexed, warm Dulbecco's modified Eagle medium (DMEM)/10% fetal calf serum (FCS) (360 μl) is added, and the contents are mixed again. To avoid the danger of FTS precipitation, the aforementioned procedure must be carried out in the correct order; it is also important that the added medium be warm and contain serum (the effect of the serum is discussed later). The resulting clear solution of FTS contains 2.5 mM FTS and 10% DMSO. This work solution is then used for standard cell culture experiments. For example, following the addition of 10 μl of solution to a well (e.g., in 24-well plates) containing 1 ml of medium, the mixture is immediately subjected to gentle rotation to yield 25 μM FTS/ 0.1% DMSO. Control cells receive 10 μl of the vehicle solution (DMEM/ 10% FCS/10% DMSO). We usually do not prepare a stock of FTS-containing medium to replace the medium in which the cells were kept, as results obtained in that case are more variable; we therefore prefer to add concentrated FTS solution to the wells as described previously. Lower concentrations of FTS are obtained by diluting the aforementioned 2.5 mM solution of FTS/10% FCS/10% DMSO with DMEM/10% FCS/10% DMSO. To obtain higher FTS concentrations, we use the 20- or 40-μl stock FTS/chloroform solution and follow the procedures described earlier. The resulting work solutions will contain, respectively, 5 and 10 mM FTS (for the higher concentration it is best to prepare and maintain these solutions at $37°$). Rather than storing any residual FTS work solutions, we discard them and prepare a fresh work solution for each experiment.

Two minor technical points should be noted in connection with the preparation of FTS solutions. First, the chloroform might gradually evaporate from the stock FTS solutions even if kept at $-70°$. Although in most cases the FTS will remain intact, it will be less soluble (presumably because the powder is wet) and sometimes not soluble at all. In such a case it is best to take another tube of the stock, add 50 μl of chloroform, mix, and then evaporate the chloroform. We have obtained perfectly soluble FTS solutions using the aforementioned procedure. Second, FTS is also soluble in ethanol. It is therefore possible to use ethanol instead of chloroform and DMSO. We do not have much experience with the behavior of ethanolic FTS solutions in storage. We do, however, use such solutions for animal experiments, as described in the next section.

The effects of FTS on cell growth, differentiation, or both might depend on cell density (discussed later). In Ras-transformed fibroblasts, good results were obtained in 24-well plates at densities of 5000 cells per well. It should be noted that densities of the various cell types tested in these plates were found to range between 2500 and 10,000 and must therefore be examined for each cell type separately. In the standard treatment protocol, cells are plated on day -1 and FTS is added on day 0, that is, about 24 h after plating. We found, however, that FTS can also be added a few hours after plating, once the cells have settled. There is usually no need for medium changes or a second addition of FTS. Cell numbers are estimated after 3 to 7 days in culture. It should be noted that some cell types do not survive in the presence of FTS if the medium does not contain serum, as discussed later. In our experience, moreover, the best way to quantify the effects of FTS on cell growth is by direct cell counting. Mitochondrial dyes can also be used to assess the effects of FTS on cell growth and cell death (Marciano *et al.*, 1995); although because of the impact of FTS on metabolism we prefer to use such dyes mostly for studying how FTS affects cell survival. The effects of FTS on cell cycle and cell growth are easy to determine by [³H]thymidine incorporation assays and fluorescence-activated cell sorter analysis (Blum *et al.*, 2006b). In those experiments the incubation periods with FTS are relatively short, ranging between 12 and 48 h. Likewise, relatively short incubation periods (12–24 h) are needed when examining the effects of FTS on cytoskeleton and cell shape (Egozi *et al.*, 1999; see also Fig. 32.1) .

2.2. Preparation of FTS solutions for animal experiments

2.2.1. Solutions for intraperitoneal FTS administration

For experiments *in vivo* we use two alternative procedures. The first is fairly simple and can be used routinely for administration of FTS at relatively low doses (5–40 mg/kg). Standard (0.1 M) stock FTS/chloroform (28 μl) is placed in a tube and the chloroform is evaporated. The FTS is then dissolved in 4 μl of warm ethanol. Warm 1 N NaOH (7 μl) is then added, followed by the addition of 980 μl of warm phosphate-buffered saline ( ). The solution is titrated to pH ≈ 8.0 with 0.1 N HCl, yielding a solution of 1 mg FTS/ml. Under this condition of mild basic PBS, the FTS will remain in solution for at least several hours. For experiments in mice weighing 25 to 35 g, we administer 0.1 to 0.2 ml of this solution (corresponding to

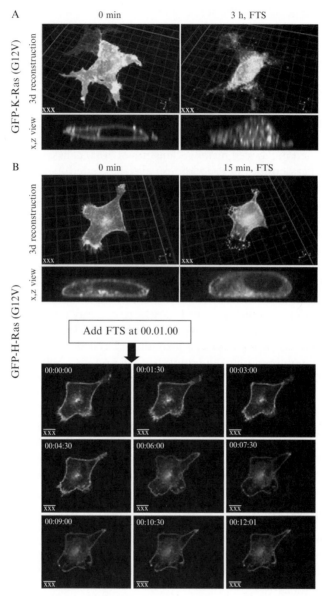

Figure 32.1 Rapid dislodgement of oncogenic Ras proteins by FTS in live cells. Mouse brain endothelial cells (Bend 3) transiently transfected with GFP-K-RasG12V using Lipofectamine 2000 (A) or stably expressing GFP-H-RasG12V (B) were treated in the presence of serum, respectively, with 50 and 80 μM FTS. The same cells were imaged before drug treatment (zero time) and followed over a period of 3 h employing a spinning disk confocal microscope (Zeiss 200M) equipped with a spinning disk confocal head (Yokogawa) and a temperature-controlled chamber. Three-dimensional renditions of cells performed with SlideBook software and x,z projection of the z stack

≈5 mg/kg) intraperitoneally. For higher doses, we either increase the injection volume up to 0.3 ml or prepare a 2× concentrated work stock solution (by dissolving FTS in 8 μl of warm ethanol). Control mice receive the same volume of the vehicle, prepared as described earlier but without FTS.

The procedure just given is useful for animal models in which the effective dose appears to be within the aforementioned range, but not if FTS doses higher than 40 mg/kg are required. For such cases we recommend a second procedure, employing the nontoxic methylated β-cyclodextrin, which appears as a white powder and is water soluble (Jindrich et al., 1995). FTS is dissolved in ethanol (2 μl/mg), and sufficient 1 N NaOH is added to convert the FTS to a sodium salt. This basic FTS solution is stirred into a PBS solution containing sufficient cyclodextrin to yield a final mixture containing an FTS to cyclodextrin ratio of 1:10. This solution is titrated with 0.1 N HCl to pH ≈8.0. The same procedure is used to prepare the vehicle control solution except that 1 N HCl is used for titration. Solutions are stored at 4°.

The following describes a typical example of the preparation of FTS in cyclodextrin. PBS (6 ml) is added to 600 mg of cyclodextrin and stirred with a magnetic stirrer. FTS (60 mg) is dissolved in 120 μl of warmed ethanol (heated by keeping the tube over warm water), and 1 N NaOH (180 μl) is then added to obtain a clear FTS solution. This solution is added, with continuous stirring, to the cyclodextrin solution. The FTS/cyclodextrin solution is then titrated with 0.1 N HCl (approximately 48 μl) to pH ≈8.0, covered with foil, and kept at 4°. The solution, containing 7.8 mg FTS/ml and 85.7 mg cyclodextrin/ml, will remain stable for at least a month. With these concentrated soluble FTS preparations, doses of up to 80 mg/kg FTS intraperitoneally can be reached easily in mice. This method of preparation may also be suitable for intravenous administration in models of human diseases, such as head trauma (Shohami et al., 2003) or stroke.

2.2.2. Solutions for oral FTS administration

For oral administration, we recommend using a formula based on carboxymethylcellulose (CMC; Sigma) (Haklai et al., 2007). To prepare a solution of 0.5% CMC, double-distilled water (≈60 ml) is added to 0.5 g of CMC. After mixing for 4 h with a magnetic stirrer, the solution is transferred to a

---

(acquired with a spacing of 0.35 μm between planes) of the same cells are depicted in A and B (upper panels). Note that FTS induced a robust dislodgment of GFP-Ras from the cell membrane. The lower panels in B depict the treatment process, which was followed continuously (30-s interval between planes) and imaged at a middle confocal plane. The panels depict selected time points before and after FTS treatment (initiated 1 min after beginning of acquisition). FTS treatment leads to a very rapid disappearance of GFP-H-Ras(G12V)-enriched filopodia and lamellipodia, followed by a transient change in cell morphology concomitant with the dissociation of GFP-H-Ras(G12V) from the membrane.

volumetric flask and the volume is adjusted to 100 ml. The clear solution is stable at 4°. To prepare a suspension of 5 mg/ml FTS in 0.5% CMC, FTS (150 mg) is placed in a mortar and ground with a pestle. A few drops of the 0.5% CMC solution are added and worked in with the pestle. More CMC solution is gradually added until the mixture acquires a consistency that can be poured. The suspension, having the appearance of white particles in the aqueous CMC solution, is then poured into a volumetric flask. The mortar and pestle are rinsed a few times with the CMC solution in the volumetric flask, and the volume is adjusted to 30 ml. The suspension is then stirred with a stirring bar for 1 h, and 3-ml aliquots are poured into vials while stirring. FTS in the aqueous CMC is stable at 4° for at least a month. Before the drug is administered, the suspension is warmed to room temperature and stirred continuously while the syringe is filled. The control vehicle solution is prepared in the same way but without FTS. The oral dose of FTS/CMC solution that the mouse receives is 20 to 60 mg/kg (e.g., a mouse weighing 25 g receives 0.1–0.3 ml of FTS/CMC). For higher dosages, FTS/CMC solutions of 10 and 20 mg/ml FTS are prepared as described earlier. The CMC formula is nontoxic, and the efficiency bioavailability of the oral FTS in this formula is 69.5% (Haklai *et al.*, 2007). This formula may be suitable for experiments in which oral dosing and long-term treatments are desired, such as in models of human diseases requiring long-term treatment, for example, neurofibromatosis type 1 (Barkan *et al.*, 2006).

3. Disruption of Ras Localization as a Potential Method for Cancer Treatment

3.1. Location is critical for Ras signaling and biological activity

Ras protein signaling depends on specific combinations of Ras activation at the plasma membrane, endomembranes, or both (Hancock and Parton, 2005), dynamic lateral segregation of Ras within the plasma membrane (Niv *et al.*, 1999, 2002; Roy *et al.*, 2005), and translocation of Ras from the plasma membrane to intracellular compartments (Hancock, 2003; Hancock and Parton, 2005; Philips, 2005). Ras proteins signal from the plasma membrane, Golgi, and perhaps also from mitochondrial membranes (Bivona *et al.*, 2006; Hancock, 2003; Hancock and Parton, 2005; Philips, 2005). The contribution of membrane localization to the signal output has been best characterized for Ras on the plasma membrane. Different Ras isoforms occupy different membrane microdomains (Hancock, 2003; Plowman *et al.*, 2005) and drive the formation of spatially distinct nanoclusters (Harding *et al.*, 2005; Tian *et al.*, 2007). Ras nanoclusters are the sites to which cytoplasmic effectors such as Raf are recruited and activated (Harding *et al.*, 2005; Tian *et al.*, 2007). Galectin-1, which can act as a Ras escort protein, has been shown to be both a structural component and a regulator of H-Ras nanoclusters (Ashery *et al.*, 2006a; Elad-Sfadia *et al.*, 2002; Paz *et al.*, 2001; Prior *et al.*, 2003; Rotblat *et al.*, 2004a). All of these processes are directed by the GDP/GTP-loading state of Ras, as well as by the C terminus polybasic farnesyl domain in K-Ras 4B and the cysteine-palmitoylated C terminus farnesyl domains in H-Ras and N-Ras (the K-Ras 4B isoform has no palmitoylated cysteines). H-Ras undergoes GTP-dependent lateral segregation between different types of nanoclusters (Plowman *et al.*, 2005; Prior *et al.*, 2003), and depalmitoylation/repalmitoylation of H-Ras and N-Ras proteins promotes their cellular redistribution and signaling (Baker *et al.*, 2003; Dong *et al.*, 2003; Linder and Deschenes, 2003; Rocks *et al.*, 2005; Roy *et al.*, 2005) by mechanisms that are not yet known and might involve chaperons. Palmitoylation of H-Ras and N-Ras also promotes their association with "rasosomes," randomly diffusing nano-particles that apparently provide a means by which multiple copies of activated Ras and its signal can spread rapidly (Ashery *et al.*, 2006a,b; Rotblat *et al.*, 2006). Ubiquitination of H-Ras evidently targets it to endosomes (Jura *et al.*, 2006), another reported site of Ras signaling (Jiang and Sorkin, 2002). The polybasic farnesyl domain of K-Ras 4B acts as a target for Ca^{2+}/calmodulin, which sequesters the active protein from the plasma membrane, thereby facilitating its trafficking to Golgi and early endosomes (Fivaz and Meyer, 2005; Villalonga *et al.*, 2001). Protein kinase C-dependent phosphorylation of S181 in K-Ras 4B provides a regulated farnesyl-electrostatic switch on K-Ras 4B, which promotes its translocation to the mitochondria and its transforming activity (Bivona *et al.*, 2006). The location of Ras proteins, governed by specific sequences and post-translational modifications, thus appears to be critical for Ras signaling and biological activity.

3.2. Targeting the localization and function of transforming Ras proteins

The C-terminal farnesylcysteine carboxymethyl ester common to all Ras proteins is an absolute requirement for their proper location and signaling dynamics (Casey *et al.*, 1989; Gutierrez *et al.*, 1989; Hancock, 2003; Kloog and Cox, 2000; Philips, 2005). This modified amino acid residue acts as a recognition unit promoting specific binding of Ras to lipids and proteins that support its membrane interactions (Hancock, 2003; Kloog and Cox, 2000; Linder and Deschenes, 2003; Philips, 2005; Silvius *et al.*, 2006). Transformation by Ras proteins is indeed critically dependent on membrane association mediated by lipid/protein interactions between the farnesyl groups of Ras and prenyl-binding pockets in their regulatory proteins, such as galectin-1, which regulates H-Ras, and galectin-3, which regulates K-Ras (Elad-Sfadia *et al.*, 2002, 2004). Recognition of farnesyl-binding domains as functionally important elements of Ras proteins (Casey *et al.*, 1989; Cox and Der, 1992; Gutierrez *et al.*, 1989) thus opens up new possibilities for targeting the localization and functions of transforming Ras proteins. Ras proteins can be efficiently extracted from membranes and their downstream signaling inhibited by even relatively nonspecific prenyl analogs, such as FTS (Gana-Weisz *et al.*, 1997; Haklai *et al.*, 1998; Marom *et al.*, 1995) and *N*-acetyl *S*-farnesyl-L-cysteine (Chiu *et al.*, 2004). Appreciation of the fine differences in membrane localization between active and inactive Ras isoforms (Niv *et al.*, 2002; Prior *et al.*, 2001, 2003) has helped explain the surprising selectivity of FTS action for oncogenic Ras.

3.3. Mode of action of FTS: selective inhibition of active Ras proteins

Farnesylthiosalicylic acid mimics the carboxy-terminal farnesyl cysteine carboxymethyl ester of Ras and dislodges the active Ras protein from cellular membranes (Elad *et al.*, 1999; Goldberg and Kloog, 2006; Haklai *et al.*, 1998; Yaari *et al.*, 2005). FTS is readily taken up by cells, and once inside the cell it specifically disrupts the association of active forms of all Ras proteins (H-Ras, K-Ras, and N-Ras) with the inner surface of the cell membrane and with other cellular membranes (Elad *et al.*, 1999; Haklai *et al.*, 1998; Jansen *et al.*, 1999). Figure 32.1 presents examples depicting the effect of FTS on the localization of green fluorescent protein (GFP)-tagged oncogenic Ras proteins in live cells. The rapid dislodgement (observable within 3 h) of GFP-tagged oncogenic K-Ras (see Fig. 32.1A) and H-Ras (see Fig. 32.1B) proteins from the cell membranes by FTS, as shown in these three-dimensional images, demonstrates the drug in action. A middle plane of a cell expressing GFP-H-Ras(G12V), which was treated with FTS and then observed in a time-lapse series acquired with a spinning disk confocal microscope, shows dislodgement of the Ras protein in real time (see Fig. 32.1B lower panels and Supplementary Movie 1).

The remarkable selectivity of FTS for active forms of Ras proteins, including oncogenic Ras isoforms and growth factor-mediated chronically active Ras proteins (Clarke *et al.*, 2003; Gana-Weisz *et al.*, 1997; Haklai *et al.*, 1998; Haring *et al.*, 1998; Khwaja *et al.*, 2005; Reif *et al.*, 1999), can be explained by the known different microlocalities of active (GTP-bound) and inactive (GDP-bound) Ras proteins (Niv *et al.*, 1999, 2002; Prior *et al.*, 2001), the stronger association of Ras-GDP than of Ras-GTP with the cell membrane (Niv *et al.*, 1999, 2002; Rotblat *et al.*, 2004b), and the accessibility of the farnesyl group of Ras to Ras escort proteins such as galectin-1 and galectin-3. FTS appears to compete with the binding only of activated Ras to these proteins (Ashery *et al.*, 2006a; Elad-Sfadia *et al.*, 2002, 2004; Paz *et al.*, 2001; Rotblat *et al.*, 2004a), as described later.

The inactive H-Ras-GDP protein associates with and forms nanoclusters in cholesterol-dependent microdomains (Hancock, 2006; Harding *et al.*, 2005; Niv *et al.*, 2002; Rotblat *et al.*, 2004b; Tian *et al.*, 2007). In contrast, the active H-Ras-GTP protein segregates into and forms distinct nanoclusters in cholesterol-independent microdomains (Harding *et al.*, 2005). Nanoclustering of H-Ras-GTP is regulated by galectin-1, which specifically binds this active Ras isoform in a farnesyl-dependent manner (Elad-Sfadia *et al.*, 2002; Paz *et al.*, 2001; Prior *et al.*, 2003). Recent work simulating the interaction of H-Ras with 1,2-dimyristoylglycero-3-phosphocholine bilayers demonstrated that H-Ras exists in two distinct conformational states, recognizable by their different modes of membrane interaction (Gorfe *et al.*, 2007). In model 1, all H-Ras lipid chains are stretched and deeply inserted into the lipid bilayer. In model 2 the lipid chains are more flexible and less deeply embedded (Gorfe *et al.*, 2007). The simulation suggested that H-Ras-GDP is more readily associated with model 1 and H-Ras-GTP with model 2 (see model in Fig. 32.2). This would promote a strong association of H-Ras-GDP with the cell membrane, while precluding interactions between the farnesyl group of H-Ras and escort proteins such as galectin-1

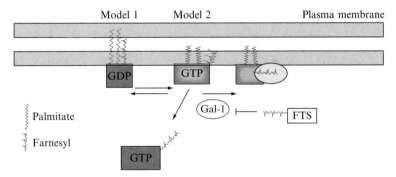

Figure 32.2 Selectivity of FTS toward activated Ras. A schematic representation of the interactions of H-Ras (depicted as squares) with a lipid bilayer, based on molecular dynamics simulation models (Gorfe *et al.*, 2007), is shown. H-Ras-GDP acquires a conformational state characterized by extended, deeply embedded lipid chains (model 1) and, upon GTP loading, acquires a conformational state (model 2) with more flexible lipid chains. In the latter conformation the farnesyl group is also less deeply inserted into the bilayer, allowing it to interact with the farnesyl-binding pocket of Gal-1 (circle). We postulate that the farnesyl cysteine mimetic FTS blocks the farnesyl-binding pocket of Gal-1, thereby preventing the interaction between Gal-1 and H-Ras-GTP needed for stable membrane association of the active Ras protein (Paz *et al.*, 2001; Rotblat *et al.*, 2004a). As a consequence, H-Ras-GTP is dislodged from the plasma membrane, presumably with a concomitant loss of its palmitoyl groups. H-Ras-GDP remains untouched, as its membrane association is independent of Gal-1 (Paz *et al.*, 2001; Rotblat *et al.*, 2004a).

(see model in Fig. 32.2). The lipid groups of the active H–Ras–GTP do not appear to be deeply inserted in the lipid bilayer (see Fig. 32.2). This is probably one of the reasons why active H–Ras–GTP is less avidly attached than H–Ras–GDP to the cell membrane (Niv *et al.*, 2002; Rotblat *et al.*, 2004b). Nonetheless, the farnesyl group of the H–Ras–GTP (model 2) is more accessible than that of H–Ras–GDP (model 1) to galectin-1 and can therefore bind this moiety of Ras and strengthen its interaction with cholesterol-independent membrane microdomains (Elad-Sfadia *et al.*, 2002; Hancock, 2003; Parton and Hancock, 2004; Rotblat *et al.*, 2004a). The H–Ras–GTP/galectin-1 nanoclusters formed in the cell membrane appear to be highly active (Elad-Sfadia *et al.*, 2002; Hanzal–Bayer and Hancock, 2007; Nicolau *et al.*, 2006). Accordingly, the Ras inhibitor FTS, which binds to galectin-1 (J. Hirsh and Y. Kloog, unpublished data) and can thus block its putative farnesyl-binding pocket (Rotblat *et al.*, 2004a), prevents H–Ras–GTP/galectin-1 complex formation and actually reduces the association of the active H–Ras isoform with the cell membrane (see Fig. 32.2). The inactive form of H–Ras, whose farnesyl group appears to be deeply inserted in the cell membrane (Gorfe *et al.*, 2007; see Fig. 32.2), cannot bind galectin-1 (Elad-Sfadia *et al.*, 2002; Paz *et al.*, 2001) and therefore is "protected" from the membrane dislodgement induced by FTS. This can explain the lower sensitivity to FTS of the wild-type H–Ras at rest than that of the oncogenic H–Ras (Haklai *et al.*, 1998) and is consistent with the low toxicity of FTS reported in normal cells (Haklai *et al.*, 1998; Marom *et al.*, 1995), in animals (Barkan *et al.*, 2006; Halaschek–Wiener *et al.*, 2000; Weisz *et al.*, 1999), and in phase I clinical trials of FTS (Borthakur *et al.*, 2007).

An analogous but nevertheless distinctive scenario has been documented for K-Ras. This isoform, unlike H-Ras or N-Ras, is not palmitoylated and binds to the cell membrane through its farnesyl group and the adjacent positively charged lysine residues that associate with acidic phospholipids (Hancock *et al.*, 1991). Active and inactive K-Ras each reside in a distinct membrane microdomain, both of which are different from those with which H-Ras proteins associate (Hancock, 2006; Hanzal–Bayer and Hancock, 2007; Parton and Hancock, 2004; Tian *et al.*, 2007). Active K-Ras-GTP (but not the inactive form) binds galectin-3, and this interaction strengthens its membrane localization and signaling (Elad-Sfadia *et al.*, 2004). The analogy between the preferential interactions of galectin-1 with active H-Ras (Elad-Sfadia *et al.*, 2002; Paz *et al.*, 2001) and those of galectin-3 with active K-Ras (Elad-Sfadia *et al.*, 2004) strongly suggests that FTS blocks the interaction of K-Ras-GTP with galectin-3. Like galectin-1, galectin-3 possesses a putative farnesyl-binding pocket (Ashery *et al.*, 2006a) that could bind either the farnesyl group of K-Ras or FTS. Point mutation analysis demonstrates that the integrity of the putative farnesyl-binding pockets of galectin-1 (Rotblat *et al.*, 2004a) and galectin-3 (Shalom, Rotblat, and Kloog, unpublished data) is critical for their proper functioning as Ras escort proteins.

Taken together, results of the experiments described earlier provide a mechanistic explanation for the selective inhibition of active Ras isoforms by FTS.

3.4. Ras inhibition by FTS affects Ras signaling

Ras proteins control diverse cellular behaviors because of the ability of all Ras isoforms to activate a wide-ranging common set of effectors, including Raf-1, PI3-K, RalGEFs, phospholipase Cε, RIN, and Tiam-1 (Bar-Sagi and Hall, 2000; Downward, 2003a; Hancock, 2003; Shields et al., 2000). Early tissue culture experiments performed with Ras-transformed rodent fibroblasts showed that FTS downregulates active H-Ras (Haklai et al., 1998; Marom et al., 1995), K-Ras (Elad et al., 1999), and N-Ras (Jansen et al., 1999), with concomitant inhibition of the canonical Raf/Mek/Erk pathway. These results are consistent with the finding that chronically active Ras can induce hyperactivation of Erk, which is blocked when active Ras proteins are inhibited by FTS (Cox and Der, 2002; Gana-Weisz et al., 1997). Similar results were obtained in a number of human cancer cell lines expressing oncogenic Ras, for example, panc-1 (Gana-Weisz et al., 2002; Weisz et al., 1999) and melanoma (Jansen et al., 1999), or in cell lines with hyperactivated wild-type Ras, for example, NF-1 cells (Barkan et al., 2006), glioblastoma (Blum et al., 2005, 2006a,b; Goldberg and Kloog, 2006), and neuroblastoma (Yaari et al., 2005). Other experiments showed that growth factor-stimulated activation of Ras and of Erk, both in untransformed cells (Gana-Weisz et al., 1997; Khwaja et al., 2005; Reif et al., 1999) and in transformed cells (Barkan et al., 2006), was strongly inhibited by FTS. In addition, activation of the Ras/PI3-K pathway was found to be inhibited by FTS, for example, in glioblastoma (Blum et al., 2005, 2006a,b; Goldberg and Kloog, 2006), in neuroblastoma (Yaari et al., 2005), and in NF-1 cells (Barkan et al., 2006). FTS also inhibited Ral activation in NF-1 cells (Barkan et al., 2006).

Taken together, the results of these studies are consistent with the classic linear paradigm of hyperactivation by chronically active Ras of its downstream effectors, and hence of inhibition of the activation of the Raf/Mek/Erk, PI3K/Akt, and RalGEF/Ral pathways by FTS-induced blockage of the active Ras proteins. Thus, these experiments provide proof of principle for the FTS-induced inhibition of Ras-signaling pathways. It is important to bear in mind, however, that the various Ras isoforms can exhibit different biological activities, which depend on the cellular context and may differentially activate one or another of the Ras effectors. It should therefore not be expected that all Ras pathways will be inhibited by FTS in any given cancer cell.

One example of distinct biological functions of Ras isoforms is given by Ras knockout mice. Knockout of K-Ras, but not of H-Ras or N-Ras, is lethal to mouse embryos (Johnson et al., 1997; Umanoff et al., 1995). Ras regulation of matrix metalloproteinase-2 transcription was abolished in K-Ras but not in N-Ras knockout fibroblasts (Liao *et al.*, 2003), and earlier experiments showed that activated Ras elicits transformation in immortal rodent cells but induces senescence in primary rodent cells (Serrano *et al.*, 1997). In addition, the types of effectors required for Ras transformation of rodent cells differ from those in human cells (Hamad *et al.*, 2002). An analogous distinct effector-type requirement was documented in PC12 cells, in which Raf and PI3-K signaling promoted cell cycle arrest, whereas RalGEF signaling promoted cell proliferation (Goi *et al.*, 1999). Another important observation was that activated K-Ras in human pancreatic tumor cell lines does not correlate with persistent activation of Erk (Yip-Schneider *et al.*, 1999). Other studies demonstrated cross talk between the various Ras pathways and showed that activation of one pathway can result in inhibition of another under certain circumstances (Wennstrom and Downward, 1999). More recently, in a mouse model of colon cancer in which mutationally activated K-Ras was expressed in the colonic epithelium, there was a surprising lack of activation of Erk in the proliferating epithelial cells, despite its activation in the differentiated cells (Haigis *et al.*, 2006). Consistently with these observations, no Erk activation was detectable in undifferentiated adenocarcinoma expressing activated K-Ras (Haigis *et al.*, 2006). It thus appears that Ras pathways other than the Raf/Mek/Erk, possibly comprising as yet unidentified Ras effectors, mediate K-Ras-driven hyperproliferation of colonic cancer cells.

The aforementioned examples of the diversity and complexity of Ras signaling in cancer cells would appear to rule out straightforward linear relationships between Ras inhibition and blockage of all Ras downstream signals. Thus, for example, in cancer cells with no Ras-driven activation of the Raf/Mek/Erk cascade, inhibition of the constitutively active Ras by FTS would not be expected to affect this pathway. If, however, the tumor is driven by chronically active Ras, its growth is likely to be inhibited by FTS concomitantly with inhibition of the active Ras protein. Experiments with FTS have indeed demonstrated that the inhibitor decreases the amounts of activated Ras both *in vitro* (Barkan *et al.*, 2006; Gana-Weisz *et al.*, 2002; Goldberg and Kloog, 2006; Jansen *et al.*, 1999; Weisz *et al.*, 1999; Yaari *et al.*, 2005) and *in vivo* (Haklai *et al.*, 2007), with resulting inhibition of tumor growth.

3.5. Dose-dependent inhibition of cell growth and tumor growth by FTS: *in vitro* results correlate well with antitumor potency *in vivo*

In line with its anti-Ras activity, FTS *in vitro* strongly inhibits Ras-driven growth (Elad *et al.*, 1999; Haklai *et al.*, 1998; Marom *et al.*, 1995), transformation (Goldberg and Kloog, 2006; Weisz *et al.*, 1999; Zundelevich *et al.*, 2007), survival (Shalom-Feuerstein *et al.*, 2004), and migration (Goldberg and Kloog, 2006; Reif *et al.*, 1999). The range of FTS concentrations required to inhibit Ras-controlled cellular behavior *in vitro* varies from as low as 2 to 10 μM (Barkan *et al.*, 2006; Goldberg and Kloog, 2006; Khwaja *et al.*, 2005; Reif *et al.*, 1999) to as high as 75 to 100 μM (Blum *et al.*, 2005; Halaschek-Wiener *et al.*, 2000; Weisz *et al.*, 1999; Yaari *et al.*, 2005). The observed sensitivity to FTS *in vitro* depends on at least three factors: cell type, cell density, and cell growth conditions (e.g., presence or absence of serum). Issues of cell type are related to non-Ras genetic aberrations on which the cancer cell depends, including other oncogenes and tumor suppressors typical of each cancer cell line. We have observed significant variations in sensitivity to growth inhibition by FTS within a given type of cancer cell line that expresses activated K-Ras, such as certain pancreatic carcinoma cell lines [panc-1 cells, for example, are more sensitive than MiaPaca cells (Weisz *et al.*, 1999)] or lung cancer cell lines [A549 are more sensitive than H23 cells (Zundelevich *et al.*, 2007)]. We also observed differences between types of tumor cells expressing oncogenic K-Ras, for example, panc-1 cells are more sensitive than SW480 colon cancer cells to FTS (Halaschek-Wiener *et al.*, 2000, 2004; Weisz *et al.*, 1999). Cancer cell types also vary in their sensitivity to FTS-induced apoptosis; the differences probably reflect different genetic backgrounds. Examples of such differences include glioblastoma (Blum *et al.*, 2005, 2006a,b; Goldberg and Kloog, 2006) and prostate (Erlich *et al.*, 2006) cancer cells, which undergo apoptosis when treated with FTS in the presence as well as in the absence of serum. Cancer cells that do not undergo apoptosis when treated with FTS in the presence of serum include pancreatic (Weisz *et al.*, 1999), colon (Halaschek-Wiener *et al.*, 2000, 2004), and lung cancer cell lines (Zundelevich *et al.*, 2007).

Cell density issues are related to factors such as saturation density and autocrine loops. When cells are plated at relatively high densities they reach growth saturation more rapidly than when the plating density is relatively low; growth inhibition by FTS is therefore less apparent at higher densities. Under such conditions the concentrations of growth factors secreted by the cancer cells are also high, counteracting the FTS-induced growth inhibition. However, a major common factor affecting the sensitivity of all cell types to FTS is the presence or absence of serum. Even though most cancer cells produce and secrete growth factors and cytokines that enhance their proliferation, the best way to maintain such cells in culture is by using serum-enriched media, for example, fetal calf serum. The serum provides growth factors and a variety of proteins, amino acids, and lipids, which together promote faster growth. The serum, however, also affects the sensitivity of the cells to FTS because of its high concentration of proteins (particularly serum albumin) and lipids, which can bind the drug. Experiments with mouse, rat, dog, and human plasma have demonstrated the high capacity of plasma proteins to bind FTS (>81 and $>89\%$ binding, respectively, at 1 and 10 μM FTS). In other words, when cells grown in the presence of serum are treated with FTS, a large amount of the drug binds to the serum proteins, and the concentration of free drug that can enter the cells is therefore lower than the nominal (total) drug concentration in the dish. In all cases that we have examined using the same cell line, cell sensitivity to FTS was indeed more pronounced when the cells were grown in the absence of serum. Early studies showed, for example, that 10 μM FTS inhibits thymidine incorporation into DNA-induced rat-1 fibroblasts by serum, by epidermal growth factor, or by thrombin (Gana-Weisz *et al.*, 1997). In the presence of serum, growth of these cells was not inhibited even at 100 μM FTS (Marom *et al.*, 1995). As another example, inhibition of liver hematopoietic stem cell migration in the absence of serum occurred at FTS concentrations as low as 2.0 to 2.5 μM (Reif *et al.*, 1999). Similarly, in the absence of serum, FTS at the low concentration of 12.5 μM inhibited the migration of human glioblastoma cell lines and reduced their Ras-GTP levels and Ras signaling (Goldberg and Kloog, 2006). In the presence of serum, however, both anchorage-dependent (Blum *et al.*, 2005) and anchorage-independent (Goldberg and Kloog, 2006) growth of these cells was inhibited only at 50 to 75 μM FTS. The relatively high sensitivity of cells to FTS in the absence of serum was also apparent in the proapoptotic effect of the drug, which canceled the survival signals of activated Ras proteins (Shalom-Feuerstein *et al.*, 2004). In the absence of serum, FTS induces apoptosis of glioblastoma (Blum *et al.*, 2005, 2006a), NF1 (Barkan *et al.*, 2006), and other cells possessing chronically active Ras at the low micromolar range. The 5- to 10-fold higher sensitivity to FTS in the absence of serum than in its presence, as well as the knowledge that serum proteins bind the drug, clearly indicate that the "true" affinity of FTS to Ras escort proteins such as galectin-1 or galectin-3 would be 5- to 10-fold higher than the apparent IC_{50} of the drug determined under normal growth conditions.

On the basis of these observations, the effect of FTS *in vivo* is expected to become apparent at doses that resemble the "true" affinity of FTS for Ras-binding partners, namely in the range of 2 to 10 μM, provided that the continuous flow of FTS with the blood allows it to accumulate in the tumor. Pharmacokinetic data support the notion that at doses of FTS that effectively inhibit tumor growth in animals, FTS levels in the plasma are clearly in the micromolar concentration range (Haklai *et al.*, 2007). Evidently, in such models the amounts of Ras in the tumor are strongly reduced and tumor growth is inhibited (Barkan *et al.*, 2006; Haklai *et al.*, 2007).

3.6. Evaluation of Salirasib as a Ras inhibitor for the potential treatment of human cancers: clinical trials

The aforementioned findings pointing to a fairly specific interference by FTS with the spatiotemporal localization of active Ras proteins are supported by several independent gene expression profiling experiments *in vitro*. Use of a ranking-based procedure, combined with functional analysis and promoter sequence analysis, deciphered the commonest and most prominent patterns of the transcriptional responses of five different human cancer cell lines to FTS (Blum *et al.*, 2007). Remarkably, the analysis identified a distinctive Ras-related core transcriptional response to FTS that was common to all cancer cell lines tested. Moreover, this signature fitted well with an independent set of experiments describing the deregulated Ras pathway signature in Ras-transformed human cell lines (Bild *et al.*, 2006b), which by itself predicted sensitivity to FTS (Bild *et al.*, 2006a). These studies strongly supported the conclusion that FTS specifically reregulates defective Ras pathways in human tumor cells, including repression of E2F- and NF-Y-regulated genes and the transcription factor FOS (Blum *et al.*, 2007).

In view of these promising results, it is now of utmost importance to confirm the mode of action of FTS (Salirasib) as a Ras inhibitor in humans and to critically evaluate its effects on Ras-driven tumors in clinical trials. Given the possibility that this simple farnesyl derivative might also interfere with the action of other prenylated proteins, it is encouraging that phase I trials point to a safe profile of Salirasib with minimal or no side effects (Borthakur *et al.*, 2007). Those trials were carried out in healthy volunteers and in patients with solid tumors, as well as in patients with hematological disorders (Borthakur *et al.*, 2007). In the latter open-label dose–escalation study, FTS was reported to possess promising activity in hematological malignancies and to be safe up to at least 800 mg twice a day (Borthakur *et al.*, 2007). Pivotal trials of Salirasib (phase I study) are currently underway (http://www.concordiapharma.com) in patients with pancreatic cancer (in combination with gemcitabine), as well as a single drug treatment in patients with nonsmall cell lung cancer with and without identified K-Ras mutations (phase II study). An auspicious outcome of these trials might indicate whether interference with the spatiotemporal localization of oncogenic Ras proteins, as well as the use of prenyl-binding domains as targets for anticancer drugs (Kloog and Cox, 2000; Magee and Seabra, 2003), are worth developing as potential methods for cancer treatment.

ACKNOWLEDGMENTS

YK is the incumbent of The Jack H. Skirball Chair for Applied Neurobiology. We thank S.R. Smith for editorial assistance. This work was supported in part by grants to YK from the Israel Science Foundation (Grant 912/06) and the Wolfson Foundation.

Supplementary Movie 1   Spinning disk confocal time-lapse images of a GFP-H-Ras (G12V)-expressing cell treated with FTS. Mouse brain endothelial cells (Bend 3) stably expressing GFP-H-Ras(G12V) were treated with 80 μM FTS as described in Fig. 32.1 and followed under a spinning disk confocal microscope in a time-lapse series centered on a middle confocal plane of the cells as in Fig. 32.1.

REFERENCES

Ashery, U., Yizhar, O., Rotblat, B., Elad-Sfadia, G., Barkan, B., Haklai, R., and Kloog, Y. (2006a). Spatiotemporal organization of ras signaling: Rasosomes and the galectin switch. *Cell. Mol.Neurobiol.* **26**, 469–493.

Ashery, U., Yizhar, O., Rotblat, B., and Kloog, Y. (2006b). Nonconventional trafficking of Ras associated with Ras signal organization. *Traffic* **7**, 1119–1126.

Baines, A. T., Lim, K. H., Shields, J. M., Lambert, J. M., Counter, C. M., Der, C. J., and Cox, A. D. (2005). Use of retrovirus expression of interfering RNA to determine the contribution of activated k-ras and ras effector expression to human tumor cell growth. *Methods Enzymol.* **407**, 556–574.

Baker, T. L., Zheng, H., Walker, J., Coloff, J. L., and Buss, J. E. (2003). Distinct rates of palmitate turnover on membrane-bound cellular and oncogenic H-ras. *J. Biol. Chem.* **278**, 19292–19300.

Barkan, B., Starinsky, S., Friedman, E., Stein, R., and Kloog, Y. (2006). The Ras inhibitor farnesylthiosalicylic acid as a potential therapy for neurofibromatosis type 1. *Clin. Cancer Res.* **12**, 5533–5542.

Bar-Sagi, D., and Hall, A. (2000). Ras and Rho GTPases: A family reunion. *Cell* **103**, 227–238.

Bild, A. H., Potti, A., and Nevins, J. R. (2006a). Linking oncogenic pathways with therapeutic opportunities. *Nat. Rev. Cancer* **6**, 735–741.

Bild, A. H., Yao, G., Chang, J. T., Wang, Q., Potti, A., Chasse, D., Joshi, M. B., Harpole, D., Lancaster, J. M., Berchuck, A., Olson, J. A., Jr., Marks, J. R., *et al.* (2006b). Oncogenic pathway signatures in human cancers as a guide to targeted therapies. *Nature* **439**, 353–357.

Bivona, T. G., Quatela, S. E., Bodemann, B. O., Ahearn, I. M., Soskis, M. J., Mor, A., Miura, J., Wiener, H. H., Wright, L., Saba, S. G., Yim, D., Fein, A., *et al.* (2006). PKC regulates a farnesyl-electrostatic switch on K-Ras that promotes its association with Bcl-XL on mitochondria and induces apoptosis. *Mol. Cell* **21**, 481–493.

Blum, R., Elkon, R., Yaari, S., Zundelevich, A., Jacob-Hirsch, J., Rechavi, G., Shamir, R., and Kloog, Y. (2007). Gene expression signature of human cancer cell lines treated with the Ras inhibitor Salirasib (S-farnesylthiosalicylic acid). *Cancer Res.* **67**, 3320–3328.

Blum, R., Jacob-Hirsch, J., Amariglio, N., Rechavi, G., and Kloog, Y. (2005). Ras inhibition in glioblastoma down-regulates hypoxia-inducible factor-1alpha, causing glycolysis shutdown and cell death. *Cancer Res.* **65**, 999–1006.

Blum, R., Jacob-Hirsch, J., Rechavi, G., and Kloog, Y. (2006a). Suppression of survivin expression in glioblastoma cells by the Ras inhibitor farnesylthiosalicylic acid promotes caspase-dependent apoptosis. *Mol. Cancer Ther.* **5**, 2337–2347.

Blum, R., Nakdimon, I., Goldberg, L., Elkon, R., Shamir, R., Rechavi, G., and Kloog, Y. (2006b). E2F1 identified by promoter and biochemical analysis as a central target of glioblastoma cell-cycle arrest in response to Ras inhibition. *Int. J. Cancer* **119**, 527–538.

Borthakur, G., Ravandi, F., O'Brien, S., Williams, B., Bauer, V., and Giles, F. (2007). Phase I study of S-trans, trans-farnesylthiosalicylic acid (FTS), a noval oral Ras inhibitor, in patients with refractory hematologic malignancies. *In* "ASCO Annual Meeting Proceeding Part I," Vol. 25.

Casey, P. J., Solski, P. A., Der, C. J., and Buss, J. E. (1989). p21ras is modified by a farnesyl isoprenoid. *Proc. Natl. Acad. Sci. USA* **86,** 8323–8327.

Chin, L., Tam, A., Pomerantz, J., Wong, M., Holash, J., Bardeesy, N., Shen, Q., O'Hagan, R., Pantginis, J., Zhou, H., Horner, J. W., 2nd, Cordon-Cardo, C., et al. (1999). Essential role for oncogenic Ras in tumour maintenance. *Nature* **400,** 468–472.

Chiu, V. K., Silletti, J., Dinsell, V., Wiener, H., Loukeris, K., Ou, G., Philips, M. R., and Pillinger, M. H. (2004). Carboxyl methylation of Ras regulates membrane targeting and effector engagement. *J. Biol. Chem.* **279,** 7346–7352.

Clarke, H. C., Kocher, H. M., Khwaja, A., Kloog, Y., Cook, H. T., and Hendry, B. M. (2003). Ras antagonist farnesylthiosalicylic acid (FTS) reduces glomerular cellular proliferation and macrophage number in rat thy-1 nephritis. *J. Am. Soc. Nephrol.* **14,** 848–854.

Cox, A. D., and Der, C. J. (1992). Protein prenylation: More than just glue? *Curr. Opin. Cell Biol.* **4,** 1008–1016.

Cox, A. D., and Der, C. J. (2002). Ras family signaling: Therapeutic targeting. *Cancer Biol. Ther.* **1,** 599–606.

Dong, X., Mitchell, D. A., Lobo, S., Zhao, L., Bartels, D. J., and Deschenes, R. J. (2003). Palmitoylation and plasma membrane localization of Ras2p by a nonclassical trafficking pathway in. *Saccharomyces cerevisiae. Mol. Cell Biol.* **23,** 6574–6584.

Downward, J. (2003a). Role of receptor tyrosine kinases in G-protein-coupled receptor regulation of Ras: Transactivation or parallel pathways? *Biochem. J.* **376,** e9–e10.

Downward, J. (2003b). Targeting RAS signalling pathways in cancer therapy. *Nat. Rev. Cancer* **3,** 11–22.

Egozi, Y., Weisz, B., Gana-Weisz, M., Ben-Baruch, G., and Kloog, Y. (1999). Growth inhibition of ras-dependent tumors in nude mice by a potent ras-dislodging antagonist. *Int. J. Cancer* **80,** 911–918.

Elad, G., Paz, A., Haklai, R., Marciano, D., Cox, A., and Kloog, Y. (1999). Targeting of K-Ras 4B by S-trans,trans-farnesyl thiosalicylic acid. *Biochim. Biophys. Acta* **1452,** 228–242.

Elad-Sfadia, G., Haklai, R., Ballan, E., Gabius, H. J., and Kloog, Y. (2002). Galectin-1 augments Ras activation and diverts Ras signals to Raf-1 at the expense of phosphoinositide 3-kinase. *J. Biol. Chem.* **277,** 37169–37175.

Elad-Sfadia, G., Haklai, R., Balan, E., and Kloog, Y. (2004). Galectin-3 augments K-Ras activation and triggers a Ras signal that attenuates ERK but not phosphoinositide 3-kinase activity. *J. Biol. Chem.* **279,** 34922–34930.

Erlich, S., Tal-Or, P., Liebling, R., Blum, R., Karunagaran, D., Kloog, Y., and Pinkas-Kramarski, R. (2006). Ras inhibition results in growth arrest and death of androgen-dependent and androgen-independent prostate cancer cells. *Biochem. Pharmacol.* **72,** 427–436.

Fivaz, M., and Meyer, T. (2005). Reversible intracellular translocation of KRas but not HRas in hippocampal neurons regulated by Ca^{2+}/calmodulin. *J. Cell Biol.* **170,** 429–441.

Gana-Weisz, M., Haklai, R., Marciano, D., Egozi, Y., Ben Baruch, G., and Kloog, Y. (1997). The Ras antagonist S-farnesylthiosalicylic acid induces inhibition of MAPK activation. *Biochem. Biophys. Res. Commun.* **239,** 900–904.

Gana-Weisz, M., Halaschek-Wiener, J., Jansen, B., Elad, G., Haklai, R., and Kloog, Y. (2002). The Ras inhibitor S-trans,trans-farnesylthiosalicylic acid chemosensitizes human tumor cells without causing resistance. *Clin. Cancer Res.* **8,** 555–565.

Goi, T., Rusanescu, G., Urano, T., and Feig, L. A. (1999). Ral-specific guanine nucleotide exchange factor activity opposes other Ras effectors in PC12 cells by inhibiting neurite outgrowth. *Mol. Cell Biol.* **19,** 1731–1741.

Goldberg, L., and Kloog, Y. (2006). A Ras inhibitor tilts the balance between Rac and Rho and blocks PI3-kinase-dependent glioblastoma cell migration. *Cancer Res.* **66,** 11709–11717.

Gorfe, A. A., Hanzal-Bayer, M., Abankwa, D., Hancock, J. F., and McCammon, J. A. (2007). Structure and dynamics of the full-length lipid-modified H-Ras protein in a 1,2-dimyristoylglycero-3-phosphocholine bilayer. *J. Med. Chem.* **50,** 674–684.

Gutierrez, L., Magee, A. I., Marshall, C. J., and Hancock, J. F. (1989). Post-translational processing of p21ras is two-step and involves carboxyl-methylation and carboxy-terminal proteolysis. *EMBO J.* **8,** 1093–1098.

Hahn, W. C., Counter, C. M., Lundberg, A. S., and Weinberg, R. L. B. (1999). Creation of human tumour cells with defined genetic elements. *Nature* **400,** 464–468.

Haigis, K., Appleton, P., Nathke, I., and Jacks, T. (2006). Pleiotropic effects of activating K-ras in colonic epithelium. *In* "Beatson International Cancer Conference," p. 77 Glasgow, Scotland.

Haklai, R., Elad-Sfadia, G., Egozi, Y., and Kloog, Y. (2007). Orally administered FTS (Salirasib) inhibits human pancreatic tumor growth in nude mice. *Cancer Chemother. Pharmacol.* **61,** 89–96.

Haklai, R., Gana-Weisz, G., Elad, G., Paz, A., Marciano, D., Egozi, Y., Ben Baruch, G., and Kloog, Y. (1998). Dislodgment and accelerated degradation of Ras. *Biochemistry* **37,** 1306–1314.

Halaschek-Wiener, J., Wacheck, V., Kloog, Y., and Jansen, B. (2004). Ras inhibition leads to transcriptional activation of p53 and down-regulation of Mdm2: Two mechanisms that cooperatively increase p53 function in colon cancer cells. *Cell Signal* **16,** 1319–1327.

Halaschek-Wiener, J., Wacheck, V., Schlagbauer-Wadl, H., Wolff, K., Kloog, Y., and Jansen, B. (2000). A novel Ras antagonist regulates both oncogenic Ras and the tumor suppressor p53 in colon cancer cells. *Mol. Med.* **6,** 693–704.

Hamad, N. M., Elconin, J. H., Karnoub, A. E., Bai, W., Rich, J. N., Abraham, R. T., Der, C. J., and Counter, C. M. (2002). Distinct requirements for Ras oncogenesis in human versus mouse cells. *Genes Dev.* **16,** 2045–2057.

Hanahan, D., and Weinberg, R. A. (2000). The hallmarks of cancer. *Cell* **100,** 57–70.

Hancock, J. F. (2003). Ras proteins: Different signals from different locations. *Nat. Rev. Mol. Cell. Biol.* **4,** 373–384.

Hancock, J. F. (2006). Lipid rafts: Contentious only from simplistic standpoints. *Nat. Rev. Mol. Cell. Biol.* **7,** 456–462.

Hancock, J. F., Cadwallader, K., and Marshall, C. J. (1991). Methylation and proteolysis are essential for efficient membrane binding of prenylated p21K-ras(B). *EMBO J.* **10,** 641–646.

Hancock, J. F., and Parton, R. G. (2005). Ras plasma membrane signalling platforms. *Biochem. J.* **389,** 1–11.

Hanzal-Bayer, M. F., and Hancock, J. F. (2007). Lipid rafts and membrane traffic. *FEBS Lett.* **581,** 2098–2104.

Harding, A., Tian, T., Westbury, E., Frische, E., and Hancock, J. F. (2005). Subcellular localization determines MAP kinase signal output. *Curr. Biol.* **15,** 869–873.

Haring, R., Fisher, A., Marciano, D., Pittel, Z., Kloog, Y., Zuckerman, A., Eshhar, N., and Heldman, E. (1998). Mitogen-activated protein kinase-dependent and protein kinase C-dependent pathways link the m1 muscarinic receptor to beta-amyloid precursor protein secretion. *J. Neurochem.* **71,** 2094–2103.

Jansen, B., Schlagbauer-Wadl, H., Kahr, H., Heere-Ress, E., Mayer, B. X., Eichler, H., Pehamberger, H., Gana-Weisz, M., Ben-David, E., Kloog, Y., and Wolff, K. (1999). Novel Ras antagonist blocks human melanoma growth. *Proc. Natl. Acad. Sci. USA* **96,** 14019–14024.

Jiang, X., and Sorkin, A. (2002). Coordinated traffic of Grb2 and Ras during epidermal growth factor receptor endocytosis visualized in living cells. *Mol. Biol. Cell* **13,** 1522–1535.

Jindrich, J., Pitha, J., Lindberg, B., Seffers, P., and Harata, K. (1995). Regioselectivity of alkylation of cyclomaltoheptaose (beta-cyclodextrin) and synthesis of its mono-2-O-methyl, -ethyl, -allyl, and -propyl derivatives. *Carbohydr. Res.* **266,** 75–80.

Johnson, L., Greenbaum, D., Cichowski, K., Mercer, K., Murphy, E., Schmitt, E., Bronson, R. T., Umanoff, H., Edelmann, W., Kucherlapati, R., and Jacks, T. (1997). K-ras is an essential gene in the mouse with partial functional overlap with N-ras. *Genes Dev.* **11,** 2468–2481.

Jura, N., Scotto-Lavino, E., Sobczyk, A., and Bar-Sagi, D. (2006). Differential modification of ras proteins by ubiquitination. *Mol. Cell* **21,** 679–687.

Khwaja, A., Sharpe, C. C., Noor, M., Kloog, Y., and Hendry, B. M. (2005). The inhibition of human mesangial cell proliferation by S-trans, trans-farnesylthiosalicylic acid. *Kidney Int.* **68,** 474–486.

Kloog, Y., and Cox, A. D. (2000). RAS inhibitors: Potential for cancer therapeutics. *Mol. Med. Today* **6,** 398–402.

Liao, J., Wolfman, J. C., and Wolfman, A. (2003). K-ras regulates the steady-state expression of matrix metalloproteinase 2 in fibroblasts. *J. Biol. Chem.* **278,** 31871–31878.

Linder, M. E., and Deschenes, R. J. (2003). New insights into the mechanisms of protein palmitoylation. *Biochemistry* **42,** 4311–4320.

Magee, A. I., and Seabra, M. C. (2003). Are prenyl groups on proteins sticky fingers or greasy handles? *Biochem. J.* **376,** e3–e4.

Marciano, D., Ben Baruch, G., Marom, M., Egozi, Y., Haklai, R., and Kloog, Y. (1995). Farnesyl derivatives of rigid carboxylic acids-inhibitors of ras-dependent cell growth. *J. Med. Chem.* **38,** 1267–1272.

Marom, M., Haklai, R., Ben Baruch, G., Marciano, D., Egozi, Y., and Kloog, Y. (1995). Selective inhibition of Ras-dependent cell growth by farnesylthiosalisylic acid. *J. Biol. Chem.* **270,** 22263–22270.

Nicolau, D. V., Jr., Burrage, K., Parton, R. G., and Hancock, J. F. (2006). Identifying optimal lipid raft characteristics required to promote nanoscale protein-protein interactions on the plasma membrane. *Mol. Cell. Biol.* **26,** 313–323.

Niv, H., Gutman, O., Henis, Y. I., and Kloog, Y. (1999). Membrane interactions of a constitutively active GFP-K-Ras 4B and their role in signaling: Evidence from lateral mobility studies. *J. Biol. Chem.* **274,** 1606–1613.

Niv, H., Gutman, O., Kloog, Y., and Henis, Y. I. (2002). Activated K-Ras and H-Ras display different interactions with saturable nonraft sites at the surface of live cells. *J. Cell Biol.* **157,** 865–872.

Parton, R. G., and Hancock, J. F. (2004). Lipid rafts and plasma membrane microorganization: Insights from Ras. *Trends Cell Biol.* **14,** 141–147.

Paz, A., Haklai, R., Elad-Sfadia, G., Ballan, E., and Kloog, Y. (2001). Galectin-1 binds oncogenic H-Ras to mediate Ras membrane anchorage and cell transformation. *Oncogene* **20,** 7486–7493.

Philips, M. R. (2005). Compartmentalized signalling of Ras. *Biochem. Soc. Trans.* **33,** 657–661.

Philips, M. R., and Cox, A. D. (2007). Geranylgeranyltransferase I as a target for anti-cancer drugs. *J. Clin. Invest.* **117,** 1223–1225.

Plowman, S. J., Muncke, C., Parton, R. G., and Hancock, J. F. (2005). H-ras, K-ras, and inner plasma membrane raft proteins operate in nanoclusters with differential dependence on the actin cytoskeleton. *Proc. Natl. Acad. Sci. USA* **102,** 15500–15505.

Prior, I. A., Harding, A., Yan, J., Sluimer, J., Parton, R. G., and Hancock, J. F. (2001). GTP-dependent segregation of H-ras from lipid rafts is required for biological activity. *Nat. Cell Biol.* **3,** 368–375.

Prior, I. A., Muncke, C., Parton, R. G., and Hancock, J. F. (2003). Direct visualization of Ras proteins in spatially distinct cell surface microdomains. *J. Cell Biol.* **160,** 165–170.

Reif, S., Weis, B., Aeed, H., Gana-Weis, M., Zaidel, L., Avni, Y., Romanelli, R. G., Pinzani, M., Kloog, Y., and Bruck, R. (1999). The Ras antagonist, farnesylthiosalicylic acid (FTS) inhibits experimentally induced liver cirrhosis in rats. *J. Hepatol.* **31,** 1053–1061.

Rocks, O., Peyker, A., Kahms, M., Verveer, P. J., Koerner, C., Lumbierres, M., Kuhlmann, J., Waldmann, H., Wittinghofer, A., and Bastiaens, P. I. (2005). An acylation cycle regulates localization and activity of palmitoylated Ras isoforms. *Science* **307,** 1746–1752.

Rotblat, B., Niv, H., Andre, S., Kaltner, H., Gabius, H. J., and Kloog, Y. (2004a). Galectin-1(L11A) predicted from a computed galectin-1 farnesyl-binding pocket selectively inhibits Ras-GTP. *Cancer Res.* **64,** 3112–3118.

Rotblat, B., Prior, I. A., Muncke, C., Parton, R. G., Kloog, Y., Henis, Y. I., and Hancock, J. F. (2004b). Three separable domains regulate GTP-dependent association of H-ras with the plasma membrane. *Mol. Cell. Biol.* **24,** 6799–6810.

Rotblat, B., Yizhar, O., Haklai, R., Ashery, U., and Kloog, Y. (2006). Ras and its signals diffuse through the cell on randomly moving nanoparticles. *Cancer Res.* **66,** 1974–1981.

Roy, S., Plowman, S., Rotblat, B., Prior, I. A., Muncke, C., Grainger, S., Parton, R. G., Henis, Y. I., Kloog, Y., and Hancock, J. F. (2005). Individual palmitoyl residues serve distinct roles in h-ras trafficking, microlocalization, and signaling. *Mol. Cell. Biol.* **25,** 6722–6733.

Serrano, M., Lin, A. W., McCurrach, M. E., Beach, D., and Lowe, S. W. (1997). Oncogenic ras provokes premature cell senescence associated with accumulation of p53 and p16INK4a. *Cell* **88,** 593–602.

Shalom-Feuerstein, R., Lindenboim, L., Stein, R., Cox, A. D., and Kloog, Y. (2004). Restoration of sensitivity to anoikis in Ras-transformed rat intestinal epithelial cells by a Ras inhibitor. *Cell Death Differ.* **11,** 244–247.

Shields, J. M., Pruitt, K., McFall, A., Shaub, A., and Der, C. J. (2000). Understanding Ras: 'it ain't over 'til it's over'. *Trends Cell Biol.* **10,** 147–154.

Shohami, E., Yatsiv, I., Alexandrovich, A., Haklai, R., Elad-Sfadia, G., Grossman, R., Biegon, A., and Kloog, Y. (2003). The Ras inhibitor S-trans, trans-farnesylthiosalicylic acid exerts long-lasting neuroprotection in a mouse closed head injury model. *J. Cereb. Blood Flow Metab.* **23,** 728–738.

Silvius, J. R., Bhagatji, P., Leventis, R., and Terrone, D. (2006). K-ras4B and prenylated proteins lacking "second signals" associate dynamically with cellular membranes. *Mol. Biol. Cell* **17,** 192–202.

Tian, T., Harding, A., Inder, K., Plowman, S., Parton, R. G., and Hancock, J. F. (2007). Plasma membrane nanoswitches generate high-fidelity Ras signal transduction. *Nat. Cell Biol.* **9,** 905–914.

Umanoff, H., Edelmann, W., Pellicer, A., and Kucherlapati, R. (1995). The murine N-ras gene is not essential for growth and development. *Proc. Natl. Acad. Sci. USA* **92,** 1709–1713.

Villalonga, P., Lopez-Alcala, C., Bosch, M., Chiloeches, A., Rocamora, N., Gil, J., Marais, R., Marshall, C. J., Bachs, O., and Agell, N. (2001). Calmodulin binds to K-Ras, but not to H- or N-Ras, and modulates its downstream signaling. *Mol. Cell. Biol.* **21,** 7345–7354.

Weisz, B., Giehl, K., Gana-Weisz, M., Egozi, Y., Ben-Baruch, G., Marciano, D., Gierschik, P., and Kloog, Y. (1999). A new functional Ras antagonist inhibits human pancreatic tumor growth in nude mice. *Oncogene* **18,** 2579–2588.

Wennstrom, S., and Downward, J. (1999). Role of phosphoinositide 3-kinase in activation of ras and mitogen-activated protein kinase by epidermal growth factor. *Mol. Cell. Biol.* **19,** 4279–4288.

Yaari, S., Jacob-Hirsch, J., Amariglio, N., Haklai, R., Rechavi, G., and Kloog, Y. (2005). Disruption of cooperation between Ras and MycN in human neuroblastoma cells promotes growth arrest. *Clin. Cancer Res.* **11,** 4321–4330.

Yip-Schneider, M. T., Lin, A., Barnard, D., Sweeney, C. J., and Marshall, M. S. (1999). Lack of elevated MAP kinase (Erk) activity in pancreatic carcinomas despite oncogenic K-ras expression. *Int. J. Oncol.* **15,** 271–279.

Zundelevich, A., Elad-Sfadia, G., Haklai, R., and Kloog, Y. (2007). Suppression of lung cancer tumor growth in a nude mouse model by the Ras inhibitor salirasib (farnesylthio-salicylic acid). *Mol. Cancer Ther.* **6,** 1765–1773.

CHAPTER THIRTY-THREE

QUANTITATIVE HIGH-THROUGHPUT CELL-BASED ASSAYS FOR INHIBITORS OF ROCK KINASES

Andrew J. Garton, Linda Castaldo, *and* Jonathan A. Pachter

Contents

Abstract

The serine/threonine kinases ROCK1 and ROCK2 are direct targets of activated rho GTPases, and aberrant rho/ROCK signaling has been implicated in a number of human diseases. We have developed novel methods for high-throughput assays of ROCK inhibitors that provide for quantitative evaluation of the ability of small molecules to inhibit the function of ROCK kinases in intact cells. Conditions for extraction of known phosphorylated substrates of ROCK were identified, and the involvement of ROCK in phosphorylation of these substrates was evaluated using small interfering RNA (siRNA). Of the potential substrates tested, MYPT1 was identified as a substrate whose phosphorylation was reduced markedly in the combined absence of ROCK1 and ROCK2 proteins, and ELISA methods were developed to allow quantitative measurement of the degree of

OSI Pharmaceuticals, Farmingdale, New York

Methods in Enzymology, Volume 439
ISSN 0076-6879, DOI: 10.1016/S0076-6879(07)00433-8

phosphorylation of MYPT1 at residue T853 in cells grown in 96-well plates. These methods are amenable to high-throughput assays for identification of ROCK inhibitors within libraries of small molecules and can be used to compare compound potencies to prioritize compounds of interest for additional evaluation. These methods should be useful in drug discovery efforts directed toward identifying potent ROCK inhibitors for potential treatment of cancer, hypertension, or other diseases involving rho/ROCK signaling.

 1. INTRODUCTION

ROCK1 and ROCK2 (rho-associated coiled-coil containing kinase-1 and -2, also known as Rokβ/p160ROCK and Rokα, respectively) are closely related members of the AGC subfamily of enzymes that are activated downstream of activated rho in response to a number of extracellular stimuli, including growth factors, integrin activation, and cellular stress (Riento and Ridley, 2003). The ROCK enzymes play key roles in multiple cellular processes, including cell morphology, stress fiber formation and function, cell adhesion, cell migration and invasion, epithelial–mesenchymal transition, transformation, phagocytosis, apoptosis, neurite retraction, cytokinesis, and cellular differentiation (Riento and Ridley, 2003). As such, ROCK kinases represent potential targets for the development of inhibitors to treat a variety of disorders, including cancer, hypertension, vasospasm, asthma, preterm labor, erectile dysfunction, glaucoma, atherosclerosis, myocardial hypertrophy, and neurological diseases (Mueller *et al.*, 2005; Sahai and Marshall, 2002; Wettschurek and Offermanns, 2002).

Many potential substrates of ROCK have been suggested as a result of studies in a variety of cellular or *in vivo* systems. However, these substrates are also potentially phosphorylated by several other protein Ser/Thr kinases, such that the degree of phosphorylation observed in cell extracts is not necessarily an accurate reflection of the activity of the ROCK enzymes. For example, the regulatory light chain component of myosin (MLC) can be phosphorylated at Ser[19] by ROCK (Amano *et al.*, 1996; Wilkinson *et al.*, 2005), but under certain conditions is also a substrate for additional kinases, including myosin light chain kinases (MLCK) (Fazal *et al.*, 2005; Katoh *et al.*, 2001), myotonic dystrophy-related and cdc42-activated kinases (MRCK) (Leung *et al.*, 1998), and ZIPK (Niiro and Ikebe, 2001). The myosin-binding subunit of protein phosphatase 1, MYPT1 (also known as MBS, MLCP), is thought to be a physiological substrate for ROCK under certain conditions and can be phosphorylated by ROCK kinases at two major sites (Thr[696] within the sequence RRSTQGV and T[853] within the sequence RRSTGVS) (Feng *et al.*, 1999; Kimura *et al.*, 1996; Velasco *et al.*, 2002; Wilkinson *et al.*, 2005). However, one or both of these sites are also recognized by several other enzymes, including myotonic dystrophy-related and cdc42-activated kinases (MRCK) (Tan *et al.*, 2001; Wilkinson *et al.*, 2005), ZIPK (MacDonald *et al.*, 2001), and DMPK (Muranyi *et al.*, 2001; Wansink *et al.*, 2003). Therefore, establishing the level of ROCK activity in cells, including the effects of potential inhibitors, has been problematic. Furthermore, methods for determining ROCK substrate phosphorylation levels in cells in a quantitative and high-throughput manner have not been described.

Evaluation of the ability of novel compounds to inhibit their target proteins within the context of an intact cell is an important component in determining the utility of such compounds for the treatment of disorders involving intracellular target proteins such as the Ser/Thr kinases ROCK1 and ROCK2. Ideally the method used to establish cell-based activity should be specific for the target protein of interest, amenable to testing a large number of potential inhibitors, and quantitative to allow identification of preferred compounds. We have developed specific assay methods that can rapidly and quantitatively determine the level of ROCK kinase activity in cells, either in tissue culture systems or *in vivo*, based on the phosphorylation status of MYPT1. These methods should be useful in efforts to discover potent ROCK inhibitors for various potential therapeutic applications.

2. Materials and Methods

2.1. Antibodies and ROCK inhibitors

The following antibodies are used for immunoblotting analysis or for antigen capture or detection in the ELISA assay: MLC (Sigma, St. Louis, MO; clone MY-21), MYPT1 (BD Transduction, San Jose, CA), pMLC(S19) (Cell Signaling Technology, Danvers, MA), pMYPT1(T696) (US Biologicals, Swampscott, MA), and pMYPT1(T853) (US Biologicals). The ROCK inhibitors are obtained from commercial sources: HA1077 (Fasudil) is from Sigma and H-1152 and Y27632 are from Calbiochem (San Diego, CA).

2.2. Cell lines and siRNA transfection

Panc-1 cells (obtained from ATCC) are grown in Dulbecco's modified Eagle's medium with 2 mM L-glutamine adjusted to contain 3.7 g/liter sodium bicarbonate, 4.5 g/liter glucose, and 10% fetal bovine serum. Cells are seeded into six-well culture dishes at 300,000 cells/well and allowed to attach overnight in 2 ml growth medium prior to transfection. siRNA oligonucleotides are diluted to 20 μM in RNase-free water and then mixed with Lipofectamine 2000 (Invitrogen, Carlsbad, CA) prior to transfection. For 10 nM final [siRNA], 25 pmol of the oligonucleotide duplex is diluted in 250 μl serum-free culture medium and then added to 250 μl serum-free medium containing 5 μl Lipofectamine 2000. The mixture is then incubated for 20 min prior to addition to cells growing in 2 ml medium. Following incubation of transfected cells for 48 h, cultures are harvested by the addition of 150 μl of 1× SDS-PAGE sample buffer, and the lysates are analyzed by immunoblotting.

2.3. Cell lysis

Lysates are prepared from cultured cells, following a brief rinse in phosphate-buffered saline (PBS), using several different buffers. Buffer A (50 mM Tris-HCl, pH 7.4, 150 mM NaCl, 10% glycerol, 1% Triton X-100, 0.5 mM EDTA, 1 mM Na$_4$VO$_3$, 1 mg/ml leupeptin, 1 mg/ml aprotinin) or buffer B (RIPA buffer: containing protease and phosphatase inhibitor cocktails; Sigma) are added to the cells in culture dishes, the cells are scraped from the dish, and the lysates are transferred to Eppendorf tubes and incubated at 4° for 20 min. Lysates are then clarified by centrifugation at 10,000 g, and the resulting supernatants are stored at −80° prior to the addition of SDS-PAGE sample buffer and analysis by immunoblotting. Alternatively, lysates are prepared by the direct addition of SDS-PAGE sample buffer to the cultures, and samples are prepared for analysis by sonication for 5 min followed by heating to 100° for 5 min. Lysates may also be prepared by the addition of Proteoextract solution (complete mammalian proteome extraction kit, Calbiochem) including 16.5 IU of Benzonase (Novagen, Inc., Madison, WI) followed by shaking for 30 min at room temperature. Samples prepared by this method are either diluted directly into SDS-PAGE sample buffer for analysis by immunoblotting or are diluted 10-fold in buffer A for analysis in antibody capture ELISA assays.

2.4. ELISA assay for ROCK inhibitors

The assay capture plate is precoated with 50 ng/well of an antibody to the MYPT1 protein (e.g., BD Transduction Laboratories, mouse monoclonal antibody) in 100 μl of 0.1 M sodium bicarbonate buffer (pH 9) by incubation at 4° overnight, followed by blocking with 100 μl/well of 3% bovine serum albumin (BSA) in PBS/Tween 20 (PBS/T) for 1 h at room temperature and washing with 200 μl/well of PBST.

Cells are seeded at 15,000 cells/well in 96-well culture plates in 90 μl of the appropriate growth medium and are allowed to incubate overnight at 37°, 5% CO$_2$, 95% humidity. Compounds are serially diluted to 10× their final concentrations in dimethyl sulfoxide (1%) and growth medium, and then 10-μl aliquots are added to each culture well followed by incubation for 1 h at 37°. Cells are then washed with PBS at room temperature and lysed by the addition of 20 μl of Proteoextract solution (Calbiochem), including

16.5 IU of Benzonase. Lysis is performed by incubation with shaking for 30 min at room temperature. Lysates are then diluted 10-fold with buffer A, and 180 μl is transferred to a 96-well capture ELISA plate. Following incubation of lysates in the capture plate overnight at 4°, assay wells are washed four times with 200 μl/well of PBST prior to the addition of 100 μl of phospho-specific detection antibody [e.g., US Biologicals, pMYPT1 (T853) rabbit polyclonal antibody] diluted in 3% BSA in PBST and incubation for 2 h at room temperature. Assay wells are then washed four times with 200 μl/well of PBST prior to the addition of 100 μl of the secondary detection reagent (e.g., HRP-conjugated antirabbit IgG, Jackson Laboratories) and incubation for 1 h at room temperature. The assay wells are then washed five times with 200 μl/well of PBST prior to the addition of appropriate detection reagent (e.g., 50 μl of Supersignal femto luminescent HRP substrate; Pierce, Rockford, IL) and quantitation using an appropriate plate reader.

Comparison of the assay signals obtained in the presence of compound with those of controls (no compound added) allowed the reduction in phosphorylation of MYPT1 to be determined over a range of compound concentrations, and these inhibition values were fitted to a sigmoidal dose–response inhibition curve to determine the IC_{50} values (i.e., the concentration of compound that reduces the level of MYPT phosphorylation to 50% of the level observed in untreated cells). These IC_{50} values represent the potency of the ROCK inhibitory activity of the compounds.

3. RESULTS AND DISCUSSION

3.1. Extraction of cytoskeletal substrates of ROCK1/ROCK2 from Panc-1 cells

The predominant function of ROCK kinase activity is to regulate cellular morphology and migration through controlling various aspects of cytoskeletal function. The substrates of ROCK are therefore frequently tightly associated with the insoluble fraction of cell extracts that contains cytoskeletal elements. In order to extract such proteins to evaluate the degree of protein phosphorylation it is necessary to use relatively harsh conditions (e.g., high concentrations of SDS, followed by boiling), which are incompatible with the use of such extracts in standard ELISA-based methods for determining protein phosphorylation levels. In order to develop a high-throughput mechanistic assay for ROCK inhibition in intact cells, potential cellular extraction methods were evaluated for their ability to solubilize MYPT1 and MLC (Fig. 33.1). These proteins were effectively extracted by boiling in SDS-PAGE sample buffer or by the commercially available reagent Proteoextract (Calbiochem), whereas much lower amounts of the

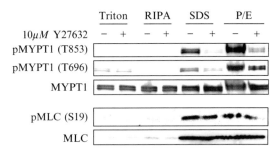

Figure 33.1 Extraction of cytoskeletal substrates of ROCK1/ROCK2 from Panc1 cell extracts. Panc-1 cells were incubated in the presence and absence of 10 μM ROCK inhibitor Y27632 for 1 h followed by lysis in the indicated buffers. Equal fractions of cell lysate were analyzed by immunoblotting for pMYPT1 (T696 or T853), total MYPT1, pMLC (S19), and total MLC.

phosphorylated proteins were released from the insoluble fraction when standard cell lysis buffers were used (see Fig. 33.1). Treatment of the cells with a ROCK inhibitor (Y27632) prior to lysis significantly reduced the level of MYPT1 phosphorylation observed without significantly affecting the total MYPT1 protein content of the extracts (see Fig. 33.1).

3.2. Evaluation of ROCK1/ROCK2 dependence of cytoskeletal signaling events in Panc-1 cells using siRNA

In order to evaluate the extent to which cytoskeletal signaling events are dependent on ROCK function, three distinct siRNA sequences targeting each of the ROCK enzymes were introduced into Panc-1 pancreatic carcinoma cells either individually or in combination. Each siRNA tested reduced significantly the expression level of ROCK1 or ROCK2 (as appropriate to the target sequence).

Reduction of ROCK1 or ROCK2 protein expression alone had minimal effect on the phosphorylation state of MYPT1, MLC, or cofilin, whereas targeting both ROCK1 and ROCK2 together significantly reduced MYPT1 phosphorylation at both T853 and T696 phosphorylation sites within 48 h of addition of siRNA (data not shown). In contrast, there was less of an effect on the phosphorylation of MLC or cofilin under these conditions. These data suggest that MYPT1 phosphorylation represents a predominantly ROCK-dependent signaling event in Panc-1 cells, which can be reduced effectively by targeting both ROCK1 and ROCK2 enzymes. Furthermore, monitoring the phosphorylation state of MYPT1 in Panc-1 cells provides a method for establishing the relative activity of ROCK kinases under different conditions (e.g., in the presence of different potential ROCK inhibitors).

3.3. Development of a quantitative ELISA assay for inhibition of MYPT1 phosphorylation by ROCK inhibitors

Three literature-validated ROCK inhibitors were evaluated for their ability to reduce MYPT1 phosphorylation levels. All three compounds effectively reduced MYPT1 phosphorylation levels at both T853 and T696 phosphorylation sites (Fig. 33.2), although the potency against the T696 phosphorylation site appeared to be lower than that observed for T853. Importantly, the relative potency of the effects observed with three known ROCK inhibitors [H1152>Y27632>fasudil (HA1077)] correlated well with the reported potency of these compounds in kinase inhibition assays performed with the purified ROCK enzyme *in vitro* (Mueller *et al.*, 2005). Immunoblotting analysis of samples from time-course experiments revealed that the steady-state phosphorylation of MYPT1 (T853) in Panc-1 cells was reduced maximally within 30 to 60 min of compound addition, indicating that this phosphorylation site is highly dynamic and that a 1-h incubation with compounds is likely to be sufficient for evaluation of the cellular potency of ROCK inhibitors (data not shown).

The ability of compounds to inhibit ROCK activity in intact cells was determined using a 96-well antibody capture ELISA assay to detect the degree of phosphorylation of the ROCK-specific substrate MYPT1 at amino acid residue T853. Cells were cultured in 96-well plates and incubated with compounds for 1 h prior to lysis in Proteoextract buffer. The phosphorylated MYPT1 protein solubilized by the Proteoextract reagent

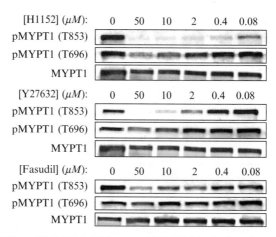

Figure 33.2   Effect of ROCK inhibitors on MYPT1 (T853) phosphorylation in Panc-1 cells. Panc-1 cells were incubated in the presence of the indicated concentrations of ROCK inhibitors H1152, Y27632, or fasudil (HA1077) for 1 h, followed by lysis in SDS-PAGE sample buffer. Equal fractions of cell lysate were then analyzed by immunoblotting for pMYPT1 (T696 or T853) and total MYPT1 protein.

was not detected readily using standard antibody capture ELISA techniques (Fig. 33.3A, undiluted sample). However, a significant (e.g., 10-fold) dilution of the sample in a Triton X-100-containing buffer yielded a sample that retained the extracted MYPT1 as a soluble protein, while also permitting efficient capture and detection using antibody-coated 96-well ELISA plates (see Fig. 33.3A). The signal/background ratio observed in this assay, obtained by comparing the signal strength in wells containing cell lysates to that in wells to which no lysate was added, was approximately 10- to 30-fold.

Quantitative comparison of three commercially available ROCK inhibitors (see Fig. 33.3B) revealed, as expected from the immunoblotting analysis (see Fig. 33.2), that the rank order of compound potency in this assay correlated well with that described in the literature for inhibition of the purified ROCK kinase enzyme *in vitro*. Furthermore, comparison of IC_{50} values obtained for greater than 60 small molecules in the cell-based assay with those from biochemical assays of ROCK kinase inhibition also demonstrated a similar rank order of compound potency within the two assay systems (data not shown). The potency of compounds determined in the pMYPT1 (T853) ELISA assay also correlated well with the potency of

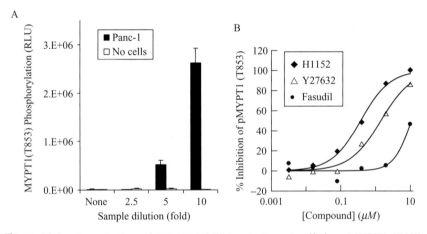

Figure 33.3 Quantitation of ROCK inhibition in Panc-1 cells by pMYPT1 (T853) ELISA. (A) Ten-microliter aliquots of Panc-1 cell lysate prepared in Proteoextract buffer (or buffer alone for control wells) were diluted in buffer A as indicated prior to addition to MYPT1 antibody-coated ELISA assay plate wells. Quantitation of the extent of MYPT1 (T853) phosphorylation was performed in triplicate assay wells as described in the text. Results plotted are mean signal intensities; error bars represent standard deviations. (B) Panc-1 cells were grown in 96-well culture plates and incubated for 1 h with the indicated concentrations of ROCK inhibitors H1152, Y27632, or fasudil (HA1077). Lysates were then prepared in Proteoextract buffer and analyzed in the pMYPT1 (T853) ELISA assay as described in the text. Results plotted are mean signal intensities; error bars represent standard deviations.

compounds in functional assays evaluating inhibition of cell migration, a process strongly believed to be regulated by ROCK kinase activity (data not shown).

In general, data outlined earlier from studies in Panc-1 cells appear to be generally applicable to other cell types, as comparable assays performed across a panel of cancer cell lines revealed similar compound potencies in the majority of cell lines tested (HCT116, PC3, DU-145, A375). However, one cell line (MDA-MB-231) appeared to be relatively insensitive to ROCK inhibition, as judged by the level of phosphorylation of MYPT1 (T853), perhaps indicating that MYPT1 phosphorylation is regulated somewhat differently in this cell type (data not shown).

The cell-based assay described here appears to be an accurate indicator of compound potency against ROCK enzymes within intact cells and is amenable to quantitative comparison of large numbers of compounds in support of drug discovery programs. In addition, similar methods can be applied to tissue samples from experiments performed *in vivo* in order to assess the ability of compounds to inhibit ROCK activity in appropriate tissues following compound dosing (data not shown). Such assays are potentially useful to guide drug discovery efforts directed toward identifying novel ROCK inhibitors for a variety of therapeutic applications, including cancer, hypertension, vasospasm, asthma, glaucoma, atherosclerosis, and neurological diseases.

REFERENCES

Amano, M., Ito, M., Kimura, K., Fukata, Y., Chihara, K., Nakano, T., Matsuura, Y., and Kaibuchi, K. (1996). Phosphorylation and activation of myosin by Rho-associated kinase (Rho-kinase). *J. Biol. Chem.* **271**, 20246–20249.

Fazal, F., Gu, L., Ihnatovych, I., Han, Y., Hu, W., Antic, N., Carreira, F., Blomquist, J. F., Hope, T. J., Ucker, D. S., and de Lanerolle, P. (2005). Inhibiting myosin light chain kinase induces apoptosis *in vitro* and *in vivo*. *Mol. Cell. Biol.* **25**, 6259–6266.

Feng, J., Ito, M., Ichikawa, K., Isaka, N., Nishikawa, M., Hartshorne, D. J., and Nakano, T. (1999). Inhibitory phosphorylation site for Rho-associated kinase on smooth muscle myosin phosphatase. *J. Biol. Chem.* **274**, 37385–37390.

Katoh, K., Kano, Y., Amano, M., Onishi, H., Kaibuchi, K., and Fujiwara, K. (2001). Rho-kinase-mediated contraction of isolated stress fibers. *J. Cell Biol.* **153**, 569–584.

Kimura, K., Ito, M., Amano, M., Chihara, K., Fukata, Y., Nakafuku, M., Yamamori, B., Feng, J., Nakano, T., Okawa, K., Iwamatsu, A., and Kaibuchi, K. (1996). Regulation of myosin phosphatase by Rho and Rho-associated kinase (Rho-kinase). *Science* **273**, 245–248.

Leung, T., Chen, X. Q., Tan, I., Manser, E., and Lim, L. (1998). Myotonic dystrophy kinase-related Cdc42-binding kinase acts as a Cdc42 effector in promoting cytoskeletal reorganization. *Mol. Cell. Biol.* **18**, 130–140.

MacDonald, J. A., Borman, M. A., Muranyi, A., Somlyo, A. V., Hartshorne, D. J., and Haystead, T. A. (2001). Identification of the endogenous smooth muscle myosin phosphatase-associated kinase. *Proc. Natl. Acad. Sci. USA* **98**, 2419–2424.

Mueller, B. K., Mack, H., and Teusch, N. (2005). Rho kinase, a promising drug target for neurological disorders. *Nat. Rev. Drug. Discov.* **4,** 387–398.

Muranyi, A., Zhang, R., Liu, F., Hirano, K., Ito, M., Epstein, H. F., and Hartshorne, D. J. (2001). Myotonic dystrophy protein kinase phosphorylates the myosin phosphatase targeting subunit and inhibits myosin phosphatase activity. *FEBS Lett.* **493,** 80–84.

Niiro, N., and Ikebe, M. (2001). Zipper-interacting protein kinase induces Ca(2+)-free smooth muscle contraction via myosin light chain phosphorylation. *J. Biol. Chem.* **276,** 29567–29574.

Riento, K., and Ridley, A. J. (2003). ROCKS: Multifunctional kinases in cell behaviour. *Nat. Rev. Mol. Cell. Biol.* **4,** 446–456.

Sahai, E., and Marshall, C. J. (2002). Rho-GTPases and cancer. *Nat. Rev. Cancer.* **2,** 133–142.

Tan, I., Ng, C. H., Lim, L., and Leung, T. (2001). Phosphorylation of a novel myosin binding subunit of protein phosphatase 1 reveals a conserved mechanism in the regulation of actin cytoskeleton. *J. Biol. Chem.* **276,** 21209–21216.

Velasco, G., Armstrong, C., Morrice, N., Frame, S., and Cohen, P. (2002). Phosphorylation of the regulatory subunit of smooth muscle protein phosphatase 1M at Thr850 induces its dissociation from myosin. *FEBS Lett.* **527,** 101–104.

Wansink, D. G., van Herpen, R. E., Coerwinkel-Driessen, M. M., Groenen, P. J., Hemmings, B. A., and Wieringa, B. (2003). Alternative splicing controls myotonic dystrophy protein kinase structure, enzymatic activity, and subcellular localization. *Mol. Cell. Biol.* **23,** 5489–5501.

Wettschureck, N., and Offermanns, S. (2002). Rho/Rho-kinase mediated signaling in physiology and pathophysiology. *J. Mol. Med.* **80,** 629–638.

Wilkinson, S., Paterson, H. F., and Marshall, C. J. (2005). Cdc42-MRCK and Rho-ROCK signalling cooperate in myosin phosphorylation and cell invasion. *Nat. Cell Biol.* **7,** 255–261.

CHAPTER THIRTY-FOUR

Tandem Affinity Purification of the BBSome, a Critical Regulator of Rab8 in Ciliogenesis

Maxence V. Nachury

Contents

Abstract

Bardet–Biedl syndrome (BBS) is a hereditary disorder whose symptoms include obesity, retinal degeneration, and kidney cysts. Intriguingly, the cellular culprit of BBS seems to lie in the primary cilium, a "cellular antenna" used by a number of signaling pathways. Yet, despite the identification of 12 BBS genes, a consistent molecular pathway for BBS had so far remained elusive. The recent discovery of a stable complex of seven BBS proteins (the BBSome) considerably simplifies the apparent molecular complexity of BBS and provides a clear insight into the molecular basis of BBS. Most tellingly, the BBSome associates with Rabin8, the guanine nucleotide exchange factor for the small GTPase Rab8, and Rab8-GTP enters the primary cilium to promote extension of the ciliary membrane.

Department of Molecular and Cellular Physiology, Stanford University School of Medicine, Stanford, California

Methods in Enzymology, Volume 439
ISSN 0076-6879, DOI: 10.1016/S0076-6879(07)00434-X

Thus, BBS is likely caused by defects in vesicular transport to the primary cilium. This chapter describes methods used to purify the BBSome using a tandem affinity purification method and presents a variation of this technique to demonstrate the existence of a stable complex of BBS proteins by sucrose gradient fractionation. When combined with state-of-the art mass spectrometry, these methods can provide a nearly complete BBSome interactome containing factors such as Rabin8.

1. INTRODUCTION

The primary cilium is a hair-like extension present at the surface of nearly every vertebrate cell that consists of a microtubule core enclosed within a membrane sheath. Although topologically continuous with the plasma membrane, the ciliary membrane contains a specific complement of lipids and proteins and excludes plasma membrane proteins, thus making the primary cilium a *bona fide* cellular organelle (Vieira *et al.*, 2006). With this realization comes a flurry of questions pertaining to vesicular transport to the primary cilium. What are the actors, the molecular mechanisms, the trafficking routes, the regulators, and so on? These questions have become all the more clinically relevant with the emergence of primary cilium dysfunction as the central pathomechanism in a range of inherited disorders characterized by obesity, brain malformations, retinal degeneration, kidney cysts, and skeletal abnormalities (Fliegauf *et al.*, 2007). One disorder that crystallizes all these symptoms is Bardet–Biedl syndrome (BBS). Given the depth of BBS genetics (12 different genes identified in 5 years), BBS proteins provide a number of prime entry points into the molecular understanding of primary cilium biogenesis.

The challenge of building from these entry points to a solid biochemical foundation was answered by the powerful tandem affinity purification (TAP) technique initiated in 1997 by the laboratory of Bertrand Seraphin (Rigaut *et al.*, 1999) and recently perfected for mammalian cell systems by Cheeseman and Desai (2005). As its name implies, the TAP purification relies on two distinct affinity tags, and a specific protease cleavage site between those tags provides a straightforward elution method to move from one affinity matrix to the next one. The net result is that the two enrichment factors are multiplied and the final eluate is essentially pure of contaminants. In the case of the mammalian TAP version, the affinity tags are the green fluorescent protein (GFP) and the S•tag, the protease site is recognized by TEV (Fig. 34.1), and the corresponding affinity matrices are anti-GFP antibody beads and S-protein agarose.

This chapter presents an application of the TAP protocol to the purification and identification of BBS4-associated proteins. In addition, it can be

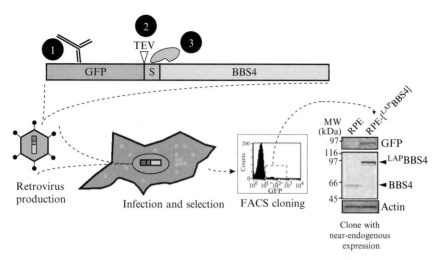

Figure 34.1 Work flow for the generation of a stable, clonal cell line with a near-endogenous level of expressions of TAP-BBS4. The TAP tag allows for purification on anti-GFP beads (1) followed by highly specific elution with the TEV protease (2) and final capture on S-protein agarose beads (3). (See color insert.)

useful to categorize interacting proteins into stable complexes and more transient interactors. For this purpose, this chapter presents a method to separate the BBSome from its associated factors on a sucrose density gradient (Nachury *et al.*, 2007).

2. MATERIALS

2.1. Reagents

Fugene: Roche; polybrene (hexadimethrine bromide): Sigma; Affi-Prep protein A beads: Bio-Rad; DMP (dimethyl pimelimidate·2HCl): Pierce; S-protein agarose: Novagen; 2× Tris-glycine SDS sample buffer: Invitrogen; HPLC-grade H_2O: Sigma; HPLC-grade acetonitrile: Alfa Aesar

2.2. Consumables

96-well flat bottom cell culture plates: Corning; glass bottom 24-well plates: Greiner; *InVitro* PETG small and extra large roller bottles: Nunc; vented caps: Nunc; 500-ml conical bottle: Corning; 5-ml polystyrene tubes: Falcon; 8-ml screw-cap tubes: Falcon; individually wrapped 0.5- and 1.5-ml microtubes: Eppendorf

2.3. Solutions and media

1. Medium for RPE cells (also use for Phoenix-A): Mix 500 ml of Dulbecco's modified Eagle's medium/F12 medium (Gibco), 50 ml fetal bovine serum (FBS; Sigma), 17.3 ml 7.5% sodium bicarbonate (Gibco), 5 ml 200 mM glutamine (Gibco), and 5 ml 10,000 U/ml penicillin/streptomycin (Gibco). Conditioned medium is made by harvesting a culture supernatant from cells that have grown for 2 days and sterilizing it by passing through a 0.22-μm filter.
2. Trypsin/EDTA solution: Make 1× with phosphate-buffered saline (PBS) from 10× stock (Gibco)
3. PBS: 10 mM sodium phosphate, pH 7.0, 150 mM NaCl
4. PBS/FBS: Mix 95 ml sterile PBS and 5 ml FBS
5. Protease inhibitor cocktail (1000× stock in dimethyl sulfoxide): 1 M AEBSF·HCl; 0.8 mM aprotinin; 50 mM bestatin; 15 mM E-64; 20 mM leupeptin; 10 mM pepstatin A; 1 M benzamidine
6. PBST: PBS containing 0.1% Tween-20
7. Coupling buffer: 0.2 M sodium borate, pH 9.0
8. Quenching buffer: 0.2 M ethanolamine, 0.2 M NaCl, pH 8.5
9. PE buffer: 3.5 M MgCl$_2$; 20 mM Tris, pH 7.4
10. LAP0: 50 mM HEPES·KOH, pH 7.4; 1 mM EGTA; 1 mM MgCl$_2$; 10% glycerol.
11. LAP100: 50 mM HEPES·KOH, pH 7.4; 100 mM KCl; 1 mM EGTA; 1 mM MgCl$_2$; 10% glycerol
12. LAP200: 50 mM HEPES·KOH, pH 7.4; 200 mM KCl; 1 mM EGTA; 1 mM MgCl$_2$; 10% glycerol
13. LAP300: 50 mM HEPES·KOH, pH 7.4; 300 mM KCl; 1 mM EGTA; 1 mM MgCl$_2$; 10% glycerol
14. SUC150: 50 mM HEPES·KOH, pH 7.4; 150 mM KCl; 1 mM EGTA; 1 mM MgCl$_2$; 5% (w/w) sucrose

The following abbreviations are used throughout: LAP^N is LAP containing 0.05% NP-40; LAP^D is LAP containing 0.5 mM DTT; LAP^{PI} is LAP containing 1× protease inhibitors; LAP^{3PI} is LAP containing 3× protease inhibitors; and LAP^{CD} is LAP containing 10 μg/ml cytochalasin D.

2.4. Cell lines

Phoenix-A: ATCC. These are the packaging cells for the production of retrovirus particles that have the ability to infect human cells.

RPE1-hTERT: ATCC. All cells are grown in an incubator at 37° and 5% CO$_2$.

2.5. Anti-GFP antibody

If this TAP method is to be used in the laboratory on a regular basis, it makes sense to generate a polyclonal antibody by injecting two rabbits with His-GFP. The antibodies can then be affinity purified on a GST-GFP column. Alternatively, one can use a commercial anti-GFP antibody, but in the author's experience, very few of them are reliable for immunoprecipitation. The author has had the best experience with Invitrogen monoclonal antibody clone 3E6.

3. PROCEDURES

3.1. Generation of a stable clonal RPE-[^{TAP}BBS4] cell line

In order to study cellular processes relevant to primary cilium assembly, the "quasi-primary" cell line RPE-hTERT is used (Jiang et al., 1999). A unique feature of this human cell line is that it never underwent crisis and transformation but instead bypassed senescence through introduction of the catalytic subunit of telomerase (hTERT). The advantage of these cells is that they are easy to grow and ciliate under well-defined conditions.

In order to gain the full benefits of these biochemical purification techniques, it is crucial to generate a stable clonal cell line expressing TAP-BBS4 at near endogenous levels. This way, artifactual interactions as a consequence of protein overexpression can be avoided.

Here a well-established retroviral infection method is used to introduce the transgene into cells and integrate it in the cell genome in a very stable manner (Kinsella and Nolan, 1996). After antibiotic selection, the heterogeneous population is sorted into single cells by a fluorescence-activated cell sorter (FACS) machine. Although quite tedious, this process is critical to the ultimate success of the biochemical purification, and the resulting clonal cell line can be quite useful for imaging studies thanks to the GFP tag. It should be noted that these steps may now be circumvented using the Flp-In system commercialized by Invitrogen that allows for the generation of isogenic cell lines with single site integration without the need for single cell cloning or viral infection. However, one limitation for now resides in the limited number of Flp-In cell lines available.

3.1.1. Cloning of BBS4 into a retroviral expression vector

BBS4 is amplified by a polymerase chain reaction with oligonucleotides 5′-ctagactagtatggctgaggagagagtcgcg-3′(forward) and 5′-cttcaagctttatttctctc-ttatttgttctgatgtttcagttgg-3′ (reverse) from clone Image 4824276. The purified product is then cut with *Spe*I and *Hin*dIII and ligated into *Spe*I/*Hin*dIII-restricted pEGFP-TEV-S (Cheeseman and Desai, 2005). The TAP•BBS4 cassette is then excised with *Afe*I and *Eco*RI and ligated into *Sna*BI/*Eco*RI-restricted pBabe-Puro.

3.1.2. Infection of cells with amphotropic retrovirus

On day 0, plate 1.8×10^6 Phoenix-A cells in a 6-cm dish, paying special attention to avoid clumping. It is important to use healthy cells that have been growing for at least a week to get good transfection efficiency. On day 1, transfect the Phoenix-A cells with pBABEpuro-TAP•BBS4 using 4 μg DNA and 6 μl Fugene in 200 μl OPTI-MEM according to the manufacturer's instructions. On day 2, replace the Phoenix-A culture supernatant with fresh medium and switch cells to 32° (the optimal temperature for virus stability). Note that for the next 7 days, BL2$^+$ guidelines must be followed, as you will be handling infectious particles. The transfection efficiency can now be checked by imaging the dish of Phoenix-A cells on an inverted fluorescence microscope. Depending on the sensitivity of the microscope, there should be from 20% up to 90% GFP-positive cells. Meanwhile, seed two 35-mm dishes with 100,000 RPE cells each. On day 3, harvest the culture supernatant containing the virus from the transfected Phoenix-A cells, filter into a fresh tube using a 0.45-μm syringe filter, add polybrene to 30 μg/ml final, and replace 1 ml of RPE cell culture supernatant with the virus/polybrene mix. Leave the other dish free of virus as a negative control. Switch RPE cells to 32° and discard Phoenix-A cells. On day 4, replace the RPE culture supernatant with fresh medium and switch cells back to 37°.

On day 5, passage each dish into a 6-cm dish and add 3 μg/ml puromycin to the medium. On day 7, replace the culture medium with fresh medium containing 3 μg/ml puromycin. Most of the uninfected cells should start dying and may have detached. Replace medium every 4 days with selective medium until cells are 80% confluent. You will see massive cell death of uninfected cells starting at day 8 and complete death before day 15. On day 15, split infected cells into a 10-cm dish with selective medium and seed one 10-cm dish with 300,000 parental RPE cells. At this step, it is a good idea to freeze about 100,000 infected cells to keep for future use in case of contamination.

3.1.3. Cloning of GFP-positive cells by FACS

On day 16, prepare two 96-well plates with 100 μl of conditioned media per well. Keep the plates in the incubator for 24 h until FACS cloning. On day 17, transfer the culture supernatant from each dish into a 15-ml conical tube, wash the dish with PBS, and detach cells with 1 ml trypsin solution for 2 min at 37°. Create a single cell suspension by pipetting cells up and down 10 times with a P1000 and transfer the cells to the tube filled with medium. Pellet cells at 200 g, resuspend in 10 ml PBS/FBS, count cells with a hemocytometer, and pellet again. Resuspend cells at 10^6 cells/ml in PBS/FBS and transfer to 5-ml tubes. A trained FACS operator and a FACS machine with a cloning module (BD FACSAria or FACSVantage, DAKO Cytomation MoFlo) are now needed in order to array the cells in 96-well plates with 1 cell per well. Use parental RPE cells to set the background fluorescence. It is a good idea to gate the sorting on the 50% lower levels of fluorescence for one plate and the 50% higher levels for the other plate. On day 19, replace the medium in each well with 100 μl of conditioned medium using a multichannel pipettor. Then, change medium every 4 days until cells reach confluence. The medium will begin to change color in the wells where cells start approaching confluence. Choose 24 clones that are growing well from each 96-well plate and passage into each well of a glass-bottom 24-well plate. Around day 29, examine the cells by fluorescence on an inverted microscope. The brightest expressers may not show proper protein localization. Others may have undetectable levels of expression. Pick six clones in which a consistent and clear localization is observed. If you have an antibody against the endogenous protein, probe the selected clones by Western blot to find the clone whose expression is closest to the endogenous protein after expanding those clones. You can now freeze a few ampoules of your best clone.

3.2. Large-scale growth of adherent cells

One caveat of the RPE cells is their low cell volume and the consequent need to grow them on a much larger scale than one would need for HEK or HeLa cells in order to perform biochemical purification. Therefore, to minimize the number of handling steps, it is preferable to use culture vessels with the largest surface available. The author has had the best experience with *InVitro* roller bottles (Nunc). The largest bottles have a surface area corresponding to 28 times the area of a 15-cm dish. The rest of the chapter focuses on the growth and purification of complexes from RPE-[^{TAP}BBS4] cells but it is preferable to perform parallel procedures with the parental RPE cell line for a powerful control.

On day 1, seed a 15-cm dish with 800,000 RPE-[^{TAP}BBS4] cells and grow until 80% confluent. On day 4, passage the cells into a small roller bottle containing 250 ml of medium and close with a vented cap. Transfer to a temperature- and CO_2-controlled roller bottle incubator (e.g., HeraCell 240). Let the cells adhere for 2 h at 6 rph and then increase the speed to 20 rph. Grow the cells until 80% confluency. On day 8, split cells into four extra-large bottles containing 500 ml of medium each. Adhere cells as just described and grow to 80% confluency. On day 12, harvest the cells, proceeding with two bottles at a time. All manipulations can be performed on the bench, as sterility does not need to be maintained anymore.

Pour medium off into a 2-liter beaker on ice, wash cells with 500 ml PBS per bottle, and pour off wash into the same beaker. Add 150 ml of trypsin solution to each bottle and return the roller bottle to 37° for 5 min. Detach cells by hitting the bottles lengthwise against a solid wall 16 times, rotating the bottle one-eighth turn every two hits. Transfer cells into a 500-ml conical bottle on ice. Collect the remaining cells by pouring 120 ml of medium/PBS from the beaker into the bottle, gently shaking and collecting into the 500-ml conical bottle. Repeat three times to fill up the 500-ml bottle. Pellet cells at 800 g for 5 min at 4° in a clinical centrifuge (e.g., Sorvall Legend RT). Resuspend cells from one 500-ml bottle in 50 ml of ice-cold PBS, transfer to a 50-ml conical tube, and pellet as described earlier. Resuspend cells from one 50-ml conical tube into 1 ml ice-cold PBS, transfer to a 1.5-ml tube, and pellet for 5 min at 800 g in a cold microcentrifuge. Wash cells with 1.5 ml ice-cold PBS and freeze pellets in liquid nitrogen. You can expect 450 μl of packed cell volume (PCV) per extra-large bottle.

3.3. TAP purification of macromolecular complexes

The entire procedure (purification and gel running) is performed over 20 h and includes an overnight incubation. All pelleting of beads is done in a benchtop picofuge (e.g., Labnet) for 10 s.

3.3.1. Preparation of anti-GFP antibody beads

Transfer 160 μl of well-resuspended slurry of Affi-Prep protein A beads into 1 ml PBST in a 1.5-ml tube. Wash three times with 1 ml PBST. You should get about 110 μl of packed beads. Resuspend the beads in 500 μl PBST, add 55 μg of affinity-purified GFP antibody, and mix for 1h. Wash the beads twice with 1 ml PBST, twice with 1 ml coupling buffer, and resuspend in 1 ml of coupling buffer containing 20 mM of the cross-linker DMP. Rotate tube for 30 min. Pellet beads and resuspend in quenching buffer. Rotate for 30 min and wash with quenching buffer. Keep beads in quenching buffer at 4° until use.

3.3.2. Preparation of cell extract

Note: From this point onward, pass all buffers through a 0.22-μm filter and do all manipulations at 4° or on ice.

Resuspend 800 μl of PCV into 4 ml LAP300$^{D/PI/CD}$. Mix well by pipetting and add 130 μl of 10% NP-40. Mix by inversion and incubate on ice for 10 min. Spin in a microcentrifuge at 20,000 g for 10 min, collect the crude supernatant, and dilute with 2.2 ml (0.5 volume) of LAP0^N to bring the salt concentration down to 200 mM. This salt concentration constitutes the sweet spot for the following biochemical purification.

Less salt and the yield and purity of the immunoprecipitation drop, more salt and weakly associated factors may be lost. Spin the diluted supernatant at 100,000 g for 1 h and collect the clarified supernatant into a tube on ice. The protein concentration should be around 8 mg/ml.

3.3.3. TAP purification step #1: GFP immunoprecipitation

Preelute 100 μl of anti-GFP antibody beads by washing three times with 1 ml of PE buffer. Wash the beads three times with 1 ml of LAP200^N, resuspend the beads with 1 ml of clarified extract, transfer to a 8-ml tube, and add the remaining extract (the tube should be almost full). Allow the TAP–BBS4 complexes to bind to the antibody beads by rotating the tube for 2 h at 4°. Then wash the beads five times with 1 ml LAP200$^{N/D/PI}$ and then twice with 1 ml LAP200$^{N/D}$ (be sure to omit protease inhibitors from the last wash).

3.3.4. TAP purification step #2: TEV cleavage

Resuspend the beads into 500 μl LAP200^N and transfer to a 0.6-ml tube. Add 3 μg of TEV protease and allow for cleavage to proceed by rotating the tube overnight at 4°.

From this point until cutting bands from the gel, it is highly preferable to perform all steps in a laminar flow hood and keep a stock of consumables (pipette tips, tubes, and buffers) dedicated for this use and to be opened under the hood only (Fig. 34.2). This will drastically reduce the amount of

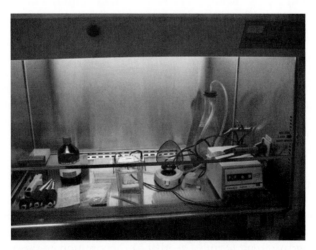

Figure 34.2 Biological safety cabinet for proteomic manipulations. All the equipment for the biochemical purification procedure, gel running, and gel processing is kept inside the hood. This way, the TEV eluate is brought into the hood and the sample does not leave the enclosure until the gel band has been excised and placed in a closed tube. (See color insert.)

dust particles finding their way in the solutions, inside the tube, onto the gel surface, and so on. Those dust particles consist in large part of microscopic flakes of dead skin and hair that are shed by the mammals inhabiting the laboratory space (Holmes, 2001). Consequently, dust contamination of the protein sample will introduce large amounts of keratins, which will inevitably be detected by the mass spectrometer. Although this may not represent a problem for the identification of abundant proteins, keratin abundance may overwhelm the analysis capacity of the liquid chromatography–mass spectrometry (LC-MS) system for low abundance proteins and obscure the peptides of interest. Performing work in a laminar flow hood will reduce keratin contamination to the absolute minimum and increase the likelihood that substoichiometric partners of the BBSome may be detected by the mass spectrometer. If possible, the use of individually wrapped microcentrifuge tubes is highly recommended.

Pellet beads and transfer supernatant to a fresh tube. Rinse beads twice with 200 μl LAP200$^{N/D/3PI}$. Pool all supernatants.

3.3.5. TAP purification step #3: S-protein agarose affinity

Wash 20 μl S-protein agarose (40 μl slurry) three times with 0.5 ml LAP200^N. Add TEV-eluted supernatant to S-protein agarose beads and rotate for 2 h at 4°. Wash beads three times with LAP200$^{N/D/PI}$ and then twice with LAP100. Add 25 μl of 2× SDS sample buffer to the beads. Heat beads on a 95° dry block for 5 min. Pellet beads and transfer supernatant to a fresh tube using a gel-loading tip. Repeat SDS elution and pool the two eluates. Keep at −80° until ready to run the gel.

3.3.6. Running the gel and sample processing

Given the low complexity of the TAP–BBS4 eluate, the entire sample can be analyzed by LC-MS/MS (liquid chromatography in line with tandem mass spectrometry) without the need for fractionation. One option is to concentrate the eluate with organic solvents (acetone, trichloroacetic acid, or methanol/chloroform) and let the mass spectrometry facility handle the dried sample. However, the submission of bands from SDS-PAGE gels is performed more routinely by most facilities and often results in lower analysis costs without significantly reducing sensitivity. When run for a very short amount of time, the gel can also concentrate the sample into a small slice volume. The author typically runs 90% of the eluate in one lane for a very short run and 10% on a different gel for full resolution and silver staining. It is important to find a gel that can accommodate the 45-μl sample (80-mm^3 well volume) and does not have a stacking gel. The Invitrogen Midi-Gels 12 + 2 wells fit these criteria nicely.

Run the gel at 100 V until the sample fully enters the gel (about 5 min). Stain the gel with Simply Blue stain (Invitrogen) in a 15-cm tissue culture dish following the manufacturer's instruction. Cut the band containing the protein (first 3 mm of the lane) and transfer to a 1.5-ml tube (Fig. 34.3A). Wash the gel slice in the tube two times with 50% acetonitrile in water. Each wash consists of 1.0 ml of 50% HPLC grade acetonitrile/water for 2 min with gentle shaking, discarding the supernatant after each wash.

Send the sample to a reliable microchemistry facility for protein identification by tandem mass spectrometry. The author has had the best experience with the Harvard Microchemistry Facility (http://www.mcb.harvard.edu/Microchem/).

Once the first set of protein identification has been received, it may be valuable to fractionate the sample by running the gel until the front dye migrates 25 mm (so called ge LC-MS technique). Excise the stained bands and the blank spaces between the bands and send all 16 samples for protein identification by mass spectrometry (see Fig. 34.3B). This strategy results in the identification of 10 additional BBSome-associated factors compared to single band submission.

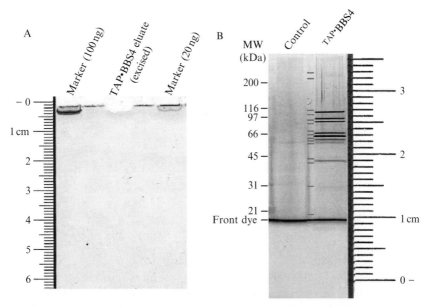

Figure 34.3 SDS-PAGE gel processing of the TAP•BBS4 eluate. (A) The TAP•BBS4 eluate was loaded onto a 10% Novex NuPAGE gel, run for 5 min at 100 V, and stained with Coomassie G-250. These parameters are sufficient for samples to fully enter the gel, and the first 3.5 mm of the lane was excised and sent for protein identification by LC-MS/MS. (B) The TAP•BBS4 eluate was loaded onto a 10% Novex NuPAGE gel and run at 100 V until the front dye migrated 25 mm into the gel. This gel was silver stained, an identical gel was loaded with nine equivalents of eluate and resolved under identical conditions, and 16 bands were excised according to the aquamarine lines drawn on the gel. For the identity of each stained band, see Nachury *et al.* (2007). (See color insert.)

3.3.7. Alternative protocol: Isokinetic sucrose gradient sedimentation of the TEV eluate

Once the full complement of BBS4-interacting factors has been identified, it is useful to distinguish between the most stable interactions and the more peripheral interactions. Sucrose gradient fractionation provides a relatively straightforward mean to fractionate the partially purified BBSome fraction that results from elution of the anti-GFP antibody beads with the TEV protease. At this stage, the BBSome has been enriched over 10,000-fold from the total extract and is still in a soluble state, which could allow for functional assays of BBSome function to be performed.

After anti-GFP immunoprecipitation, wash the beads five times with 1 ml LAP200$^{N/D/PI}$ and then twice with 1 ml SUC150$^{N/D}$. Resuspend the beads into 250 μl SUC150$^{N/D}$, add 3 μg of TEV protease, and rotate overnight at 4°. Collect the supernatant (TEV eluate). Pour the 10 to 40% (w/w) sucrose gradient in 14 × 89-mm polyallomer tubes (Beckman) using a Gradient Master (Biocomp) and sucrose solutions made in SUC150$^{N/D}$ buffer. Carefully apply 200 μl of TEV eluate to the top of the gradient and spin tubes in a SW41 rotor for 18 h at 40,000 rpm at 4°. Open the tubes under the laminar flow hood and collect 400-μl fractions from the top of the gradient. Concentrate the fractions by methanol/chloroform precipitation (Wessel and Flügge, 1984) under the hood using newly opened bottles of methanol, chloroform, and water and resuspend

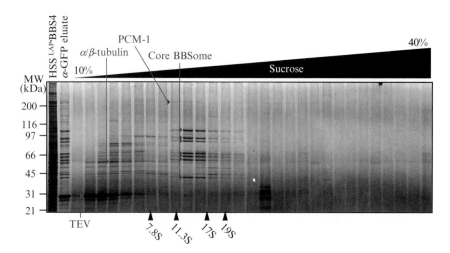

Figure 34.4 Fractionation of the BBSome on a sucrose density gradient. The TEV eluate of TAP•BBS4 was fractionated by sedimentation velocity on a linear 10 to 40% sucrose density gradient. Fractions were resolved on 4 to 12% NuPAGE gels and silver stained. Sedimentation coefficient markers were run simultaneously on an identical gradient. Reproduced with permission from Nachury *et al.* (2007). (See color insert.)

precipitates in 25 μl of SDS sample buffer. Given the concentration of such dilute samples, it is crucial that keratin contamination be kept to an absolute minimum in order to yield clean results after silver staining the gels. After boiling the samples, add 100 mM iodoacetamide (4 M stock in methylformamide) to reduce point streaking ("rainy" lane) after silver staining. Load samples immediately on a gradient SDS-PAGE gel and stain the gel with ammoniacal silver (Chevallet *et al.*, 2006). Bands corresponding to the BBSome are found in the fractions corresponding to a sedimentation coefficient of 12–14S (Fig. 34.4).

ACKNOWLEDGMENTS

I am grateful to Iain Cheeseman for his help in the initial stages of using the TAP tag and to Alex Loktev for his technical assistance in the experiments presented here.

REFERENCES

Cheeseman, I. M., and Desai, A. (2005). A combined approach for the localization and tandem affinity purification of protein complexes from metazoans. *Sci. STKE* **2005,** pl1.

Chevallet, M., Luche, S., and Rabilloud, T. (2006). Silver staining of proteins in polyacrylamide gels. *Nat. Protoc.* **1,** 1852–1858.

Fliegauf, M., Benzing, T., and Omran, H. (2007). When cilia go bad: Cilia defects and ciliopathies. *Nat. Rev. Mol. Cell. Biol.* **8,** 880–893.

Holmes, H. (2001). "The Secret Life of Dust: From the Cosmos to the Kitchen Counter, the Big Consequences of Little Things." Wiley, New York.

Jiang, X. R., Jimenez, G., Chang, E., Frolkis, M., Kusler, B., Sage, M., Beeche, M., Bodnar, A. G., Wahl, G. M., Tlsty, T. D., and Chiu, C. P. (1999). Telomerase expression in human somatic cells does not induce changes associated with a transformed phenotype. *Nat. Genet.* **21,** 111–114.

Kinsella, T. M., and Nolan, G. P. (1996). Episomal vectors rapidly and stably produce high-titer recombinant retrovirus. *Hum. Gene Ther.* **7,** 1405–1413.

Nachury, M. V., Loktev, A. V., Zhang, Q., Westlake, C. J., Peranen, J., Merdes, A., Slusarski, D. C., Scheller, R. H., Bazan, J. F., Sheffield, V. C., and Jackson, P. K. (2007). A core complex of BBS proteins cooperates with the GTPase Rab8 to promote ciliary membrane biogenesis. *Cell* **129,** 1201–1213.

Rigaut, G., Shevchenko, A., Rutz, B., Wilm, M., Mann, M., and Seraphin, B. (1999). A generic protein purification method for protein complex characterization and proteome exploration. *Nat. Biotechnol.* **17,** 1030–1032.

Vieira, O. V., Gaus, K., Verkade, P., Fullekrug, J., Vaz, W. L., and Simons, K. (2006). FAPP2, cilium formation, and compartmentalization of the apical membrane in polarized Madin-Darby canine kidney (MDCK) cells. *Proc. Natl. Acad. Sci. USA* **103,** 18556–18561.

Wessel, D., and Flügge, U. I. (1984). A method for the quantitative recovery of protein in dilute solution in the presence of detergents and lipids. *Anal. Biochem.* **138,** 141–143.

Author Index

W

Wacheck, V., 478, 481
Wagers, A. J., 367, 368
Wagner, U., 26
Wahl, G. M., 505
Walch-Solimena, C., 315, 316, 318
Waldmann, H., 488
Walker, J., 475
Wallace, R., 17
Wallis, T. S., 146
Walmsley, M. J., 236, 240, 371, 372, 373
Walsh, C. A., 256, 287
Walsh, J. E., 345
Walsh, T., 104
Waltenberger, J., 195, 402
Walter, M., 162
Walzl, G., 206
Wan, H., 208
Wan, P. T., 26
Wang, B., 105
Wang, F., 375
Wang, J., 368
Wang, J. C., 18
Wang, J. Y., 112, 380
Wang, L., 18, 76, 183, 226, 375, 415, 462
Wang, Q., 162, 248, 483
Wang, R., 329
Wang, W., 220
Wang, Y., 220
Wang, Z., 269, 275
Wansink, D. G., 493
Ward, E., 452
Ward, J. M., 22
Warne, P. H., 92
Warner, A., 305, 306
Warner, J. K., 18
Warren, J. S., 305
Warren, S., 368
Warshaw, A. L., 452
Wasik, M., 380
Wassarman, D. A., 428
Watanabe, G., 220
Watanabe, N., 136, 182, 183, 184
Watanabe, T., 240
Waterfield, M. D., 280
Waterman-Storer, C. M., 406
Watson, C. J., 248
Wattel, E., 382
Way, M., 146, 380
Weaver, V. M., 414
Webb, D. J., 414
Webb, I. J., 367
Webb, M. R., 115, 118
Weber, C., 375
Weber, D. S., 280
Weber, K. S., 375
Weber, P. C., 375

Wee, S., 105
Weel, J. F., 154
Wehland, J., 146
Wei, L., 183
Wei, L. L., 225
Wei, W., 104
Wei, Y., 105
Weibel, E. R., 304
Weijer, K., 260
Weil, R., 172
Weinberg, R. A., 6, 40, 446, 453, 468, 469
Weinberg, R. L. B., 468, 469
Weinrich, S. L., 454
Weinshenker, D., 429
Weir, E. K., 192
Weis, B., 476, 479, 480, 481, 482
Weisdorf, D., 369
Weiss, E., 112
Weissleder, R., 75
Weissman, I., 17
Weissman, I. L., 367, 368
Weisz, B., 471, 478, 479, 480, 481, 482
Welch, D. R., 221, 225, 231
Welcsh, P., 104
Welker, E., 298
Wellbrock, C., 26
Welsh, C. F., 376
Welsh, M. J., 208
Welti, R., 330
Wengler, G., 379
Wennerberg, K., 55, 150, 380
Wenning, M. J., 372
Wennstrom, S., 480
Werb, Z., 207, 370
Werness, B. A., 459
Wes, P. D., 340
Wesolowski, G., 400
Wessel, D., 512
Wessler, S. R., 375
West, I., 269
Westbury, E., 474, 475, 477
Westerberg, L., 379
Westlake, C. J., 354, 503
Wettschureck, N., 492
Wewers, M. D., 305
White, G. C. II, 458
White, G. E., 396
White, H., 372
Whitworth, A. J., 340
WHO, 206
Wiener, H., 97, 98, 476
Wiener, H. H., 89, 92, 96, 474, 475
Wieringa, B., 493
Wiestler, O. D., 381
Wigler, M. H., 104
Wijnands, E., 260
Wikstroem, P., 280
Wilde, C., 136

Subject Index

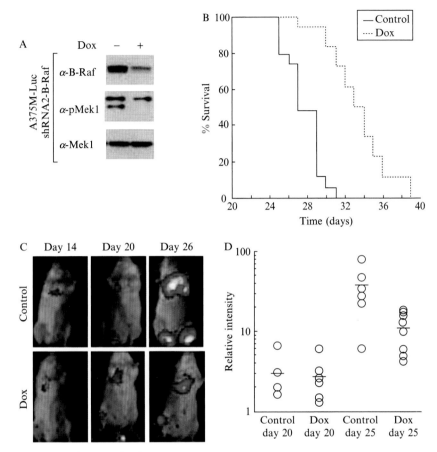

Klaus P. Hoeflich _et al.,_ Figure 3.4 Reduction of A375M systemic tumor growth by B-Raf shRNA knockdown. (A) Western blot analysis showing expression of B-Raf and phosphorylation of Mek1 in uninduced cells (lane 1) and cells treated with 2 mg/ml Dox for 72 h (lane 2). Total Mek1 serves as an internal control to show equal loading. (B) Kaplan–Meier survival data of _scid-beige_ mice injected intravenously with 4×10^5 A375M-luc/shRNA-B-Raf cells and receiving drinking water containing 5% sucrose only (control) or sucrose with 1 mg/ml Dox. Animals were monitored for tumor onset and illness until they reached a terminal stage and were euthanized. Each group consisted of at least 10 mice. The reduction in tumor growth conferred by Dox-mediated B-Raf knockdown is significant according to the log-rank test, $p < 0.0001$. (C) Representative _in vivo_ bioluminescence imaging of mice and (D) quantification of tumor burden of mice receiving Dox versus sucrose-treated control mice. Homogeneous cohorts of mice with established tumor lesions were divided into treatment groups 2 weeks after injection of A375M-luc/shRNA-B-Raf cells. Bioluminescence is represented relative to intensity at day 14 for each animal (adapted from Hoeflich _et al.,_ 2006).

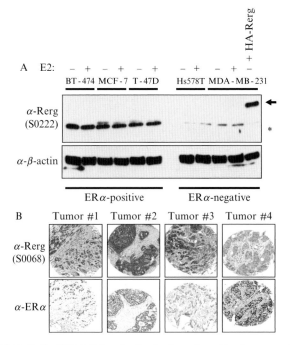

Ariella B. Hanker et al., Figure 5.1 Antibody detection of endogenous Rerg expression in breast cancer cell lines and tumor tissue. (A) Detection of endogenous Rerg expression in breast carcinoma cell lines. Endogenous Rerg and ectopically expressed HA-tagged Rerg were detected in breast cancer cells by Western blot analysis using the S0222 anti-Rerg antibody. The indicated cell lines were incubated in complete growth media, either with or without 10 nM 17β-estradiol (E2, Sigma) for 24 h. A strong band at ≈26 kDa (asterisk) was seen in all Rerg mRNA-positive cell lines (BT-474, MCF-7, and T-47D), whereas a band at ≈30 kDa (arrow) was detected in MDA-MB-231 cells stably infected with the pBabe-puro HAII Rerg plasmid. Blot analysis for β-actin (Sigma) was used as a loading control. (B) Rerg expression levels correlate with ERα expression in breast cancer tissue. Rerg and ERα protein expression in primary invasive breast tissues were detected by IHC analyses. Parallel staining is shown for Rerg (S0068) and ERα from four representative samples.

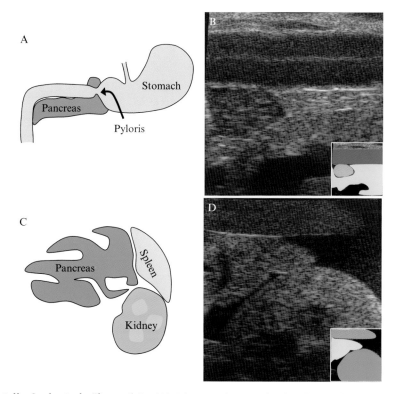

Natalie Cook *et al.*, Figure 6.1 (A) Diagram showing landmarks used to locate the head of the pancreas by ultrasound. (B) Ultrasound image of the head of the pancreas showing a longitudinal section of the proximal duodenum (coded blue in the inset key), pancreas (yellow), and a axial section of a more distal segment of intestine (orange). (C) Diagram of the landmarks used to locate the tail of the pancreas by ultrasound. (D). Ultrasound image of the tail of the pancreas showing the proximity of the pancreas (coded yellow in the inset key) to the left kidney (green) and spleen (lavender).

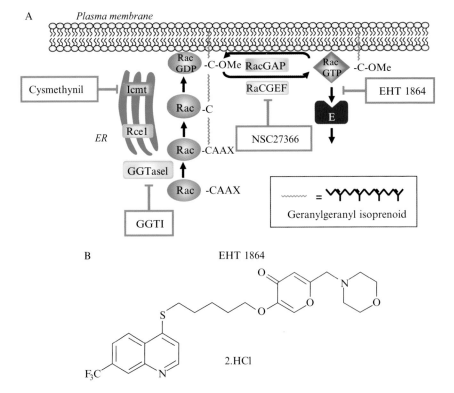

Cercina Onesto *et al.*, Figure 9.1   Inhibition of Rac GTPase function. (A) Approaches for blocking Rac function. Various small molecule inhibitors of Rac function have been described or considered. These include inhibitors of Rac post-translational modification. Rac terminates in a CAAX tetrapeptide sequence (C = Cys, A = aliphatic amino acid, X = Leu). This CAAX motif signals for three sequential post-translational modifications that convert the cytosolic, inactive Rac GTPase to a plasma membrane-associated protein. Geranylgeranyltransferase I (GGTaseI) catalyzes addition of the C20 geranylgeranyl isoprenoid to the Cys residue of the CAAX motif, followed by Rac converting enzyme 1 (Rce1)-catalyzed proteolytic removal of the AAX residues, and isoprenylcysteine carboxyl methyltransferase (Icmt)-catalyzed carboxyl methylation of the now terminal geranylgeranylated cysteine residue. Rac cycles between an inactive GDP-bound and an active GTP-bound state that is regulated by GTPase activating proteins (RhoGAPs) and guanine nucleotide exchange factors (RhoGEFs). Rac-GTP binds preferentially to a large spectrum of functionally diverse effectors (E) that regulate cytoplasmic signaling networks. GGTaseI inhibitors (GGTIs) block all CAAX-signaled modifications, rendering Rac cytosolic and inactive. Cysmethynil blocks the final CAAX modification step by inhibiting Icmt. NSC27366 inhibits RacGEF activation of Rac, whereas EHT 1864 impairs Rac-GTP formation and prevents Rac binding and activation of downstream effectors. (B) Structure of EHT 1864.

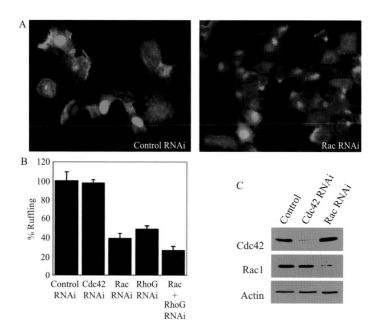

Jayesh C. Patel and Jorge E. Galán, Figure 11.2 *Salmonella*-induced actin remodeling requires Rac and RhoG. (A) Henle-407 cells were cotransfected with GFP (to detect transfected cells) and RNAi constructs targeting Cdc42, Rac, or RhoG as indicated. Two days post-transfection, cells were infected with wild-type *S. typhimurium* for 30 min and stained with TRITC-conjugated phalloidin to visualize remodeling of the actin cytoskeleton. (B) The percentage of transfected (GFP positive) cells displaying actin rearrangements as a consequence of bacterial infection was enumerated and standardized considering the number of transfected cells exhibiting actin-rich ruffles in RNAi control transfections to be 100%. Results represent the mean ± SD of three independent experiments. (C) Western blot showing RNAi efficacy. Cell lysates were prepared 2 days post-transfection with RNAi constructs or a mock control, and protein levels were analyzed using antibodies directed against endogenous Cdc42, Rac, and actin.

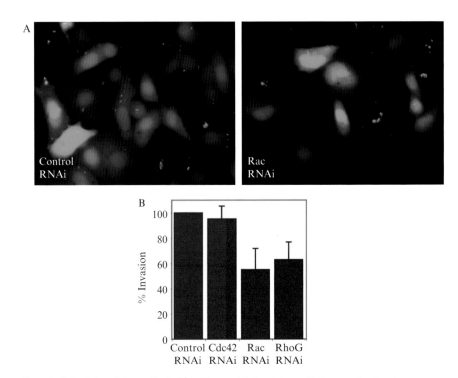

Jayesh C. Patel and Jorge E. Galán, Figure 11.3 *Salmonella* internalization into non-phagocytic cells requires Rac and RhoG. (A) Intestinal Henle-407 cells were cotrans-fected with GFP (to detect transfected cells) and RNAi constructs targeting Cdc42, Rac, or RhoG as indicated. Two days post-transfection, cells were infected with wild-type DsRED *S. typhimurium* as described in the text. DsRED-positive *Salmonella* repre-sent internalized bacteria only, as the expression of DsRED is induced once external bacteria are killed by gentamicin treatment. (B) The percentage of transfected (GFP positive) cells displaying internalized DsRED bacteria was enumerated and standar-dized considering the number of transfected cells with internalized bacteria in RNAi control transfections to be 100%. Results represent the mean ± SD of three indepen-dent experiments.

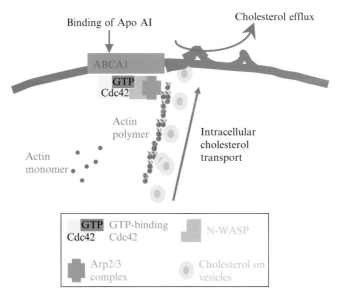

Ken-ichi Hirano *et al.*, Figure 12.1   Hypothetical scheme for Cdc42-mediated cholesterol efflux from cells. The binding of apo AI with ABCA1 on the cell surface activates
Cdc42. The GTP form of Cdc42 triggers intracellular events, such as assembly and
rearrangement of the actin cytoskeleton and activations of some kinases. Although we
still do not know the detailed mechanism, we believe that cholesterol moves as vesicles
from the intracellular site to the plasma membrane along with arrangement of the actin
cytoskeleton. On the cell surface, ABCA1 assembles apo AI and cholesterol for its
efflux from cells.

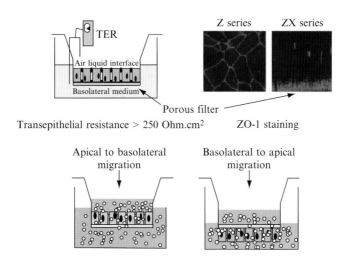

Joanna C. Porter, Figure 16.1   Assay for transepithelial migration of human T cells
across a confluent monolayer of human bronchial epithelial cells. (Top) Human bronchial epithelial cells (HBE) can be grown on either side of a Costar Transwell filter and
polarized with the TJ marker ZO-1 positioned correctly; ZO-1 is in red on the Z series
and ZX series, and the porous filter is stained green. (Bottom) T-cell migration can
be measured in the apical-to-basal direction (epithelial cells grown on the topside of the
filter) and in the more physiological basal-to-apical direction (epithelial cells are grown
on the underside of the filter).

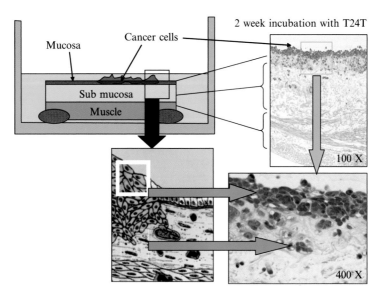

Matthew D. Nitz *et al.*, Figure 17.1 Bladder organ culture (BOC) invasion assay. Invasion is a critical step in bladder cancer metastasis. The BOC invasion assay replicates this step on mouse bladders *in vitro*. (Top left) Basic setup of the BOC assay. (Bottom left) Invasion of the submucosa that the BOC assay is designed to test. (Right) Actual H&E sections of a BOC assay using T24T cells. At higher magnification (400×) it is quite clear that these cells have breached the mucosal layer and are invading the submucosa.

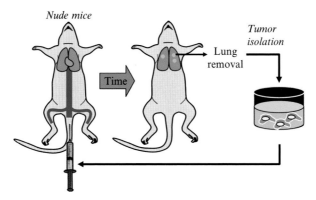

Matthew D. Nitz *et al.*, Figure 17.2 Procedure for selecting increasingly metastatic cell lines. Briefly, cells were harvested and injected into the mouse lateral tail vein for lung metastases. After tumors have formed, the lungs were harvested and new cell lines were created from metastatic nodules. The process was repeated until the desired metastatic capacity was reached.

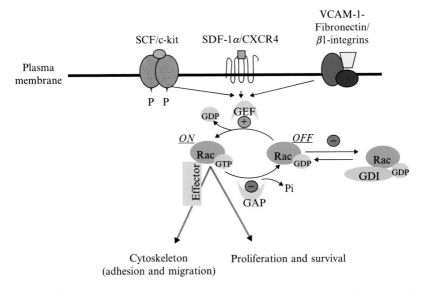

David A. Williams *et al.*, Figure 27.1 Rac GTPases integrate signals from multiple surface receptors involved in HSC engraftment and retention.

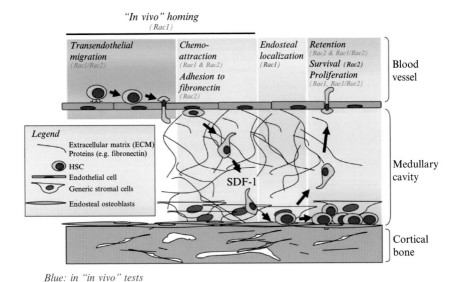

David A. Williams *et al.*, Figure 27.2 Unique and overlapping roles of Rac1 versus Rac2 in hematopoiesis. Putative utilization of Rac1 and Rac2 in processes involved in homing, migration, endosteal localization, retention, and proliferation/survival is shown.

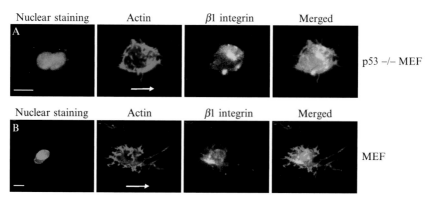

Nuclear staining	Actin	β1 integrin	Merged

A

p53 –/– MEF

Nuclear staining	Actin	β1 integrin	Merged

B

MEF

Stéphanie Vinot *et al.*, Figure 29.3 Immunofluorescence of cells cultured in a 3D Matrigel matrice network. (A) Example of rounded migrating cells. Deficiency in p53 leads to rounded migration as exemplified by p53-/- MEF. The merged image depicts the covisualization of nuclear staining (green), actin (red), and integrin β1 (yellow). (B) Unstressed cell harboring wild-type p53 exhibits an elongated-type migration. Visualizations of nucleus, actin, and integrin β_1 are identical to A. Arrows indicate the direction of cell migration. Bars: 10 μm.

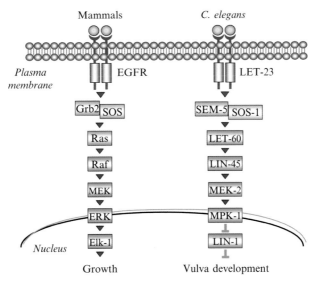

David J. Reiner *et al.*, Figure 30.1 A conserved Ras>Raf>ERK MAPK pathway specifies growth in mammalian cells and vulval cell fate in *C. elegans*. In the worm, an EGFR–ligand-like signal, LIN-3 (released from the anchor cell in the gonad; see Fig. 30.2), binds to the LET-23 receptor tyrosine kinase, which becomes activated and binds the adaptor/exchange factor complex SEM-5/SOS-1 (Grb-2/SOS) to activate the Ras small GTPase LET-60. Activated Ras then triggers the equivalent of the Raf>MEK>ERK MAPK cascade of kinases. ERK/MPK-1 enters the nucleus to phosphorylate and activate or inactivate, respectively, Ets family transcription factors such as LIN-1, which negatively regulates the induction of vulval precursor cells.

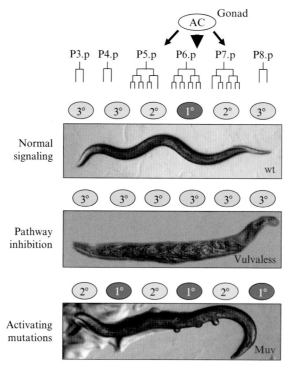

David J. Reiner et al., Figure 30.2   Normal or aberrant vulval formation in *C. elegans* is determined by the exposure of vulval precursor cells to graded levels of activation of the Ras>Raf> MEK>ERK MAPK signaling pathway. Under "normal signaling" conditions (top worm photograph), the anchor cell (AC) of the gonad releases signals (*e.g.,* LIN-3 ligand) to activate the Ras pathway in adjacent VPCs (P6.p, P7.p, etc.). In wild-type animals, VPCs adjacent to the AC receive the strongest signal. These cells assume 1° and 2° fates and together develop into a functional vulva. The more distal VPCs are normally exposed to lower levels of signaling inputs from the Ras pathway. These cells remain uninduced and therefore assume a 3° fate and fuse to the hypodermal syncytium rather than contributing to vulval formation. Under conditions of "pathway inhibition" (middle worm photograph), whether because of loss-of-function mutations or drug treatments, blockade of critical elements of the pathway leads to the absence of inductive signals. All VPCs therefore remain undifferentiated and assume a 3° cell fate that results in a failure to develop a functional vulva (vulvaless phenotype) and hence the inability to lay eggs. As a consequence, the mature eggs hatch inside the parent, causing the "bag-of-worms" phenotype shown. Conversely, constitutive signaling ("activating mutations," bottom worm photograph) leads to hyperinduction of all VPCs, which adopt only 1° and 2° fates, resulting in the formation of both a single functional vulva and additional protruding nonfunctional vulva-like structures called pseudovulvae. Worms with such pseudovulvae are described as exhibiting a multivulva (Muv) phenotype.

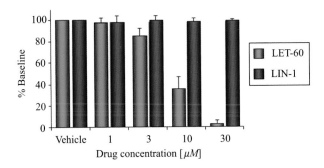

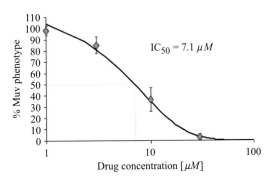

David J. Reiner et al., Figure 30.5 The MEK inhibitor U0126 dose dependently reduces the Muv phenotype induced by constitutively activated Ras in *let-60* (*n1046*gf) worms (top, green), but does not affect the Muv phenotype induced by loss of the downstream Ets family transcription factor in *lin-1*(*null*) animals (top, red). "Percent baseline" refers to the percentage of animals displaying a Muv phenotype in the presence of inhibitor compared to vehicle-only control. The dose–response curve (bottom) demonstrates that the IC50 for Muv inhibition in *let-60*(gf) worms is approximately 7.1 μM, and essentially maximal inhibition can be achieved at the 30 μM dose to which *lin-1*(*null*) worms are insensitive.

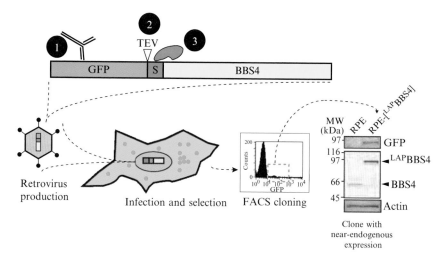

Maxence V. Nachury, Figure 34.1 Work flow for the generation of a stable, clonal cell line with a near-endogenous level of expressions of TAP-BBS4. The TAP tag allows for purification on anti-GFP beads (1) followed by highly specific elution with the TEV protease (2) and final capture on S-protein agarose beads.

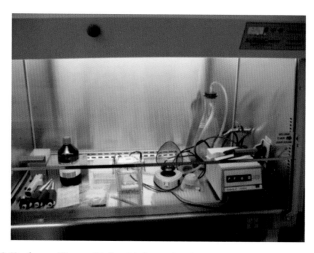

Maxence V. Nachury, Figure 34.2 Biological safety cabinet for proteomic manipulations. All the equipment for the biochemical purification procedure, gel running, and gel processing is kept inside the hood. This way, the TEV eluate is brought into the hood and the sample does not leave the enclosure until the gel band has been excised and placed in a closed tube.

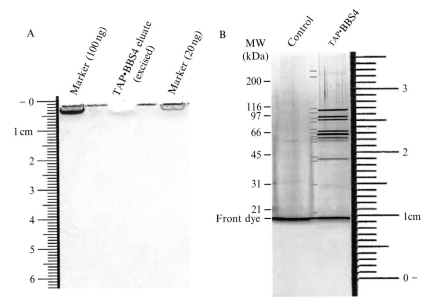

Maxence V. Nachury, Figure 34.3 SDS-PAGE gel processing of the TAP•BBS4 eluate. (A) The TAP•BBS4 eluate was loaded onto a 10% Novex NuPAGE gel, run for 5 min at 100 V, and stained with Coomassie G-250. These parameters are sufficient for samples to fully enter the gel, and the first 3.5 mm of the lane was excised and sent for protein identification by LC–MS/MS. (B) The TAP•BBS4 eluate was loaded onto a 10% Novex NuPAGE gel and run at 100 V until the front dye migrated 25 mm into the gel. This gel was silver stained, an identical gel was loaded with nine equivalents of eluate and resolved under identical conditions, and 16 bands were excised according to the aquamarine lines drawn on the gel. For the identity of each stained band, see Nachury *et al.* (2007).

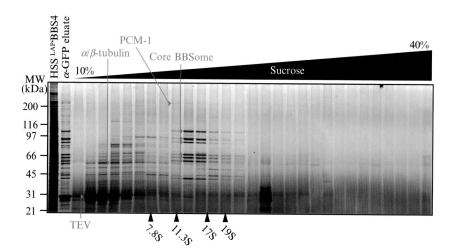

Maxence V. Nachury, Figure 34.4 Fractionation of the BBSome on a sucrose density gradient. The TEV eluate of TAP•BBS4 was fractionated by sedimentation velocity on a linear 10 to 40% sucrose density gradient. Fractions were resolved on 4 to 12% NuPAGE gels and silver stained. Sedimentation coefficient markers were run simultaneously on an identical gradient. Reproduced with permission from Nachury *et al.* (2007).